The Emergence of the Acheulean in East Africa and Beyond

Vertebrate Paleobiology and Paleoanthropology Series

Edited by

Eric Delson
Vertebrate Paleontology, American Museum of Natural History
New York, NY 10024, USA
delson@amnh.org

Eric J. Sargis
Anthropology, Yale University
New Haven, CT 06520, USA
eric.sargis@yale.edu

Focal topics for volumes in the series will include systematic paleontology of all vertebrates (from agnathans to humans), phylogeny reconstruction, functional morphology, Paleolithic archaeology, taphonomy, geochronology, historical biogeography, and biostratigraphy. Other fields (e.g., paleoclimatology, paleoecology, ancient DNA, total organismal community structure) may be considered if the volume theme emphasizes paleobiology (or archaeology). Fields such as modeling of physical processes, genetic methodology, nonvertebrates or neontology are out of our scope.

Volumes in the series may either be monographic treatments (including unpublished but fully revised dissertations) or edited collections, especially those focusing on problem-oriented issues, with multidisciplinary coverage where possible.

More information about this series at http://www.springer.com/series/6978

The Emergence of the Acheulean in East Africa and Beyond

Contributions in Honor of Jean Chavaillon

Edited by

Rosalia Gallotti

Dipartimento di Scienze dell'Antichità, Sapienza Università di Roma, Rome, Italy

Université Bordeaux, UMR 5199 PACEA-PPP, Pessac, France

Margherita Mussi

Dipartimento di Scienze dell'Antichità, Sapienza Università di Roma, Rome, Italy

Springer

Editors
Rosalia Gallotti
Dipartimento di Scienze dell'Antichità
Sapienza Università di Roma
Rome
Italy

and

Université Bordeaux
UMR 5199 PACEA-PPP
Pessac
France

Margherita Mussi
Dipartimento di Scienze dell'Antichità
Sapienza Università di Roma
Rome
Italy

ISSN 1877-9077 ISSN 1877-9085 (electronic)
Vertebrate Paleobiology and Paleoanthropology Series
ISBN 978-3-030-09371-6 ISBN 978-3-319-75985-2 (eBook)
https://doi.org/10.1007/978-3-319-75985-2

Cover illustration: The upper Awash Valley at Melka Kunture, with the Awash River in the foreground and the Wochacha volcano in the background; inset, obsidian massive scraper (early Acheulean, Garba IVD, Melka Kunture) (photos by Rosalia Gallotti).

Printed on acid-free paper

This Springer imprint is published by the registered company Springer International Publishing AG part of Springer Nature
The registered company address is: Gewerbestrasse 11, 6330 Cham, Switzerland

This volume is dedicated to the memory of Jean Chavaillon (March 25, 1925–December 21, 2013), the leading archaeologist and Quaternary geologist who researched with unfailing enthusiasm the African Pleistocene and directed from 1965 to 1995 the French Archaeological Mission at Melka Kunture.

Jean Chavaillon (front row center) at Melka Kunture in 1969, with workers and collaborators. Next to him his wife Nicole, also an archaeologist, with the little Florence, one of their daughters. Next to Nicole another archaeologist, Françoise Hivernel, and slightly behind and between them, Jean Gire, in charge of the archaeological drawings

Foreword

Jean Chavaillon: The Scientist, The Teacher, and The Colleague

Very few professionals have marked their paths on the scene of prehistoric research as vividly as Professor Jean Chavaillon. Prehistoric research in Africa was in its infancy, and very few sites were known in East Africa when this scientist joined the quest for knowledge about early man's environment and cultures. In his earlier professional career in Africa, his interest was focused on North Africa, where he studied the Quaternary formations of the northwestern Sahara, culminating in a 393-page monograph published in 1964, i.e., just before he turned his gaze toward Ethiopia.

The discovery of the Melka Kunture paleoanthropological site in 1963 by G. Dekker and results of a reconnaissance survey in the following months by G. Bailloud showed the great potential of the site. Subsequently, Jean Chavaillon took charge and assumed the task of planning and undertaking a multiyear research project at Melka Kunture; where from 1965 until 1995 he led a multidisciplinary paleoanthropological research program.

Professor Jean Chavaillon was also among the first prehistorians to work in the Lower Omo paleoanthropological sites. In the Omo Shungura sites, he conducted geological and archaeological investigations between 1967 and 1976. There he discovered, with his team, the world's oldest stone artifacts known at the time, dated to ~2.3 million years old (Ma). In doing so, he pushed back the antiquity of stone tool making by a half million years vis-a-vis the previously known oldest discoveries, dated to 1.8 Ma at Olduvai Gorge in Tanzania. During these years, he was working for the Institut d'Archéologie in Addis Ababa and was appointed by the Ethiopian Government as coordinator of the Omo International Research Group which was composed of French, American, and Kenyan research teams. The French contingent was initially led by Professor Camille Arambourg who was later replaced by Yves Coppens. F. Clark Howell organized the American team, and Richard Leakey (representing his father Louis Leakey) led the Kenyan team. The Omo paleoanthropological research laid the foundation for later multidisciplinary approaches in paleoanthropology. Successful results of this multicomponent approach served as yardsticks for biochronological dating at sites where secure radiometric dates are lacking. The discoveries made in the Lower Omo remain among the most important scientific milestones in human biological and cultural evolution, as well as the paleoenvironmental contexts of these important discoveries. Jean Chavaillon's contribution in this regard is paramount. The discoveries he made in the Lower Omo were published in several scientific journals and still serve as major references for understanding the behavioral evolution of our ancestors.

While working in the Omo, Professor Jean Chavaillon was also deeply engaged in paleoanthropological research at Melka Kunture. He organized there a multidisciplinary research team which also included young students of Quaternary research. Since 1965, this resulted in the discovery of multiple layers of human occupation at various localities, among which the most famous are Karre, Garba IV, Gombore I and II, and Simbiro. Discoveries from these sites have opened up new discussions among prehistorians, providing additional knowledge as well as a paradigm shift in our understanding of both the beginnings and evolution of stone tool technologies and early human behavior in relation to the paleoenvironment.

Jean Chavaillon demonstrated at Melka Kunture that all known stages of prehistoric stone technologies are represented in a well-defined and dated stratigraphic sequence. The main occurrences of these technologies are: the Oldowan from Karre I, Gombore I, Gombore Iγ, and Garba IVG-E; Acheulean lithic production from Garba IVD, Garba XII, Simbiro III, Gombore II, Garba I, Garba IIIE-B; the Middle Stone Age from Garba IIIA-B; and the Late Stone Age from Wofi II, Wofi III, and Kella I. In addition to these, the Balchit site was a quarry source for obsidian in both prehistoric and historical times. The Melka Kunture site is unique in the world in presenting such an extended sequence of human stone technologies and continuous site use over the last 1.8 Myr. This became evident in the archaeological record, thanks to the unreserved and continuous research efforts of Jean Chavaillon and his team. In addition to the evolution of the various stages of stone technologies and geological sequences, his research at Melka Kunture has also produced evidence on space management and land use by prehistoric people.

During his early years of research at Melka Kunture, Jean Chavaillon was encouraging young students to work on the Ethiopian Quaternary. Among many, he supported the geological work of Maurice Taieb in the Awash Basin, which eventually resulted, on top of the vast knowledge generated about the basin itself, in the discovery of the Hadar, Middle Awash, and Gona paleoanthropological sites, which became crucial locations for understanding our biological and cultural evolution.

A modest man who never advertised his great accomplishments, honest to his career, and in love with the site in which he worked, Jean Chavaillon was a great site protector. On several occasions, he fought against the actions of clandestine fossil collectors. In 1972, he advocated for the delineation of the Melka Kunture site within an area of 800 hectares of Archaeological Park. This was finally achieved in the 1990s (although not all of the area he envisaged to be protected was included in the park). In his plan, he proposed a site museum at Melka Kunture to showcase the finds, which came into reality in collaboration with his colleague Marcello Piperno late in the 1990s.

Jean Chavaillon had also helped save the Melka Kunture site from irreversible damage by bringing the issue to the attention of the relevant decision makers in 1972–1973. His interest was not limited just to the protection of the site. In 1979, after discussions with the relevant Ethiopian authorities, and following their approval, he raised funds for the construction in Addis Ababa of a repository for paleoanthropological findings and also built a laboratory facility where the materials were stored and studied. This facility, which was put in place and organized by Chavaillon, has served the task for which it was designed during more than 30 years. I was honored to work in this facility, using his old office, as Head of Archaeology and Paleontology of the Ministry of Culture and Sports (now Culture and Tourism) of Ethiopia for 15 years. Most importantly, the establishment of this facility inspired other researchers (led by the late J. D. Clark) to build and organize additional facilities. This culminated in the construction of the new "state-of-the-art" research facility built by the Ethiopian Government on the premises of the National Museum, where all of the paleoanthropological and archaeological findings from across the country are housed.

The results of Chavaillon's monumental work at Melka Kunture have been published in more than 60 articles and books and known worldwide. Students of archaeology have benefited and continue to benefit from knowledge acquired through his research endeavors. In Ethiopia, his findings were included in the educational curriculum; and during my high school days, I was one of those students who benefited from learning the results of his impressive research which were fit into the curriculum. My undergraduate archaeology courses in prehistory were enriched by the results of his research at Melka Kunture and Omo. Much later, in the early 1980s, I had the benefit of learning not only from his work but personally from him in the National Museum of Ethiopia collection rooms; even later, I formally attended his lectures while a student at the Institut de Paleontologie Humaine in Paris. He was always keen on his advice and openhearted, providing me with invaluable advice while I was working on my Ph.D. thesis. As a member of my Examining Board, he honestly and professionally commented on my work. Students who benefited from his wisdom and research works are currently thriving all over the world.

Professor Jean Chavaillion's scientific works were not limited to Melka Kunture or Omo. He also worked in several other areas including Gotera (a Middle Stone Age site in Southern Ethiopia) and late Acheulean sites around Lake Ziway (again in Ethiopia). Jean Chavaillon was an unflagging prehistorian of great intelligence. He was farsighted in his planning for the sites he loved so much. This selfless great prehistorian gallantly handed over the task of the leadership of the research work of his beloved Melka Kunture in 1996 to his longtime friend and colleague Professor Marcello Piperno. Despite that, he continued to work year after year with the new team leader, and together they undertook in 2004 the publication of a monumental monograph on Melka Kunture. And again, after the transfer of the research leadership at Melka Kunture to Professor Margherita Mussi, he was always supportive of her efforts, until unfortunately he left us for the last time. His work will continue to inspire us all for many more years to come.

The international workshop on "The Emergence of the Acheulean in East Africa" organized by Margherita Mussi and Rosalia Gallotti to commemorate the 50th anniversary of Melka Kunture and to celebrate the lifetime achievements of Professor Jean Chavaillon has brought together scholars working on the East African Acheulean at Università di Roma Sapienza on September 12–13, 2013. I would like to seize this opportunity to thank Margherita Mussi and Rosalia Gallotti for inviting me to take part in this important workshop and to celebrate the life of this great prehistorian.

Yonas Beyene
Association for Research and Conservation of Culture (ARCC)
and French Center for Ethiopian Studies (CFEE), Ethiopia

Preface

In 2013 an international roundtable was held in Rome, discussing "The Emergence of the Acheulean in East Africa" to celebrate the 50th anniversary of the discovery of Melka Kunture (Upper Awash valley, Ethiopia).

The theme had been carefully selected. During the second half of the last century, the archaeological research at Melka Kunture directed by Jean Chavaillon, head of the French archaeological mission, had led to the discovery of Oldowan sites and of an impressive sequence of Acheulean layers—plus some important Middle Stone Age and Late Stone Age sites. From 1999 to 2010, the Italian archaeological mission, under the direction of Marcello Piperno, further focused on the Oldowan of Garba IV, also opening for display to the general public a new Acheulean area, Gombore II OAM. Since 2011, under the direction of one of the editors (MM), new fieldwork has been aimed at updating and completing previous research.

Accordingly, both the Oldowan and the Acheulean are extremely well documented at Melka Kunture. However, while workshops specifically addressing the earliest developments in lithic technology had been held since the beginning of the century, such as the "First Hominid Technology Workshop" (Bellaterra, Spain, 2003) and the "Conference on Early Stone Tools and Cognitive Evolution" (Stanford University, USA, 2010), the origin of the Acheulean in East Africa, and its relationships with the Oldowan, had not been collectively discussed in a decade. Thanks to the Wenner-Gren Foundation, which generously sponsored the meeting (grant no. CONF-626), we were able to fill this gap. Researchers who were working on the earliest Acheulean were asked to present recent results and share their experiences, allowing fruitful discussion. The program, list of participants, and abstracts of the communications are available at http://melkakunture.it/research/fifty_years.

A volume of proceedings was the obvious outcome of this collective effort. The Vertebrate Paleobiology and Paleoanthropology Series—which already included "Interdisciplinary Approaches to the Oldowan" edited by Erella Hovers and David R. Braun—was the perfect option. We gladly acknowledge the support given to this project ever since the beginning by the Series Editors, Eric Delson and Eric Sargis. This new volume reflects fairly well the roundtable of 2013, but there are also differences. For various reasons some of the original participants were eventually unable to produce a paper, as always happens with proceedings. Vice versa, we expanded the volume with some chapters on the preceding Oldowan, on the African fauna, on the Acheulean in Asia and, eventually, on the Acheulean in Europe, where it develops later than elsewhere.

In doing so, we contacted tens of colleagues who were asked to review the papers, definitely improving the quality of the final versions. While they will remain anonymous, they must be assured that we are most grateful to them for their time and dedication. We also thank Università di Roma Sapienza, and especially Dipartimento di Scienze dell'Antichità, which provided the venue where the 2013 meeting was held.

For the opening of the roundtable in September 2013, Jean Chavaillon sent a touching letter in his own hand, ending with the following words "Chers Amis, bon courage, belles et fructueuses découvertes. Avanti!" At the time he was a frail, 88-year-old gentleman, but the enthusiasm of this great prehistorian for archaeological research was unshaken. Sadly, he died the same year, just 3 months later. This volume is dedicated to his memory, as an outstanding researcher who focused most of his work on Melka Kunture.

Rome, Italy

Margherita Mussi
Italian Archeological Mission at Melka Kunture and Balchit

Chers amis, bon courage, belles et fructueuses découvertes.

Avanti !

Jean Chavaillon

Contents

Contributors

* indicates preferred addresses for contact

Nick Ashton

Department of Britain, Europe and Prehistory, British Museum, Franks House, 56 Orsman Road, London, N1 5QJ, UK

Enrique Baquedano

*Institute of Evolution in Africa (I.D.E.A.), Museo de San Isidro, Plaza de San Andrés 2, 28005 Madrid, Spain; Museo Arqueológico Regional, Plaza de las Bernardas s/n, 28801 Alcalá de Henares, Madrid, Spain

Raymonde Bonnefille

CNRS, Université Aix-Marseille, UMR 7330, Cerege BP 80, 13545 Aix-en-Provence, France

Isabel Cáceres

*Universitat Rovira i Virgili (URV), Avinguda de Catalunya 35, 43002 Tarragona, Spain; IPHES, Institut Català de Paleoecologia Humana i Evolució Social, Zona Educacional 4, Campus Sescelades URV (Edifici W3), 43007 Tarragona, Spain

Robin Dennell

Department of Archaeology, University of Exeter, Exeter EX4 4QE, UK

Fernando Diez-Martín

Department of Prehistory and Archaeology, University of Valladolid, Plaza del Campus s/n, 47011 Valladolid, Spain

Manuel Domínguez-Rodrigo

*Department of Prehistory, Complutense University, Prof. Aranguren s/n, 28040 Madrid, Spain; Institute of Evolution in Africa (I.D.E.A.), Museo de San Isidro, Plaza de San Andrés 2, 28005 Madrid, Spain

Rosalia Gallotti

*Dipartimento di Scienze dell'Antichità, Sapienza Università di Roma, Piazzale Aldo Moro 5, 00185 Rome, Italy; Université Bordeaux, UMR 5199 PACEA-PPP, Bâtiment B18, Allée Geoffroy Saint-Hilaire, CS 50023, F 33615 PESSAC CEDEX, France

Denis Geraads

Centre de Recherche sur la Paléobiodiversité et les Paléoenvironnements, Sorbonne Universités, MNHN, CNRS, UPMC, CP 38, 8 rue Buffon, 75231 PARIS Cedex 05, France

Amanda C. Leiss

Department of Anthropology, Yale University, 10 Sachem Street, New Haven, CT 06511-3707, USA

Luis de Luque

Fundación Conjunto Paleontológico de Teruel-Dinópolis, Avenida de Sagunto s/n, 44002 Teruel, Spain

Audax Z. P. Mabulla

Archaeology Unit, University of Dar es Salaam, P.O. Box 35050, Dar es Salaam, Tanzania

Rita Melis

Dipartimento di Scienze chimiche e geologiche, Università di Cagliari, Via Trentino 51, 09127 Cagliari, Italy

Marie-Hélène Moncel

Département de Préhistoire, UMR 7194 CNRS, Muséum national d'Histoire naturelle, 1 rue René Panhard, 75013 Paris, France

Margherita Mussi

Dipartimento di Scienze dell'Antichità, Sapienza Università di Roma, Piazzale Aldo Moro 5, 00185 Rome, Italy

Joaquín Panera

Centro Nacional de Investigación sobre la Evolución Humana (CENIEH), Sierra de Atapuerca 3, 09002 Burgos, Spain

Alfredo Pérez-González

Centro Nacional de Investigación sobre la Evolución Humana (CENIEH), Sierra de Atapuerca 3, 09002 Burgos, Spain

Michael J. Rogers

Department of Anthropology, Southern Connecticut State University, 501 Crescent Street, New Haven, CT 06515-1355, USA

Raquel Rojas-Mendoza

Asociación "El Hombre y el Medio", Santo Tomás de Aquino, 21, 28982 Madrid, Spain

Susana Rubio-Jara

Centro Nacional de Investigación sobre la Evolución Humana (CENIEH), Sierra de Atapuerca 3, 09002 Burgos, Spain

Policarpo Sánchez Yustos

Department of Prehistory and Archaeology, University of Valladolid, Plaza del Campus s/n, 47011 Valladolid, Spain

Manuel Santonja

Centro Nacional de Investigación sobre la Evolución Humana (CENIEH), Sierra de Atapuerca 3, 09002 Burgos, Spain

Sileshi Semaw

*Centro Nacional de Investigación sobre la Evolución Humana (CENIEH), Sierra de Atapuerca 3, 09002 Burgos, Spain; CRAFT, Stone Age Institute, Indiana University, 1392 W. Dittemore Rd., Gosport, IN 47433, USA

Dietrich Stout

Department of Anthropology, Emory University, 1557 Dickey Drive, Atlanta, GA 30322, USA

Pierre-Jean Texier

Aix Marseille Univ, CNRS, Minist Culture, LAMPEA, UMR 7269, 5 rue du Château de l'horloge-BP 647, F-13094 Aix-en-Provence Cedex 2, France

Chapter 1
The Emergence of the Acheulean in East Africa: Historical Perspectives and Current Issues

Rosalia Gallotti and Margherita Mussi

Abstract We review below the Acheulean of East Africa from two perspectives: the history of research and the current state of the art. The definition of Acheulean industries has changed considerably over 150 years and since the earliest research in Africa. A brief presentation of the main discoveries, of the many theories, and of the various methods used in Acheulean archaeological research will help in understanding the current debate and the topics addressed in this volume.

Keywords History of archaeological research • Current issues • Typology • Technology

The Acheulean developed over more than 1.5 million years and is the longest lasting Palaeolithic culture. It is also the one with the widest geographical distribution, spreading over Africa and Eurasia.

Gabriel de Mortillet recognized the Acheulean as such in Europe, at the end of the nineteenth century. He was a geologist who later became a leading archaeologist (Nicole 1901). Following a geological methodology, he used tool types as index fossils, relating them to the best-known and most typical locality. This allowed him to characterize prehistoric periods and to put them in a chronological sequence. In 1872, in his "Classification des diverses périodes de l'âge de la pierre", he described a number of prehistoric lithic collections from northern France, taking St. Acheul as a type site. Accordingly, he defined an "Époque de St. Acheul", with a characteristic implement, or index fossil: the "coup-de-poing" (de Mortillet 1872). "Biface", which refers to the same tool type, was first used later by Vayson de Pradenne (1920).

This iconic tool had even attracted attention before de Mortillet's time, and well before Boucher de Perthes (1847) and Lyell (1863) established the antiquity of humans in Europe. In 1797 John Frere sent a letter to the Society of Antiquaries of London describing "...weapons of war, fabricated and used by a people who had not the use of metals... The situation in which these weapons were found may tempt us to refer them to a very remote period indeed, even beyond that of the present world...". The letter came with two handaxes from Hoxne (Suffolk), which are now on display at the British Museum, and was published in Frere (1800). Admittedly, the finds of John Frere did not attract much attention at the time, and de Mortillet remains the founding father of the Acheulean (Mussi 2014).

Ten years after the first publication, G. de Mortillet choose Chelles as type site and changed his original nomenclature: the term "Chellean" was introduced (de Mortillet 1883). The term "Acheuléen", translated in English as Acheulean or Acheulian, was introduced in the 1920s. Through time, Acheulean superseded previous terminologies and came to include "Chellean" and "Abbevillian". The latter names had been in use for some time for industries with rougher and "more primitive" bifacial tools, apparently belonging to an earlier stage of human development—stages which were eventually found to lack stratigraphic consistency (Déchelette 1924; Breuil 1932). While other parts of de Mortillet's nomenclature became obsolete, the "Époque de St Acheul", renamed "Acheulean", has ever since remained in full use (Mussi 2014).

R. Gallotti (✉) · M. Mussi
Dipartimento di Scienze dell'Antichità, Sapienza Università di Roma, Piazzale Aldo Moro 5, 00185 Rome, Italy
e-mail: rosalia.gallotti@uniroma1.it

M. Mussi
e-mail: margherita.mussi@uniroma1.it

R. Gallotti · M. Mussi
Italian Archaeological Mission at Melka Kunture and Balchit

R. Gallotti
Université Bordeaux, UMR 5199 PACEA-PPP, Bâtiment B18 allée Geoffroy Saint-Hilaire, CS 50023, 33615 Pessac Cedex, France

R. Gallotti and M. Mussi (eds.), *The Emergence of the Acheulean in East Africa and Beyond: Contributions in Honor of Jean Chavaillon*, Vertebrate Paleobiology and Paleoanthropology,
https://doi.org/10.1007/978-3-319-75985-2_1

Around 1880, some years after the introduction of the "Époque de St Acheul", the first African handaxes were found near Temassinine, in Algeria, and Rivière (1896) made formal reference to a Lower Paleolithic industry from the site of Boul Baba, in Tunisia. Several "Chellean biface sites" were then discovered in North Africa at the beginning of the twentieth century (Flamand and Laquière 1906; Chantre 1908; Pallary 1911; Seligman 1921; Breuil 1930; Roffo 1934; Biberson 1961). In East Africa, collections including handaxes were similarly made since the last decade of the nineteenth century, when they were shipped back to European museums (Leakey 1931; Mussi 1973). Around 1905, handaxes, compared to those of France, were also collected at Stellenbosch, close to Cape Town (Péringuey 1911).

In East Africa, Gregory (1896, 1921), a trained geologist, discovered Olorgesailie, a major Acheulean site, but its location was lost for a further forty years. Since 1926, Louis Leakey made expeditions that set the foundations of East African prehistory. As a result of the first two expeditions, he published the handaxes from Kariandusi (Leakey 1931). From 1931 to 1951 he worked at Olduvai, attracted there by the discovery made by Reck of mammalian fossil assemblages (Reck 1914, 1926). His research aimed at defining the evolutionary cultural stages within the exposed geological horizons of the gorge (Leakey 1965). By 1932 he was able to report to the first IUPPS congress (held in London) a sequence from Olduvai ranging from the "pre-Chellean" to the "Aurignacian" (Leakey 1934, 1936). This is when he introduced the name "Oldowan" to identify the "pre-Chellean" industries. During the World War II, together with his wife Mary, he rediscovered Olorgesailie (Gowlett 1990). Then in 1947 he organized in Nairobi the First Pan-African Congress, which marked the beginning of "the time of the Acheulean" in Africa. Several communications provided support to this theme, e.g., those of Leakey himself on Olorgesailie and Kariandusi; of J. Desmond Clark on Acheulean sites from the Somalilands; and of H. Breuil on the survey of raised beaches all around Africa (Breuil 1952; Clark 1952; Leakey 1952). As already summarized at the time by van Riet Lowe (1952: 167) "While it is now widely held that the essential home of the Hand-Axe Culture is to be sought in Africa, we find, when we set out in the search of its roots that as soon as we leave this continent we flounder in mists of uncertainty. If, on the other hand, we remain here, we find that here—and here only—we have a long series of earlier well-stratified cultures which led us naturally and directly to the establishment of the Hand-Axe Culture".

The Pan-African Congresses gradually started to address issues related to terminology and typology, prompting useful debates. At first, and throughout the first half of the twentieth century, developments in African archaeological research were strongly linked to those in Europe. Attempts were made to discard the imported terminology and forge a local nomenclature, most notably in South Africa (Goodwin and Van Riet Lowe 1929; Van Riet Lowe 1952), where "Stellenbosch" was introduced in 1925 as an alternative label. In East Africa, Leakey (1931) described a "Kenya Chellean" (later to be dismissed, just as was the European "Chellean") and a "Kenya Acheulean" as well. However, at the time of the Fourth Pan-African Congress of Prehistory in 1959, Kleindienst (1962: 81) makes clear that "work in Africa is an outgrowth of the European tradition of prehistory".

In 1953, J.D. Clark had started excavating at Kalambo Falls (Clark 1969, 1974). The Isimila site complex was also discovered (Howell et al. 1962). The lithic assemblages from these two sites were pivotal in a major study by Kleindienst (1961, 1962).

In North Africa too this was a time of chronological and typological refinements, and of new discoveries. Investigations started in the impressive costal sequence around Casablanca and in the Acheulean site of Ternifine (Arambourg 1955; Balout et al. 1967). In 1961, Biberson proposed to subdivide the Acheulean of North Africa into eight phases (Biberson 1961). His model of typological progression was a reference for L.S.B. Leakey in East Africa. Balout (1955) and Tixier (1956) furthermore carried out major studies of bifacial tool typology (especially cleavers). The outcome of scientific activity all over Africa was published in three monumental works on the prehistory of East, South, and North Africa, respectively (Cole 1954; Clark 1959; McBurney 1960).

In the 1960s, archaeological teams routinely included researchers from a number of countries, involved in large interdisciplinary projects, as is well reflected in the collective volume "Background to Evolution in Africa", edited by W. W. Bishop and J.D. Clark in 1967. It included formal recommendations in order to update the African terminology, notably abandoning the obsolete "Chellean", and the improperly defined "Hand-Axe Culture". "Acheulian" was instead recommended.

This was also the time when a major advance in field research became widely accepted: the punctual data recording of archaeological excavations, championed in Europe by Leroi-Gourhan (1950). Although the importance of careful stratigraphic study had long been recognized, recording artifact distribution was just beginning. When in 1960 Mary Leakey started large-scale excavations at Olduvai, she was paying attention to the study of living floors. Further theoretical and methodological advances rest on the work of Glynn Isaac at Olorgesailie, also in the early 1960s. In the Olorgesailie monograph (Isaac 1977), he revised and simplified Kleindienst's typology, adding a metrical approach to artifact analysis. He also introduced a landscape approach and was the first to pay attention to site formation processes.

New field activity and new dating techniques began to establish the antiquity of the African Acheulean. The Leakeys led expeditions on the west side of Lake Natron at Peninj (Gowlett 1990). This new project was followed by R. Leakey and Gl. Isaac. In 1964, an age of 1.4 million years was assessed for the very early Acheulean at Peninj (Isaac 1967; Isaac and Curtis 1974). In 1963, Mary Leakey had excavated EF-HR at Olduvai, where a similar age was established for the early Acheulean (Leakey 1971, 1975). In L.S.B. Leakey's opinion the Acheulean had emerged from the Oldowan (Leakey 1936), while Mary Leakey regarded it instead as an intrusive phenomenon. In 1967 she proposed a model based on large-scale excavations in four sites of Bed I and in nine of Bed II. The interpretation of Bed I remained largely unchanged, but the sequence of Bed II was thoroughly modified (de la Torre and Mora 2014). In Middle and Upper Bed II, she differentiated the Lower Acheulean from the Developed Oldowan, basing her observations on handaxe frequencies (Leakey 1971, 1975). In her opinion, those were two different but coexisting cultures. This model was most probably influenced by Breuil's (1932) in Europe, i.e., by the supposed parallel evolution there of Clactonian and Chellean. While in L.S.B. Leakey's model (1951) the first evidence of specific tools or techniques were the main proxy for cultural change in an evolutionary sequence, Mary Leakey's (1971) hypothesis was based on the frequencies of types. As the term suggests, the Developed Oldowan was a local evolution of the Oldowan. The Lower Acheulean appeared as intrusive and unrelated to any preceding lithic complex. Leakey (1967) also speculated on links between cultures and hominins, suggesting the equation Oldowan = *Homo habilis* and Acheulean = *Homo erectus*.

M. Leakey subsequently divided the Developed Oldowan into Developed Oldowan A (DOA) and Developed Oldowan B (DOB). They were stratigraphically located respectively below and above Tuff IIB (Leakey 1971). The DOA included mostly Oldowan artifacts, although the frequency of spheroids, subspheroids, and light-duty tools increased. In the DOB the main difference is the addition of some handaxes. She also established that handaxes had to be c. 40% of the tool types in any Acheulean assemblage. Then, in 1975, after excavating in Beds III, IV, and in the Masek Beds (1968–1971), she came to the conclusion that at Olduvai there was no evidence of handaxes becoming more refined through time, and that the Oldowan persisted in Bed IV as a parallel tradition (Developed Oldowan C, or DOC; Leakey and Roe 1994).

M. Leakey's model became an issue of discussion as soon as proposed. Research in Olduvai's Beds I and II became the milestone for all subsequent investigations on the Early Stone Age.

In 1969, Isaac put forward a functional/ecological explanation for the coexisting Developed Oldowan and Acheulean (Isaac 1969). He pointed out that Developed Oldowan sites were located close to the Olduvai paleolake margins, while Acheulean sites apparently were in a fluvial environment—as Hay's research further supported some years later (1976, 1990). Besides, in Isaac's opinion, the main Acheulean innovation was the ability to detach large flake blanks for handaxe manufacture. The need of accessing large boulders for flaking regulated the landscape distribution of Developed Oldowan/Lower Acheulean sites. Accordingly, the Developed Oldowan was just a facies of the Lower Acheulean.

In 1963, J. Dekker discovered Melka Kunture and in 1964 G. Bailloud started investigating there. From the next year on, J. Chavaillon carried out large-scale excavations, revealing an impressive Early-Middle Pleistocene sequence and producing a different scenario (Bailloud 1965; Chavaillon et al. 1979; Chavaillon and Piperno 2004). Chavaillon et al. (1979) stated that the Acheulean had emerged there at c. 1.0 Ma, i.e., later than elsewhere in East Africa (Leakey 1971, 1975) and divided the local sequence into four stages: ancient (1.0 Ma), middle (0.8–0.5 Ma), upper (0.4–0.3 Ma), and final Acheulean (0.25–0.15 Ma). They concluded that the dichotomy Developed Oldowan/Lower Acheulean suggested by Leakey (1971, 1975) at Olduvai did not exist on the Ethiopian plateau. The cultural change happened locally as a gradual evolution of the technical equipment within a unilineal sequence from Oldowan to Acheulean (Chavaillon 1980; Chavaillon and Chavaillon 1980).

In the mid-1970s archaeological investigation flourished in the Ethiopian Rift, where Hadar and Gadeb were discovered (Clark and Kurashina 1979; Kalb et al. 1982; Clark et al. 1984). In the meantime, R. Leakey was working at Koobi Fora in collaboration with Gl. Isaac. Further south, L. C. King and W.W. Bishop were researching around Lake Baringo, discovering Chesowanja and Kilombe (Bishop et al. 1978; Harris and Gowlett 1980; Gowlett et al. 1981; Gowlett 1991, 1993). Discoveries made in South Africa did not have the same impact as those in East Africa, mostly because of the lack of datable volcanic deposits, but remarkable work was conducted at Amanzi Springs, Montagu Cave, and Cave of Hearths (Mason 1962, 1966; Deacon 1970, 1975; Keller 1973). Large series of Acheulean artifacts were also studied by Stiles (1979a, b) and compared to those of Olduvai Bed II.

In the 1970–1990s, several scholars revised Olduvai assemblages using typological and metrical approaches. Stiles (1977, 1980, 1991) studied those from the Middle and Upper Bed II concluding that all of them were early Acheulean. Jones (1979, 1994) focused on Bed IV artifacts, coupling his typological study with an intense experimental program. He eventually agreed with Leakey (1975) that no diachronic evolution in handaxe refinement existed from Middle Bed II to Bed IV. Nevertheless, he recognized higher reduction intensity in the handaxe shaping of DOC assemblages compared to those of the Bed IV Acheulean, probably because of more resharpening (Jones 1994). In Bed IV he differentiated DOC and Acheulean: DOC sites were devoted to varied functional activities, while Acheulean sites corresponded to discard areas.

More revision was based on published data. Davis (1980) supported the validity of M. Leakey's original model and the Developed Oldowan as a distinctive industry, criticizing Stiles (1977) for focusing instead on a single tool type. Gowlett (1988) identified disparities between the DOB and the Acheulean handaxes through statistical analysis, although he argued that morphometric dissimilarities did not necessarily correspond to two cultural phyla. To the contrary, both Callow (1994) and Roe (1994) underlined that there were substantial metrical differences when comparing DOB/DOC and Acheulean handaxes. This, in their opinion, validated the distinction between Developed Oldowan and Acheulean. Although M. Leakey's cultural model was discussed again and again, her methodology and typology were widely accepted. However, Toth (1982), Gl. Isaac (1986), Potts (1991) and later I. de la Torre and Mora (2005) as well as Semaw et al. (2009) all revised her typology.

A different theoretical approach was introduced after the excavations in 1983 at Isenya on the Kenyan highlands (Roche et al. 1988). The study of the artifact assemblages is a hallmark in East African archaeological research. P.-J. Texier and H. Roche carried out a systematic analysis following the technological approach based on the chaîne opératoire concept, developed in France since the early 1960s (Roche and Texier 1991; Texier and Roche 1995a, b; Texier 1996). This approach supersedes the study of the final state of the artifact, analyzing all of the technical sequences performed as well as the technical and cognitive skills involved in tool production (Leroi-Gourhan 1964, 1971; Pelegrin 1985; Geneste 1989, 1991; Perlès 1991; Inizan et al. 1999). This multiplies the observable production patterns and allows the researcher to investigate variations at different levels.

Although the technological approach played a minor role in 1990s, in recent years it has been frequently used for both Oldowan and Acheulean complex studies. A technological approach characterizes the revision of lithic collections excavated in previous decades at Peninj, Olduvai, Melka Kunture, and Gadeb (de la Torre et al. 2003, 2008; de la Torre and Mora 2005; de la Torre 2009, 2011; Gallotti et al. 2010, 2014; Gallotti 2013; Diez-Martín et al. 2014a, b; Gallotti and Mussi 2017; Sánchez Yustos et al. 2017) as well as the study of new assemblages from Konso, Gona, and West Turkana (Quade et al. 2004, 2008; Lepre et al. 2011; Chevrier 2012; Beyene et al. 2013, 2015).

The research developments of the last two decades provided much of the rationale for this volume. Many topics were discussed in the 2013 workshop "The Early Acheulean in East Africa" in Rome, from which this book stems. The workshop included research projects addressed to study the nature of the early Acheulean as a lithic production system(s) in a technological perspective and at microregional scale.

Over the last two decades, much effort was also aimed at redating the earliest Acheulean, whose age increased at Kokiselei 4, West Turkana (1.76 Ma; Lepre et al. 2011); KGA6-A1 (1.75 Ma) and KGA4-A2 (1.6 Ma) in Konso (Beyene et al. 2013; Suwa et al. 2015); FLK West at Olduvai ($\sim$1.7 Ma; Diez-Martín et al. 2015); BSN-12 and OGS-12 at Gona ($\sim$1.6 Ma; Quade et al. 2004); and Garba IVD at Melka Kunture ($\sim$1.6 Ma; Gallotti and Mussi 2018). Currently, solid geochronological data place the early Acheulean in East Africa between 1.76 and $\sim$1.30 Ma (Leakey 1971; Asfaw et al. 1992; Katoh et al. 2000; Beyene 2003; Roche et al. 2003; Quade et al. 2004, 2008; Nagaoka et al. 2005; de la Torre et al. 2008; Semaw et al. 2009; Lepre et al. 2011; Beyene et al. 2013; Gallotti 2013; Diez-Martín et al. 2015). Accordingly, the emergence of the Acheulean gets closer to the late Oldowan, supporting the idea that the Oldowan–Acheulean transition corresponds to a rapid change rather than the outcome of evolutionary trends (e.g., Semaw et al. 2018). This also rekindled the debate about the existence of a Developed Oldowan, as well as about the paradigm that the early Acheulean is the cultural product of *Homo erectus sensu lato*.

The first contribution of this volume is devoted to the Oldowan techno-complexes: it is a report on the current state of our knowledge in East Africa. It will help evaluating if the early Acheulean originates, or not, from earlier technologies (Gallotti 2018). Reviewing the outcome of fifteen years of techno-economic studies allows the author to identify two main Oldowan chronological horizons, an earlier one (2.6–2.3 Ma) and a later one (2.0–1.6 Ma), thus separated by a gap of 300 thousand years. In both periods, Oldowan lithic productions show a high intra- and inter-site variability, which are the outcome of multiple experiments aimed at finding the technical solutions allowing to properly exploit the available lithic resources. The various attempts happen to be alike or diverse at different levels. Furthermore, according to multiple factors they are linked to different paleoenvironments and subsistence strategies. Accordingly, Gallotti (2018) finds little empirical support for notions such as "technological stasis" and "uniformity", or for a progressive development. However, she underlines that the late Oldowan shows more intra-site as well as inter-site flaking method variability.

Unfortunately, very few East African sequences have yielded both late Oldowan and early Acheulean assemblages, complicating a detailed comparative evaluation of the technical innovations and/or traditions defining the emerging Acheulean and the status of DOB.

Texier (2018) assesses that in West Turkana the late Oldowan industry of Kokiselei 5 (KS5; 1.87 Ma) shows the full technical control of three-dimensional space, which is the prerequisite of the bifacial shaping concept later developed in the early Acheulean of Kokiselei 4 (KS4; 1.76 Ma). The diversification of flaking methods identified at KS5 confirmed that the late Oldowan knappers of West Turkana were able to exploit a wider range of raw materials than those of Lokalalei 2C (2.34 Ma). The same variability of flaking methods characterizes both the late Oldowan (Garba IVE-F, ~1.7 Ma) and the early Acheulean (Garba IVD, ~1.6 Ma) at Melka Kunture, as well as the oldest Acheulean industry of Olduvai (Diez-Martín et al. 2015). Nevertheless, the knappers of Garba IVD provide evidence of a technological leap: they acquired an incipient ability to configure the raw material geometry thanks to the preparation of the striking platform, the management of volume/convexity during flaking, and the setting of a hierarchy among surfaces. The same innovations appear both in small–medium and in large flake extraction, representing the central technical advancement of the early Acheulean in the Ethiopian highlands. Cores showing a radial or centripetal exploitation are also frequent at earlier sites, such as DK in Olduvai (de la Torre and Mora 2005) and Gona (Stout et al. 2010). However, this happens within methods where core volume and convexity configuration are not fully managed, and there is no hierarchy between a flaking surface and a prepared striking platform.

Prepared methods appear systematically around 1.6–1.3 Ma, associated or not with large tool productions. At Nyabusosi (~1.5 Ma), the centripetal exploitation of one surface from a natural or prepared striking platform is the only flaking method for small flake production (Texier 2005). At Gadeb 2E, there are examples of well-structured exploitation sequences, such as hierarchical centripetal and discoid methods (de la Torre 2011). At Olduvai BK and TK Upper Floor (~1.35 Ma), in some cases the centripetal hierarchical method is implemented, which corresponds to a change in the flaking modalities in the Olduvai sequence (de la Torre and Mora 2005). As argued by de la Torre et al. (2008), the adoption of these flaking criteria is relevant in cultural terms, because the same technological knowledge seems shared by knappers at Peninj both in the ST complex and in the Escarpment, regardless of the presence of large tools. Thus, this technical feature was used by de la Torre et al. (2008) to assign the ST complex to the early Acheulean, while it was previously classified as Oldowan (de la Torre et al. 2003). Nevertheless, the reanalysis (Diez-Martín et al. 2012, 2018) of some of the cores studied by de la Torre et al. (2003) questions the existence of the bifacial hierarchical centripetal exploitation at Type Section. New criteria are suggested to assign this assemblage to the early Acheulean, i.e., the presence of large flakes and of resharpening/configuration flakes from large tool shaping.

Core preparation as a proxy for the emergence of the Acheulean has been recently reinvestigated in Olduvai at SHK (~1.5 Ma) and BK. Sánches Yustos et al. (2017) recognize that core preparation allows producing serviceable striking angles, when the latter are not available in the original core blank. Summing up, core preparation, volume management in a three-dimensional space, and hierarchy among surfaces all occur in a variety of operative schemes. By commanding these technical solutions and gaining the ability to exploit a wider range of geometries, the toolmakers freed themselves for the first time from the constraints of the natural blanks. This advancement is present both in DOB and in early Acheulean assemblages, suggesting a close relationship between them.

The ambiguous status of the DOB has been a matter of debate ever since its first definition. Several authors criticized the dichotomy between DOB and early Acheulean and proposed instead to assign DOB industries lacking handaxes to the early Acheulean. This means that the two industries are interrelated within the same cultural tradition (de la Torre and Mora 2005, 2014; de la Torre et al. 2008; Semaw et al. 2009, 2018; de la Torre 2011, 2016; Diez-Martín and Eren 2012; Gallotti 2013). Nevertheless, although a technological approach has been systematically adopted in the last two decades, early technology paradigms are often rooted in previous typological postulates. This epistemological contradiction constrained the DOB/early Acheulean debate (Sánchez-Yustos et al. 2017), while there is not yet a formal redefinition of the DOB status in technological terms. Besides, the DOA is not properly discriminated from the early Acheulean. At Olduvai the DOA is stratigraphically located below and above Tuff IIB, dated to ~1.6 Ma, while the early Acheulean has been recently discovered at FLK West (~1.7 Ma; Diez-Martín et al. 2015).

The focus on small débitage in the last few years is a relevant analytical development of early technology research. It modifies previously established paradigms and partially supersedes the handaxe "abuse" in Acheulean studies. For decades, analyses focusing on this single component had bolstered the idea of a uniform and static Acheulean, lacking innovation over hundreds of thousands of years and across a number of varied environmental settings (e.g., Nowell and White 2010). The typological features and degree of refinement of handaxes were used to group together lithic assemblages far apart in space and time and then to chart the supposed evolution of Acheulean technology. But, although the outcomes of technical processes might be typologically similar, there are many ways of combining raw material selection and acquisition patterns, percussion motions, and technical sequences (Gallotti 2018). Accordingly, the recent technological studies also shed new light on the high variability of the large tools in the early Acheulean.

Large tool production definitely is an innovation signaling the emergence of the Acheulean. The presence and frequency of the new chaînes opératoires within 1.76–1.3 Ma East African assemblages show relevant intra- and inter-site variability. Raw material provisioning is frequently based on local secondary sources, as in the case of small débitage (de la Torre et al. 2008; Harmand 2009; Gallotti 2013; Díez-Martín et al. 2018; Gallotti and Mussi 2018; Santonja et al. 2018; Texier 2018), but also on primary sources when specific raw materials are looked for (de la Torre and Mora 2005; de la Torre 2011). Nevertheless, the selection of morphologies and lithotypes is not a new acquisition; although the published evidence is limited, this behavior has been documented at earlier sites (Plummer et al. 1999; Hovers et al. 2002; Stout et al. 2005; Goldman-Neuman and Hovers 2009; Harmand 2009; Gallotti 2018).

Large tools were manufactured on various blanks, such as large cobbles, tabular clasts, and large flakes. The ability to detach large flakes has been considered as the very distinctive trait of the early Acheulean (e.g., Isaac 1969). Unfortunately, the related large cores are very rare or altogether lacking. Most inferences on flaking methods derive from the observation of large tools. At KS4, large flakes were obtained by splitting large cobbles (Texier 2018). At FLK West, large flakes were detached by bifacial or multifacial exploitation when other small- and medium-size flake series were also produced (Diez-Martín et al. 2015). In this case large flakes do not seem the outcome of a chaîne opératoire distinct from small débitage. At Melka Kunture, in contrast, there is marked metrical discontinuity between small–medium and large cores. The Garba IVD early Acheulean site yielded cores documenting a specific flaking process to produce large flakes to be turned into large tools. Such flaking methods involve the systematic preparation of the flaking surface and volume management according to the discoid concept (Gallotti 2013; Gallotti and Mussi 2018). Similar technological patterns have been suggested by de la Torre et al. (2008) at Peninj. On the other hand, Diez-Martín et al. (2018) argue that at Peninj Escarpment sites the technical patterns of large tools, and unmodified large flakes suggest bifacial models based on the orthogonal intersections of the removals. In other instances, as at Olduvai TK, the criterion that guided the selection of slabs, i.e., two natural symmetrical and parallel surfaces, is comparable to that of the flakes, making bifacial reduction easier (Santonja et al. 2018). Overall, the technical parameters followed to select or produce large blanks were inconsistent, just as the retouch/shaping processes. On large tools, unifacial or bifacial retouching is often quite limited and definitely not invasive. It never aims at managing the whole volume, just at modifying edges. If there is shaping at all, it is limited to part of the volume, creating a pointed tip (Diez-Martín et al. 2018; Gallotti and Mussi 2018; Santonja et al. 2018; Semaw et al. 2018; Texier 2018). Such large tools coexist, even in the same assemblage, with highly symmetrical and bifacially flaked large tool types, showing a full management of the blank volume (Diez-Martín et al. 2018; Santonja et al. 2018). In this scenario, cleavers (sensu Tixier 1956) are very rare and do not show the predetermined aspects of the blanks typical of later assemblages (e.g., Gallotti and Mussi 2018; Texier 2018).

In summary, between 1.76 and 1.3 Ma in terms of both final form and the stages of the chaînes opératoires, the large tools are more diverse and variable than usually assumed in the literature.

The contributors to this volume do not speak in one voice, as each chapter discusses the emergence of the Acheulean in a specific site context. The review of past and present perspectives suggests that, after more than a century of research on the East African Acheulean, a large amount of data is now available on its emergence. Nevertheless, efforts are still needed to establish a comprehensive chronostratigraphic framework. Other topics also remain open issues. From a technological perspective, the origin and end, unity or variability at intra and inter-site scale, definition, and even the classical equation Acheulean = handaxe (or in some cases Acheulean = bifacial phenomenon) are all matters of discussion.

The factors that caused the technical innovations leading to the origin of the Acheulean might be linked to changes in biotic/abiotic resources, in climate and in paleoenvironment. Two papers in this volume explore the paleobotanical context of Melka Kunture in the Early/Middle Pleistocene (Bonnefille et al. 2018) and the faunal composition (Geraads 2018) at the Oldowan–Acheulean transition. Bonnefille et al. (2018) argue that in the Ethiopian highlands *Homo erectus sensu lato* adapted to mountain climatic conditions with marked daily temperature contrast. This happened when *Homo erectus* produced both late Oldowan and early Acheulean industries. Geraads (2018) reviews the large mammal record of several major eastern African sites. He concludes that the Oldowan–Acheulean transition, which is a key event in human evolution, does not correspond to any major turnover of large mammal faunas. Both results do not support the paradigm that the cultural and biological changes in human evolution recorded in East Africa between 1.9 and 1.5 Ma are contemporaneous with and possibly fostered by modifications of the natural environment.

Besides, while much research focused on the chronostratigraphic limit of the Acheulean emergence and on its relationship with the Oldowan, the last appearance datum of the early Acheulean and the developments after ~1.3 Ma are still poorly understood. From then up to 0.7 Ma, long stratigraphic sequences are known in East Africa at a limited number of sites, and several actually are of uncertain age (e.g., Hay 1976; Isaac and Isaac 1997; Roche et al. 2003; Quade et al. 2008; de la Torre 2011; Beyene et al. 2013; Gallotti and Mussi 2017).

In West Turkana, no sites are recovered between Kokiselei 4, dated to 1.76 Ma, and Nadung'a 4, dated to ~0.7 Ma (Roche et al. 2003). A gap of 0.3 Ma is recorded in the Busidima Formation at Gona between ~1.3 Ma and ~1.0 Ma (Quade et al. 2008). At Konso, KGA12-A1 is dated to ~1.25 Ma. After a hiatus of the archaeological record lasting 0.4 Myr, KGA20-A1 and KGA20-A2 are both dated to ~0.85 Ma (Beyene et al. 2013). The Gadeb sites occur within a long and loosely defined time interval between 1.45 and 0.7 Ma, without a specific age for each occurrence (Clark and Kurashina 1979; de la Torre 2011). At Olduvai, Bed II is stratigraphically complex, with facies changes, faulting, and unconformities. Further investigations are definitely required to improve the chrono-stratigraphic framework of the archaeological sites (McHenry et al. 2016). Additionally, sites in Bed III are mostly dated only by paleomagnetism and sedimentation rates (Hay 1994; but Kimbel 1997 reports a date of 1.33 for basal Tuff III-1 from Manega 1993). The age of the contact between Beds II and III has been estimated at ~1.2 Ma (McHenry et al. 2016), but Delson and Van Couvering (2000) suggested an age closer to 1.35 Ma. The age of the top of Bed III was estimated at ca. 0.8 Ma (Hay 1994), but Tamrat et al. (1995) imply a much older date (Delson and Van Couvering 2000). Only one site with a substantial concentration of fauna and stone artifacts is currently known in Bed III, i.e., Juma's Korongo (Pante 2013). Further investigations are definitely required to improve the chrono-stratigraphic framework of the archaeological sites (McHenry et al. 2016). In the Dawaitoli Formation of the Middle Awash a tuff at the base of Member U-2 has been dated to 0.64 Ma. Acheulean sites are located both below and above this tuff, but precise chronometric dates are not available (Schick and Clark 2003).

In contrast with the dearth of evidence in areas with long-established sequences, new areas were settled for the first time around 1 Ma. This is notably the case of Isenya, Olorgesailie, Kariandusi, and Kilombe (Isaac 1977; Gowlett 1993; Gowlett and Crompton 1994; Durkee and Brown 2014).

The reason why the archaeological evidence decreases over time in East Africa, or is altogether lacking, before this fresh surge in the number of sites has not yet been investigated in any detail. In large-scale syntheses, this gap in the record is generally overlooked. To make it even more difficult to properly understand the ongoing changes, human fossils are very rare in this interval (e.g., Manzi 2012; Ghinassi et al. 2015; Profico et al. 2016). This happens in a crucial period of human evolution, when *Homo ergaster/erectus* evolves and disappears, while *Homo heidelbergensis* emerges.

After ~1.5–1.2 Ma a global climate change also occurs. This is the Early/Middle Pleistocene Transition, when the dominant periodicity of glacial/interglacial cycles shifts from 41,000 to 100,000 years, and ice-caps start accumulating at northern latitudes. The change is well documented in the northern hemisphere, between broadly MIS 36 (~1.2 Ma) and MIS 13 (~0.54–0.46 Ma) (e.g., Lisiecki and Raymo 2005). In the Mediterranean, the deposition of hematite-rich aeolian dust from the Sahara increases after ~0.95 Ma, positively suggesting aridity in Africa at 0.87 Ma (MIS 22) (Larrasoaña et al. 2003). More recently, Trauth et al. (2009) evidenced increasing aridity in Africa after ~1.5 Ma, a trend matched south of the Sahara by a shift from C_3 to C_4 vegetation between ~1.5 and ~0.7 Ma, as evidenced by the stable carbon isotope record (Ségalen et al. 2007). As Bonnefille et al. (2018) underlined, at Melka Kunture a dramatic change in vegetation at ~0.8 Ma points to a climate much cooler than today and than ever before. The high elevation, above 2000 m asl, grants more visibility to climatic variation than at lower elevation in the Rift Valley.

The gap in late Early/early Middle Pleistocene record had a causal role in conventionally fixing at ~1.3 Ma the "end point" of the early Acheulean. Sharon (2007, 2010) identifies two stages within the Acheulean techno-complex, based on the systematic use of large flakes as large tool blanks: (1) an early Acheulean, predating 1 Ma, when large flakes are not a primary technological praxis and cleavers are absent; and (2) a subsequent Large Flake Acheulean (LFA), a "distinct segment in the Acheulian techno-complex that is technologically and typologically distinguishable from others" (Sharon 2010: 228). Texier as well as Gallotti and Mussi address this topic in this volume. These authors assess that at ~1.0 Ma a marked change occurs. The predetermination of large tool blank technical aspects, the standardization of large tool types, the systematic provisioning at primary sources, and the consequent fragmentation of the large tool chaînes opératoires are a giant leap in technical productions during the late Early Pleistocene. At Melka Kunture, these technical innovations occur at the time of the gradual emergence of a new and more encephalized type of hominin: *Homo heidelbergensis* (Profico et al. 2016; Gallotti and Mussi 2017, 2018).

Finally, in the last thirty years, a general consensus emerged that "the Acheulian lithic technology was transported out of Africa" (Santonja and Villa 2006: 467). Based on this axiom, several out of Africa models have been suggested and discussed (e.g., Bar-Yosef 1987; Carbonell et al. 1999, 2010; Bar-Yosef and Belfer-Cohen 2001; Rightmire 2001; Mithen and Reed 2002; Kozlowski 2005; Santonja and Villa 2006; Lycett and von Cramon-Taubadel 2008; Despriée et al. 2010; Gallotti 2016). Accordingly, the last two chapters of this volume are devoted to a review of the oldest Acheulean evidence beyond Africa, i.e., in Asia (Dennell 2018) and in Western Europe (Moncel and Ashton 2018).

Acknowledgements We would like to express deep thanks to Marie-Hélène Moncel, to Pierre-Jean Texier, and to Eric Delson for their useful suggestions.

References

Arambourg, C. (1955). Le gisement de Ternifine et l'Atlanthropus. *Bulletin de la Société Préhistorique Française, 52,* 90–95.

Asfaw, B., Beyene, Y., Suwa, G., Walker, R. C., White, T. D., WoldeGabriel, G., et al. (1992). The earliest Acheulean from Konso-Gardula. *Nature, 360,* 732–735.

Bailloud, G. (1965). *Le gisement paléolithique de Melka-Konturé.* Addis Abeba: Institut Ethiopien d'Archéolgie, Cahier n.1.

Balout, L. (1955). *Préhistoire de l'Afrique du Nord, essai de chronologie.* Paris: Arts et Métiers Graphiques.

Balout, L., Biberson, P., & Tixier, J. (1967). L'Acheuléen de Ternifine, gisement de l'Atlanthrope. *L'Anthropologie, 71,* 217–235.

Bar-Yosef, O. (1987). Pleistocene connexions between Africa and Southwest Asia: an archaeological perspective. *The African Archaeological Review, 5,* 29–38.

Bar-Yosef, O., & Belfer-Cohen, A. (2001). From Africa to Eurasia—early dispersals. *Quaternary International, 75,* 19–28.

Beyene, Y. (2003). The emergence and development of the Acheulean at Konso. *Anthropological Science, 111,* 58.

Beyene, Y., Katoh, S., WoldeGabriel, G., Hart, W. K., Uto, K., Sudo, M., et al. (2013). The characteristics and chronology of the earliest Acheulean at Konso, Ethiopia. *Proceedings of the National Academy of Sciences, 110,* 1584-1591.

Beyene, Y., Asfaw, B., Katsuhiro, S., & Suwa, G. (2015). *Konso-Gardula Research Project, Vol. 2. Archaeological Collections: Background and the Early Acheulean Assemblages.* Tokyo: The University Museum, The University of Tokyo.

Biberson, P. (1961). *Le Paléolithique inférieur du Maroc atlantique.* Rabat: Publications du Service des Antiquités du Maroc, Fascicule 16.

Bishop, W. W., & Clark, J. D. (1967). *Background to Evolution in Africa.* Chicago: Chicago University Press.

Bishop, W. W., Hill, A., & Pickford, M. (1978). Chesowanja: A revised geological interpretation. In W. W. Bishop (Ed.), *Geological Background to Fossil Man* (pp. 309–327). Edinburgh: Scottish Academic Press.

Boucher de Perthes, J. (1847). *Antiquités celtiques et antédiluviennes, mémoire sur l'industrie primitive et les arts à leur origine*, I vol. Paris.

Bonnefille, R., Melis, R. T., & Mussi, M. (2018). Variability in the Mountain Environment at Melka Kunture Archaeological Site, Ethiopia, During the Early Pleistocene (~1.7 Ma) and the Mid-Pleistocene Transition (0.9–0.6 Ma). In R. Gallotti & M. Mussi (Eds.), *The Emergence of the Acheulean in East Africa and Beyond. Contributions in Honor of Jean Chavaillon.* Cham: Springer.

Breuil, H. (1930). L'Afrique préhistorique. *Cahiers d'Art, 8–9,* 449–500.

Breuil, H. (1932). Les industries à éclats du paléolithique ancien, I, le Clactonien. *Préhistoire, 1,* 125–190.

Breuil, H. (1952). Raised marine beaches around the African continent and their relation to Stone Age cultures. In L. S. B. Leakey & S. Cole (Eds.), *Proceedings of the Pan-African Congress on Prehistory*, Nairobi 1947 (pp. 91–93). Oxford: Basil Blackwell.

Callow, P. (1994). The Olduvai bifaces: technology and raw materials. In M. D. Leakey & D. A. Roe (Eds.), *Olduvai Gorge. Volume 5. Excavations in Beds III, IV and the Masek Beds, 1968–1971* (pp. 235–253). Cambridge: Cambridge University Press.

Carbonell, E., Mosquera, M., Rodríguez, X. P., Sala, R., & Made, J. V. D. (1999). Out of Africa: The Dispersal of the Earliest Technical Systems Reconsidered. *Journal of Anthropological Archaeology, 18,* 119–136.

Carbonell, E., Sala, R., Rodríguez, X. P., Mosquera, M., Ollé, A., Vergés, J. M., et al. (2010). Early hominid dispersals: A technological hypothesis for "out of Africa". *Quaternary International, 223–224,* 36–44.

Chantre, E. (1908). L'âge de la pierre dans la Berberie orientale, Tripolitaine et Tunisie. *37e Congrès Association française pour l'Avancement des Sciences*, Clermont Ferrand, 2e partie (pp. 686–688).

Chavaillon, J. (1980). Chronologie archéologique de Melka-Kunturé (Ethiopie). In R. E. Leakey & B. A. Ogot (Eds.), *Proceedings VIII Panafrican Congress of Prehistory and Quaternary Studies, 1977, Nairobi* (pp. 200–201). Nairobi: The International Louis Leakey Memorial Institute for African Prehistory.

Chavaillon, J., & Chavaillon, N. (1980). Evolution de l'Acheuléen à Melka-Kunturé (Ethiopie). *Anthropologie, XVIII*(2/3), 153–159.

Chavaillon, J., & Piperno, M. (2004). *Studies on the Early Paleolithic site of Melka Kunture, Ethiopia.* Florence: Origines.

Chavaillon, J., Chavaillon, N., Hours, F., & Piperno, M. (1979). From the Oldowan to the Middle Stone Age at Melka-Kunture (Ethiopia). Understanding cultural changes. *Quaternaria, XXI,* 1–26.

Chevrier, B. (2012). Les assemblages à pièces bifaciales au Pléistocene inférieur et moyen ancien en Afrique de l'Est et au Proche-Orient. Nouvelle approche du phénomène bifacial appliquée aux problématiques de migrations, de diffusion et d'évolution locale. Ph.D. Dissertation, Université Paris Ouest Nanterre La Défense.

Clark, J. D. (1952). Recent prehistoric research in the Somalilands. In L. S. B. Leakey & S. Cole (Eds.), *Proceedings of the Pan-African Congress on Prehistory, Nairobi 1947* (pp. 146–164). Oxford: Basil Blackwell.

Clark, J. D. (1959). *The Prehistory of Southern Africa.* Harmondsworth: Penguin.

Clark, J. D. (1969). *Kalambo Falls Prehistoric Site, Vol. I: The geology, palaeoecology and detailed stratigraphy of the excavations.* Cambridge: Cambridge University Press.

Clark, J. D. (1974). *Kalambo Falls Prehistoric Site, Vol. II: The late prehistoric remains.* Cambridge: Cambridge University Press.

Clark, J. D., & Kurashina, H. (1979). An analysis of earlier Stone Age bifaces from Gadeb (Locality 8E), Northern Bale Highlands, Ethiopia. *The South African Archaeological Bulletin, 34,* 93–109.

Clark, J. D., Asfaw, B., Assefa, G., Harris, J. W. K., Kurashina, H., Walter, R. C., et al. (1984). Palaeoanthropological discoveries in the Middle Awash Valley, Ethiopia. *Nature, 307,* 423–428.

Cole, S. (1954). *The Prehistory of East Africa.* Harmondworth: Penguin.

Davis, D. D. (1980). Further consideration of the developed Oldowan at Olduvai Gorge. *Current Anthropology, 21,* 840–843.

Deacon, H. J. (1970). The Acheulian occupation at Amanzi Springs, Uitenhage District, Cape Province. *Annals of the Cape Provincial Museums (Natural History), 8,* 89–189.

Deacon, H. J. (1975). Demography, subsistence and culture during the Acheulian in southern Africa. In K. W. Butzer & G. Ll. Isaac (Eds.), *After the Australopithecines* (pp. 543–570). The Hague: Mouton.

Déchelette, J. (1924). *Manuel d'archéologie préhistorique, celtique et gallo-romaine.* Paris: Alphonse Picard et fils.

de Mortillet, G. (1872). Classification des diverses périodes de l'âge de la pierre. *Revue d'Anthropologie, 1,* 432–435.

de Mortillet, G. (1883). *Le Préhistorique, Antiquité de l'homme.* Paris: Reinwald.

de la Torre, I. (2009). Technological Strategies in the Lower Pleistocene at Peninj (West of Lake Natron, Tanzania). In K. Schick & N. Toth (Eds.), *The Cutting Edge: New Approaches to the Archaeology of Human Origins* (pp. 93–113). Gosport, IN: The Stone Age Institute Publication Series 3.

de la Torre, I. (2011). The Early Stone Age lithic assemblages of Gadeb (Ethiopia) and the developed Oldowan/early Acheulean in East Africa. *Journal of Human Evolution, 60,* 768–812.

de la Torre, I. (2016). The origins of the Acheulean: Past and present perspectives on a major transition in human evolution. *Philosophical Transactions B, 371,* 20150245.

de la Torre, I., & Mora, R. (2005). *Technological Strategies in the Lower Pleistocene at Olduvai Beds I & II.* Liege: ERAUL.

de la Torre, I., & Mora, R. (2014). The transition to the Acheulean in East Africa: An assessment of paradigms and evidence from Olduvai Gorge (Tanzania). *Journal of Archaeological Method and Theory, 21,* 781–823.

de la Torre, I., Mora, R., Domínguez-Rodrigo, M., De Luque, L., & Alcalá, L. (2003). The Oldowan industry of Peninj and its bearing on the reconstruction of the technological skills of Lower Pleistocene hominids. *Journal of Human Evolution, 44,* 203–224.

de la Torre, I., Mora, R., & Martínez-Moreno, J. (2008). The early Acheulean in Peninj (Lake Natron, Tanzania). *Journal of Anthropological Archaeology, 27,* 244–264.

Delson, E., & Van Couvering, J. A. (2000). Composite stratigraphic chart of Olduvai Gorge. In E. Delson, I. Tattersall, J. A. Van Couvering, & A. S. Brooks (Eds.), *Encyclopedia of Human Evolution and Prehistory, Second Edition* (p. 488). New York and London: Taylor & Francis Group.

Dennell, R. (2018). The Acheulean Assemblages of Asia: A Review. In R. Gallotti & M. Mussi (Eds.), *The Emergence of the Acheulean in East Africa and Beyond. Contributions in Honor of Jean Chavaillon.* Cham: Springer.

Despriée, J., Voinchet, P., Tissoux, H., Moncel, M.-H., Arzarello, M., Robin, S., et al. (2010). Lower and Middle Pleistocene human settlements in the middle Loire River Basin, Centre Region, France. *Quaternary International, 223–224,* 345–359.

Diez-Martín, F., & Eren, M. I. (2012). The Early Acheulean in Africa: Past paradigms, current ideas, and future directions. In M. Domínguez-Rodrigo (Ed.), *Stone tools and fossil bones. Debates in the archaeology of human origins* (pp. 310–357). Cambridge: Cambridge University Press.

Diez-Martín, F., Cuartero, F., Sánchez Yustos, P., Baena, J., Domínguez-Rodrigo, M., & Rubio, D. (2012). Testing cognitive skills in Early Pleistocene hominins: An analysis of the concepts of hierarchization and predetermination in the lithic assemblages of Type Section (Peninj, Tanzania). In M. Domínguez-Rodrigo (Ed.), *Stone tools and fossil bones. Debates in the archaeology of human origins* (pp. 245–309). Cambridge: Cambridge University Press.

Diez-Martín, F., Sánchez Yustos, P., Gómez de la Rúa, D., Gómez González, J. A., de Luque, L., & Barba, R. (2014a). Early Acheulean technology at Es2-Lepolosi (ancient MHS-Bayasi) in Peninj (Lake Natron, Tanzania). *Quaternary International, 322–323,* 209–236.

Diez-Martín, F., Sánchez Yustos, P., Gómez González, J. A., Luque, L., Gómez de la Rúa, D., & Domínguez-Rodrigo, M. (2014b). Reassessment of the Early Acheulean at EN1-Noolchalai (Ancient RHS-Mugulud) in Peninj (Lake Natron, Tanzania). *Quaternary International, 322–323,* 237–263.

Diez-Martín, F., Sánchez Yustos, P., Uribelarrea, D., Baquedano, E., Mark, D.F., Mabulla, A., et al. (2015). The Origin of the Acheulean: The 1.7 Million-Year-Old Site of FLK West, Olduvai Gorge (Tanzania). *Scientific Reports*, *5*, 17839.

Diez-Martín, F., Sánchez Yustos, P., & de Luque L. (2018). The East African Early Acheulean of Peninj (Lake Natron, Tanzania). In R. Gallotti & M. Mussi (Eds.), *The Emergence of the Acheulean in East Africa and Beyond. Contributions in Honor of Jean Chavaillon.* Cham: Springer.

Durkee, H., & Brown, F. H. (2014). Correlation of volcanic ash layers between the early Pleistocene Acheulean sites of Isinya, Kariandusi, and Olorgesailie, Kenya. *Journal of Archaeological Science, 49,* 510–517.

Flamand, G. B. M., & Laquière, E. (1906). Nouvelles recherches sur la Préhistoire dans le Sahara et le Haut pays oranais. *Revue Africaine, 50,* 204–243.

Frere, J. (1800). Account of flint weapons discovered at Hoxne in Suffolk, in a letter to the Rev. John Brand, Secretary. *Archaeologia, 13,* 204–205.

Gallotti, R. (2013). An older origin for the Acheulean at Melka Kunture (Upper Awash, Ethiopia). Techno-economic behaviors at Garba IVD. *Journal of Human Evolution, 65,* 594–620.

Gallotti, R. (2016). The East African origin of the Western European Acheulean technology: Fact or paradigm? *Quaternary International, 411,* 9–24.

Gallotti, R. (2018). Before the Acheulean in East Africa: An Overview of the Oldowan Lithic Assemblages. In R. Gallotti & M. Mussi (Eds.), *The Emergence of the Acheulean in East Africa and Beyond. Contributions in Honor of Jean Chavaillon.* Cham: Springer.

Gallotti, R., & Mussi, M. (2017). Two Acheuleans, two humankinds: From 1.5 to 0.85 Ma at Melka Kunture (Upper Awash, Ethiopian highlands). *Journal of Anthropological Sciences, 95,* 137–181.

Gallotti, R., & Mussi, M. (2018). Before, During, and After the Early Acheulean at Melka Kunture (Upper Awash, Ethiopia): A Techno-economic Comparative Analysis. In R. Gallotti & M. Mussi (Eds.), *The Emergence of the Acheulean in East Africa and Beyond. Contributions in Honor of Jean Chavaillon.* Cham: Springer.

Gallotti, R., Collina, C., Raynal, J.-P., Kieffer, G., Geraads, D., & Piperno, M. (2010). The Early Middle Pleistocene site of Gombore II (Melka Kunture, Upper Awash, Ethiopia) and the issue of Acheulean Bifacial Shaping Strategies. *African Archaeological Review, 27,* 291–322.

Gallotti, R., Raynal, J.-P., Geraads, D., & Mussi, M. (2014). Garba XIII (Melka Kunture, Upper Awash, Ethiopia): A new Acheulean site of the late Lower Pleistocene. *Quaternary International, 343,* 17–27.

Geneste, J. M. (1989). Économie des ressources lithiques dans le Moustérien du sud ouest de la France. In M. Otte (Ed.), *L'Homme de Néanderthal, La Subsistance* (Vol. 6, pp. 75–97). Liege: ERAUL.

Geneste, J. M. (1991). Systèmes techniques de production lithique : Variations technoéconomiques dans les processus de réalisation des outillages paléolithiques. *Technique et Culture, 17*(18), 1–35.

Geraads, D. (2018). Faunal Change in Eastern Africa at the Oldowan—Acheulean Transition. In R. Gallotti & M. Mussi (Eds.), *The Emergence of the Acheulean in East Africa and Beyond. Contributions in Honor of Jean Chavaillon.* Cham: Springer.

Ghinassi, M., Oms, O., Papini, M., Scarciglia, F., Carnevale, G., Sani, F., et al. (2015). An integrated study of the *Homo*-bearing Aalat stratigraphic section (Eritrea): An expanded continental record at the Early-Middle Pleistocene transition. *Journal of African Earth Sciences, 112,* 163–185.

Goldman-Neuman, T., & Hovers, E. (2009). Methodological issues in the study of Oldowan raw material selectivity: Insights from A.L. 894 (Hadar, Ethiopia). In E. Hovers & D. R. Braun (Eds.), *Interdisciplinary Approaches to the Oldowan* (pp. 71–84). Dordrecht: Springer.

Goodwin, A. J. H., & Van Riet Lowe, C. (1929). The Stone Age cultures of South Africa. *Annals of the South African Museum, 27,* 1–289.

Gowlett, J. A. J. (1988). A case of developed Oldowan in the Acheulean? *World Archaeology, 20,* 13–26.

Gowlett, J. A. J. (1990). Archaeological studies of human origins & early prehistory in Africa. In P. Robertshaw (Ed.), *A History of African Archaeology* (pp. 13–38). London: James Currey Ltd.

Gowlett, J. A. J. (1991). Kilombe—Review of an Acheulian site complex. In J. D. Clark (Ed.), *Cultural beginnings* (pp. 129–136). Bonn: Dr. Rudolf Habelt Gmbh.

Gowlett, J. A. J. (1993). Le site acheuléen de Kilombe : stratigraphie, géochronologie, habitat et industrie lithique. *L'Anthropologie, 97,* 69–84.

Gowlett, J. A. J., & Crompton, R. H. (1994). Kariandusi: Acheulian morphology and the question of allometry. *The African Archaeological Review, 12,* 3–42.

Gowlett, J. A. J., Harris, J. W. K., Walton, D., & Wood, B. A. (1981). Early archaeological sites, hominid remains and traces of fire from Chesowanja, Kenya. *Nature, 294,* 125–129.

Gregory, J. W. (1896). *The Great Rift Valley*. London: Frank Cass.

Gregory, J. W. (1921). *The Rift Valleys and Geology of East Africa.* London: Seeley, Service.

Harmand, S. (2009). Variability in Raw Material Selectivity at the Late Pliocene site of Lokalalei, West Turkana, Kenya. In E. Hovers & D. R. Braun (Eds.), *Interdisciplinary Approaches to the Oldowan* (pp. 85–97). Dordrecht: Springer.

Harris, J. W. K., & Gowlett, J. A. J. (1980). Evidence of early stone industries at Chesowanja, Kenya. In R. E. Leakey & B. A. Ogot (Eds.), *Proceedings VIII Panafrican Congress of Prehistory and Quaternary Studies, 1977, Nairobi* (pp. 208–212). Nairobi: The International Louis Leakey Memorial Institute for African Prehistory.

Hay, R. L. (1976). *Geology of the Olduvai Gorge*. Berkeley: University of California Press.

Hay, R. L. (1990). Olduvai Gorge: a case history in the interpretation of hominid paleoenvironments in East Africa. In L. F. Laporte (Ed.), *Establishment of a Geologic Framework for Paleoanthropology* (Vol. 242, pp. 23–37). Boulder: Geological Society of America.

Hay, R. (1994). Geology and dating of Beds III, IV, and the Masek Beds. In M. D. Leakey & D. A. Roe (Eds.), *Olduvai Gorge Excavations in Beds III, IV, and the Masek Beds 1968–1971* (pp. 8–14). Cambridge: Cambridge University Press.

Hovers, E., Schollmeyer, K., Goldman, T., Eck, G., Reed, K. E., Johanson, D. C., et al. (2002). Late Pliocene Archaeological Sites in Hadar, Ethiopia. *Journal of Human Evolution, 42,* 1–17.

Howell, F. C., Cole, G. H., & Kleindienst, M. R. (1962). Isimila, an Acheulean occupation site in the Iringa Highlands, Southern Highlands Province, Tanganyika. In G. Mortelmans & Nenquin J. (Eds.), *Proceedings IV Panafrican Congress of Prehistory* (pp. 43–80). Tervuren: Musée royal de l'Afrique central.

Inizan, M. L., Reduron-Ballinger, M., Roche, H., & Tixier, J. (1999). *Technology and Terminology of Knapped Stone (Préhistoire de la Pierre taillée 5).* Nanterre: CREP.

Isaac, G. Ll. (1967). The stratigraphy of the Peninj Group—early Middle Pleistocene Formations west of Lake Natron, Tanzania. In W. W. Bishop & J. D. Clark (Eds.), *Background to Evolution in Africa* (pp. 229–258). Chicago: University of Chicago Press.

Isaac, G. Ll. (1969). Studies of early culture in East Africa. *World Archaeology, 1,* 1–28.

Isaac, G. Ll. (1977). *Olorgesailie. Archeological Studies of a Middle Pleistocene Lake Basin in Kenya.* Chicago: University of Chicago Press.

Isaac, G. Ll. (1986). Foundation stones: Early artefacts as indicators of activities and abilities. In G. N. Bailey, P. Callow, & C. B. M. McBurney (Eds.), *Stone Age Prehistory: Studies in Memory of Charles McBurney* (pp. 221–241). Cambridge: Cambridge University Press.

Isaac, G. Ll., & Curtis, G. H. (1974). Age of the Acheulean industries from the Peninj Group, Tanzania. *Nature, 249,* 624–627.

Isaac, G. Ll., & Isaac, B. (1997). *Koobi Fora Research Project, Plio-Pleistocene Archaeology* (Vol. 5). Oxford: Clarendon Press.

Jones, P. R. (1979). Effects of raw materials on biface manufacture. *Science, 204,* 835–836.

Jones, P. R. (1994). Results of experimental work in relation to the stone industries of Olduvai Gorge. In M. D. Leakey & D. A. Roe (Eds.), *Olduvai Gorge. Volume 5. Excavations in Beds III, IV and the Masek Beds, 1968–1971* (pp. 254–298). Cambridge: Cambridge University Press.

Kalb, J. E., Jolly, C. J., Mebrate, A., Tebedge, S., Smart, C., Oswald, E. B., et al. (1982). Fossil mammals and artefacts from the Middle Awash Valley, Ethiopia. *Nature, 298,* 25–29.

Katoh, S., Nagaoka, S., WoldeGabriel, G., Renne, P., Snow, M. G., Beyene, Y., et al. (2000). Chronostratigraphy and correlation of the Plio-Pleistocene tephra layers of the Konso Formation, southern Main Ethiopian Rift, Ethiopia. *Quaternary Science Review, 19,* 1305–1317.

Keller, C. M. (1973). *Montagu Cave in Prehistory: A Descriptive Analysis.* Berkeley: University of California Press.

Kimbel, W. H. (1997). Hominid Speciation and Pliocene Climatic Change. In E. S. Vrba, G. H. Denton, T. C. Partridge, & L. H. Burckle (Eds.), *Paleoclimate and Evolution, with Emphasis on Human Origins* (pp. 425–437). New Haven and London: Yale University Press.

Kleindienst, M. R. (1961). Variability within the late Acheulean assemblage in eastern Africa. *South African Archaeological Bulletin, 16,* 35–52.

Kleindienst, M. R. (1962). Components of the East African Acheulian assemblage: An analytic approach. In G. Mortelmans & J. Nenquin (Eds.), *Actes du IVe Congrès Panafricain de Préhistoire* (pp. 81–105). Tervuren: Annales du Musée Royal d'Afrique Centrale.

Kozlowski, J. K. (2005). Les premières migrations humaines et les premières étapes du peuplement de l'Europe. *Diogène, 211,* 9–25.

Larrasoaña, J. C., Roberts, A. P., Rohling, E. J., Winklhofer, M., & Wehausen, R. (2003). Three million years of monsoon variability over the northern Sahara. *Climate Dynamics, 21,* 689–698.

Leakey, L. S. B. (1931). *The Stone Age Cultures of Kenya Colony.* London: Cambridge University Press.

Leakey, L. S. B. (1934). The Olduvai Cultural Sequence. *Proceedings of the 1st Congress of Prehistoric and Protohistoric Sciences, London 1932* (pp. 73–74).

Leakey, L. S. B. (1936). *Stone Age Africa: An Outline of Prehistory in Africa.* London: Oxford University Press.

Leakey, L. S. B. (1951). *Olduvai Gorge.* Cambridge: Cambridge University Press.

Leakey, L. S. B. (1952). Olorgesailie. In L. S. B. Leakey & S. Cole (Eds.), *Proceedings of the Pan-African Congress on Prehistory, Nairobi 1947* (p. 209). Oxford: Basil Blackwell.

Leakey, L. S. B. (1965). The new Olduvai hominid discoveries and their geological setting. *Proceedings of the Geological Society of London, 1617,* 104–109.

Leakey, M. D. (1967). Preliminary survey of the cultural material from Beds I and II, Olduvai Gorge, Tanzania. In W. W. Bishop & J. D. Clark (Eds.), *Background to Evolution in Africa* (pp. 417–446). Chicago: Chicago University Press.

Leakey, M. D. (1971). *Olduvai Gorge. Volume 3. Excavations in Bed I and II, 1960–1963.* Cambridge: Cambridge University Press.

Leakey, M. D. (1975). Cultural Patterns in the Olduvai Sequence. In K. W. Butzer & G. Ll. Isaac (Eds.), *After the Australopithecines. Stratigraphy, Ecology, and Cultural Change in the Middle Pleistocene* (pp. 477–493). Chicago: Mouton.

Leakey, M. D. & Roe, D. A. (1994). *Olduvai Gorge, vol. 5: Excavations in Beds III, IV and the Masek Beds, 1968–1971.* Cambridge University Press, Cambridge.

Lepre, C. J., Roche, H., Kent, D. V., Harmand, S., Quinn, R. L., Brugal, J.-P., et al. (2011). An earlier origin for the Acheulean. *Nature, 477,* 82–85.

Leroi-Gourhan, A. (1950). *Les fouilles préhistoriques : techniques et méthodes.* Paris: Picard.

Leroi-Gourhan, A. (1964). *Le geste et la parole. Technique et langage.* Paris: Albin Michel.

Leroi-Gourhan, A. (1971). *Evolution et Technique. L'Homme et la Matière*. Paris: Albin Michel.

Lisiecki, L. E., & Raymo, M. E. (2005). A Pliocene-Pleistocene stack of 57 globally distributed benthic $\delta^{18}O$ records. *Paleoceanography, 20,* PA1003.

Lycett, S. J., & von Cramon-Taubadel, N. (2008). Acheulian variability and hominin dispersal: A model-bound approach. *Journal of Archaeological Science, 35,* 553–562.

Lyell, C. (1863). *The Geological Evidences of the Antiquity of Man with Remarks on Theories of the Origin of Species by Variation*. London: John Murray.

McBurney, C. B. M. (1960). *The Stone Age of Northern Africa*. Harmondsworth: Penguin.

McHenry, L. J., Njau, J. K., de la Torre, I., & Pante, M. C. (2016). Geochemical "fingerprints" for Olduvai Gorge Bed II tuffs and implications for the Oldowan–Acheulean transition. *Quaternary Research, 85,* 147–158.

Manega, P. C. (1993). Geochronology, Geochemistry, and Isotopic Study of Plio-Pleistocene Hominid Sites and the Ngorongoro Volcanic Highlands in Northern Tanzania. Ph.D. Dissertation, University of Colorado.

Manzi, G. (2012). On the trail of the genus *Homo* between archaic and derived morphologies. *Journal of Anthropological Sciences, 90,* 1–18.

Mason, R. J. (1962). *The Prehistory of the Transvaal*. Johannesburg: Witwatersrand University Press.

Mason, R. J. (1966). The excavation of Doornlaagte Earlier Stone Age camp, Kimberley district. *Actas del V Congreso Panafricano de Prehistoria y del estudio del Cuaternario, Tenerife, 1963*(2), 187–188.

Mithen, S., & Reed, M. (2002). Stepping out: A computer simulation of hominid dispersal from Africa. *Journal of Human Evolution, 43,* 433–581.

Moncel, M.-H., & Ashton, N. (2018). From 800 to 500 ka in Western Europe. The Oldest Evidence of Acheuleans in Their Technological, Chronological, and Geographical Framework. In R. Gallotti & M. Mussi (Eds.), *The Emergence of the Acheulean in East Africa and Beyond. Contributions in Honor of Jean Chavaillon*. Cham: Springer.

Mussi, M. (1973). La question de l'Acheuléen de la Somalie. *Origini, VII,* 7–33.

Mussi, M. (2014). The Outstanding Universal Value of sites related to human evolution in Asia: The Acheulean as a case study for cooperation. *World Heritage Papers, 39,* 24–34.

Nagaoka, S., Katoh, S., WoldeGabriel, G., Sato, H., Nakaya, H., Beyene, Y., et al. (2005). Lithostratigraphy and sedimentary environments of the hominid-bearing Pliocene-Pleistocene Konso Formation in the southern Main Ethiopian Rift, Ethiopia. *Palaeogeography, Palaeoclimatology, Palaeoecology, 216,* 333–357.

Nicole, M. (1901). Eloge de Gabriel de Mortillet. *Bulletin de la Société d'Anthropologie de Paris, 2,* 559–572.

Nowell, A., & White, M. (2010). Growing up in the Middle Pleistocene: Life History Strategies and Their Relationship to Acheulean Industries. In A. Nowell & I. Davidson (Eds.), *Stone Tools and the Evolution of Human Cognition* (pp. 67–81). Boulder: Colorado University Press.

Pallary, P. (1911). Les collections préhistoriques du Musée des Antiquités algériennes. *Revue Africaine, 55,* 306–326.

Pante, M. (2013). The larger mammal fossil assemblage from JK2, Bed III, Olduvai Gorge, Tanzania: Implications for the feeding behavior of *Homo erectus*. *Journal of Human Evolution, 64,* 68–82.

Pelegrin, J. (1985). Réflexions sur le comportement technique. In M. Otte (Ed.), *La signification culturelle des industries lithiques* (pp. 72–91). Liege: Studia Praehistorica Belgica 4, BAR International Series 239.

Péringuey, L. (1911). The Stone Age of South Africa. *Annals of the South African Museum, 8,* 1–218.

Perlès, C. (1991). Économie des matières premières et économie du débitage : deux conceptions opposées ? In S. Beyries, L. Meignen, & P. J. Texier (Eds.), *25 ans d'études technologiques en Préhistoire : bilan et perspectives. Proceedings XIème rencontres internationales d'Archéologie et d'Histoire d'Antibes* (pp. 35–46). Juan-les-Pins: APDCA.

Plummer, T. W., Bishop, L., Ditchfield, P., & Hicks, J. (1999). Research on late Pliocene Oldowan sites at Kanjera South, Kenya. *Journal of Human Evolution, 36,* 151–170.

Potts, R. (1991). Why the Oldowan? Plio-Pleistocene toolmaking and the transport of resources. *Journal of Anthropological Research, 47,* 153–176.

Profico, A., Di Vincenzo, F., Gagliardi, L., Piperno, M., & Manzi, G. (2016). Filling the gap. Human cranial remains from Gombore II (Melka Kunture, Ethiopia; ca. 850 ka) and the origin of *Homo heidelbergensis*. *Journal of Anthropological Sciences, 94,* 1–24.

Quade, J., Levin, N., Semaw, S., Stout, D., Renne, P., Rogers, M. J., et al. (2004). Paleoenvironments of the earliest stone toolmakers, Gona, Ethiopia. *Geological Society of America Bulletin, 116,* 1529–1544.

Quade, J., Levin, N., Simpson, S., Butler, R., McIntosh, W., Semaw, S., et al. (2008). The Geology of Gona, Afar, Ethiopia. *Geological Society of America Bulletin, Special Papers, 446,* 1–31.

Reck, H. (1914). Erste Vorläufige Mitteilung über den Fund eines fossilen Menschenskelets aus Zentral-Afrika. *Sitzungsbericht der Gesellschaft der naturforschender Freunde, 3,* 81–95.

Reck, H. (1926). Prahistorische Grab und Menschenfunde und ihre Beziehungen zur Pluvialzeit in Ostafrika. *Mitteilungen der Deutschen Schutzgebiete, 34,* 81–86.

Rightmire, P. G. (2001). Patterns of hominid evolution and dispersal in the Middle Pleistocene. *Quaternary International, 75,* 77–84.

Rivière, J. (1896). L'industrie préhistorique du silex en Tunisie. *25e Congrès Association française pour l'Avancement des Sciences, Carthage-Tunis, 1st part* (pp. 199–200).

Roche, H., & Texier, P.-J. (1991). La notion de complexité dans un ensemble lithique. Application aux séries acheuléennes d'Isenya (Kenya). In S. Beyries, L. Meignen, & P. J. Texier (Eds.), *25 ans d'études technologiques en Préhistoire : bilan et perspectives. Proceedings XIème rencontres internationales d'Archéologie et d'Histoire d'Antibes* (pp. 99–108). Juan-les-Pins: APDCA.

Roche, H., Brugal, J.-P., Delagnes, A., Feibel, C., Harmand, S., Kibunjia, M., et al. (2003). Les sites archéologiques plio-pléistocènes de la formation de Nachukui, Ouest-Turkana, Kenya : bilan synthétique 1997–2001. *Comptes Rendus Palevol, 2,* 663–673.

Roche, H., Brugal, J.-P., Lefèvre, D., & Texier, P.-J. (1988). Isenya : état des recherches sur un nouveau site acheuléen de l'Afrique orientale. *African Archaeological Review, 6,* 27–55.

Roe, D. A. (1994). A metrical analysis of selected sets of handaxes and cleavers from Olduvai Gorge. In M. D. Leakey & D. A. Roe (Eds.), *Olduvai Gorge. Volume 5. Excavations in Beds III, IV and the Masek Beds, 1968–1971* (pp. 146–234). Cambridge: Cambridge University Press.

Roffo, P. (1934). Les civilisations du Mzab, Alger, Heintz. Compte rendu par R. Vaufrey. *L'Anthropologie, 45,* 417.

Sánchez Yustos, P., Diez-Martín, F., Domínguez-Rodrigo, M., Duque, J., Fraile, C., Baquedano, E., et al. (2017). Diversity and significance of core preparation in the Developed Oldowan technology—Reconstructing the flaking processes at SHK and BK (Middle-Upper Bed II, Olduvai Gorge, Tanzania). *Boreas, 46,* 874–893.

Santonja, M., & Villa, P. (2006). The Acheulean of Western Europe. In N. Goren Inbar & G. Sharon (Eds.), *Axe Age, Acheulian Tool-making from Quarry to Discard* (pp. 429–478). London: Equinox Publishing.

Santonja, M., Rubio-Jara, S., Panera, J., Pérez-González, A., Rojas-Mendoza, R., Domínguez-Rodrigo, M., et al. (2018). Bifacial

Shaping at the TK Acheulean Site (Bed II, Olduvai Gorge, Tanzania): New Excavations 50 Years After Mary Leakey. In R. Gallotti & M. Mussi (Eds.), *The Emergence of the Acheulean in East Africa and Beyond. Contributions in Honor of Jean Chavaillon*. Cham: Springer.

Schick, K., & Clark, J. D. (2003). Biface technological development and variability in the Acheulean industrial complex in the Middle Awash region of the Afar Rift, Ethiopia. In M. Soressi & H. L. Dibble (Eds.), *Multiple Approaches to the Study of Bifacial Technologies* (pp. 1–30). Philadelphia: University of Pennsylvania, Museum of Archaeology and Anthropology.

Ségalen, L., Lee-Thorp, J. A., & Cerling, T. (2007). Timing of C_4 grass expansion across sub Saharan Africa. *Journal of Human Evolution, 53,* 549–559.

Seligman, C. G. (1921). The older Palaeolithic age in Egypt. *Journal of the Royal Anthropological Institute, 51*, 115–144.

Semaw, S., Rogers, M., & Stout, D. (2009). The Oldowan-Acheulian transition: Is there a 'Developed Oldowan' artifact tradition? In M. Camps & P. R. Chauhan (Eds.), *Sourcebook of Paleolithic Transitions. Methods, Theories and Interpretations* (pp. 173–193). New York: Springer.

Semaw, S., Rogers, M., Stout, D., Cáceres, I., & Leiss, A. C. (2018). The Early Acheulean ~1.6–1.5 Ma from Gona, Ethiopia: Issues on the Emergence of the Acheulean in Africa. In R. Gallotti & M. Mussi (Eds.), *The Emergence of the Acheulean in East Africa and Beyond. Contributions in Honor of Jean Chavaillon*. Cham: Springer.

Sharon, G. (2007). *Acheulean large flake industries: Technology, chronology, and significance*. BAR (British Archaeological Reports) International Series 1701. Oxford: Archaeopress.

Sharon, G. (2010). Large flake Acheulean. *Quaternary International, 223–224,* 226–233.

Stiles, D. (1977). Acheulian and Developed Oldowan. The Meaning of Variability in the Early Stone Age. *Mila, 6,* 1–35.

Stiles, D. (1979a). Recent archaeological findings at the Sterkfontein site. *Nature, 277,* 381–382.

Stiles, D. (1979b). Early Acheulian and Developed Oldowan. *Current Anthropology, 20,* 126–129.

Stiles, D. (1980). Industrial Taxonomy in the Early Stone Age of Africa. *Anthropologie, XVIII,* 189–207.

Stiles, D. (1991). Early hominid behaviour and culture tradition: Raw material studies in Bed II, Olduvai Gorge. The African Archaeological Review, 9, 1–19.

Stout, D., Quade, J., Semaw, S., Rogers, M., & Levin, N. (2005). Raw material selectivity of the earliest stone toolmakers at Gona, Afar, Ethiopia. *Journal of Human Evolution, 48,* 365–380.

Stout, D., Semaw, S., Rogers, M. J., & Cauche, D. (2010). Technological variation in the earliest Oldowan from Gona, Afar, Ethiopia. *Journal of Human Evolution, 58,* 474–491.

Suwa, G., Beyene, Y., & Asfaw, B. (2015). *Konso-Gardula Research Project, Vol. 1*. Tokyo: The University Museum, The University of Tokyo.

Tamrat, E., Thouveny, N., Taieb, M., & Opdyke, N. D. (1995). Revised magnetostratigraphy of the Plio-Pleistocene sedimentary sequence of the Olduvai Formation (Tanzania). *Palaeogeography, Palaeoclimatology, Palaeoecology, 114,* 273–283.

Texier, P.-J. (1996). L'Acheuléen d'Isenya (Kenya), approche expérimentale des principales chaînes opératoires lithiques. In *La vie préhistorique* (pp. 58–63). Paris: Société préhistorique française, Faton.

Texier, P.-J. (2005). L'Oldowayen dans le Grand Rift occidental: le site NY18 à Nyabusosi (Ouganda). In M. Sahnouni (Ed.), *Le Paléolithique en Afrique, l'Histoire la Plus Longue* (pp. 83–98). Paris: Artcom.

Texier, P.-J. (2018). Technological Assets for the Emergence of the Acheulean? Reflections on the Kokiselei 4 Lithic Assemblage and Its Place in the Archaeological Context of West Turkana, Kenya. In R. Gallotti & M. Mussi (Eds.), *The Emergence of the Acheulean in East Africa and Beyond. Contributions in Honor of Jean Chavaillon*. Cham: Springer.

Texier, P.-J., & Roche, H. (1995a). The impact of predetermination on the development of some Acheulian chaînes opératoires. In Bermudez, J. (Ed.), *Paper Presented at the Conference Human Evolution in Europe and the Atapuerca Evidence, 2* (pp. 403–420). Castillo de la Motta, Medina del Campo, Valladolid 1992. Junta de Castilla y Leon.

Texier, P.-J., & Roche, H. (1995b). Polyèdre, sub-sphéroïde, sphéroïde et bola : des segments plus ou moins longs d'une même chaîne opératoire. *Cahier Noir, 7,* 31–40.

Tixier, J. (1956). Le hachereau dans l'Acheuléen nord-africain. Notes typologiques. In *Proceedings of the XV session of the Congrès préhistorique de France, July 15–22, 1956, Poitiers-Angoulême* (pp. 914–923).

Toth, N. (1982). The Stone Technologies of Early Hominids at Koobi Fora, Kenya, an Experimental Approach. Ph.D. Dissertation, University of California.

Trauth, M. H., Larrasoaña, J. C., & Mudelsee, M. (2009). Trends, rhythms and events in Plio-Pleistocene African climate. *Quaternary Science Reviews, 28,* 399–411.

Van Riet Lowe, C. (1952). The development of the hand-axe culture in South Africa. In L. S. B. Leakey & S. Cole (Eds.), *Proceedings of the Pan-African Congress on Prehistory, Nairobi 1947* (pp. 167–177). Oxford: Basil Blackwell.

Vayson de Pradenne, A. (1920). La plus ancienne industrie de Saint-Acheul. *L'Anthropologie, XXX,* 441–496.

Chapter 2
Before the Acheulean in East Africa: An Overview of the Oldowan Lithic Assemblages

Rosalia Gallotti

Abstract In 2009, Hovers and Braun published in Springer's Vertebrate Paleobiology and Paleoanthropology Series the volume "Interdisciplinary Approaches to the Oldowan," stemming from the symposium of the 2006 SAA meeting in Puerto Rico. Many contributors focused on the description of the Oldowan as a lithic production system, showing the high technical variability of the techno-complexes. As pointed out by Braun and Hovers (2009: 4), even if most or all scholars agree that the study of Oldowan behaviors is fundamental to understand early hominin evolution, "not all would agree on a definition of the Oldowan." Forty years after it was first defined (Leakey 1971, 1975), many sites scattered over approximately one million years are labelled as "Oldowan" in large-scale syntheses. While the available data are highly fragmented both in time and space, and the study of lithic assemblages follows different theoretical and methodological approaches, major overviews simply take for granted that a correlation among the East African assemblages is inescapable. However, the term Oldowan is still a vague concept, lacking a comprehensive definition of what an Oldowan technology is. Additionally, who were the authors of the Oldowan stone tools remains an open question.

Nine years after the publication of Hovers and Braun's volume, this is a short overview and update of the current state of our knowledge of the Oldowan technical behaviors recorded in East Africa, to put in the proper perspective specific sites with "emerging" Acheulean.

Keywords Oldowan • Lithic techno-economy • Technological stasis/development • Early Pleistocene hominins

2.1 Introduction

In Leakey's original definition, the Oldowan is a cultural entity within East Africa dating between ca. 1.8 and 1.5 Ma. This definition pertains exclusively to the material culture and is grounded in the chrono-stratigraphic and cultural framework of the Olduvai sequence as described in Leakey's report (1971).

Following the methodological approach developed in Europe by Bordes (1961), Leakey (1971) established the fundamental characters of the Oldowan industries of Olduvai. She proposed the terminology and associated descriptive criteria allowing to define them. She further defined in 1975 the guidelines of the typological evolution of the Olduvai industries suggesting a homogeneous nomenclature and producing a unilinear evolutionary cultural synthesis. Leakey's seminal definition has been the reference ever since, over more than forty years, for the Early Pleistocene prehistory of East Africa. Nevertheless, some later analyses, albeit based on Leakey's typological list, focused on the characterization of specific types of artifacts in a diachronic perspective, often extracting them out of their relations with the other elements of the same complex. In the end, the perception of the intra-site synchronism was missed (Bower 1977; Dies and Dies 1980; Wynn 1981; Willoughbly 1987; Sahnouni 1991).

R. Gallotti (✉)
Dipartimento di Scienze dell'Antichità, Sapienza Università di Roma, Piazzale Aldo Moro 5, 00185 Rome, Italy
e-mail: rosalia.gallotti@uniroma1.it

R. Gallotti
Italian Archaeological Mission at Melka Kunture and Balchit

R. Gallotti
Université Bordeaux, UMR 5199 PACEA-PPP, Bâtiment B18 allée Geoffroy Saint-Hilaire, CS 50023, 33615 Pessac Cedex, France

R. Gallotti and M. Mussi (eds.), *The Emergence of the Acheulean in East Africa and Beyond: Contributions in Honor of Jean Chavaillon*, Vertebrate Paleobiology and Paleoanthropology,
https://doi.org/10.1007/978-3-319-75985-2_2

A few years after Leakey, Isaac proposed an alternative system for the study of the Oldowan industries. Isaac's model linked each lithic component to a specific production stage and suppressed the functional connotations of each morpho-type found in M. Leakey's perspective (Isaac 1977, 1984, 1986). Isaac also insisted on the need of relating stone artifacts to economic and ecological patterns, in order to evaluate if some degree of variability was the outcome (Isaac 1977). The differences highlighted between the KBS and Karari industries of Koobi Fora on the one hand, and the Oldowan of Olduvai Gorge on the other, made clear that with two different methodological approaches it was truly a challenge to evaluate inter-site Oldowan technical variability (Isaac and Isaac 1997).

At the end of the 1970s, Chavaillon (1979) introduced a detailed typology of the percussion material, which is abundant at Melka Kunture as well as at Olduvai, but only rarely found at Koobi Fora. Chavaillon (1979) underlined for the first time that percussion activities, overlooked in Leakey's and Isaac's models, rather were a structural element of early technologies.

In the 1980s, several primatologists and some archaeologists suggested that Oldowan industries were sets of tools that did not require skills beyond those observed in apes (Wynn and McGrew 1989; Foley 1991). The Oldowan was seen as an expedient toolkit. In the same years, Toth's (1982, 1985) reinterpretation of the Oldowan led to a somewhat different picture: the Oldowan was reassessed as the result of a more complex technological behavior than documented among non-human primates. However, even if more elaborate, this technology was seen as definitely simpler and more expedient than in later Acheulean industries (Martínez-Moreno et al. 2003). Toth's (1982, 1985) experimental studies furthermore demonstrated that the flakes, considered by M. Leakey as waste, actually were end products, as already argued by Isaac (1977).

As it had been in the case of Leakey's studies, Isaac and Toth's contributions opened the path to new research, mainly by American scholars. Unfortunately, as de la Torre and Mora pointed out (2009: 17), this next generation of analyses lacked "the theoretical and methodological approaches that had guided these authors", i.e., respectively, the ecological perspective and the experimental analysis (e.g., Potts 1988, 1991; Semaw 1997, 2000; Kimura 1999, 2002; Ludwigh 1999; Noll 2000; de la Torre et al. 2003).

The discovery of sites older than 2 Ma elsewhere in East Africa, notably Gona/Hadar and Omo in Ethiopia, and West Turkana in Kenya, promoted the idea of an even simpler and less developed technology (Chavaillon 1976; Roche and Tiercelin 1980; Roche 1989; Kibunjia 1994; de Lumley and Beyene 2004). Initially, this earlier evidence was assigned to the Oldowan, giving further weight to an assumed one-million-year-long stasis in technological development (Semaw 2000; Semaw et al. 2003). Between the end of the 1980s and the beginning of the 1990s, the hypothesis started to emerge that more archaic artifacts, predating 2 Ma, should be classified separately. A new terminology was introduced, including "pre-Oldowan," "Shungura facies," or "Nachukui industry" (Howell et al. 1987; Piperno 1987; Roche 1989, 1996; Kibunjia 1994, 1998; Roche et al. 1999, 2003).

The detailed analysis of artifacts available at the 2.34 Ma site of Lokalalei 2C (West Turkana, Kenya) suggests that early hominids displayed distinct technical competencies and techno-economic behaviors more sophisticated than expected. Accordingly, it was argued that intra-site complexity and inter-site variability were not accounted for by the existing chrono-cultural classifications (Delagnes et al. 2005). This recognized that higher degree of variability and complexity is not just the outcome of the analysis of Lokalalei 2C. It is also the effect of a new theoretical and methodological approach to the study of stone artifacts, i.e., of the chaîne opératoire approach. The chaîne opératoire concept allows to place each object in a precise technical context, identifying all the technical processes performed for its production, from raw material procurement modalities to the manufacture and use phases, ending with final abandonment (Leroi-Gourhan 1964, 1971; Lemonnier 1976; Pelegrin 1985; Geneste 1989, 1991; Boëda 1991; Perlès 1991; Inizan et al. 1999). While a typological approach considers exclusively the final state of the technical operations, making comparisons and recognizing likeness or differences, the technological approach takes into account the whole history of a lithic object and the various modalities of manufacturing. This allows for a broad range of variability, well beyond the finished object.

Right now, the chaîne opératoire approach is widely used in lithic studies of early technologies and many assemblages have been recently reviewed following this concept. Even if the level and the details of analyses vary from one site to another, they generated a common methodological background allowing comparisons.

The next sections will explore the techno-economic behaviors of any Oldowan assemblage analyzed through this methodology (Table 2.1) in order to micro-analyze (to the degree possible) the lithic productions in order to comment on the issues of complexity, diversity, and flexibility of behaviors. Based on Isaac and Isaac's typological analysis (1997), the Early Pleistocene[1] assemblages from the KBS and Okote members at Koobi Fora (1.9–1.6 Ma) are considered here as a local variant of the Oldowan from Olduvai. The Koobi Fora assemblages, however, are not discussed here, because of the lack of a systematic and comprehensive

[1]I use in this paper the revised timescale approved by IUGS, in which the base of the Pleistocene is defined by the GSSP of the Gelasian Stage at 2.588 (2.6) Ma (Gibbard et al. 2010).

Table 2.1 East African Oldowan sites discussed in this paper

Site	Sub-site	Age (Ma)	m²	Hominins	Fauna	Lithic assemblage	Percussion elements	Small débitage					References
								Raw material provisioning	Raw material morphology	Flaking exploitation	Retouched flakes (% of all the flakes)	Technique	
Gona	EG10	2.6–2.5	13	Absent	Absent	2236	Absent	Local	Medium-sized rounded cobbles	• Unifacial (mainly) simple/unidirectional/centripetal • Bifacial partial • Multidirectional irregular	2.5	Free hand	Semaw et al. (2003), Stout et al. (2005, 2010), Semaw (2006)
	EG12		8	Absent	Absent	754					4		
	OGS-7		2.6	Absent	Present	~700		Local/non-local		• Unifacial simple/unidirectional/centripetal • Bifacial partial and multidirectional irregular (mainly)	Absent		
Hadar	AL849	2.4–2.3	20	*Homo* aff. *habilis*	Present	4828	Absent	Local	Medium-sized angular cobbles	Mainly unifacial	Absent	Free hand	Hovers (2003, 2009), Goldman-Neuman and Hovers (2009, 2012)
	AL666		2?		Present	224		Local/non-local (?)		Mainly unifacial		Free hand Bipolar	
West Turkana	LA1	2.34	60	*Homo* aff. *habilis*	Present	445	?	Local	Large-sized angular and globular cobbles	• Unifacial (mainly)/bifacial/multifacial multidirectional • Bifacial • Multifacial	?	Free hand	Roche et al. (2003), Delagnes and Roche (2005), Harmand (2009)
	LA2C		17	*Paranthropus aethiopicus*	Present	2614	Present		Large-medium-sized blocks and cobbles	• Simple • Unifacial (mainly) multidirectional • Bifacial • Multifacial	2.5		
	KS5	1.87	65	–	Present	1727	Present	Local	Large-sized angular, flat and globular cobbles	• Unifacial • Bifacial partial • Multifacial multidirectional/orthogonal • Centripetal	1.2	Free hand	Texier et al. (2006), Texier (2018)
Omo	Omo 57	2.34	?	*Homo sp.*	Absent	498	Absent	Local	Small-sized angular cobbles	Unifacial unidirectional	0.8	Free hand	de la Torre (2004)
	Omo 123		?	*Paranthropus aethiopicus*		1314					Absent	Bipolar (only two cores)	
Kanjera South	Ex. 1, 2, 5, 6	>2.0	4–169	Absent	Present	4474	?	Local/non-local	Small-medium-sized rounded cobbles	• Unifacial unidirectional • Bifacial centripetal • Multifacial multidirectional	?	Free hand Bipolar	Braun et al. (2008a, b, 2009a, b, c), Lemorini et al. (2014)
Fejej	FJ-1a	~1.9	85	*Homo* aff. *habilis* *Paranthropus boisei*	Present	2610	Present	Local	Medium-sized rounded and angular cobbles	• Unifacial unipolar/bipolar/multipolar/centripetal • Bifacial unipolar/bipolar/centripetal • Multifacial multidirectional	0.8	Free hand Bipolar (very few)	de Lumley et al. (2004b), Barsky et al. (2011)

(continued)

Table 2.1 (continued)

Site	Sub-site	Age (Ma)	m^2	Hominins	Fauna	Lithic assemblage	Percussion elements	Small débitage					References
								Raw material provisioning	Raw material morphology	Flaking exploitation	Retouched flakes (% of all the flakes)	Technique	
Olduvai	DK	>1.8	380[a]	*Homo habilis*	Present	1180	Present	Local	Medium-sized cobbles Tabular blocks	• Unifacial unidirectional/peripheral • Bifacial partial/peripheral • Multifacial multidirectional	1.6	Free hand	Leakey (1971), de la Torre and Mora (2005), Mora and de la Torre (2005) Only for Levels 1–2 see also Diez-Martín et al. (2010)
	FLK Zinj	>1.8	~300	*Paranthropus boisei*		2663				• Unifacial unidirectional/peripheral • Bifacial partial/peripheral	1.5		
	FLK North Level 6	>1.8	? / 6–8[b]	Absent		130				Unifacial/bifacial partial	Absent		
	FLK North Level 5	>1.8	?	*Homo habilis?*		132 / 168[c]				Unifacial/bifacial partial/peripheral	2		
	FLK North Level 4	>1.8	264	Absent		83				• Unifacial/bifacial partial • Multifacial multidirectional	Absent		
	FLK North Level 3	>1.8	100?			214				• Unifacial/bifacial partial/peripheral • Multifacial multidirectional	1.3		
	FLK North Levels 2–1	>1.8	100?			1456				• Unifacial unidirectional/centripetal • Bifacial unidirectional/ multidirectional/ • centripetal • Trifacial	5	Free hand Bipolar	
	FLK North *D.* Level	>1.66	?			36		Only 3 cores					
	FLK North SC	~1.6	?			248		Local	Medium-sized cobbles Tabular blocks	Unifacial/bifacial partial/peripheral	5		
Melka Kunture	Garba IVE	>1.7	34	*Homo erectus s.l.*	Present	1222	Absent	Local	Small-medium-sized rounded and angular cobbles	• Multifacial multidirectional • Multifacial multidirectional with one preferential flaking surface • Unifacial unidirectional • Unifacial centripetal/radial • Unifacial/bifacial partial	15	Free hand	Gallotti and Mussi (2015)
	Garba IVF		12	Absent		193					10		

[a]DK was excavated in 4 sectors: DK IA (27 m^2), DK I Strips 1-111 (218.7 m^2), DK IB (45 m^2), and DK IC (89.1 m^2)
[b]Two test trenches excavations were performed in 2007–2008 by The Olduvai Palaeoanthropology and Palaeoecology Project (TOPPP)
[c]Diez-Martín et al. (2010) studied 168 items from Levels 1 to 5. There is not a count per level. Flaking methods identified by Diez-Martín et al. (2010) mainly concern Levels 1–2, but include also specimens from Levels 3–5

reassessment in a technological perspective, except for some studies focused on specific aspects of the lithic production (e.g., Braun et al. 2008b, c, 2009a). The current overview aims at producing a detailed dataset for a broader evaluation of the continuity/discontinuity technological patterns from the Oldowan to the early Acheulean, also including an intra- and inter-site perspective.

2.2 "Older Than the Oldowan": The Lomekwian and the Origin of Stone Technology

After the discovery of the most ancient Oldowan assemblages at Gona (~2.6 Ma; Semaw et al. 1997, 2003; Semaw 2000), several authors speculated about an earlier phase of stone knapping. Gona artifacts appeared to be too well made to have been the first experiments in producing sharp-edged stone flakes. It was also argued that percussive activities other than knapping, well documented in extant taxa of non-human primates, could have been structural components of hominin stone tool use (e.g., Roche et al. 1999; Semaw 2000; Panger et al. 2002; Semaw et al. 2003; Davidson and McGrew 2005; Marchant and McGrew 2005; Mora and de la Torre 2005; Carvalho et al. 2008; Rogers and Semaw 2009; de la Torre 2011; Carvalho and McGrew 2012).

The cut-marked bones from Dikika (Ethiopia; McPherron et al. 2010, 2011; Thompson et al. 2015), dated to >3.39 Ma, indirectly added fuel to speculations on pre-2.6 Ma stone tool use. Other researchers questioned the find, attributing the marks to natural wear and tear such as trampling (Domínguez-Rodrigo et al. 2010b, 2011, 2012).

The recent discovery of stone artifacts dated to ~3.3 Ma at Lomekwi 3 (LOM 3) in West Turkana validated the hypotheses of an older origin for stone tool technology. The lithic assemblage consists of 149 surfaces and in situ artifacts, used for flaking and percussion activities (Harmand et al. 2015). Cores, significantly larger than those discovered in the Oldowan sites, are made mainly from very large-sized cobbles of basalts and phonolite, selected among cobbles and blocks of all sizes available in paleochannels at less than 100 m from the site. Cores are mainly unifacial with unidirectional superposed and contiguous removals. The knappers were also able to laterally rotate the cores, flaking one surface through multidirectional removals, and to flip them over for bifacial exploitation. However, the poorly controlled percussive motion combined with a tough raw material availability resulted in repetitive failed blows, hinged removals, and step fractures (Harmand et al. 2015).

Few cores display very short small scars along an edge, which could be the outcome of using them as tools, as well as of the technique. Replication experiments suggest that flaking activities were performed using passive hammers and/or bipolar techniques (Harmand et al. 2015), rarely identified in the Oldowan (e.g., Mora and de la Torre 2005; Diez-Martín et al. 2009, 2010; de la Torre and Mora 2010). This hypothesis is possibly confirmed by pieces weighing up to 15 kg bearing similar wear and fractures, which can be interpreted as anvils or passive elements. These techniques allow understanding how the knappers were able to flake huge blanks, hardly knapped by direct freehand percussion. On medium-sized cobbles battering marks and fractured surfaces document the coexistence of hand-held or active hammerstones. A combination of flaking and percussion is further documented by some flakes with natural dorsal faces showing battered areas. As pointed out by Harmand et al. (2015: 313): "The use of individual objects for several distinctive tasks reflects a degree of technological diversity both much older than previously acknowledged and different from the generally unipurpose stone tools used by primates." Nevertheless, Hovers (2015: 295) suggests that it is also possible that "at the beginning of each discrete episode of its use, each artifact was perceived merely as available raw material."

According to Harmand et al. (2015), at Lomekwi 3 the techniques seem closer to those involved in nut-cracking by non-human primates, than to the direct freehand percussion recorded in Oldowan assemblages. This further underlines the central role played by percussion activities at the dawn of technology, as it had already been suggested (e.g., Matsuzawa 1996; Mora and de la Torre 2005; Diez-Martín et al. 2009, 2010; Haslam et al. 2009; de la Torre and Mora 2010; Visalberghi et al. 2013; Bril et al. 2015; Hayashi 2015). Although direct comparisons between Lomewiki 3 and Oldowan assemblages are difficult, given the different levels of analysis, "the technological and morphological differences between the LOM 3 and early Oldowan assemblages are significant enough that amalgamating them would mask important behavioral and cognitive changes occurring among hominins over a nearly 2-million-year timespan" (Harmand et al. 2015: 314). For this reason, and because of the long time gap separating LOM 3 from the Oldowan sites, Harmand et al. (2015) introduced the name "Lomekwian".

2.3 The Oldowan Assemblages

2.3.1 *Gona (Ethiopia)*

At ~2.6 Ma, i.e., 700,000 years later, stone artifacts are documented at Gona (Ethiopia). Stone artifacts of great antiquity were known at Gona ever since the early 1970s

(Johanson et al. 1978, 1982). Low-density scatters of surface artifacts had been discovered east of the Kada Gona River (Corvinus 1976; Corvinus and Roche 1976, 1980; Roche and Tiercelin 1977, 1980). Subsequent excavations at West Gona allowed discovering low density in situ artifacts (Harris 1983; Harris and Semaw 1989). Systematic excavations and geochronological analyses started in 1992–1994. More localities along the Kada and Ounda Gona rivers yielded surface and in situ artifacts, mostly recovered from EG10, EG12, and OGS-7. EG10 and EG12 are stratigraphically located below the AST-2.75 Tuff providing a minimum age of 2.52 Ma, and just above the 2.6 Ma Gauss-Matuyama polarity transition (McDougall et al. 1992; Semaw et al. 1997). The same polarity transition occurs below the archaeological level at OGS-7, sealed by the Gonash-14 Tuff, dated in turn to 2.53 ± 0.15 Ma (Semaw et al. 2003). Two archaeological levels at EG10 and one level at EG12 yielded high densities of stone artifacts with no associated bones (Semaw 2006). OGS-7 is the only —and oldest—site where artifacts were discovered with fossil bones, a few of them possibly showing human modification (Semaw et al. 2003). Modified bones of a similar age showing cutmarks and percussion marks made by stone tools were discovered at Bouri in the Middle Awash, but without any associated artifacts (de Heinzelin et al. 1999).

The three lithic assemblages are minimally disturbed. They are exclusively focused on small-medium flake production from local raw materials available in paleochannels. Comparing the geological and archaeological samples points to a high degree of raw material selectivity, based on rock type, phenocrysts, and groundmass (Stout et al. 2005). Additionally, lithotype frequencies significantly differ between East Gona sites and OGS-7. While at EG10-12 the assemblages are dominated by trachyte and rhyolite, which are largely available in local conglomerates, the OGS-7 series are mainly composed of high-quality aphanitic and vitreous volcanic materials. Such lithotypes are scarcely found in local conglomerates, and accordingly they had to be transported into the site (Stout et al. 2010).

Knapping technique is direct hard hammer percussion, with no evidence of bipolar, anvil, or throwing techniques. Five flaking methods were observed: simple unifacial, centripetal unifacial, unidirectional, partial bifacial, and irregular multifacial. Nevertheless, the EG10-12 débitage is dominated by unifacial cores. At OGS-7 the cores instead are bifacial or multifacial, with a higher number of flake scars indicating a relatively intense reduction. According to Stout et al. (2010), although the existence of distinct techno-cultural traditions cannot be ruled out, this variability has been interpreted as possibly originated by environmental diversity. OGS-7 formed on a channel bank or channel margin, while both EG-10 and EG-12 were located on a proximal floodplain (Stout et al. 2010).

2.3.2 Hadar (Ethiopia)

Early Pleistocene archaeological occurrences were also discovered in the upper Kada Hadar Member of the Hadar Formation in the Makaamitalu Basin, where the A.L. 666 and A.L. 894 sites are overlain by the 2.33 ± 0.07 Ma BKT-3 Tuff (Kimbel et al. 1996; Campisano 2012).

A.L. 666 yielded a very restricted lithic assemblage of flakes and angular fragments, and it represents the oldest spatiotemporal co-occurrence of tools, fauna, and hominids, i.e., a maxilla of *Homo* aff. *habilis* (Kimbel et al. 1996, 1997).

At A.L. 894 a few hundred bone fragments and several thousand lithic artifacts were discovered in the silty-clay deposits of the floodplain of a low-energy stream. The rodent fauna provides important evidence for a faunal shift in the small-mammal community—coincident with a similar faunal shift in the large-mammal community (Reed 2008)—between 3.2 Ma and 2.4 Ma. This shift documents increasing aridity, but not a dramatic change in the local paleoenvironment. This shift coincides also with the local extinction of *Australopithecus afarensis* and the emergence of the genus *Homo* at Hadar (Reed and Geraads 2012). The lithic assemblage possibly represents the palimpsest of several occupations. Taphonomic interpretation of the well-preserved bone assemblage shows lack of functional association between artifacts and bones, underscoring the complexity of site formation processes (Domínguez-Rodrigo and Martínez-Navarro 2012). Nevertheless, the numerous refits among lithic items suggest that burial happened relatively fast and with minimal geological disturbance (Hovers 2009). Most of the artifacts are complete or broken flakes (83%), with cores in small numbers (1%). Angular fragments make the remaining part of the assemblage. Rather than linked to anthropic flaking, they might well have been the result of post-depositional mechanisms (Hovers 2003).

The artifacts mainly are on cobbles of volcanic rocks, primarily rhyolite, basalt, and trachyte. Cobble size suggests transport from a source located at least several meters away. They were accurately selected in the nearby conglomerate, especially in the case of rhyolites. However, at A.L. 666 the rocks are more homogeneous and more fine-grained than at A.L. 894. Quartz and chert were also exploited. These rock types are found, even if scarcely, in the conglomerate, but transport from an unknown source cannot be ruled out. These patterns suggest not only selectivity, but possibly also the search for the best quality raw materials out of the available lithic resources. Besides, some cobbles were imported to the site partly decorticated, after the knappers had tested them at source (Goldman-Neuman and Hovers 2009, 2012).

Flaking at A.L. 894 was mainly unipolar, while at A.L. 666 bipolar flaking is documented for quartz and chert exploitation. Step and hinge scars are frequent, taking place when the knappers were unable to maintain appropriate knapping angles. When flawed flake terminations occurred, the error could not be removed by further flaking from the same direction. However, cores with large surfaces were rotated in an attempt of rectifying the error and continuing to flake before the knapper made a decision to remove a "cleaning flake" bearing hinge and step scars (Hovers 2009; Goldman-Neuman and Hovers 2012).

2.3.3 *West Turkana (Kenya)*

Lithic artifacts comparable in age to the Hadar assemblages were discovered in the Nachukui Formation (West Turkana, Kenya) at Lokalalei 1 (LA1) and Lokalalei 2C (LA2C). The archaeological levels are capped by a mollusk-packed sandstone, a marker bed dated to 2.35 Ma which is the local boundary between the Lokalalei and Kalochoro members (Harris et al. 1988). The Kokiselei and Ekalalei tuffs (2.40 $\pm$ 0.05 and 2.34 $\pm$ 0.04 Ma, respectively) lying below the Lokalalei 1 and Lokalalei 2C are correlated with tuffs E and F-1 of the Shungura Formation (Harris et al. 1988; Feibel et al. 1989). Accordingly, the Lokalalei sites have an estimated age of 2.34 $\pm$ 0.05 Ma (Roche et al. 1999). However, after the revision of the local lithostratigraphy, LA2C could be marginally younger than LA1 (Brown and Gathogo 2002).

The archaeological levels also yielded faunal remains, which are poorly preserved and devoid of any recognizable trace of human action, except from a single cutmark on a bone fragment from the surface (Roche et al. 1999; Brugal et al. 2003). A tooth attributed to an early *Homo* was found at Lokalalei 1a, in the same lithostratigraphic unit as LA1 (Prat et al. 2005). Accordingly, early *Homo* is a candidate as a knapper of the lithic assemblages, even if *Paranthropus aethiopicus* was also present at 2.5 Ma in West Turkana (Walker et al. 1986).

Lithic items were knapped from lavas (phonolite, trachyte, basalt, and rhyolite) available in paleochannels at a maximum distance of 50 m from the site (Harmand 2009).

At LA1, flaking is produced mainly on large cobbles, which are rounded and therefore lacking any flat surfaces or suitable natural striking platforms. The resulting knapping sequences are opportunistic and heterogeneous. The cores bear evidence of frequent knapping accidents and of repeated impact damage from failed percussions (Delagnes and Roche 2005). On the opposite, LA2C technical strategies are based on the implementation of constant technical rules, documenting advanced manual dexterity. Although remains are distributed in a thickness of 50 cm, the feeble impact of post-depositional disturbance on this assemblage is demonstrated by the high number of refits scattered horizontally and vertically throughout the entire sand deposit, by the high ratio of elements <1 cm, as well as by the fresh physical state (Delagnes and Roche 2005). A set of cores shows one-to-three unorganized flake scars (simple flaking), as if they had been tested and discarded. Another set displays evidence of organized flaking on fine-grained phonolite clasts. Whole cobbles and angular blocks were knapped without any preparation, even if the deliberate breakage off-site of the largest specimens into large fragments can occur. In two cases, refitting groups testify that a preliminary phase of flaking was also conducted off-site. The cores displaying organized flaking were mainly flaked on a single surface which is the largest available one. Several series of flakes were extracted from natural or rectified platforms. The numerous changes of flaking direction ensured that the surface remained reasonably flat and regular. In some cases, few removals on another face can occur at a final reduction stage. Alternate flaking is also documented: in order to produce several series of flakes, some cores display two or three surfaces used in succession as flaking surfaces and as striking platforms. As the knappers used and maintained angles among surfaces which were already available, the outcome are cores with a final shape similar to that of the natural blank. Additionally, cores do not show any impact damage from failed or repetitive percussions. A few cores bear evidence of retouch on edges with an angle close to 90° for further use as tools after flaking. Retouch also occurs on a very few flakes, along either one or, more rarely, two edges. Several cobbles show thickly pitted areas due to percussion. They are of the same medium-grained phonolite as unworked specimens, but they are heavier. Therefore it is likely that during raw material procurement, the knappers selected specimens better suited for percussion or broke them to obtain fragments suitable for recurrent flaking (Delagnes and Roche 2005).

After a hiatus of $\sim$0.5 Ma, Oldowan artifacts reappear at Kokiselei 5, where an archaeological level is slightly younger than the KBS Tuff (1.87 Ma; Texier 2018). Cobbles of local rocks were collected in alluvial deposits near the site and selected on the base of their knapping suitability and morphology. Although the initial geometry of the blank conditioned the exploitation modalities, cores show the ability to reconfigure the core and to rearrange the striking platform (Texier et al. 2006; Texier 2018). A large split cobble of phonolite/trachyte displays a point partially shaped by unifacial removals (Texier et al. 2006).

2.3.4 *Omo (Ethiopia)*

Sites of $\sim$2.3 Ma were discovered four decades ago in the Omo Valley of southern Ethiopia, in members E and F of the Shungura Formation (Chavaillon 1970, 1975, 1976; Merrick et al. 1973; Merrick and Harry 1976; Chavaillon and Boisaubert 1977). Member E is dated between 2.40 ± 0.05 and 2.324 ± 0.020 Ma (Feibel et al. 1989; McDougall and Brown 2008), and Member F between 2.324 ± 0.020 Ma and 2.271 ± 0.041 Ma (McDougall and Brown 2008; McDougall et al. 2012). According to Chavaillon's typological description (1976), the Omo collection was composed of small-sized quartz items, i.e., fragments, flakes, and cores, and of abundant wastes. The latter were considered either as natural residues, detached during flaking, or as hammerstone fragments produced by intense percussion. After this description, the constraints imposed by small-sized quartz cobbles allowed simple smashing only, and no proper knapping and flake production. Accordingly, the Omo industries were evidence of expedient technology. Because of the lack of lithic types as defined at Olduvai, and because of the predominant tiny fragments and irregular quartz pieces, Chavaillon (1976) introduced the term "Shungura facies" for the Omo industries.

Howell et al. (1987) discussed the context and stratigraphic position of Omo 71 and Omo 84, the two sites belonging to Member E. Later, de la Torre (2004) questioned the very evidence of intentionality in stone artifacts, assessing that lithics from the two sites (24 objects from Omo 71 and 200 from Omo 84) must rather be interpreted as natural assemblages. This was confirmed during recent research in the Shungura Formation (Delagnes et al. 2011). Thus there is no consensual evidence of stone knapping in Member E.

In Member F, five archaeological sites were discovered, notably FtJi 1, 2, 5, Omo 57, and Omo 123. Merrick and Merrick (1976) divided them into sites in primary position (FtJi 2 and Omo 123), and sites in secondary contexts from the fluvial channel (FtJi 1, 5, and Omo 57). According to the brief, non-exhaustive description, the quartz industries from FtJi 1, 2, and 5 include few whole and broken flakes and high numbers of angular fragments (Merrick and Merrick 1976). According to Chavaillon (1976), in addition to flakes and angular fragments, the Omo 57 and 123 assemblages also contained cores. This was confirmed by de la Torre (2004), who pointed out, however, that many so-called artifacts actually are unworked objects. He recognized the systematic unifacial unidirectional exploitation of the cores, and occasionally even a rotation aimed at making the most of the knapping surfaces. Given that the flakes show previous removals of the dorsal face, more than one series of flakes was detached from cores. However, the small size of the angular quartz pebbles that are used as core blanks did not allow a high flake productivity. The natural angles and available surfaces are used until exhaustion, without any attempt of rejuvenation. Notwithstanding the small size of the blanks, direct knapping with a hammerstone was widely used and the bipolar technique adopted in two cases only. Very few knapping accidents, such as split fracture, are recorded, suggesting some control of the strength used in impacting the cores (de la Torre 2004).

New investigations in the Shungura Formation (Boisserie et al. 2008, 2010; Delagnes et al. 2011) allowed discovering several new localities in Member F, ranging from small concentrations with few pieces to sites with abundant material. The artifacts are almost exclusively quartz flakes and fragments. Systematic surveys demonstrated that the quartz pebbles available in paleochannels near the archaeological sites do not outnumber any other rock types, as lavas, granites, and cryptocrystalline rocks. Vice versa, quartz dominates in archaeological collections. This means that quartz was deliberately selected for knapping purposes, most probably for angular morphologies which facilitated the initial flaking phases (Delagnes et al. 2011).

2.3.5 Kanjera South (Kenya)

Oldowan occurrences were found at Kanjera South, on the northern margins of the Homa Mountain Carbonatite Complex (Homa Peninsula, southwestern Kenya). They belong to Beds KS-1 to KS-3 of the Southern Member of the Kanjera Formation, which predate the base of the Olduvai subchron at 1.95 Ma (Behrensmeyer et al. 1995; Plummer et al. 1999; Bishop et al. 2006).

Paleoenvironmental data provide the early documentation of hominin activities in an open habitat within a grassland-dominated ecosystem at Kanjera South (Plummer et al. 2009b). Multiple lines of evidence suggest that Kanjera "was a lightly-wooded to open grassland habitat, with a lake to the North and presumably bushes and woods lining nearby hills and perhaps some drainages. Wash from the foothills of the Homa Mountain drained toward the lake, burying faunal and lithic materials on a generally low-energy alluvial plain during KS-1 through KS-3 deposition" (Plummer et al. 2009a: 158).

A detailed analysis of raw material provisioning is available. Knapped rocks derived from a wide variety of geographically distinct primary and secondary sources. Lithologies incorporated in the assemblages include igneous rocks, sedimentary rocks, metamorphic rocks, and metasomatized rocks. Approximately 28% of the artifacts were made with non-local raw materials, i.e., rocks available outside outcrops and drainage systems of the Homa Peninsula (Braun et al. 2008a, b, 2009b, c).

Several flaking methods were in use. Centripetal flaking on two surfaces is the best solution when rounded cobbles are available. Two opposed convex flaking surfaces are exploited without any hierarchy, and the maintenance of convexities happens rarely. Many of them are bipyramidal. Centripetal débitage can also occur on the ventral face of large flakes, reused as core blanks. In the case of multifacial multidirectional cores, continuous core rotation allows maintaining acute angles and the production of several generations of flakes. Some such cores are of Homa limestone, otherwise rarely used. The reduction process changes progressively the original volume structure of the blank. A unifacial unidirectional method is applied on coarse-grained rhyolite/dacites. Bipolar technique also occurs, sometimes mixed to hand-held percussion.

The lack of long series of removals might simply reflect the small size of the original cobbles. Core reduction seems to covary with the availability and quality of available raw materials. Raw materials from farther away were actually reduced more extensively than those available nearby (Braun et al. 2008a, b, 2009b, c; Lemorini et al. 2014).

2.3.6 Fejej (Ethiopia)

The Fejej region is located in the Ethiopian sector of the African Rift system, in southwestern Ethiopia, and at the northernmost extremity of the Koobi Fora Formation, which was identified on the east side of Lake Turkana (Asfaw et al. 1991; de Lumley and Beyene 2004). Archaeological sites were discovered in the FJ-1 locality, where sector FJ-1a was selected for an excavation. The age of the archaeological layer is constrained between 1.95 ± 0.03 Ma, the onset of the Olduvai subchron, and 1.869 ± 0.021 Ma, the ^{39}Ar–^{40}Ar age of KBS Tuff (Cande and Kent 1995; de Lumley et al. 2004a; McDougall and Brown 2006). Three hominin teeth from FJ-1a were ascribed to a young adult *Homo* aff. *habilis.* A distal humerus fragment with affinities to *Australopithecus boisei* was discovered in an older unit (de Lumley and Marchal 2004).

The stone assemblage from Fejej FJ-1a was knapped from local raw materials collected nearby in the alluvial deposits of a small river. Quartz cobbles, mainly of small size, are dominant (91%) over basalt cobbles (7%), while just a few more artifacts were knapped from other rock types. Sampling in the FJ-1 conglomerate highlighted that quartz and basalt cobbles and pebbles are only ca. 35% of rock types, showing that quartz was purposefully selected. Thick and oval cobbles were often used as knapping hammerstones, but percussion marks also occur on cores, suggesting that they happened to be used as multipurpose tools. Flake production was often carried out on angular cobbles taking advantage of the available angles. The industry is predominantly composed of flakes and angular fragments, cores are few, and retouched flakes scarce. Artifacts are small, given the size of the available blanks.

Reduction sequences are short. Flaking is mostly unifacial unidirectional, multidirectional, bipolar, or centripetal. When centripetal exploitation occurs, it is to adapting to the original cobble morphology. Multifacial and bifacial reduction strategies are rarely observed. Striking platforms are usually natural. Flake negative scars were occasionally used as platforms, producing multifacial orthogonal cores. Direct, hard hammer percussion is most frequently observed, although some of the cores were flaked using controlled bipolar percussion on an anvil (de Lumley et al. 2004b; Barsky et al. 2011).

2.3.7 Olduvai (Tanzania)

The Oldowan sites of Olduvai are located in Bed I, Lower Bed II, and Middle Bed II up to Tuff IIB (Leakey 1971). De la Torre and Mora (2005) completely reassessed some of the lithic assemblages from Beds I and II, which had been excavated in the 1960s, from a technological perspective. In their analysis, the lithic collections from three sites were assigned to the Oldowan, i.e., those from DK, FLK Zinj, and FLK North.

The DK site is the oldest one at Olduvai. It lies just below Tuff IB, dated to 1.848 ± 0.003 Ma (Habermann et al. 2016). FLK Zinj is 6 m below Tuff IF, dated to 1.803 ± 0.002 Ma (Habermann et al. 2016). FLK North has a more complex stratigraphic development, with eight archaeological levels. Levels 6–1 belong to Upper Bed I below Tuff IF. The *Deinotherium* level lies below Tuff IIA (1.66 Ma) at the base of Lower Bed II. Sandy Conglomerate (SC) level is located in the lower part of the Middle Bed II, below Tuff IIB dated to ~1.6 Ma (Manega et al. 1993).

At all sites, lavas and quartz were mainly used. The provisioning area is no more than 4 km from any given site and, in most cases, probably at as little as 2 km from the Gorge. As a rule, lavas derive from stream cobbles, whereas the majority of quartz originates from tabular blocks transported from the source at Naibor Soit. Knappers of FLK North focused more on exploiting high-quality raw materials, mainly phonolites from a nearby stream that had not existed earlier. At FLK North SC, chert, which is available at specific time spans in the Olduvai sequence, was also collected in the vicinity (Stiles et al. 1974; Hay 1976; Blumenschine and Peters 1998; Stiles 1998; Kyara 1999; Kimura 2002; de la Torre and Mora 2005).

Percussion activities play an important role in the Oldowan assemblages of Olduvai. Active hammerstones for knapping are rounded river cobbles mainly of lavas, rather uniform in size and with ergonomic shapes that enable their use for hand-held percussion. Some were also used as cores. They show areas with thick pitting and, when a fracture occurs, the cobble is rotated or discarded. The selection of lavas was most probably linked to their weigh as well as to the need of rounded shapes, given that quartz is usually available as tabular clasts. A different type of active hammerstones are cobbles or blocks used in heavy percussion activities other than knapping, generating fracture angles in a large area of the blank. Passive percussion elements, i.e., anvils with two opposite battered surfaces, are mainly of quartz, probably because of the convenient tabular morphology, granting stability during the process (de la Torre and Mora 2005, 2010; Mora and de la Torre 2005; Diez-Martín et al. 2009).

Although percussion occurs in all Oldowan sites, it is the main technical activity at FLK North Levels 6–1 and the only one in the *Deinotherium* level. De la Torre and Mora (2005) argue that in the two instances knapping activities were residual. They draw a scenario in which hominins were mainly involved in the intense percussion of a variety of organic materials, particularly long bone shafts. However, no evidence of hammerstone-broken bones has been found in most of the FLK North levels (Domínguez-Rodrigo et al. 2007).

Small-medium flake production was the only knapping activity as for the other Oldowan assemblages. Many flakes show remains of natural surfaces, suggesting limited intensity of exploitation and little recurrence in core flaking. Inter-site metrical similarities suggest a rather homogeneous metric module. Very few flakes were retouched, modifying one or two edges, without creating specific morphologies.

Flaking is unifacial, bifacial, and multifacial. Unifacial exploitation is usually unidirectional from natural striking platforms. The resulting flakes are elongated and were produced from a flaking surface usually restricted to a portion of the blank and just occasionally involving its periphery. In a few instances, after the flaking surface is exhausted, the core is rotated to pursuit the same method on a new surface. Bifacial cores are also abundant. They can be bifacial abrupt, when there is an interchange between striking and flaking surfaces; bifacial peripheral with the intersection of two asymmetrical flaking surfaces; and bifacial simple partial, when removals involve only a portion of two flaking surfaces and are used alternatively as platform for flaking. Few cores from DK are evidence of multifacial exploitation: the core surfaces were alternatively flaked through multidirectional removals without a clear organization of the reduction process and without any specific platform preparation. These cores were abandoned when exhausted (de la Torre and Mora 2005).

Recent investigations at FLK North Levels 6–1 unearthed a new archaeological assemblage. Levels 1–2 yielded the majority of lithic objects (Domínguez-Rodrigo et al. 2010a). Battering is the main activity inferred again at the site, just as in the previous assessment (de la Torre and Mora 2005). Flaking, however, displays more complex patterns than in the assemblage excavated by Leakey and revised by de la Torre and Mora (2005). Freehand direct percussion is largely used; bipolar technique is also present, but restricted to the knapping of quartz slabs. Flaking modalities are (Diez-Martín et al. 2010):

(1)
- unifacial unipolar, when only one striking direction or platform is visible;
- unifacial bipolar, when two opposed and parallel striking platforms show negative scars in two directions.

(2)
- bifacial unipolar, when two surfaces have been exploited from a single striking platform;
- bifacial bipolar opposed, when the exploitation occurs on two opposed parallel striking surfaces on two different planes;
- bifacial multipolar orthogonal, when negative scars occur on two surfaces and the negative scar sequences are arranged orthogonally. Bifacial cores show more intense exploitation and volume reduction, although no such core has been exploited to exhaustion.

(3)
- trifacial, when the exploitation is carried out on three different surfaces and from multiple striking platforms.

(4)
- centripetal, both unifacial and bifacial, ending with intense reduction.

2.3.8 Melka Kunture (Ethiopia)

Two Oldowan assemblages were recovered at Garba IV levels E and F in the basal part of the Melka Kunture Formation, stratigraphically located just below the Grazia Tuff, dated to <1.719 ± 0.199 (Piperno et al. 2004, 2009; Raynal et al. 2004; Morgan et al. 2012; Gallotti and Mussi 2015). Tamrat et al. (2014) include both levels in a normal polarity interval (N1), which they interpret as the end of the Olduvai subchron.

The assemblages share similar technical patterns and include all the phases of small débitage chaînes opératoires. Approximately 80% of the artifacts are of obsidian, followed by aphyric lavas and by few items of porphyritic and microdoleritic basalts. All these rocks were abundantly available nearby in paleochannels, but obsidian cobbles and pebbles are found there in small proportion when compared to aphyric lavas (Kieffer et al. 2002, 2004). The large proportion of high-quality raw material artifacts is clearly the outcome of strict selection by the knappers.

In addition to a few simple cores, several structured flaking methods were recognized, closely dependent on blank geometry. The most frequent flaking method is irregular multifacial multidirectional exploitation. Cobbles with several flat surfaces were flaked irregularly using any available flaking angle, without preparing a striking platform, to produce the largest feasible number of flakes. Besides, a few cores show major flaked surface(s) with unidirectional long flake scars exploited at the beginning of the reduction process. Multifacial multidirectional cores were usually abandoned when their size had been considerably reduced.

Unifacial unidirectional exploitation of the longest available surface of elongated cobbles produced flakes with a length exceeding the width. The natural convex surface of cobbles was chosen for detaching flakes by the peripheral unidirectional method over the maximal available extension. The striking platform was a naturally flat surface.

Centripetal/tangential exploitation was mostly performed on a flaking surface from a natural peripheral platform, or from a striking platform rectified by only a few removals. Volume and convexity configurations were not managed, recurrence and preparation are absent, and there is no evidence of hierarchy. When the lack of convexities (as on flat cobbles) made it possible to detach only short flakes, the resulting cores are unifacial or bifacial partial ones.

Retouch frequently occurs exclusively on obsidian flakes. Side-scrapers and notches, as well as small points, were manufactured at Garba IVE-F. They can be grouped in two sets. The first one, identified only in layer E, consists of flakes whose edges were modified by a retouch that did not transform the original blank into any standard form. The retouch is continuous but highly variable, ranging from marginal to invasive. The resulting tools display large dimensional and morphological variability. In contrast, the second set (41 tools) displays a retouch process aimed specifically at producing a small point, modifying the distal part of the blank though standardized procedures expressed by the simultaneous occurrence of: (1) a repetitive intention to shape the distal portion of the flake into a tip; (2) a repetitive intention to create a convergence; and (3) a recurrent search for a small and homogeneous size (Gallotti and Mussi 2015).

2.4 Discussion

The evidence for Oldowan stone knapping falls into two chronological sets divided by a gap of approximately 0.3 Myr.[2] The first set, the early Oldowan, includes archaeological sites dated between 2.6 and 2.3 Ma, located in the northern part of the Rift Valley with two main clusters: the Middle Awash River (Gona and Hadar) and the Lake Turkana Basin (Omo and West Turkana). Except West Turkana, these long sequences did not yield later Oldowan assemblages and a relevant chronological gap further separates the Oldowan from the early Acheulean appearing in the area at approximately 1.8–1.6 Ma (Quade et al. 2004, 2008; Lepre 2011; Beyene et al. 2013). The second set dates between 2.0 and 1.6 Ma and includes the later Oldowan sites of Olduvai Gorge, of Melka Kunture in the Ethiopian highlands, of Kanjera South in the Lake Victoria Basin, and of Fejej in the Omo-Turkana Basin at the northern limit of the Koobi Fora Formation. At Olduvai, some late Oldowan sites temporally overlap with the earliest Acheulean (Diez-Martín et al. 2015; de la Torre 2016).

We synthesize the main distinctive characters of the Oldowan lithic assemblages as follows: (1) stone tools were mainly cutting tools belonging to flaking processes, i.e., flakes; (2) stone knapping shows an advanced knowledge of stone fracture mechanics and was performed with a controlled hand-held percussion technique, rarely with bipolar percussion; (3) percussion devoted to activities other than knapping is scarcely documented.

Interestingly, the frequency of hammerstones and of other percussion/pounding tools in the early Oldowan is extremely low (e.g., Kibunjia 1994; Roche et al. 1999; Semaw 2000; Delagnes and Roche 2005). Percussion elements lack at Gona, Hadar, and LA1 in West Turkana. When found, as at LA2C, they document highly controlled percussion motions for hand-held knapping. The hammerstones display a high density of impact scars in circumscribed areas, which were produced by a precise and recurrent use, according to stable motor habits. Accordingly, the cores do not display impact damage from failed percussion, such as might be caused by an inaccurate appreciation of the required strength, or by inadequate manipulating or gripping (Delagnes and Roche 2005).

At Olduvai, percussion elements are abundant in all Oldowan sites and part of them are devoted to activities other than knapping, i.e., probably to food processing (de la Torre and Mora 2005; Mora and de la Torre 2005). This does not seem to reflect ecological differences between Olduvai and sites in Gona, West Turkana, Hadar, or Kanjera South, as the Oldowan sites are located in a mosaic of different landscapes, ranging from wooded areas to arid ones and open settings (e.g., Plummer et al. 1999, 2009a, b; Aronson et al. 2008; López-Sáez and Domínguez-Rodrigo 2009; Quinn et al. 2013). Hovers (2007) hypothesized that pounding activities documented in Olduvai do not represent the retention of earlier behavior, and that pounding possibly was rather "reinvented" during technological evolution. Nevertheless, pounding activities are not present in the sub-contemporaneous sites of Kanjera South, Fejej, and Melka Kunture. This means that, because of multiple but

[2]Recently, Harmand et al. (2016) announced the discovery of 2.3–2.0 Ma sites in the Nasura Complex (West Turkana), but detailed data are not yet available.

unknown factors, at Olduvai such activities might correspond to some kind of "resurrection" of a retained behavior —a behavior previously invented, experienced, and mastered for thousands of years.

At the dawn of lithic technology, knapper activities were associated with pounding on anvil rather than with hand-held core reduction. The small collection of stone tools from LOM3 is unlike any from Oldowan localities, mainly including flakes. On the opposite, most of the LOM3 artifacts are hammerstones, anvils, cores, and worked cobbles. Harmand et al. (2015) pointed out that LOM3 assemblage possibly represents an intermediate stage between a pounding-oriented stone tool use—as documented in modern non-human primates—and the flaking-oriented knapping behavior of Oldowan toolmakers.

The use of bipolar technique is not documented in all Oldowan sites. When used, it is often for flaking specific raw materials, such as quartz usually occurring as small cobbles (de la Torre 2004; de la Torre and Mora 2005; Diez-Martín et al. 2010; Delagnes et al. 2011; Goldman-Neuman and Hovers 2012). Nevertheless, even if difficulties in manipulating small-sized core blanks probably led to favoring bipolar technique as an alternative, as at Kanjera South, Fejej, and Olduvai, at Omo sites this technique is documented by just two quartz cores. The Omo knappers were so much in command of fine gripping skills, that tiny cobbles were not simply smashed but rather handled according to the same technical principles used in the exploitation of larger raw materials at Gona and West Turkana (de la Torre 2004). Just as percussion for food processing, bipolar technique for knapping, documented since 3.3 Ma at LOM3, was retained and reused as a proper solution for knapping over hundreds of thousands of year. However, the Omo evidence points to the fact that this resurgence cannot be always explained by raw material constraints.

Results from a functional approach to percussive technology recently demonstrated that stone knapping is guided by a rather specific set of mechanical constraints, more complex than those involved in nut-cracking (Bril et al. 2015). Freehand stone-knapping techniques require the ability of understanding the relationships among the parameters which define the conchoidal fracture, complex bimanual skills, and the fine motor properties of a firm precision grip. This means that the transition from percussive techniques, such as nut-cracking, to freehand knapping techniques required improved perceptual abilities, learning capacities, and a bimanual dexterity superior to that of any non-human primates. All this was experienced over approximately 700,000 years (Stout and Chaminade 2007; Stout et al. 2008).

Understanding conchoidal fracture mechanisms in parallel with systematic hand-held percussion, together with a sporadic use of bipolar technique and anvil percussion for food processing are the distinctive characters shared by the Oldowan assemblages when compared with the previous Lomekwian.

However, even if most or even all of the Oldowan assemblages known to date share the knapping principles, some intra- and inter-site variability appears and actually increases when analyzing the multiple steps of the chaînes opératoires.

In lithic resource acquisition, for example, the toolmaker selectively aimed at collecting fine-grained rocks from locally available raw materials (Stout et al. 2005; Braun et al. 2009b, c; Goldman-Neuman and Hovers 2009, 2012; Harmand 2009; Gallotti and Mussi 2015). Transport distances in the majority of cases are minimal, ranging from some hundred meters to a few kilometers. Nevertheless, an incipient exploitation of non-local raw materials involving longer distances is evidenced at OGS-7, at Kanjera South, and probably at A.L.666 (Stout et al. 2005; Braun et al. 2008a; Goldman-Neuman and Hovers 2012). We note that non-local and local provisioning systems may occur simultaneously in the same region, as at Gona. Accordingly, transport across the landscape cannot be explained in a straightforward evolutionary perspective, also because this behavior is not systematically documented at later Oldowan sites (de la Torre and Mora 2005; Barsky et al. 2011; Gallotti and Mussi 2015).

In some cases another criterion guiding raw material selection is any cobble/block angularity that facilitates the first steps of knapping. The preference for angular cobbles at Omo and Hadar sites is conceptually similar to the selection of angular cobbles at Fejej and Melka Kunture, of blocks at LA2C as well as of tabular clasts in Olduvai (de la Torre 2004; Delagnes and Roche 2005; de la Torre and Mora 2005; Barsky et al. 2011; Goldman-Neuman and Hovers 2012; Gallotti and Mussi 2015). The selection of blanks with serviceable striking angles implies that knappers were somewhat constrained by the raw material geometry. At LA2C, rounded cobbles were mainly used for simple flaking. The random production of few flakes apparently is a response to the difficult task of exploiting inappropriate shapes, rather than the outcome of poor quality in terms of grain and homogeneity. This also means that in this case the knappers did not strive to create any such angles. The importance of angular morphologies is also highlighted by the use of natural surfaces as striking platforms, which strictly speaking were never prepared, but somewhat rectified by a few removals (Delagnes and Roche 2005).

At A.L. 894, step and hinge scars were frequently produced when knappers were unable to maintain appropriate knapping angles but insisted flaking always from the same direction. Only cores with large surfaces were rotated in an attempt of rectifying the mismanagement, continuing the flaking process (Hovers 2009; Goldman-Neuman and Hovers

2012). At LA2C, the knappers insisted on the same surface producing successive series of flakes from multiple striking platforms, a practice that maintains a flat flaking surface (Delagnes and Roche 2005). On the contrary, at the sub-contemporaneous site of LA1 the knappers were apparently unable to apply this procedure. The cores bear evidence of frequent knapping accidents and repeated impact damage from failed percussions (Kibunjia 1994; Delagnes and Roche 2005). At KS5, knappers took advantage of the initial angular morphology, but they were also able to rotate the core and to create new angles (Texier et al. 2006; Texier 2018).

The perceived usefulness of a flat surface also directed raw material collection. At LA2C some large blocks were broken off-site in order to obtain a serviceable flaking surface. This behavior implies an incipient ability at fragmenting the chaîne opératoire, also visible at A.L. 894, where knappers tested cobbles at the source and imported them into the site already partly decorticated (Goldman-Neuman and Hovers 2009, 2012). Anyway, any chaîne opératoire fragmentation is unusual in the Oldowan and when specific components are lacking this is usually due to selective post-depositional mechanisms.

Although the search for raw materials with appropriate angles is recurrent at Oldowan sites, knappers also successfully flaked rounded shapes. Rounded elements were not just flaked ever since the first appearance of the Oldowan techno-complexes, but they were also flaked on more than one surface at OGS7, where bifacial and multifacial exploitations occur more frequently than unifacial ones (Stout et al. 2010).

On the opposite, core methods are mainly unifacial in the contemporaneous EG10 and EG12 sites, just as in the older Oldowan assemblages (Hovers 2003, 2009; de la Torre 2004; Delagnes and Roche 2005; Stout et al. 2010). At a final reduction stage few removals sometimes occur on a second face, documenting an occasional rotation of the core as at LA2C and Omo (de la Torre 2004; Delagnes and Roche 2005). Flake extraction from more than one surface, i.e., bifacial and multifacial exploitations implying a continuous core rotation, happens in the older Oldowan assemblages, but, except for OGS7, never represents the main flaking modality. Bifacial and multifacial flaking methods were more frequently adopted in the later Oldowan assemblages as at Kanjera, Fejej, Olduvai, Melka Kunture, and KS5, increasing the intra-site method variability (de la Torre and Mora 2005; Braun et al. 2009c; Diez-Martín et al. 2010; Barsky et al. 2011; Gallotti and Mussi 2015).

Simultaneous flaking of two or more surfaces warranties a higher productivity. Nevertheless, we underline that knappers at LA2C were able to detach a considerable number of flakes from a single surface thanks to specific procedures. The large size of LA2C core blanks definitely plays a role in this high productivity. This possibly explains why the extraction of a considerable number of flakes by creating and maintaining a single flaking surface is impossible with small cores like those from Omo (de la Torre 2004; Delagnes and Roche 2005). Conversely, elsewhere as at Gona, Kanjera South, Fejej, Garba IVE-F, and FLK North the absence of natural angles in rounded cobbles was overcome thanks to a more appropriate flaking modality: the unifacial centripetal method, which takes advantage of the core periphery as a striking platform, while the centripetal direction of flaking allows to detach more than one series of flakes with longer cutting edges (Semaw 2006; Braun et al. 2009c; Diez-Martín et al. 2010; Barsky et al. 2011; Gallotti and Mussi 2015).

However, as for unifacial cores, bifacial and multifacial exploitations correspond to an adaptation to the original cobble morphology. In the case of bifacial exploitation the notions of convexity maintenance, hierarchy between the two surfaces, and platform preparation simply do not exist. Bifacial centripetal methods are the proper solution when exploiting rounded cobbles and, when the convexities were exhausted, no attempt was made to restructure the core morphology. A multifacial multidirectional modality exploits several surfaces irregularly using any available angle <90° to produce the largest feasible number of flakes. Thus, blank morphology is respected and not modified during flaking. These practices were adopted in both early and late Oldowan assemblages, regardless of the relative frequency of the flaking methods (e.g., de la Torre and Mora 2005; Barsky et al. 2011; Gallotti and Mussi 2015).

From 2.6 Ma onward, the main objective of flaking was to maximize flake production rather than to detach flakes with specific technical aspects, notwithstanding high intra- and inter-site variability linked to variable technical solutions. Besides, flakes with one or more edges modified by retouch are only an occasional component of the Oldowan techno-complexes. When present at all, small tools are found in very small percentages and do not show any standardization (de la Torre and Mora 2005; Delagnes and Roche 2005; Semaw 2006; Barsky et al. 2011; Zaidner 2013). Nevertheless, there is also evidence of a specific technical process at Garba IVE-F, not recorded elsewhere in an Oldowan assemblage: the systematic search for small pointed forms on obsidian blanks. Most probably the small pointed tools are an occasional technological development driven by an unknown specific techno-functional purpose, certainly facilitated by the high knapping suitability of obsidian (Gallotti and Mussi 2015). This "invention" (*sensu* Hovers 2012) was not followed by any further production of similar tools in the early/middle Acheulean of Melka Kunture region (Gallotti et al. 2010, 2014; Gallotti 2013).

A fundamental question remains unanswered: who were the authors of the Oldowan stone tools? From 2.6 to 1.6 Ma, the data currently available indicate that several species belonging to various genera practiced hard stone knapping. This time period marks the transition from *Australopithecus* to *Homo* and the appearance of the robust *Australopithecus* species, in East Africa usually placed in genus *Paranthropus*.

The earliest Oldowan industries of Gona are not associated with any human fossils, but *Australopithecus garhi*, dated to 2.5 Ma, was discovered in neighboring Bouri. *Australopithecus garhi* so far is the only possible candidate as toolmaker of the oldest Oldowan industries (Asfaw et al. 1999). *Australopithecus garhi*, in turn, was sub-contemporaneous to *Paranthropus aethiopicus*, dated between 2.5 and 2.3 Ma. Until now, this robust species has been exclusively found in the Turkana Basin, i.e., not in the Awash Basin. Nevertheless, a recently recovered partial hominin mandible from the Ledi-Geraru area in the Afar region, dated to 2.8–2.75 Ma, extends the fossil record of *Homo* back in time by 0.4 Ma, making it a further candidate toolmaker of the Gona Oldowan (Villmoare et al. 2015). Besides, we know that from 2.3 to 1.5 Ma, representatives of the *Paranthropus* and *Homo* genera could have interacted in several regions. *Homo* aff. *habilis* is present at Hadar at 2.3 Ma. It had previously coexisted with *Paranthropus aethiopicus* in Omo and West Turkana, and later with *Paranthropus boisei* at Fejej and Olduvai (Leakey 1971; Walker et al. 1986; Kimbel 1997; de Lumley and Marchal 2004; Prat et al. 2005).

Since 1.9 Ma, a new *Homo* species appears, i.e., *Homo ergaster/erectus*, whose first recorded evidence is at ~ 1.9 at Koobi Fora (Lepre and Kent 2015). The recent discovery of a phalanx at Olduvai allows dating back *Homo erectus sensu lato* to >1.84 Ma in this region (Domínguez-Rodrigo et al. 2015). At Melka Kunture, ever since >1.7 Ma the Early Pleistocene fossil record points exclusively to genus *Homo*, i.e., to *Homo erectus sensu lato* (Condemi 2004; Di Vincenzo et al. 2015). *Paranthropus* and *Homo habilis* have not been discovered there. This possibly suggests that *Homo erectus* was the first and only species of the time able to adapt to mountain environments as the Ethiopian highlands (Mussi et al. 2015).

2.5 Conclusions

This review of fifteen years of techno-economic studies of Oldowan assemblages clearly demonstrates that the one-million-year-long technological stasis (Semaw et al. 1997; Semaw 2000; Stout et al. 2010) and the systematic techno-cultural homogeneity of the 2.6–2.3 Ma industries (de la Torre 2004) are both hardly supported when the analysis is detailed enough, developing beyond the basic principles of stone knapping and the simple presence/absence, or increase/decrease, of technical components.

Producing a conchoidal fracture allowing to knap stones is a single action. But ever since 2.6 Ma there are multiple ways of combining raw material selection and acquisition patterns, percussion motions, and flaking sequences. These multiple processes, in turn, are linked to manifold factors as landscape dynamics, climatic shifts, and biotic/abiotic resource availability.

In the Oldowan scene, as emerging from techno-economic analyses, we discover multiple actions intermingling without a temporal trend—i.e., no less to more "evolved" behaviors—or a continuous spatial distribution—i.e., no homogeneity and intra-site variability at Gona and West Turkana (Delagnes and Roche 2005; Stout et al. 2010). Rather than corresponding to two evolutionary technological steps, the distinction between early and late Oldowan assemblages has a chronological significance (e.g., Delagnes and Roche 2005). The only difference so far detected in late Oldowan assemblages (≤ 2.0 Ma) is the rather systematic intra-site coexistence of a panoply of flaking methods and practices, which at earlier sites were only tested and experienced at inter-site scale. This means that since 2.0 Ma knappers were able to simultaneously apply several technical solutions in order to exploit all the available lithic resources. Accordingly, it is difficult to support the idea of the one-million-year-long technological stasis (Semaw et al. 1997; Semaw 2000; Stout et al. 2010), which seems actually restricted to the 2.6–2.3 Ma sites. Intra-site variability and variation happen at Gona and West Turkana, but not at Omo and Hadar. In any case, variability and variation do not preclude stasis if there are no trends of temporal change through time. Nevertheless, an intra-site technical evolution is visible at West Turkana, where Delagnes and Roche (2005) have suggested that the differences between Lokalalei 1 (older) and Lokalalei 2C (younger) could be related to evolutionary processes. Besides, so far the Oldowan assemblages cannot be definitively attributed to a specific Early Pleistocene hominid lineage. The identity of the earliest hominid toolmakers remains elusive. The paleoanthropological diversity, coupled with the technological diversity, is not in accordance with a linear technological evolution, and it magnifies the difficulty of defining the tempos and modes of a possible technological stasis or evolution.

As a concluding remark, a clear definition of the Oldowan cannot be worked out without following back to the use of tool types as cultural markers. Vice versa, providing a comprehensive definition in a techno-economic perspective is a difficult and to some extent unresolved task. This is due to the very nature of the lithic productions of this age, rather than to the methodological approach. Even if Oldowan stone knapping shows an advanced knowledge of stone fracture mechanics, raw material constraints played such a determining role that the knappers again and again had to experiment the most proper technical solutions. The technical outcome happened to be alike or diverse at different levels and according to multiple factors. Obviously, if a definition of the Oldowan technology is unavailable in its homeland at the current state of the research, exporting this term out of East Africa to identify techno-complexes scattered in several continents over nearly two million years is just a sterile exercise, masking a strong epistemological incongruence.

Acknowledgements I am very grateful to the three anonymous reviewers for their comments which improved this manuscript. The revised version of this chapter benefited also from the detailed suggestions by Margherita Mussi.

References

Aronson, J. L., Hailemichael, M., & Savin, S. M. (2008). Hominid environments at Hadar from paleosol studies in a framework of Ethiopian climate change. *Journal of Human Evolution, 55,* 532–550.

Asfaw, B., Beyene, Y., Semaw, S., Suwa, G., White, T. D., & WoldeGabriel, G. (1991). Fejej: A new paleoanthropological research area in Ethiopia. *Journal of Human Evolution, 21,* 137–143.

Asfaw, B., White, T., Lovejoy, O., Latimer, B., Simpson, S., & Suwa, G. (1999). *Australopithecus garhi*: A new species of early hominid from Ethiopia. *Science, 284,* 629–635.

Barsky, D., Chapon-Sao, C., Bahain, J.-J., Beyene, Y., Cauche, D., Celiberti, V., et al. (2011). The Early Oldowan Stone-Tool Assemblage from Fejej FJ-1a, Ethiopia. *Journal of African Archaeology, 9,* 207–224.

Behrensmeyer, A. K., Potts, R., Plummer, T. W., Tauxe, L., Opdyke, N., & Jorstad, T. (1995). The Pleistocene locality of Kanjera, Western Kenya: Stratigraphy, chronology and paleoenvironments. *Journal of Human Evolution, 29,* 247–274.

Beyene, Y., Katoh, S., WoldeGabriel, G., Hart, W. K., Uto, K., Sudo, M., et al. (2013). The characteristics and chronology of the earliest Acheulean at Konso, Ethiopia. *Proceedings of the National Academy of Sciences, 110,* 1584–1591.

Bishop, L. C., Plummer, T. W., Ferraro, J. V., Braun, D., Ditchfield, P. W., Hertel, F., et al. (2006). Recent Research into Oldowan Hominin Activities at Kanjera South, Western Kenya. *African Archaeological Review, 23,* 31–40.

Blumenschine, R. J., & Peters, C. R. (1998). Archaeological predictions for hominid land use in the paleo-Olduvai Basin, Tanzania, during lowermost Bed II times. *Journal of Human Evolution, 34,* 565–607.

Boisserie, J. R., Delagnes, A., Beyene, Y., & Schuster, M. (2010). Reconstructing the African background to human expansions in Eurasia: New research in the Shungura Formation, Ethiopia. *Quaternary International, 223–224,* 426–428.

Boisserie, J. R., Guy, F., Delagnes, A., Hlukso, L. J., Bibi, F., Beyene, Y., et al. (2008). New palaeoanthropological research in the Plio-Pleistocene Omo Group, Lower Omo Valley, SNNPR (Southern Nations, Nationalities and People Regions), Ethiopia. *Comptes Rendus Palevol, 7,* 429–439.

Boëda, E. (1991). Approche de la variabilité des systèmes de la production lithique des industries du Paléolithique inférieur et moyen : chronique d'une variabilité attendue. *Technique et culture, 17*(18), 37–39.

Bordes, F. (1961). *Typologie du Paléolithique inférieur et moyen.* Bordeaux: Delmas.

Bower, J. R. F. (1977). Attributes of Oldowan and Lower Acheulean tools: "tradition" and design in the Early Lower Palaeolithic. *The South African Archaeological Bullettin, 32,* 113–126.

Braun, D. R., Plummer, T., Ditchfield, P., Ferraro, J. V., Maina, D., Bishop, L. C., et al. (2008a). Oldowan behavior and raw material transport: Perspectives from the Kanjera Formation. *Journal of Archaeological Science, 35,* 2329–2345.

Braun, D. R., Rogers, M. J., Harris, J. W. K., & Walker, S. J. (2008b). Landscape-scale variation in hominin tool use: Evidence from the developed Oldowan. *Journal of Human Evolution, 55,* 1053–1063.

Braun, D. R., Tactikos, J. C., Ferraro, J. V., Arnow, S. L., & Harris, J. W. K. (2008c). Oldowan reduction sequences: methodological considerations. *Journal of Archaeological Science, 35,* 2153–2163.

Braun, D. R., & Hovers, E. (2009). Introduction: Current Issues in Oldowan Research. In E. Hovers & D. R. Braun (Eds.), *Interdisciplinary Approaches to the Oldowan* (pp. 111–128). Dordrecht: Springer.

Braun, D. R., Harris, J. W. K., & Maina, D. (2009a). Oldowan Raw Material Procurement and use: Evidence from the Koobi Fora Formation. *Archaeometry, 51,* 26–42.

Braun, D. R., Plummer, J. V., Ditchfield, P., Bishop, L. C., & Ferraro, T. (2009b). Oldowan Technology and Raw Material Variability at Kanjera South. In E. Hovers & D. R. Braun (Eds.), *Interdisciplinary Approaches to the Oldowan* (pp. 99–110). Dordrecht: Springer.

Braun, D. R., Plummer, T., Ferraro, J. V., Ditchfield, P., & Bishop, L. C. (2009c). Raw material quality and Oldowan hominin tool stone preferences: evidence from Kanjera South, Kenya. *Journal of Archaeological Science, 36,* 1605–1614.

Bril, B., Parry, R., & Dietrich, G. (2015). How similar are nut-cracking and stone-flaking? A functional approach to percussive technology. *Philosophical Transactions of the Royal Society B, 370,* 20140355.

Brown, F. H., & Gathogo, P. N. (2002). Stratigraphic Relation between Lokalalei 1A and Lokalalei 2C, Pliocene Archaeological Sites in West Turkana, Kenya. *Journal of Archaeological Science, 29,* 699–702.

Brugal, J. P., Roche, H., & Kibunjia, M. (2003). Faunes et paléoenvironnements des principaux sites archéologiques plio-pléistocènes de la formation de Nachukui (Ouest Turkana, Kenya). *Comptes Rendus Palevol, 2,* 675–684.

Campisano, C. J. (2012). Geological summary of the Busidima Formation (Plio-Pleistocene) at the Hadar paleoanthropological site, Afar Depression, Ethiopia. *Journal of Human Evolution, 62,* 338–352.

Cande, S. C., & Kent, D. V. (1995). Revised calibration of the geomagnetic polarity timescale for the Late Cretaceous and Cenozoic. *Journal of Geophysical Research, B, Solid Earth and Planets, 100,* 6093–6095.

Carvalho, S., & McGrew, W. C. (2012). The origins of the Oldowan: Why chimpanzees (*Pan troglodytes*) still are good models for technological evolution in Africa. In M. Domínguez-Rodrigo (Ed.), *Stone Tools and Fossil Bones: Debates in the Archaeology of Human Origins* (pp. 222–244). Cambridge: Cambridge University Press.

Carvalho, S., Cunha, E., Sousa, C., & Matsuzawa, T. (2008). Chaînes opératoires and resource-exploitation strategies in chimpanzee (*Pan troglodytes*) nut cracking. *Journal of Human Evolution, 55,* 148–163.

Chavaillon, J. (1970). Découverte d'un niveau oldowayen dans la basse vallée de l'Omo (Ethiopie). *Bulletin de la Société Préhistorique Française, 67,* 7–11.

Chavaillon, J. (1975). Le site paléolithique ancien l'Omo 84 (Ethiopie). *Abbay, Documents Histoire Civilisation Ethiopiéenne, 6,* 9–18.

Chavaillon, J. (1976). Evidence for the technical practices of Early Pleistocene hominids, Shungura Formation, Lower Omo Valley, Ethiopia. In Y. Coppens, F. C. Howell, Gl. Isaac, & R. E. Leakey (Eds.), *Earliest man and environments in the Lake Rudolf Basin* (pp. 565–573). Chicago: University of Chicago Press.

Chavaillon, J. (1979). Essai pour une typologie du matériel de percussion. *Bulletin de la Société Géologique Française, 76,* 230–233.
Chavaillon, J., & Boisaubert, J.-L. (1977). Prospection archéologique dans le Gemu-Gofa et la basse vallée de l'Omo. *Abbay, Documents Histoire Civilisation Ethiopienne, 8,* 3–10.
Condemi, S. (2004). The Garba IV E mandible. In J. Chavaillon & M. Piperno (Eds.), *Studies on the Early Paleolithic Site of Melka Kunture, Ethiopia* (pp. 687–701). Florence: Origines.
Corvinus, G. (1976). Prehistoric Exploration at Hadar, Ethiopia. *Nature, 261,* 571–572.
Corvinus, G., & Roche, H. (1976). La préhistoire dans la région de Hadar (Bassin de l'Awash, Afar, Ethiopie): premiers résultats. *L'Anthropologie, 80,* 315–324.
Corvinus, G., & Roche, H. (1980). Prehistoric Exploration at Hadar in the Afar (Ethiopia) in 1973, 1974, and 1976. In R. E. F. Leakey & B. A. Ogot (Eds.), *Proceedings of the VIIIth Panafrican Congress of Prehistory and Quaternary Studies* (pp. 186–188). Nairobi: The International Louis Leaky Memorial Institute for African Prehistory.
Davidson, I., & McGrew, W. C. (2005). Stone tools and the uniqueness of human culture. *Journal of the Royal Anthropological Institute, 11,* 793–817.
de Heinzelin, J., Clark, J. D., White, T. W., Hart, W., Renne, P., WoldeGabriel, G., et al. (1999). Environment and behavior of 2.5-million-year-old Bouri hominids. *Science, 284,* 625–629.
de Lumley, H., & Beyene, Y. (2004). *Les sites préhistoriques de la région de Fejej, Sud-Omo, Ethiopie, dans leur contexte stratigraphique et paléontologique*. Paris: Association pour la diffusion de la pensée française (ADPF), Editions Recherche sur les civilisations.
de Lumley, H., Bahain, J. J., Beyene, Y., Chapon, C., Durandau, A., Laurent, M., et al. (2004a). Etudes stratigraphique, sédimentologique, magnétostratigraphique et géochronologique du site de Fejej FJ-1. In H. de Lumley & Y. Beyene (Eds.), *Les sites préhistoriques de la région de Fejej, Sud-Omo, Ethiopie, dans leur contexte stratigraphique et paléontologique* (pp. 129–172). Paris: Association pour la diffusion de la pensée française (ADPF), Editions Recherche sur les civilisations.
de Lumley, H., Beyene, Y., Barsky, D., Byrne, L., Camara, A., Cauche, D., et al. (2004b). L'industrie lithique préoldowayenne du site de Fejej FJ-1. In H. de Lumley & Y. Beyene (Eds.), *Les sites préhistoriques de la région de Fejej, Sud-Omo, Ethiopie, dans leur contexte stratigraphique et paléontologique* (pp. 391–563). Paris: Association pour la diffusion de la pensée française (ADPF), Editions Recherche sur les civilisations.
de Lumley, M.-A., & Marchal, F. (2004). Les restes d'hominidés du site de Fejej FJ-1. In H. de Lumley & Y. Beyene (Eds.), *Les sites préhistoriques de la région de Fejej, Sud-Omo, Ethiopie, dans leur contexte stratigraphique et paléontologique* (pp. 341–389). Paris: Association pour la diffusion de la pensée française (ADPF), Editions Recherche sur les civilisations.
de la Torre, I. (2004). Omo revisited. *Current Anthropology, 45,* 439–465.
de la Torre, I. (2011). The origins of stone tool technology in Africa: A historical perspective. *Philosophical Transactions of the Royal Society B, 366,* 1028–1037.
de la Torre, I. (2016). The origins of the Acheulean: past and present perspectives on a major transition in human evolution. *Philosophical Transactions of the Royal Society of London B, 371,* 20150245.
de la Torre, I. & Mora, R. (2005). *Technological Strategies in the Lower Pleistocene at Olduvai Beds I & II.* Liege: ERAUL.
de la Torre, I., & Mora, R. (2009). Remarks on the Current Theoretical and Methodological Approaches to the Study of Early Technological Strategies in Eastern Africa. In E. Hovers & D. R. Braun (Eds.), *Interdisciplinary Approaches to the Oldowan* (pp. 15–24). Dordrecht: Springer.
de la Torre, I., & Mora, R. A. (2010). Technological analysis of non-flaked stone tools in Olduvai Beds I & II. Stressing the relevance of percussion activities in the African Lower Pleistocene. *Paleo, special issue 2009–2010,* 13–34.
de la Torre, I., Mora, R., Domínguez-Rodrigo, M., De Luque, L., & Alcalá, L. (2003). The Oldowan industry of Peninj and its bearing on the reconstruction of the technological skills of Lower Pleistocene hominids. *Journal of Human Evolution, 44,* 203–224.
Delagnes, A., & Roche, H. (2005). Late Pliocene knapping skills: The case of Lokalalei 2C, West Turkana, Kenya. *Journal of Human Evolution, 48,* 435–472.
Delagnes, A., Boisserie, J.-R., Beyene, Y., Chuniaud, K., Guillemot, C., & Schuster, M. (2011). Archaeological investigations in the Lower Omo Valley (Shungura Formation, Ethiopia): New data and perspectives. *Journal of Human Evolution, 61,* 215–222.
Dies, K. F., & Dies, M. B. (1980). Researches on the palaeotechnology of artefacts (Developed Oldowan B and Acheulean) with mathematical statistics. In R. E. F. Leakey & B. A. Ogot (Eds.), *Proceedings of the VIIIth Panafrican Congress of Prehistory and Quaternary Studies* (pp. 41–42). Nairobi: The International Louis Leakey Memorial Institute for African Prehistory.
Diez-Martín, F., Sánchez Yustos, P., Domínguez-Rodrigo, M., Mabulla, A. Z. P., & Barba, R. (2009). Were Olduvai hominins making butchering or battering tools? Analysis of a recently excavated lithic assemblage from BK (Bed II, Olduvai Gorge, Tanzania). *Journal of Anthropological Archaeology, 28,* 274–289.
Diez-Martín, F., Sánchez Yustos, P., Domínguez-Rodrigo, M., Mabulla, A. Z. P., Bunn, H. T., Ashley, G. M., et al. (2010). New insights into hominin lithic activities at FLK North Bed I, Olduvai Gorge, Tanzania. *Quaternary Research, 74,* 376–387.
Diez-Martín, F., Sánchez Yustos, P., Uribelarrea, D., Baquedano, E., Mark, D. F., Mabulla, A., et al. (2015). The Origin of The Acheulean: The 1.7 Million-Year-Old Site of FLK West, Olduvai Gorge (Tanzania). *Scientific Reports, 5,* 17839.
Di Vincenzo, F., Rodriguez, L., Carretero, J. M., Collina, C., Geraads, D., Piperno, M., et al. (2015). The massive fossil humerus from the Oldowan horizon of Gombore I, Melka Kunture (Ethiopia, >1.39 Ma). *Quaternary Science Review, 122,* 207–221.
Domínguez-Rodrigo, M., & Martínez-Navarro, B. (2012). Taphonomic analysis of the early Pleistocene (2.4 Ma) faunal assemblage from A.L. 894 (Hadar, Ethiopia). *Journal of Human Evolution, 62,* 315–327.
Domínguez-Rodrigo, M., Barba, R., & Egeland, C. P. (2007). *Deconstructing Olduvai.* New York: Springer.
Domínguez-Rodrigo, M., Mabulla, A. Z. P., Bunn, H. T., Diez-Martín, F., Barba, R., Domínguez-Solera, S., et al. (2010a). Disentangling hominin and carnivore activities near a spring at FLK North (Olduvai Gorge, Tanzania). *Quaternary Research, 74,* 363–375.
Domínguez-Rodrigo, M., Pickering, T. R., & Bunn, H. T. (2010b). Configurational approach to identifying the earliest hominin butchers. *Proceedings of the National Academy of Sciences of the United States of America, 107,* 20929–20934.
Domínguez-Rodrigo, M., Pickering, T. R., & Bunn, H. T. (2011). Reply to McPherron et al.: Doubting Dikika is about data, not paradigms. *Proceedings of the National Academy of Sciences of the United States of America, 108,* E117–E117.
Domínguez-Rodrigo, M., Pickering, T., & Bunn, H. T. (2012). Experimental study of cut marks made with rocks unmodified by human flaking and its bearing on claims of ~3.4-million-year-old butchery evidence from Dikika, Ethiopia. *Journal of Archaeological Science, 39,* 205–214.
Domínguez-Rodrigo, M., Pickering, T. R., Almécija, S., Heaton, J. L., Baquedano, E., Mabulla, A., et al. (2015). Earliest modern human-like hand bone from a new >1.84-million-year-old site at Olduvai in Tanzania. *Nature Communications, 6,* 7987.
Feibel, C. S., Brown, F. H., & McDougall, I. (1989). Stratigraphic context of fossil hominids from the Omo group deposits: Northern

Turkana Basin, Kenya and Ethiopia. *American Journal of Physical Anthropology, 78,* 595–622.
Foley, R. (1991). *The Origins of Human Behaviour.* London: Unwin Hyman Ltd.
Gallotti, R. (2013). An older origin for the Acheulean at Melka Kunture (Upper Awash, Ethiopia): Techno-economic behaviours at Garba IVD. *Journal of Human Evolution, 65,* 594–620.
Gallotti, R., & Mussi, M. (2015). The Unknown Oldowan: ~1.7-Million-Year-Old Standardized Obsidian Small Tools from Garba IV, Melka Kunture, Ethiopia. *PlosOne, 10,* e0145101.
Gallotti, R., Collina, C., Raynal, J.-P., Kieffer, G., Geraads, D., & Piperno, M. (2010). The Early Middle Pleistocene site of Gombore II (Melka Kunture, Upper Awash, Ethiopia) and the issue of Acheulean bifacial shaping strategies. *African Archaeological Review, 27,* 291–322.
Gallotti, R., Raynal, J.-P., Geraads, D., & Mussi, M. (2014). Garba XIII (Melka Kunture, Upper Awash, Ethiopia): A new Acheulean site of the late Lower Pleistocene. *Quaternary International, 343,* 17–27.
Geneste, J. M. (1989). Économie des ressources lithiques dans le Mousterien du sudouest de la France. In M. Otte (Ed.), *L'Homme de Néanderthal, La Subsistance* (Vol. 6, pp. 75–97). Liege: ERAUL.
Geneste, J. M. (1991). Systèmes techniques de production lithique: Variations technoéconomiques dans les processus de réalisation des outillages paléolithiques. *Techique et Culture, 17*(18), 1–35.
Gibbard, P. L., Head, M. J., Walker, M. J. C., & the Subcommission on Quaternary Stratigraphy (2010). Formal ratification of the Quaternary System/Period and the Pleistocene Series/Epoch with a base at 2.58 Ma. *Journal of Quaternary Sciences, 25,* 96–102.
Goldman-Neuman, T., & Hovers, E. (2009). Methodological Issues in the Study of Oldowan Raw Material Selectivity: Insights from A.L. 894 (Hadar, Ethiopia). In E. Hovers & D. R. Braun (Eds.), *Interdisciplinary Approaches to the Oldowan* (pp. 71–84). Dordrecht: Springer.
Goldman-Neuman, T., & Hovers, E. (2012). Raw material selectivity in Late Pliocene Oldowan sites in the Makaamitalu Basin, Hadar, Ethiopia. *Journal of Human Evolution, 62,* 353–366.
Habermann, J. M., McHenry, L. J., Stollhofen, H., Tolosana-Delgado, R., Stanistreet, I. G., & Deino, A. L. (2016). Discrimination, correlation, and provenance of Bed I tephrostratigraphic markers, Olduvai Gorge, Tanzania, based on multivariate analyses of phenocryst compositions. *Sedimentary Geology, 339,* 115–133.
Harmand, S. (2009). Variability in Raw Material Selectivity at the Late Pliocene site of Lokalalei, West Turkana, Kenya. In E. Hovers & D. R. Braun (Eds.), *Interdisciplinary Approaches to the Oldowan* (pp. 85–97). Dordrecht: Springer.
Harmand, S., Lewis, J. E., Feibel, C. S., Lepre, C. J., Prat, S., Lenoble, A., et al. (2015). 3.3-million-year-old stone tools from Lomekwi 3, West Turkana, Kenya. *Nature, 521,* 310–315.
Harmand, S., Prat, P., Boes, X., Lewis, J., Feibel, C., & Roche, H. (2016). The Missing Oldowan: New 2.3–2.0 Ma Sites from the Nasura Complex, West Turkana, Kenya. *PESHE, 2016,* 117.
Harris, J. W. K. (1983). Cultural beginnings: Plio-Pleistocene archaeological occurrences from the Afar, Ethiopia. *African Archaeological Review, 1,* 3–31.
Harris, J. W. K., & Semaw, S. (1989). Pliocene archaeology at the Gona River, Hadar. *Nyame Akuma, 31,* 19–21.
Harris, J. M., Brown, F. H., & Leakey, M. G. (1988). *Stratigraphy and Paleontology of Pliocene and Pleistocene Localities West of Lake Turkana, Kenya.* Los Angeles: Natural History Museum of Los Angeles County.
Haslam, M., Hernandez-Aguilar, A., Ling, V., Carvalho, S., de la Torre, I., De Stefano, A., et al. (2009). Primate archaeology. *Nature, 460,* 339–344.
Hay, R. L. (1976). *Geology of the Olduvai Gorge.* Berkeley: University of California Press.
Hayashi, M. (2015). Perspectives on object manipulation and action grammar for percussive actions in primates. *Philosophical Transactions of the Royal Society B, 370,* 20140350.
Hovers, E. (2003). Treading carefully: Site formation processes and Pliocene lithic technology. In J. Martinez, R. Mora, & I. de la Torre (Eds.), *Oldowan: Rather More than Smashing Stones. First Hominid Technology Workshop* (pp. 145–164). Barcelona: Centre d'Estudis del Patrimoni Arqueològic de la Prehistòria, Universitat Autònoma de Barcelona.
Hovers, E. (2007). Book review of "Technological Strategies in the Lower Pleistocene at Olduvai Beds I and II" by I. de la Torre and R. Mora 2006. ERAUL 112. University of Liege: Liege. *PaleoAnthropology, 2007,* 53–57.
Hovers, E. (2009). Learning from mistakes: Flaking accidents and knapping skills in the assemblage of A.L. 894 (Hadar, Ethiopia). In K. Schick & N. Toth (Eds.), *The Cutting Edge: New Approaches to the Archaeology of Human Origins* (pp. 137–150). Gosport: Stone Age Institute Press.
Hovers, E. (2012). Invention, Reinvention and Innovation: The Makings of Oldowan Lithic Technology. In S. Elias (Ed.), *Origins of Human Innovation and Creativity* (pp. 51–68). Elsevier B.V. (Development in Quaternary science, Vol. 16).
Hovers, E. (2015). Tools go back in time. *Nature, 521,* 294–295.
Hovers, E., & Braun, D. R. (2009). *Interdisciplinary Approaches to the Oldowan.* Dordrecht: Springer.
Howell, F. C., Haesaerts, P., & de Heinzelin, J. (1987). Depositional environments, archaeological occurrences and hominids from members E and F of Shungura Formation (Omo Basin, Ethiopia). *Journal of Human Evolution, 16,* 665–700.
Inizan, M. L., Reduron-Ballinger, M., Roche, H., & Tixier, J. (1999). *Technology and Terminology of Knapped Stone (Préhistoire de la Pierre taillée 5).* Nanterre: CREP.
Isaac, G. L. (1977). *Olorgesailie. Archeological studies of a Middle Pleistocene Lake Basin in Kenya.* Chicago: University of Chicago Press.
Isaac, G. L. (1984). The Archaeology of Human Origins: Studies of the Lower Pleistocene in East Africa 1971–1981. In F. Wendorf & A. E. Close (Eds.), *Advances in World Archaeology* (pp. 1–87). Orlando: Academic Press.
Isaac, G. L. (1986). Foundation Stones: Early Artifacts as Indicators of Activities and Abilities. In G. N. Bailey & P. Callow (Eds.), *Stone Age Prehistory: Studies in Memory of Charles McBurney* (pp. 221–241). Cambridge: Cambridge University Press.
Isaac, G. Ll., & Issac, B. (1997). *Koobi Fora Research Project, Plio-Pleistocene Archaeology* (Vol. 5). Oxford: Clarendon Press.
Johanson, D. C., Taieb, M., & Coppens, Y. (1982). Pliocene hominids from the Hadar Formation, Ethiopia (1973–1977): Stratigraphic, chronological, and paleoenvironmental contexts, with notes on hominid morphology and systematics. *American Journal of Physical Anthropology, 57,* 373–402.
Johanson, D. C., White, T. D., & Coppens, Y. (1978). A new species of the genus *Australopithecus* (Primates: Hominidae) from the Pliocene of eastern Africa. *Kirtlandia, 28,* 1–14.
Kibunjia, M. (1994). Pliocene archaeological occurrences in the Lake Turkana Basin. *Journal of Human Evolution, 27,* 159–171.
Kibunjia, M. (1998). Archaeological investigation of Lokalalei 1 (GaJh5): A Late Pliocene site west of lake Turkana, Kenya. Ph.D. Dissertation, Rutgers University.
Kieffer, G., Raynal, J.-P., & Bardin, G. (2002). Cadre structural et volcanologiques des sites du Paléolithique ancien de Melka Kunture (Awash, Ethiopie): Premiers résultats. In J.-P. Raynal, C. Albore-Livadie, & M. Piperno (Eds.), *Hommes et Volcans. De l'Éruption à l'Objet* (pp. 77–92). Liege: Les Dossiers de l'Archéo-Logis 2.

Kieffer, G., Raynal, J.-P., & Bardin, G. (2004). Volcanic markers in coarse alluvium at Melka Kunture (Upper Awash, Ethiopia). In J. Chavaillon & M. Piperno (Eds.), *Studies on the Early Paleolithic Site of Melka Kunture, Ethiopia* (pp. 93–101). Florence: Origines.

Kimbel, W. H., Johanson, D. C., & Rak, Y. (1997). Systematic assessment of a maxilla of *Homo* from Hadar, Ethiopia. *American Journal of Physical Anthropology, 103,* 235–262.

Kimbel, W. H., Walter, R. C., Johanson, D. C., Reed, K. E., Aronson, J. L., Assefa, Z., et al. (1996). Late Pliocene *Homo* and Oldowan stone tools from the Hadar Formation (Kada Hadar Member), Ethiopia. *Journal of Human Evolution, 31,* 549–561.

Kimura, Y. (1999). Tool-using strategies by early hominids at Bed II, Olduvai Gorge, Tanzania. *Journal of Human Evolution, 37,* 807–831.

Kimura, Y. (2002). Examining time trends in the Oldowan technology at Beds I and II, Olduvai Gorge. *Journal of Human Evolution, 43,* 291–321.

Kyara, O. A. (1999). Lithic Raw Materials and Their Implications on Assemblage Variation and Hominid Behavior During Bed II, Olduvai Gorge, Tanzania. Ph.D. Dissertation, Rutgers University.

Leakey, M. D. (1971). *Olduvai Gorge. Excavations in Bed I and II, 1960–1963* (Vol. 3). Cambridge: Cambridge University Press.

Leakey, M. D. (1975). Cultural Patterns in the Olduvai Sequence. In K. W. Butzer & G. L. Isaac (Eds.), *After the Australopithecines. Stratigraphy, Ecology, and Cultural Change in the Middle Pleistocene* (pp. 477–493). Chicago: Mouton.

Lemonnier, P. (1976). La description des chaînes opératoires. Contribution à l'analyse des systèmes techniques. *Technique et Culture, 1,* 100–151.

Lemorini, C., Plummer, T. W., Braun, D. R., Crittenden, A. N., Ditchfield, P. W., Bishop, L. C., et al. (2014). Old stones' song: Use-wear experiments and analysis of the Oldowan quartz and quartzite assemblage from Kanjera South (Kenya). *Journal of Human Evolution, 72,* 10–25.

Lepre, C., & Kent, D. V. (2015). Chronostratigraphy of KNM-ER 3733 and other Area 104 hominins from Koobi Fora. *Journal of Human Evolution, 86,* 99–111.

Lepre, C. J., Roche, H., Kent, D. V., Harmand, S., Quinn, R. L., Brugal, J.-P., et al. (2011). An earlier origin for the Acheulian. *Nature, 477,* 82–85.

Leroi-Gourhan, A. (1964). *Le Geste et la Parole. Technique et Langage.* Paris: Albin Michel.

Leroi-Gourhan, A. (1971). *Evolution et Technique. L'Homme et la Matière.* Paris: Albin Michel.

López-Sáez, J. A., & Domínguez-Rodrigo, M. (2009). Palynology of OGS-6a and OGS-7, two new 2.6 Ma archaeological sites from Gona, Afar, Ethiopia: Insights on aspects of Late Pliocene habitats and the beginnings of stone tool use. *Geobios, 42,* 503–511.

Ludwig, B. V. (1999). A Technological Reassessment of East Africa Plio-Pleistocene Lithic Artifact Assemblages. Ph.D. Dissertation, Rutgers University.

Manega, P. (1993). Geochronology, Geochemistry and Isotopic Study of the Plio-Pleistocene Hominid Sites and the Ngorongoro Volcanic Highland in Northern Tanzania Ph.D. Dissertation, University of Colorado.

Marchant, L. F., & McGrew, W. C. (2005). Percussive technology: Chimpanzee baobab smashing and the evolutionary modelling of hominin knapping. In V. Roux & B. Bril (Eds.), *Stone knapping: The necessary conditions for a uniquely hominin behavior* (pp. 341–350). Cambridge: McDonald Institute for Archaeological Research.

Martínez-Moreno J., Mora, R., & de La Torre, I. (2003). Oldowan, rather more than smashing stones: An introduction to "The Technology of First Humans" workshop. In J. M. Moreno, R. M. Torcal, & I. de la Torre Sainz (Eds.), *Oldowan: Rather more than smashing tools. First Hominid Technology Workshop* (pp. 11–35). Bellaterra: Treballs d'Arqueologia 9.

Matsuzawa, T. (1996). Chimpanzee intelligence in nature and in captivity: Isomorphism of symbol use and tool use. In W. C. McGrew, L. F. Marchant, & T. Nishida (Eds.), *Great Ape Societies* (pp. 196–209). Cambridge: Cambridge University Press.

McDougall, I., & Brown, F. H. (2006). Precise $^{40}Ar/^{39}Ar$ Geochronology for the Upper Koobi Fora Formation, Turkana Basin, Northern Kenya. *Journal of the Geological Society, 163,* 205–220.

McDougall, I., & Brown, F. H. (2008). Geochronology of the pre-KBS Tuff sequence, Omo Group, Turkana Basin. *Journal of The Geological Society, 165,* 549–562.

McDougall, I., Brown, F. H., Cerling, T. E., & Hillhouse, J. W. (1992). A reappraisal of the Geomagnetic Polarity Time Scale to 4 Ma using data from the Turkana Basin, East Africa. *Geophysical Research Letters, 19,* 2349–2352.

McDougall, I., Brown, F. H., Vasconcelos, P. M., Cohen, B. E., Thiede, D. S., & Buchanan, M. J. (2012). New single crystal $^{40}Ar/^{39}Ar$ ages improve timescale for deposition of the Omo Group, Omo–Turkana Basin, East Africa. *Journal of The Geological Society, 169,* 213–226.

McPherron, S. P., Alemseged, Z., Marean, C. W., Wynn, J. G., Reed, D., Geraads, D., et al. (2010). Evidence for stone-tool-assisted consumption of animal tissues before 3.39 million years ago at Dikika, Ethiopia. *Nature, 466,* 857–860.

McPherron, S. P., Alemseged, Z., Marean, C. W., Wynn, J. G., Reed, D., Geraads, D., et al. (2011). Tool-marked bones from before the Oldowan change the paradigm. *Proceedings of the National Academy of Sciences of the United States of America, 108,* E116.

Merrick, H. V., & Harry, V. (1976). Recent archaeological research in the Plio-Pleistocene deposits of the Lower Omo, southwestern Ethiopia. In G. L. Isaac & E. R. McCown (Eds.), *Human origins: Louis Leakey and the East African evidence* (pp. 461–481). Menlo Park, CA: W. A. Benjamin.

Merrick, H. V., & Merrick, J. P. S. (1976). Archaeological occurrences of earlier Pleistocene age from the Shungura Formation. In Y. Coppens, F. C. Howell, Gl. Isaac, & R. E. Leakey (Eds.), *Earliest man and environments in the Lake Rudolf Basin* (pp. 574–584). Chicago: University of Chicago Press.

Merrick, H. V., de Heinzelin, J., Haesaerts, P., & Howell, F. C. (1973). Archaeological occurrences of Early Pleistocene age from the Shungura Formation, Lower Omo Valley, Ethiopia. *Nature, 242,* 572–575.

Mora, R., & de la Torre, I. (2005). Percussion tools in Olduvai Beds I and II (Tanzania): Implications for early human activities. *Journal of Anthropological Archaeology, 24,* 179–192.

Morgan, L. E., Renne, P. R., Kieffer, G., Piperno, M., Gallotti, R., & Raynal, J.-P. (2012). A chronological framework for a long and persistent archaeological record: Melka Kunture, Ethiopia. *Journal of Human Evolution, 62,* 104–115.

Mussi, M., Bonnefille, R., & Gallotti, R. (2015). Melka Kunture (Upper Awash, Ethiopian plateau): The earliest human settlement of a high mountain system. *PESHE, 2015,* 167.

Noll, M. P. (2000). Components of Acheulean lithic assemblages variability at Olorgesailie, Kenya. Ph.D. Dissertation, University of Illinois at Urbana-Champaign.

Panger, M. A., Brooks, A. S., Richmond, B. G., & Wood, B. (2002). Older Than the Oldowan? Rethinking the Emergence of Hominin Tool Use. *Evolutionary Anthropology, 11,* 235–245.

Pelegrin, J. (1985). Réflexions sur le comportement technique. In M. Otte (Ed.), *La signification culturelle des industries lithiques* (pp. 72–91). Liege: Studia Praehistorica Belgica 4, BAR International Series 239.

Perlès, C. (1991). Économie des matières premières et économie du débitage : deux conceptions opposées ? In S. Beyries, L. Meignen,

& P. J. Texier (Eds.), *25 ans d'études technologiques en Préhistoire : bilan et perspectives. Proceedings XIème rencontres internationales d'Archéologie et d'Histoire d'Antibes* (pp. 35–46). Juan-les-Pins: APDCA.

Piperno, M. (1987). Chronostratigraphic and Cultural Framework of the *Homo habilis* sites. In G. Giacobini (Ed.), *Hominidae. Proceedings of the 2nd International Congress of Human Paleontology, Torino 1987* (pp. 189–195). Milano: Jaca Book.

Piperno, M., Bulgarelli, G. M., & Gallotti, R. (2004). The site of Garba IV. The test trenches A and B and the sounding in level E. In J. Chavaillon & M. Piperno (Eds.), *Studies on the Early Paleolithic Site of Melka Kunture, Ethiopia* (pp. 581–587). Florence: Origines.

Piperno, M., Collina, C., Gallotti, R., Raynal, J.-P., Kieffer, G., Le Bourdonnec, F.-X., et al. (2009). Obsidian exploitation and utilization during the Oldowan at Melka Kunture (Ethiopia). In E. Hovers & D. R. Braun (Eds.), *Interdisciplinary Approaches to the Oldowan* (pp. 111–128). Dordrecht: Springer.

Plummer, T. W., Bishop, L. C., Ditchfield, P. W., Ferraro, J. V., Kingston, J. D., Hertel, F., et al. (2009a). The Environmental Context of Oldowan Hominin Activities at Kanjera South, Kenya. In E. Hovers & D. R. Braun (Eds.), *Interdisciplinary Approaches to the Oldowan* (pp. 149–160). Dordrecht: Springer.

Plummer, T. W., Bishop, L., Ditchfield, P., & Hicks, J. (1999). Research on late Pliocene Oldowan sites at Kanjera South, Kenya. *Journal of Human Evolution, 36,* 151–170.

Plummer, T. W., Ditchfield, P. W., Bishop, L. C., Kingston, J. D., Ferraro, J. V., Braun, D. R., et al. (2009b). Oldest Evidence of Toolmaking Hominins in a Grassland Dominated Ecosystem. *PlosOne, 4,* e7199.

Potts, R. (1988). *Early Hominid Activities at Olduvai.* New York: Aldine and Gruyter.

Potts, R. (1991). Why the Oldowan? Plio-Pleistocene toolmaking and the transport of resources. *Journal of Anthropological Research, 45,* 153–176.

Prat, S., Brugal, J. P., Tiercelin, J. J., Barrat, J. A., Bohn, M., Delagnes, A., et al. (2005). First occurrence of early *Homo* in the Nachukui Formation (West Turkana, Kenya) at 2.3–2.4 Myr. *Journal of Human Evolution, 49,* 230–240.

Quade, J., Levin, N., Semaw, S., Stout, D., Renne, P., Rogers, M. J., et al. (2004). Paleoenvironments of the earliest stone toolmakers, Gona, Ethiopia. *Geological Society of America Bulletin, 116,* 1529–1544.

Quade, J., Levin, N., Simpson, S., Butler, R., McIntosh, W., Semaw, S., et al. (2008). The Geology of Gona, Afar, Ethiopia. *Geological Society of America Bulletin, Special Paper, 446,* 1–31.

Quinn, R. L., Lepre, C. J., Feibel, C. S., Wright, J. D., Mortlock, R. A., Harmand, S., et al. (2013). Pedogenic carbonate stable isotopic evidence for wooded habitat preference of early Pleistocene tool makers in the Turkana Basin. *Journal of Human Evolution, 65,* 65–78.

Raynal, J.-P., Kieffer, G., & Bardin, G. (2004). Garba IV and the Melka Kunture Formation. A preliminary lithostratigraphic approach. In J. Chavaillon & M. Piperno (Eds.), *Studies on the Early Paleolithic Site of Melka Kunture, Ethiopia* (pp. 137–166). Florence: Origines.

Reed, K. E. (2008). Paleoecological patterns at the Hadar hominin site, Afar Regional State, Ethiopia. *Journal of Human Evolution, 54,* 743–768.

Reed, D. N., & Geraads, D. (2012). Evidence for a Late Pliocene faunal transition based on a new rodent assemblage from Oldowan locality Hadar A.L. 894, Afar Region, Ethiopia. *Journal of Human Evolution, 62,* 328–337.

Roche, H. (1989). Technological evolution in the early hominids. *OSSA, 14,* 97–98.

Roche, H. (1996). Remarques sur les plus anciennes industries en Afrique et en Europe. In F. Facchini (Ed.), *The First Humans and Their Cultural Manifestations, Section 4, XIII International Congress of Prehistoric and Protohistoric Sciences. (Forlì 8–14 September 1996), Colloquium VIII* (pp. 55–63). Forlì: A.B.A.C.O.

Roche, H., & Tiercelin, J. J. (1977). Découverte d'une industrie lithique ancienne in situ dans la formation d'Hadar, Afar central, Ethiopie. *Comptes Rendus de l'Académie des Sciences Series D, 284,* 187–174.

Roche, H., & Tiercelin, J. J. (1980). Industries lithiques de la formation Plio-Pléistocène d'Hadar: campagne 1976. In R. E. F. Leakey & B. A. Ogot (Eds.), *Proceedings of the VIIIth Panafrican Congress of Prehistory and Quaternary Studies* (pp. 194–199). Nairobi: The International Louis Leaky Memorial Institute for African Prehistory.

Roche, H., Brugal, J.-P., Delagnes, A., Feibel, C., Harmand, S., Kibunjia, M., et al. (2003). Les sites archéologiques plio-pléistocénes de la formation de Nachukui, Ouest-Turkana, Kenya: bilan synthétique 1997-2001. *Comptes Rendus Palevol, 2,* 663–673.

Roche, H., Delagnes, A., Brugal, J.-P., Feibel, C. S., Kibunja, M., Mourre, V., et al. (1999). Early hominid stone tool production and technical skill 2.34 Myr ago, in West Turkana, Kenya. *Nature, 399,* 57–60.

Rogers, M., & Semaw, S. (2009). From Nothing to Something: The Appearance and Context of the Earliest Archaeological Record. In M. Camps & P. Chauhan (Eds.), *Sourcebook of Paleolithic Transitions. Methods, Theories and Interpretations* (pp. 155–171). New York: Springer.

Sahnouni, M. (1991). Étude comparative des galets taillés polyédriques, subsphériques et sphériques des gisements d'Ain Hanech (Algérie Orientale) et d'Olduvai (Tanzanie). *L'Anthropologie, 97,* 51–68.

Semaw, S. (1997). Late Pliocene Archaeology of the Gona River Deposits, Afar, Ethiopia. Ph.D. Dissertation, Rutger University.

Semaw, S. (2000). The world's oldest stone artifacts from Gona, Ethiopia: Their implications for understanding stone technology and patterns of human evolution 2.6–1.5 million years ago. *Journal of Archaeological Science, 27,* 1197–1214.

Semaw, S. (2006). The oldest stone artifacts from Gona (2.6–2.5 Ma), Afar, Ethiopia: Implications for understanding the earliest stages of stone knapping. In N. Toth & K. Schick (Eds.), *The Oldowan: Case studies into the earliest Stone Age* (pp. 43–75). Gosport: Stone Age Institute Press.

Semaw, S., Renne, P., Harris, J. W. K., Feibel, C. S., Bernor, R. L., Fesseha, N., et al. (1997). 2.5-million-year-old stone tools from Gona, Ethiopia. *Nature, 385,* 333–336.

Semaw, S., Roger, M. J., Quade, J., Renne, P. R., Butler, R. F., Dominguez-Rodrigo, M., et al. (2003). 2.6-Million-year-old stone tools and associated bones from OGS-6 and OGS-7, Gona, Afar, Ethiopia. *Journal of Human Evolution, 45,* 169–177.

Stiles, D. (1998). Raw material as evidence for human behaviour in the Lower Pleistocene: The Olduvai case. In M. Petraglia & R. Korisettar (Eds.), *Early Human Behavior in Global Context* (pp. 133–150). London: Routledge.

Stiles, D., Hay, R. L., & O'Neil, J. R. (1974). The MNK Chert Factory Site, Olduvai Gorge, Tanzania. *World Archaeology, 5,* 285–308.

Stout, D., & Chaminade, T. (2007). The evolutionary neuroscience of tool making. *Neuropsychologia, 45,* 1091–1100.

Stout, D., Quade, J., Semaw, S., Rogers, M., & Levin, N. (2005). Raw material selectivity of the earliest stone toolmakers at Gona, Afar, Ethiopia. *Journal of Human Evolution, 48,* 365–380.

Stout, D., Semaw, S., Rogers, M. J., & Cauche, D. (2010). Technological variation in the earliest Oldowan from Gona, Afar, Ethiopia. *Journal of Human Evolution, 58,* 474–491.

Stout, D., Toth, N., Schick, K., & Chaminade, T. (2008). Neural correlates of Early Stone Age toolmaking: Technology, language and cognition in human evolution. *Philosophical Transactions of the Royal Society B, 363,* 1939–1949.

Tamrat, E., Thouveny, N., Taieb, M., & Brugal, J. P. (2014). Magnetostratigraphic study of the Melka Kunture archeological site (Ethiopia) and its chronological implications. *Quaternary International, 343,* 5–16.

Texier, P.-J. (2018). Technological Assets for the Emergence of the Acheulean? Reflections on the Kokiselei 4 Lithic Assemblage and Its Place in the Archaeological Context of West Turkana, Kenya. In R. Gallotti & M. Mussi (Eds.), *The Emergence of the Acheulean in East Africa and Beyond. Contributions in Honor of Jean Chavaillon.* Cham: Springer.

Texier, P.-J., Roche, H., & Harmand, S. (2006). Kokiselei 5, formation de Nachukui, West Turkana (Kenya) : un témoignage de la variabilité ou de l'évolution des comportements techniques au Pléistocène ancien ? In P. de Maret, E. Cornelissen, & I. Ribot (Eds.), *Actes du XIVe congrès UISPP (Liège, 2001), section 15 : Préhistoire en Afrique/African Prehistory* (pp. 11–22). Liege: Archaeopress, BAR International Series, 1522.

Thompson, J. C., McPherron, S. P., Bobe, R., Reed, D., Barr, W. A., Wynn, J. G., et al. (2015). Taphonomy of fossils from the hominin-bearing deposits at Dikika, Ethiopia. *Journal of Human Evolution, 86,* 112–135.

Toth, N. (1982). The stone technologies of early hominids at Koobi Fora, Kenya: an experimental approach. Ph.D. Dissertation, University of California.

Toth, N. (1985). The Oldowan Reassessed: A Close Look at Early Stone Artifacts. *Journal of Archaeological Science, 12,* 101–120.

Villmoare, B., Kimbel, W. H., Seyoum, C., Campisano, C. J., DiMaggio, E. N., Rowan, J., et al. (2015). Early Homo at 2.8 Ma from Ledi-Geraru, Afar, Ethiopia. *Science, 347,* 1352–1355.

Visalberghi, E., Haslam, M., Spagnoletti, N., & Fragaszy, D. (2013). Use of stone hammer tools and anvils by bearded capuchin monkeys over time and space: construction of an archeological record of tool use. *Journal of Archaeological Science, 40,* 3222–3232.

Walker, A. C., Leakey, R. E. F., Harris, J. M., & Brown, F. H. (1986). 2.5 Myr *Australopithecus boisei* from west of lake Turkana, Kenya. *Nature, 322,* 517–522.

Willoughbly, P. R. (1987). *Spheroids and Battered Stones in the African Early Stone Age* (p. 321). Oxford: BAR International Series.

Wynn, T. (1981). The Intelligence of Oldowan Hominids. *Journal of Human Evolution, 10,* 529–541.

Wynn, T., &. McGrew W. C. (1989). An ape's view of the Oldowan. *Man, 24,* 383–398.

Zaidner, Y. (2013). Adaptive flexibility of Oldowan hominins: Secondary use of flakes at Bizat Ruhama, Israel. *PLoS ONE, 8,* e66851.

Chapter 3
Technological Assets for the Emergence of the Acheulean? Reflections on the Kokiselei 4 Lithic Assemblage and Its Place in the Archaeological Context of West Turkana, Kenya

Pierre-Jean Texier

Abstract On the western side of the Turkana basin, the sedimentological members of the Nachukui Formation expose a unique succession of archaeological site complexes ranging from 0.7 to 3.3 Ma. Following the analysis of the oldest and most remarkable lithic assemblages, we propose a model clarifying the chronology and possible operative modes of the first stone knappers; the technological components which around 1.76 Ma led to a new method in stone working: shaping. It appears that they gradually substituted newly mastered technical advances for the initial selection of blocks or cobbles naturally displaying a suitable shape. The alternating of conceptual advances, first concretized in the appropriate selection of natural block shapes, then in major technical innovations, seems to have been the rhythm of a very slow and hesitant tempo, leading to the formalization of the oldest Acheulean lithic assemblages then to a new technological world from 1.0 Ma.

Keywords Oldowan • Early Acheulean • Lithic technology • Earliest technologies • Concept of tool

3.1 Pluralist Tool Makers, a Single Technological Framework

The relationship of wild chimpanzees with tools, currently observed by primatologists, is that of a single living species (*Pan troglodytes*) evolving in a natural environment where the plant component is a key element. The fossil evidence for the relationship with the stone tools of this great ape is poor and only gives us information limited to the use and the economy (transport and reuse) of a material, non-transformed except by use.

Conversely, we are interested here in the knapped stone objects that are together with faunal remains the relics abandoned by representatives of several extinct fossil species. In the Late Pliocene/Early Pleistocene context in which these ancient hominins evolved, any possible vegetable component of their equipment has not been preserved to date.

Current data indicate that from 3.3 to 1.0 Ma, several species belonging to several genera of hominins dedicated themselves to hard stone knapping. The oldest lithic assemblages we will consider are geographically relatively concentrated, but could have been produced by representatives of different species, or even of a different genus: *Kenyanthropus* and/or *Australopithecus* (Harmand et al. 2015). Besides, we know that from 2.3 to 1.5 Ma, representatives of the *Paranthropus* and *Homo* genera could have interacted on the western shores of Lake Turkana (Prat et al. 2005).

Whatever the level of expertise, hard stone knapping is only possible by combining several well-identified parameters that come within the province of solid-state physics; this forms a rigid framework to which none of these operators could deviate. Therefore, we will not discuss here the stages of a continuous technological evolution that would have resulted from a single species, but, based on concrete remains, we will discuss how the inescapable technical obstacles to the development of the first known bifacial forms have been overcome between 3.3 and 1.0 Ma, by one or more hominin species. If several of them were able to provide, with or without continuation, a technical solution to the overcoming of the first obstacles met, a single lineage, one that led to the only species present around 1.0 Ma, *Homo erectus s.l.*, was able to overcome them all.

P.-J. Texier (✉)
Aix Marseille Univ, CNRS, Minist Culture, LAMPEA, UMR 7269, 5 rue du Château de l'horloge-BP 647, F-13094 Aix-en-Provence Cedex 2, France
e-mail: pj.texier@gmail.com

R. Gallotti and M. Mussi (eds.), *The Emergence of the Acheulean in East Africa and Beyond: Contributions in Honor of Jean Chavaillon*, Vertebrate Paleobiology and Paleoanthropology,
https://doi.org/10.1007/978-3-319-75985-2_3

3.2 Origin of the Term "Acheulean"

Gabriel de Mortillet first used the term "Acheulean" in 1872 to refer to industries with handaxes (which he then referred to as "coup de poing") from the middle terrace of the Somme River, near the village of Saint-Acheul and the town of Amiens. But the term "biface" (handaxe) was coined by Vayson de Pradenne to describe the first large pieces shaped on each of their faces, with more or less pronounced bilateral and bifacial symmetry, which had been unearthed in unquestionably very ancient sediments in the terraces of the Somme River (Vayson de Pradenne 1920).

For the archaeologists of the time, handaxes represented the oldest known material evidence left by those considered as the very first protagonists of prehistory. Easy to identify, relatively easy to "read", the handaxe naturally emerged as the most representative tool of the new Acheulean culture. It remains "the first truly shaped object that is known, the first shape completely invented by mankind"[1] (Tixier and de Saint-Blanquat 1992: 8). It thus became the undisputed techno-typological marker of a culture that subsequently proved very widespread in Africa, the Middle East, Eurasia and equally widely spread over time. In Africa, the Acheulean handaxe is often associated with the cleaver on flake, another emblematic large cutting tool, whose bifacial symmetry, however, is not the first morphological characteristic. In Europe, cleavers are often lacking from the considered Acheulean lithic assemblages. We will also see that this other marker of the African Acheulean always seems to have appeared with a chronological discrepancy with regard to the handaxe. This delayed first appearance is an important factor to consider in the research on the mechanisms and chronology of the emergence of the Acheulean. The main reason is the minimum level of predetermination required to make possible the manufacturing of a cleaver on flake (Roche and Texier 1996).

So far, prehistorians have agreed on an eastern African origin of the Acheulean, but at a time that still remains unclear because sites with bifacial objects reliably dated over a million years were, and still are, particularly rare (de La Torre and Mora 2005; de la Torre et al. 2008; Gallotti 2013). The discovery and the very ancient dating of two of them, by partially filling this gap (Kato et al. 2000; Lepre et al. 2011; Beyene et al. 2013), are just restarting the debate.

The diffusion of this culture can be followed quite easily thanks to the considered typo-technological markers. Thus, it is commonly accepted that the Acheulean left Africa in several waves toward the Middle East, before gradually reaching Europe and Asia in a succession between 1.7 and 0.8 Ma (Bar Yosef and Goren-Inbar 1993; Goren-Inbar et al. 2000; Petraglia 2003). However, other researchers favor a Near Eastern re-emergence of bifacial shaping (Chevrier 2012).

We propose here to examine the determining technological milestones (Fig. 3.1) that must have punctuated the path leading in 2.3 million years from the first known knapped objects (3.3 Ma) to the first certain bifacial shapes (1.76 Ma), then to the stereotyped forms of an ancient African Acheulean (0.98 Ma), already technologically mature like that of Isenya in Kenya (Durkee and Brown 2014).

3.3 The Concept of Tool

The use and often reuse of stone tools consisting in natural and unmodified shapes have been frequently observed and well documented by numerous primatologists with regards to modern chimpanzees (Mc Grew 1992; Joulian 1996; Boesch and Tomasello 1998) and moreover demonstrated for fossil chimpanzees (Mercader et al. 2002, 2007). But intentional manufacturing of stone tools was never documented.

However, the modern knapping of flakes by chimpanzees (*P. troglodytes* or *P. paniscus*), often born and bred in captivity and that never engage in such activities in the wild, appears as an artifact introduced by modern experimenters and should therefore be considered with caution. If the experiments that have been attempted in this direction show the existence of some potential of these species in this domain, they also allow apprehending the duration of the necessary learning process and the very quickly reached limits of its technical expression (Toth et al. 1993; Texier 2012).

The production of flakes, at will and in series or the shaping of stone objects by direct or indirect percussion, remains the prerogative of one or more representatives of the sub-tribe of Hominina that includes all the species of the genera *Homo*, *Australopithecus* and *Paranthropus*. However, the identification of one or several of them as the author of these productions will always remain a problem.

Unearthed in stratigraphic context, the remains of the first lithic productions of the hominins who occupied the African Rift in the Late Pliocene and Early Pleistocene mark in an almost unchanging and "readable" manner the crossing of a milestone in the history of the human lineage:

[1]"Le premier objet vraiment façonné que l'on connaisse, la première forme totalement inventée par des hommes."

Fig. 3.1 Sketch map showing the location of the sites mentioned in this paper

the manufacturing of stone objects, hitherto unknown, using basic tools (hammers or anvils), natural but carefully chosen, and raw materials selected according to their morphologies, their mechanical properties, and their module.

It is clear through these first but already abundant productions, that the elementary principles of fragile fracturing and knapping were assimilated straight away, because the technical gesture and its consequences only become repeatable at will under this condition.

3.4 The Oldest Knapped Tools, the Early Acheulean

The oldest knapped tools currently listed were unearthed in about twenty sites of the Ethiopian or Kenyan sections of the East African Rift (Table 3.1). Their ages range from 3.3 to 2.0 Ma. These sites have yielded early Oldowan and Oldowan lithic assemblages, consisting in several dozens to several hundreds, or even thousands of objects (Roche and Tiercelin 1977; Delagnes and Roche 2005; Semaw 2005; Semaw et al. 2010; Hovers 2012).

To the west of Lake Turkana in Kenya, at the base of the Lomekwi member of the Nachukui Formation, the recent discovery of numerous knapped objects from a stratigraphic context dated to 3.3 Ma (Harmand et al. 2015) has dramatically thrown back in time by 0.7 Ma (Roche and Tiercelin 1977) the first concrete manifestation of the crossing of what some consider as the threshold of hominization. It can then be assessed by the ability to make stone objects recurrently, as simple as they are, with all the underlying ability for anticipation, minimal understanding of the conchoidal fracturing of rocks and manual dexterity.

Moreover, the great age of the Kokiselei 4 site (KS4, West Turkana, Kenya) that yielded many large unifacial and bifacial objects as well as picks shaped on cobbles or split cobbles was recently confirmed by a date at 1.76 Ma (Lepre et al. 2011).

In Ethiopia, in the vast and rich Acheulean complex of Konso Gardula, a lithic assemblage (KGA6-A1), very similar in nature and size to that of KS4 (Table 3.2), could avail of an equivalent age (Beyene et al. 2013, 2015). If this is the case, the presence of three cleavers on flakes in this assemblage suggests looking for the origin of the Acheulean of KGA6-A1 even further in time because of the technological mastery this implies.

Finally, let us remember that the experimental manufacturing of series of large cutting tools (LCTs) in phonolite allowed identifying clearly the range of available techniques and the committed know-how necessary for the working of such raw materials. These experiments also allowed establishing and characterizing the phasing of the making of these tools, evaluating the execution times and the amount of generated knapping wastes. Anticipation is clearly emerging at that level through the selection of the module and the quality of the raw materials, in the obtaining of some handaxe blanks and of all the blanks for cleavers on flakes (Roche and Texier 1991, 1996; Texier 1996; Bouthinon 2002).

Based on the spectacular dating recently conducted by researchers working in East Africa and on the contribution of the technological analysis of the new unearthed lithic assemblages, it seems legitimate to wonder about the mechanisms that have led from the very first knapped objects to the formalization of new "chaînes opératoires" oriented toward bifacial shaping. This forms a long journey

Table 3.1 Size of the assemblages and distribution of the raw materials within the main lithic groups of the Late Pliocene constituted from systematic surface and in situ collecting. Data from de la Torre 2004; Stout et al. 2005; Hovers 2009; Semaw et al. 2010; Harmand et al. 2015

	LOM3	EG10	EG12	OG7 (2000 excavation)	AL 894	LA2C	Omo 123
Age	3.3 Ma	2.6 Ma	2.6 Ma	2.6 Ma	2.36 Ma	2.34 Ma	2.3 Ma
Surface	130	1551	309	65	–	492	1014
In situ	19	685	445	188	–	2122	767
TOTAL	149	2236	754	253	4828	2614	1781
Raw materials	Phonolite 34.2% Basalt 34.9%	Trachyte 79.0% Rhyolite 11.4%	Trachyte 66.1% Rhyolite 17.7%	Trachyte 29.3% Rhyolite 26.3% Aphan. volc. 15.8%	Rhyolite 71.0% Basalt 24.0%	Phonolite 74.7% Basalt 14.2%	Quartz 96.4% Chert 2.2%
w.f. = whole flakes	Trachyphon 23.5% Others 7.4%	Basalt 7.0% Others 2.6% (w.f.)	Basalt 6.5% Others 9.7% (w.f.)	Latite 10.6% Vitr. volcanic 5,0% Basalt 3.5% Others: 9.5% (w.f.)	Trachyte 3.0% Others 2.0% (w.f.)	Trachyte 9.7% Others 1.4%	Lava 1.4%

Table 3.2 Distribution of the main categories of tools within the two oldest Acheulean lithic assemblages (aPhon. is used as a short for aphyric phonolite)

Site	Age	Cleavers	Handaxes	Picks	Others	Main raw material	Total
KS4	1.76	0	3	11	14	aPhon.	28
KGA6-A1	1.75	3	4	11	10	Basalt	28

whose stages are punctuated by the development of new concepts and decisive technical innovations. Still deeply rooted in the Oldowan, the new emerging "chaînes opératoires" marked the Early Paleolithic with their technological footprint in a lasting or recurring manner, depending on whether one chooses the model of a unique Acheulean that was to spread latter out of Africa, or whether one is a supporter of the local invention or reinvention of the bifacial object.

However, the question does not arise here as abruptly, as we propose to look at the very roots of the bifacial phenomenon and to discuss the nature and chronology of the conceptual advances and technological knowledge that have gradually made possible the manufacturing of the earliest bifacial pieces.

3.5 Detaching Flakes …

Detaching one or more flakes, even at an elementary technological level, remains an operation of great complexity, as evidenced by the difficulties encountered by many researchers and modern manufacturers to try (implied as on the edge of a block of raw material) to interpret and put this marginal phenomenon of fracturing into equations (Cottrell and Kaminga 1979; Bertouille 1989; Zarzycki 1991; Tsirk 2014).

Such an approach first requires the selection of a cobble or a block of dense and tenacious rock (to act as a hammer or anvil), whose natural convexities allow concentrating and returning on impact the energy imparted to it, on a limited area, at a preselected and then reached position.

This also implies the selection of blocks of materials identified as suitable for knapping by their homogeneity, texture, and hardness. The project can be modified depending on the sizes of the available blocks or corollary, a selection of blocks of specific morphology or module can be done according to the planned project and the technical background of the knapper.

Thus, the incidence of the trajectory followed by the hammer toward the block, or by the block toward the anvil, its weight, the speed imparted to it, the location of the point of impact on the edge of the worked block, the geometry of the block in the impact area, and the topography of the surface to be knapped are all factors that the experienced knapper should take into account simultaneously during each technical gesture, to achieve a result in line with his/her expectations.

Often, the naturally irregular shape of the worked blocks only allows knappers with an elementary technology to obtain very limited series of flakes, isolated, possibly adjacent, or alternating.

The early Oldowan, Oldowan, and early Acheulean knappers did not always have the necessary technology to carry out their projects directly. Thus at first, they favored the choice of blocks with a natural morphology that allowed them to avoid an obstacle, of which they were aware, but which was still technically insurmountable for them.

3.6 Sketching the First Stages of Knapped Stone Technology

The knapping of hard stone is applied to a volume of raw material suitable for knapping. It is expressed in the three dimensions of space. If the selected material is simply knapped, then some flakes are the desired products. They can be used as tool without any modification. They are, or have been in this case, accompanied by many by-products resulting from their preparation. Conversely, when the material is gradually shaped to manufacture one single artifact, in that case the flake only has the status of a by-product, possibly usable (Texier and Roche 1995; Roche 2007).

Schematically, the débitage may be represented by a system with three X, Y, and Z axes of the same origin. Meanwhile, bifacial shaping can be represented by two sets of arrows, possibly alternating and arranged in opposition. Thus, in the case of a débitage, we propose to refer to the X and Y axes to symbolize the part of the volume of a material that can be exploited without reworking the initial geometry of the worked block/core. The third axis is used to schematize the activation of a preexisting natural striking platform or the creation and management of an unnatural striking platform (Fig. 3.2).

In the case of bifacial shaping, the mutual arrangement of the arrows reflects the quality of the sequence of removals and potentially, the relative chronology of the series of technical events that occurred during shaping (Fig. 3.3).

Where it appears to the technological analysis that the control of the sub-volumes to work is uncertain (systematic

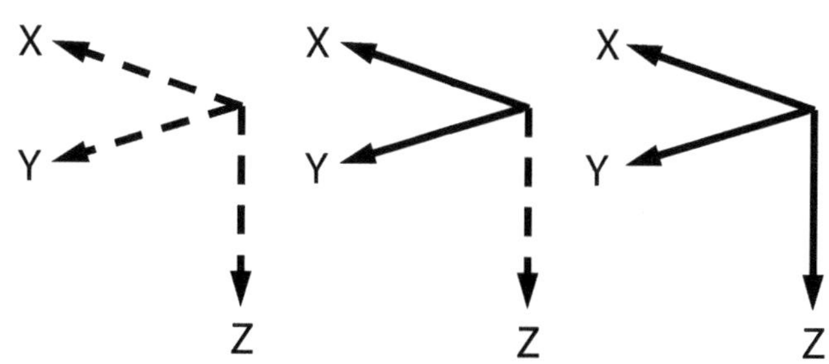

Fig. 3.2 Sketching the ability to control each of the three directions of space when flaking (dotted lines: uncontrolled direction; continuous lines: controlled direction)

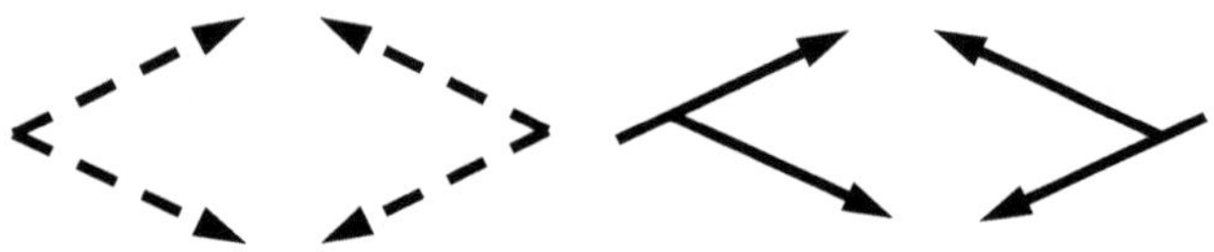

Fig. 3.3 Sketching the ability to control in space the direction and the mutual organization of the removal of flakes when shaping (dotted lines: uncertain control and uncertain organization of the flakes removal; continuous lines: controlled and mutually organized series of flakes removal)

and non-"repaired" knapping accidents), the directions that illustrate its exploitation are shown as dotted lines. Conversely, when it appears that the control of this space has become the norm, the concerned axes are then represented by a continuous line.

3.7 West Turkana: The Unity of Place

The twenty years of presence in the field of the Mission préhistorique au Kenya in the context of the West Turkana Archaeological Project[2] has brought together in a locally and chronologically tightened context, at the same time geologically and archaeologically exceptionally favorable, a unique documentation of the technological developments of the first lithic productions, benefiting in a way from the unity of place.

We suggest to look in these chronologically well-determined lithic assemblages for the diagnostic elements of the technological knowledge (and the processes of their acquisition) that enabled some knappers to venture conceptually, then concretely, in the first recorded attempts at bifacial shaping.

3.7.1 *The Nachukui Formation*

The main levels of volcanic ash (tephra) that punctuate the Nachukui Formation divide this thick sedimentary deposit (712 m) in eight separate members that bear the names of the intermittent streams that are eroding it (Roche et al. 2003). The geochemical signature of these volcanic ashes allows correlating them with some of those from other formations in the Omo group (these are directly dated tephra deposits): the Shungura Formation to the north and the Koobi Fora Formation to the east (Haileb et al. 2004). Moreover, a correlation of the Turkana basin tephra deposits (especially the KBS and Chari Tuffs) with those of the Konso Formation has been proposed (Katho et al. 2000; McDougall and Brown 2006; McDougall et al. 2012; Beyene et al. 2013).

To date, four members of the Nachukui Formation (members of Lomekwi, Kalochoro, Kaitio, and Nariokotome) have yielded 10 archaeological complexes for a total of 59 sites with ages ranging from 3.3 to 0.7 Ma (Table 3.3).

Our approach is specifically based on data from the preliminary study of the artifacts collected on the surface or in stratigraphy at Lomekwi 3 (Harmand et al. 2015), and from the in-depth technological analysis of the lithic assemblages of three sites; two Oldowan sites, Lokalalei 2C and Kokiselei 5, were excavated exhaustively (Delagnes and Roche 2005, Texier et al. 2006); at Kokiselei 4, the excavation was more limited in extension but with systematic surface collecting for the early Acheulean (Lepre et al. 2011).

The last three of these sites have notably in common comparable raw materials (aphyric phonolite) while the numerous refits that could be done confirm and refine the technological reading.

Table 3.3 Members of the Nachukui Formation (West Turkana, Kenya). Time intervals, thicknesses of sedimentary deposits, coding of the name of the sites complexes and corresponding number of sites (after Roche 2011)

Members	Age (Ma)	Thickness (m)	Archaeological complexes	Number of archaeological sites
Nariokotome	1.30–0.7	70	KL; NK; NAD	18
Natoo	1.65–1.30	75		
Kaitio	1.90–1.65	169	KS; KLD; NY	30
Kalochoro	2.35–1.90	72	LA; NAS	8
Lokalalei	2.50–2.35	42	NAS	2
Lomekwi	3.35–2.50	159	KUS; LOM	1
Kataboi	4–3.35	34		
Lonyumum	>4	91		

[2]The West Turkana Archaeological Project (WTAP) is a joint program of the National Museums of Kenya (National Museums of Kenya) and of the Mission préhistorique au Kenya (MPK). Created and directed by H. Roche from 1994 to 2013, the WTAP is now directed by S. Harmand and J. Lewis. This program yearly benefits of the institutional (Commission consultative des recherches archéologiques à l'étranger) and financial support of the French "Ministère des Affaires étrangères et du Développement international."

Other sites, older than Lokalalei 2C or contemporary, were discovered in Hadar in Ethiopia. The lithic assemblages of some of them have recently been the subject of technological studies (Hovers 2009, 2012; Stout et al. 2010). However, they can account for substantially different technical behaviors than those other knappers have shown, living around the same time a thousand kilometers away, west of Lake Turkana. To benefit in some way from the unity of place and from a narrow range of raw materials, the diachronic evolution model of the technologies presented here will refer exclusively to the sites of West Turkana.

3.7.1.1 Lomekwi 3 (LOM3)

LOM3 site was discovered during surveys carried out in 2011 by the WTAP team at the base of Lomekwi member (3.35–2.5 Ma). Some artifacts and skeletal remains, some of which in situ, were then collected. At the end of the field campaign that followed, along with the discovery of new relatively poorly preserved bone remains attributed to six species of mammals, 149 lithic pieces were collected, including 19 in an indisputable stratigraphic context (Harmand et al. 2015). They are essentially flakes or flake fragments bearing indisputable knapping traces, relatively bulky worked blocks of an average weight of about 3 kg, and elements that were used in active or passive percussion. The worked blocks mostly evidence flake scars terminating as hinge and step fractures (Fig. 3.4). The raw materials are in equivalent proportion, phonolites (35%), and basalts (34%) and to a lesser extent, trachyphonolites (23%). They are still currently available in modules compatible with the artifacts from LOM3, in gravel from the dismantling of ancient alluvial formations surrounding the site. The well-argued age of 3.3 Ma that is proposed (Harmand et al. 2015) makes it the oldest archaeological site known to date. In the current state of knowledge, this discovery finally demonstrates conclusively that hominins other than those

Fig. 3.4 Lomekwi 3. **a** Refitting surface flake and in situ unifacial core worked using passive hammer and bipolar technique (1.85 kg). These two conjoining artifacts show series of percussion marks on cortex documenting a prior use for different purpose; **b** Unifacial passive hammer core (2.04 kg); **c** and **d** Unifacial bipolar cores, respectively, 3.45 and 2.58 kg. Impacts due to the countercoups are localized on the opposite edge from the striking platform. All these cores show numerous knapping accidents due to assessing errors and to the poor quality of the raw material (Harmand et al. 2015). (Photographs courtesy of MPK-WTAP)

related to the *Homo* genus, which appears at the earliest at 2.8 Ma (Villmoare et al. 2015), also knapped hard stone. *Kenyanthropus platyops* and/or *Australopithecus afarensis* and/or *Australopithecus deyiremeda* (Haile-Selassie et al. 2015), whose presence is confirmed at Lomekwi or more widely in East Africa, could therefore be the makers.

In light of these recent findings, a thorough re-evaluation of the prehensile and manipulative abilities of these hominins from the Late Pliocene and of the gestures related to the manufacturing of stone tools is required (Harmand et al. 2015, supplementary information). Questioning the relevance of the morphological characteristics used till now to evaluate "precision", a study conducted by Pouydebat et al. (2006) demonstrated that a hand that does not seek to be precise does not necessarily lack the ability in tool making. Working on the primitive hand of *Australopithecus sediba* (Kivell et al. 2011) or on the wrist remains of *Homo floresiensis* (Tocheri et al. 2007, 2008) anthropologists came also to the conclusion that more than one type of hominin hand can be responsible for stone tools.

The technological analysis of the currently available lithic assemblage indicates that raw material blocks poorly adapted to knapping were intentionally modified and/or knapped by unifacial or semi-peripheral alternating direct percussion, with hard hammer, or by percussion of the blocks on passive hammer or again by bipolar percussion on anvil.

We can retain from the lithic productions of this exceptionally ancient site that if their knapping schemes are extremely simple, using elementary knapping techniques, their quantity, together with the butts and the characteristic traces observed on the lower face of the flakes, indicate that the basic knapping principles were sufficiently assimilated to be reproducible at will. However, in LOM3 numerous knapping accidents and percussion traces also show assessing errors of the involved parameters, the still uncertain mastery of the knapping gestures, both partially due to the poor suitability of the raw materials flaked, as well as to the ambivalence of some blocks, knapped after having been used in percussion.

Elementary knapping schemes have resulted in short series of removals or in the creation of very irregular cutting edges. Due to the low technical level shown by those first craftsmen of prehistory, the initial morphology and morphometry of the blocks were major constraints. Multiple step fractures and hinge terminations observed on the flakes or on their negatives of removal and the presence of numerous impact or percussion marks behind the core edges are all elements indicating the limited knapping capacity of the selected materials and the low control of the percussive gestures (Fig. 3.5). The most elementary knapping principles were assimilated, but the three directions of space were still awkwardly controlled when working blocks.

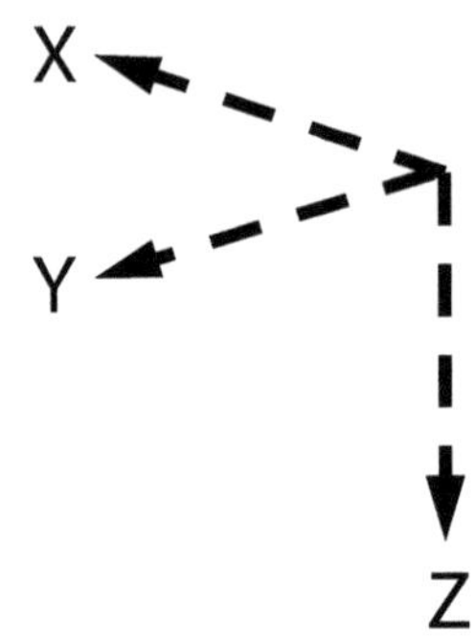

Fig. 3.5 LOM3: an awkwardly control of the three directions of space when flaking

3.7.1.2 Lokalalei 2C (LA2C)

A lack of technological elaboration was especially assumed about Lokalalei 1 site (Roche 1989), also emphasized by Kibunjia (1994, 1998). With the discovery of the neighboring LA2C site, it was later put forward that at 2.34 Ma hominin groups displayed distinct levels of skills as, at a lesser level, variations of quality in the locally available raw materials would also have had a significant role.

The Lokalalei 2C site (LA2C) was discovered during surveys conducted in the deposits of the Kalochoro member in 1998. The presence under the LA1 and LA2C sites of two tuffs, Kokiselei and Ekalalei, correlated with the tuffs E and F-1 of the Shungura Formation (Ethiopia), respectively, dated at 2.40 ± 0.05 and 2.34 ± 0.04 Ma, allowed to assign an age of 2.34 ± 0.05 Ma to the latter (Roche et al. 1999). This age is still relevant despite a reviewing of the local lithostratigraphy conducted more recently (Brown and Gathogo 2002).

The LA2C site was excavated exhaustively on about 17 m^2 that had been spared by erosion. LA2C yielded 2,614 lithic pieces associated with relatively poorly preserved skeletal remains.

Besides its age, which made it one of the oldest known archaeological sites at the time of its discovery, one of the remarkable aspects of the lithic assemblage is that over 13% of refits were possible. Often complete and sometimes combining several dozen pieces (Fig. 3.6), they allowed an exceptionally fine technological reading of the sequence of the technical gestures done by the knappers (Delagnes and Roche 2005).

This sequence can be outlined as follows:

- Choosing a specific raw material in the range of locally available volcanic materials in LA2C (mainly trachyte and phonolite).
- Choosing in these materials small blocks with a dihedral angle formed by the intersection of two surfaces, cortical

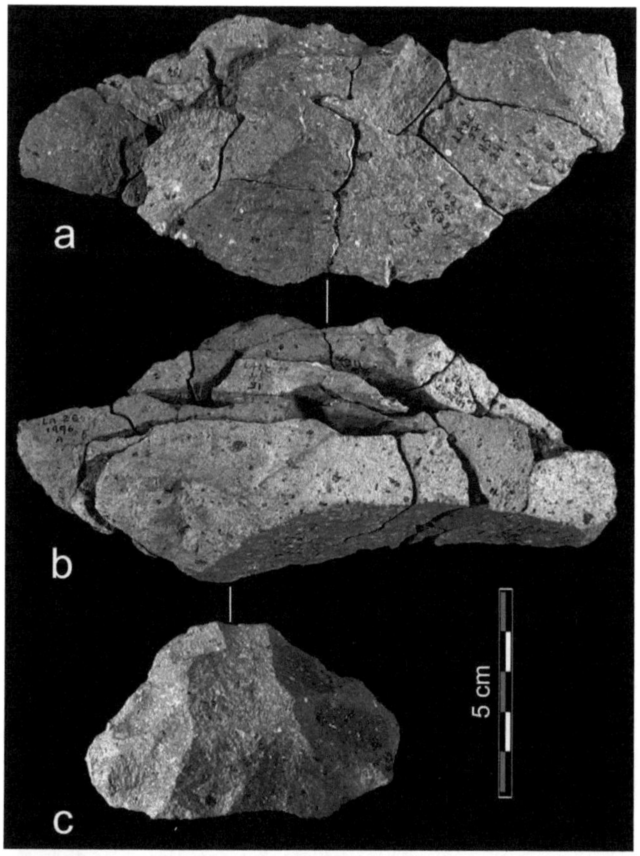

Fig. 3.6 LA2C. Upper (a) and lateral (b) view of an intermediate reconstruction of refitting group 33 reassembling 38 items. View from the flaked surface of the residual core (c). These images show the characteristic geometry of the majority of the cores at LA2C. The faces of the original dihedral that played the role of striking platforms during the exploitation of the cores result from the intentional fracturing of a fine-grained basalt block. (Photograph P-JT—MPK/WTAP)

or from fracturing, which are alternately used as natural striking platform throughout the débitage of a succession of short series of 2 to 5 flakes.

A careful selection of a material with very similar mechanical properties from one block to another and offering natural or summarily created forms (fractured blocks), according to what we know from all the available lithic assemblages on the technical background of early Oldowan and Oldowan knappers (Fig. 3.7), is what allowed the LA2C craftsmen to easily bypass the major handicap that still was their poor technical command of the third dimension of space.

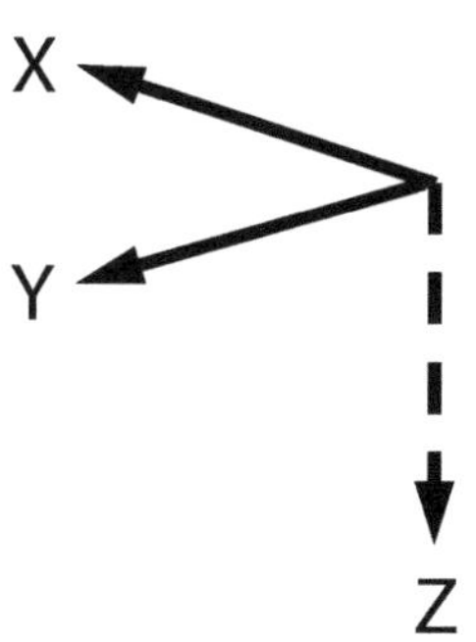

Fig. 3.7 LA2C: a good technical control of two directions of space and an indirect control of the third one obtained by flaking rigorously selected blocks

3.7.2 The Kokiselei Sites Complex

Currently, 10 sites were found in the archaeological complex of Kokiselei (KS), which takes place in its entirety in the Kaitio Member. Their stratigraphic place clearly showed that the KS1 and KS6 sites, especially, are among the oldest in the complex, with a slightly younger age than the KBS Tuff (1.87 Ma). Furthermore, the stratigraphic position of KS5 can be compared with that of the Oldowan site KS6.

Located in the flood clays that mark the top of the sequence of the complex, KS4 is significantly different from this first group of sites. An age of 1.76 Ma was calculated for the early Acheulean of KS4 (Lepre et al. 2011) that comes from sediments located 4.5 m above the Olduvai/Matuyama reversal.

KS6 and KS1 yielded typically Oldowan lithic assemblages, both typologically and technologically. The study of the Oldowan assemblage of KS5, stratigraphically very close to KS6, clearly shows, especially thanks to several refits, that its authors had already acquired the technical skills needed to deal with the same efficiency with sufficiently homogeneous and isotropic materials in all three dimensions of space. A very roughly shaped piece on a cobble could be the evidence of a still very timid attempt toward other materials modules and toward another knapping method, shaping (Texier et al. 2006).

3.7.2.1 Kokiselei 5 (KS5)

The unique archaeological level of KS5 only showed a very slight vertical dispersion of the material. It was excavated extensively on a 65 m^2 surface. It yielded a few bone remains (n = 280), relatively poorly preserved. The 1,727 pieces of the lithic assemblage that included several decisive diagnostic refits was the subject of a thorough technical analysis.

It appears from this study that the KS5 knappers used a wider variety of raw materials than at LA2C and KS1, sometimes very poor in quality.

The analysis of three refits, I, J, and F in particular (Fig. 3.8), clearly shows that the KS5 knappers had managed to cross a critical threshold in the conduct of débitage. This analysis reveals in particular that the technological level

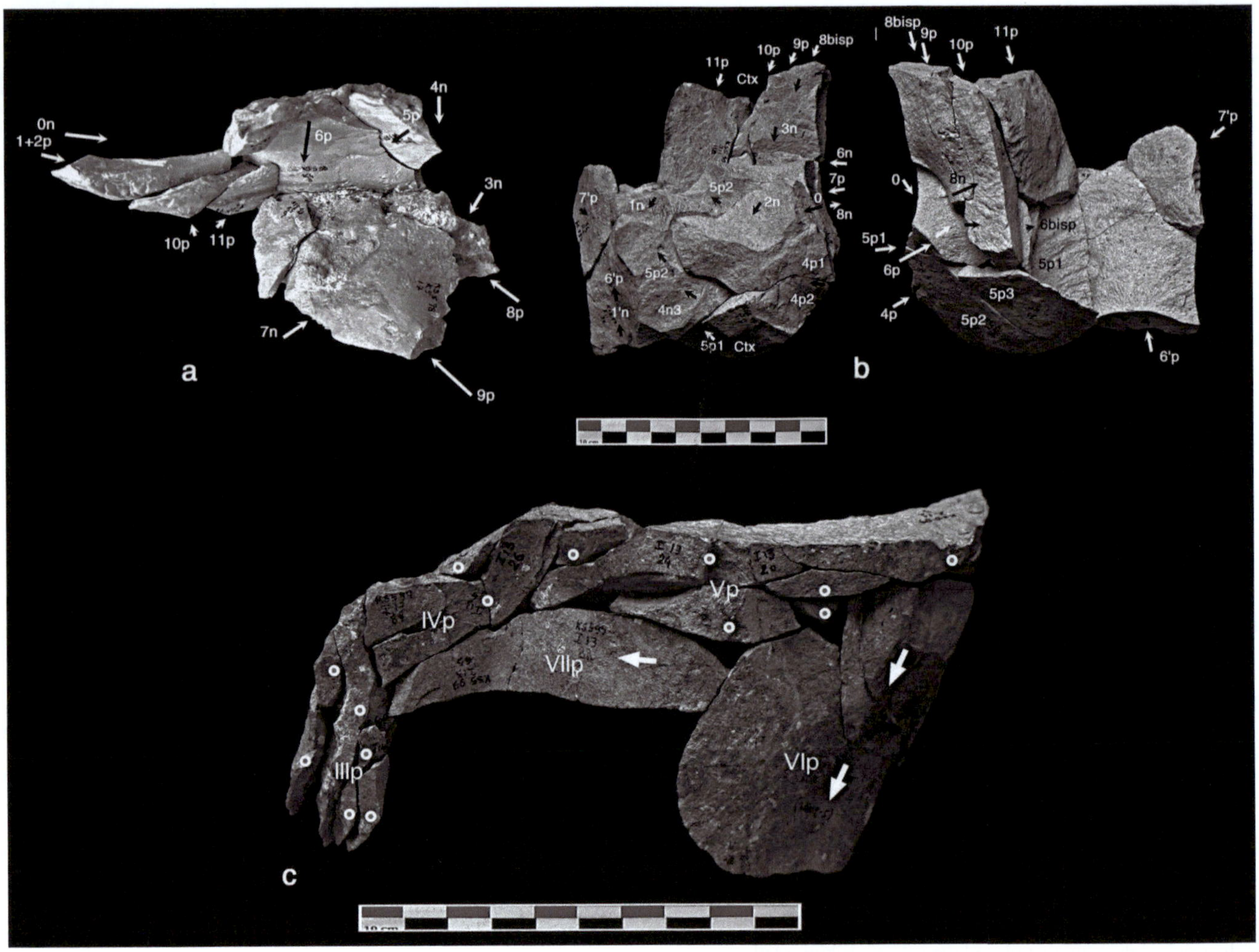

Fig. 3.8 Three refitting groups from KS5 (a = refitting I; b = refitting F; c = refitting J) documenting a negative (n) or a positive (p) way the redirection of flake removals during the cortex removal phase or the core débitage phase of various raw materials. (Photograph P-JT. MPK/WTAP)

reached allowed them to create at will and to maintain the necessary striking platforms to continue and/or reorganize the débitage (Texier et al. 2006). This is the first and most finely recorded evidence for sequences of flake removals technically and angularly outstandingly well controlled in the three dimensions of the worked volume. This advance, which is a real technological leap, enabled them to dispense in a large measure with the selection, so far unavoidable, of the morphology of the blocks to be worked. Sorting of the lithic assemblage by raw materials and by worked blocks confirms that the KS5 knappers were able to use a much wider range of materials. This predate from about 0.3 Ma the diversification of débitage methods like at Garba IVD (Melka Kunture) where it was recently demonstrated (Gallotti and Mussi 2018) that among several available methods the choice of a specific one was both influenced by raw material geometry and by technical purposes.

The knappers of KS5 had understood the importance of the role played in a débitage by the ability to create at will a new striking platform to redirect, reorganize, and continue a flake production (Fig. 3.9). Several refits show that these knappers had reached the required level of technological expertise. Therefore, any block of a raw material suitable for knapping that could be handled had now become workable.

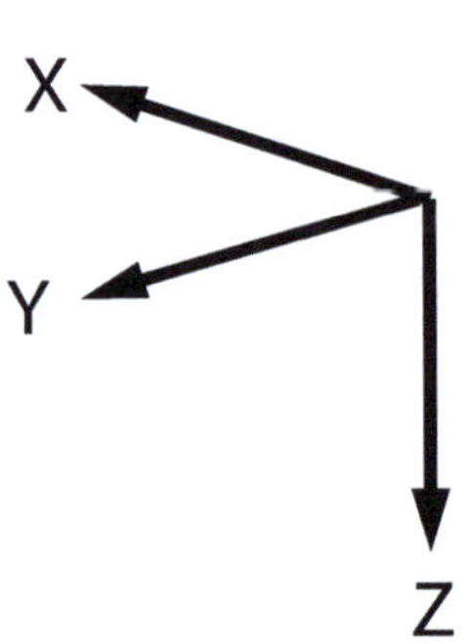

Fig. 3.9 KS5: a good control of the three directions of space reached in flaking when the ability to create at will new striking platforms is acquired

All the elements were already in place in KS5 to allow the mental construct and the formalization of a new concept: bifacial shaping. It materialized gradually (see the examples of Kokiselei 4 and Isenya below) by a succession of knapping operations whose aim was to make one single object by sculpting the raw material (Inizan et al. 1995) to create a specific morphology by successive removals of invasive, alternate, or alternating flakes.

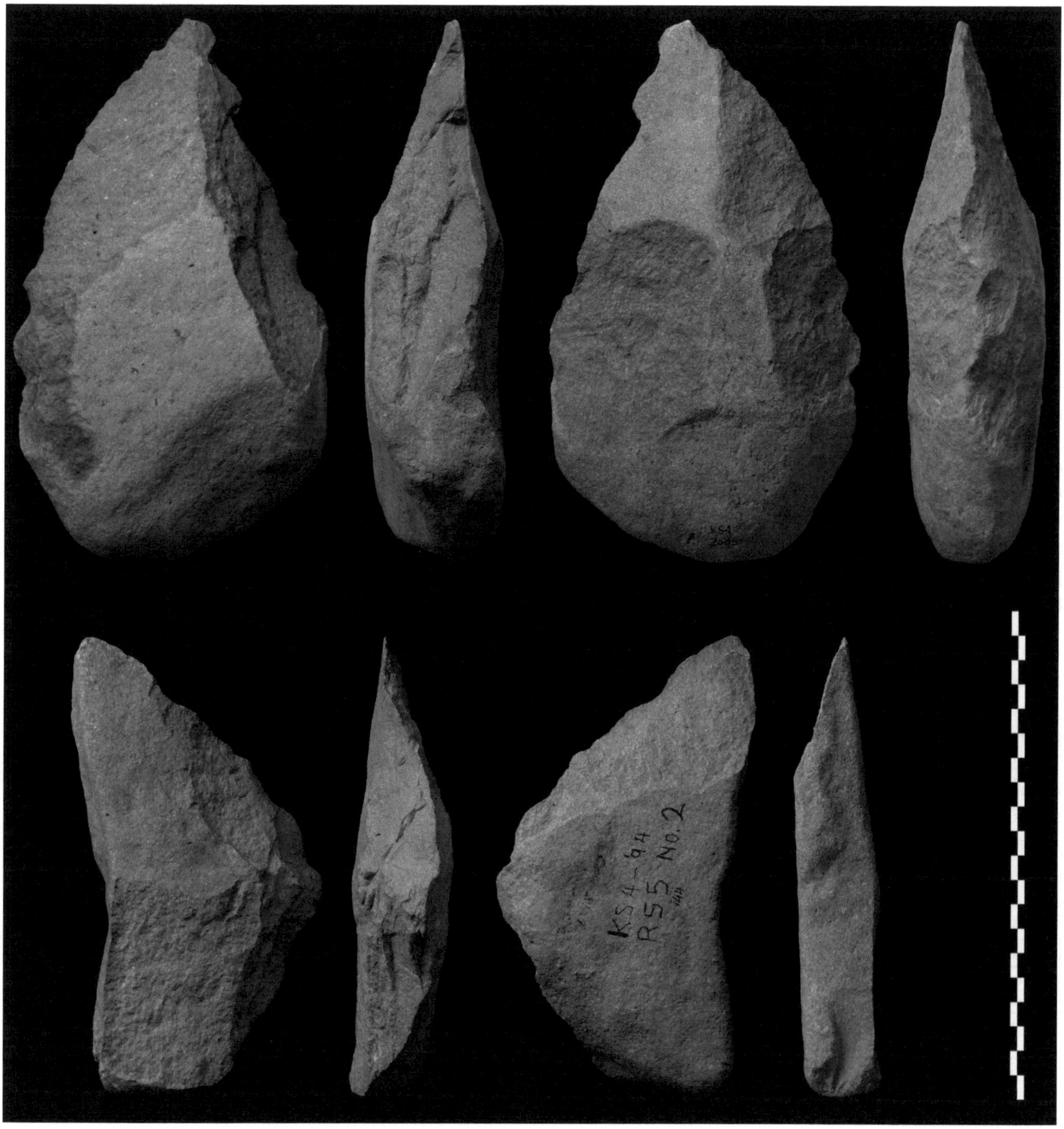

Fig. 3.10 KS4: basic bifacial shaping of a flat cobble and partial bifacial shaping of the half of a split cobble. Direct hard hammer percussion on aphyric phonolite. (Photograph P-JT. MPK/WTAP)

3.7.2.2 Kokiselei 4 (KS4)

The flood clays that yielded the artifacts from KS4 were quite largely depleted by erosion. For this reason, and also because of the low concentration of artifacts, the KS4 site could only be the subject of limited excavations. However, extensive surface collecting was regularly carried out. Numerous refits between the material collected on the surface and the material collected in stratigraphy have validated the consistency of the resulting assemblage (Lepre et al. 2011). Furthermore, these refits are contributing very significantly to the technological analysis of the assemblage, especially by allowing the reconstruction of voluminous objects and the study of their fracturing mode.

The KS4 site is only a few hundred meters away from KS5. It punctuates remarkably the end of the Kokiselei complex, for which all the sites are situated in the Kaitio Member between 1.87 and 1.76 Ma.

The KS4 lithic assemblage presently contains 202 items. Made for 85.3% of aphyric phonolite (for 6.3% of basalt and 8.4% of trachyte), it shows a certain monotony at the petrographic level. This only reflects the sought after module for the blanks, mostly present among the available phonolite cobbles (Harmand 2005, 2012). This assemblage is essentially characterized by the presence of large flakes and heavy tools. These were obtained by débitage, splitting of large aphyric phonolite cobbles and by shaping of the fracturing products, or by direct shaping, unifacial or bifacial, of flat phonolite cobbles (Fig. 3.10). Several refits tell us about the dimensions of the initially selected slabs or cobbles. Overwhelmingly in aphyric phonolite (74%), they reach or exceed 30 cm in length in their long axis (Table 3.4). Direct percussion with heavy hammer of large cobbles resting on an anvil was an effective way to split pieces of this module. The lower faces of flakes or flake fragments bearing the characteristic traces of intentional knapping evidence it, but surprisingly flat, including in areas close to the points of impact.

In the absence of hammers and alongside the heavy fashioned toolkit, the rest of the KS4 assemblage consists of cores and untreated flakes, very variable in size (up to 20 cm), showing, without particular organization, the use of materials often poor in quality.

The heavy shaped equipment of KS4, with 28 elements, includes in particular picks with trihedral (n = 8) or square (n = 3) sections, unifacial (n = 8) or roughly shaped handaxes (n = 3), as well as pieces left as rough outs (n = 6).

The metric and technical characteristics of the large shaped pieces from flat, split, or fractured cobbles on anvil put the lithic assemblage of KS4 in technological discontinuity with the Oldowan in general and with the lithic groups of the sub-contemporary or slightly older sites of the Kokiselei complex.

The technological level revealed by the refits of KS5 is perfectly compatible with that required in the production in its simplest form of a new knapping method, shaping. Thus, this event can be considered as the new milestone of technological developments taking root locally in the Oldowan and remained with or without a future.

In the new area of exploration that opened to the KS4 knappers, the raw material supply was again decisive. Indeed, we will see that the methods of acquisition and standardization of the blanks to be shaped needed tens of thousands of years to take form and establish themselves as techno-cultural markers of the Acheulean.

Thus, the need to bypass obstacles up to then technically insurmountable forced the KS4 knappers to seek new supply sources rich in large module cobbles, flat if possible, which they found in their close environment (Harmand 2005).

KS4: Techniques

The choice of large cobbles is obviously compulsory for those who want to shape tools about twenty centimeters long in their bigger axis. But the selection of large flat cobbles or

Table 3.4 Main metric attributes of the KS4 major tool types; raw materials determination after Harmand (2005)

Shaped tool types	n.	Mean length (mm)	Mean weight (g)	Length range (mm)	Weight range (g)	Raw materials
Trihedral picks	8	197.6	1310.1	160–248	857–2010	aPh2
Diamond section picks	3	202	1356	170–222	1260–1463	aPh1-aPh2
Uniface/cortical str. plat	5	198.2	1124.2	172–220	824–1390	aPh2-aTr1qz
Uniface/fracture str. plat.	3	189.3	1078	170–200	876–1423	aPh2
Bifacial rough out	6	196.6	1210.6	178–227	491–2120	aPh2
Biface	3	216.6	1388.6	195–235	766–1950	aPh2
Total or metric range	28	196–216	1078–1388	160–235	491–2120	74% aPh

the splitting up in the thickness of thick cobbles, with heavy hammer and resting on anvil, have this time allowed the KS4 knappers to overcome a major technological gap to meet the new situation created by the realization of a new concept.

The bifacial or unifacial working, even summary, of halves of thick cobbles previously split and/or fractured on anvil, or of large flat cobbles, involves the use of three variants of the same technique: direct percussion with heavy hammer, direct and passive hammer percussion, direct hard hammer percussion.

The double impact caused by percussion on passive hammer allows fracturing thick cobbles, otherwise unusable, and generates remarkably flat fracture surfaces.

KS4: Sequences

The large flat cobbles, blocks, or split cobbles are then summarily worked by direct unifacial or alternate or alternating bifacial removals of short series of flakes. The study of the heavy KS4 toolkit reflects a rudimentary sequence of removals, whose number never reaches more than 12 on the most elaborate pieces. At KS4, the very basic aspect of the shaping is due to the still deficient representation the KS4 knappers had of the order of the removals to achieve, as well as to a lack of precision in the execution of the technical gestures, and to a lesser extent, to the sometimes very average quality of the raw material worked.

In an attempt to realize a new concept, despite the limits imposed by their know-how of the moment, and the technical space in which they lived, the KS4 knappers followed a similar approach to that of the LA2C knappers. Access to new sources of raw material in terms of morphology and module, if necessary by resorting to splitting the larger cobbles on anvil with heavy hammer, enabled them to acquire and summarily shape the large tools they needed and that their technology did not yet allow them to acquire by débitage.

Technological control of débitage in the three dimensions of space could be demonstrated in KS5. Some tens of thousands of years later, the partial transformation of blanks by bifacial shaping appeared around 1.76 Ma in the neighboring site of KS4. This was another way for the knappers of the time to express technologically their appropriation of three-dimensional space. The selection of a specific module for cobbles and their possible fracturing enabled them once again to overcome their technological deficiencies (Fig. 3.11). Direct percussion with hard stone was then the only technique available at this stage of the chaîne opératoire. It is much later than the sequences were organized, then became standardized and that other techniques appeared in the chaîne opératoire of bifacial shaping.

Fig. 3.11 KS4: an uncertain control of the direction of flakes removal and of their mutual organization in bifacial shaping

3.8 Gw1-Isenya Acheulean Site

The multilayered Acheulean site of Isenya is located in the Kajiado district (Kenya), 65 km south of Nairobi, in the Pleistocene sediments overlying the Tertiary volcanic entablatures that border the left bank of the Gregory Rift. Excavated thirty years ago, the seven main levels of Acheulean occupation have yielded an abundant archaeological material. This site is taken as an example here because of the technological characteristics of its lithic assemblages, of their size and their age that has recently been distinctly increased (Durkee and Brown 2014). This recent review now makes it both one of the best-dated Acheulean sites and the most richly documented with regard to the systematic and recurrent use of direct ironwood hammer percussion for shaping large bifacial objects in phonolite(s).

Numerous LCTs (Table 3.5) were collected in seven of the main archaeological layers (Va, Vb, Vla, VIb1, VIb21, VIb22, and VIc1) that succeed each other in the 80 cm thick alluvial deposits of the Pleistocene sequence, and that are topped by recently dated volcanic ashes in the western sector of the excavation.

Based on a substantial experimental program, their technological study (Roche and Texier 1996; Bouthinon 2002) has provided many indications about the acquisition modes of the blanks of these LCTs, their degree of predetermination and the techniques used in the various phases of the representative chaînes opératoires of what can be considered as an Acheulean in full technical maturity. In particular, it was shown that the chaînes opératoires of the two major categories of LCTs widely represented in Isenya are closely intertwined at the level of the acquisition phase of the blank flakes with heavy stone hammer directly on the phonolite outcrops.

3.8.1 *Predetermination of Some Blanks*

Handaxes: when the shaping has not completely obliterated the knapping traces of the original blank and prevents its interpretation (95%), it appears from this study that the

Table 3.5 Isenya: count and metric characteristics of the LCTs from the main archaeological levels. Handaxes: mean value of the number of removals per face: dorsal (A) and ventral (B). According to data from P-J. Texier and M. Millet (lengths in mm; weight in grams)

Layer	Va	Vb	VIa	VIb1	VIb21	VIb22	VIc1	Total
N. handaxes	90	111	239	108	57	68	5	678
Handaxes mean length/weight	188/624	191/689	199/770	200/817	182/737	166/645	138/402	193/740
Mean removals A/B	9/10	9/10	9/10	9/10	7/9	6/8	4/6	8.6/9.7
N. cleavers	20	97	302	207	217	247	66	1156
Cleavers mean length/weight	167/695	177/897	172/900	168/807	176/850	176/877	165/820	173/857
Total LCT per layer	110	208	541	315	274	315	71	1834

handaxe blanks in Isenya have mostly been flakes with a partly predetermined morphology (Fig. 3.12a). These flakes are both broad and short, with a convex lower side and often hinge termination. 13.6% of them are short flakes, wide, and déjeté, whose butt was often kept in proximo-lateral position at the end of a shortened shaping time (Roche and Texier 1996).

Cleavers: the cleaver on flake cannot be made in series without the perfect control of predetermining flake removals. Thus whatever the ancient age considered, the presence of cleavers on flakes is clearly indicative of the mastery of the predetermination concept, and of a higher level of anticipation to that required to obtain a handaxe, even the most carefully executed (Roche and Texier 1991). The presence of three cleavers on flakes in the lithic assemblage of KGA6-A1 in Konso Gardula could indicate the existence in Ethiopia of an even older Acheulean than at KS4 if the age of 1.75 Ma was confirmed. The cleaver on flake is an asymmetric tool, mostly obtained at Isenya at the expense of a short flake, broad and with a convex lower face, laterally overlapping the negative of removal of a flake predetermining its future terminal bevel. Its final shape is in most cases determined by the rapid implementation by direct hard hammer percussion of a series of alternating removals to take out the plane of the supporting flake butt, then to summarily rectify its delineation (Fig. 3.12b). A distal series of often limited direct removals, also aimed at regularizing and strengthening this edge, fashion the final shape of the tool. The morphological axis of the final piece is perpendicular to the technological axis of the original blank. The bevel created by the removal of the predetermining flake is for the most part spared of any modification, but sometimes voluntarily reduced when working the edges of the tool. Direct stone percussion is the only necessary technique for producing such a tool. The shaping of 4.1% of them however was continued beyond this stage by then using the systematically implemented technique in Isenya in the later stages of the bifacial shaping: direct percussion with an organic hammer (Texier and Roche 1995).

At Isenya, all layers taken into account, on a sample of 949 cleavers, the orientation of the débitage axis of the original blank is identifiable at 91%. In 81% cases (n = 700), this axis makes an angle equal or close to the perpendicular with the morphological axis of the tool. When the orientation of the technological axis of the flake creating the bevel is identifiable (n = 252), it is always parallel or slightly convergent with the axis of the original blank. Their number and these last observations reinforce in a convincing way arguments for a production in series and predetermination.

3.8.2 Techniques and Sequences

Direct heavy hammer percussion is the only technique applied in the unique method of obtaining the LCT blanks in Isenya. It is the only technique that allows to obtain, on the very outcrop of phonolite, flakes whose weight should vary between 1 and 2 kg to make handaxes of a mean weight of 740 g (n = 621, all layers combined) and cleavers with a mean weight of 857 g (n = 948, all layers combined). A skilled knapper can manipulate with two hands with satisfactory accuracy a 5–10 kg stone hammer (Petrequin and Petrequin 1993; Madsen and Goren-Inbar 2004).

The initial phase of the shaping that consists in removing the major imperfections of the blanks (cracks, protuberances, planes, etc.) is done by hard stone hammer direct percussion. It is common to both chaînes opératoires of these two LCTs. This is an essential prerequisite to a bifacial shaping such as that carried out at Isenya because it makes possible the coming into play of the new technique, essential to the smooth running of the following phases. A bifacial balance of the volumes begins to be sketched in this first stage in the selection, direct or inverse, and the location of the very first shaping flakes.

In the presence of characteristic knapping traces of a working by organic hammer direct percussion, reinforced by the results of a large program of experimentation with local raw materials (Roche and Texier 1996; Bouthinon 2002), and in the absence in sub-Saharan Africa of Cervidae whose antlers could provide an alternative solution, percussion with ironwood is the only possible technique to interpret the fine working of these phonolite blanks previously prepared with stone.

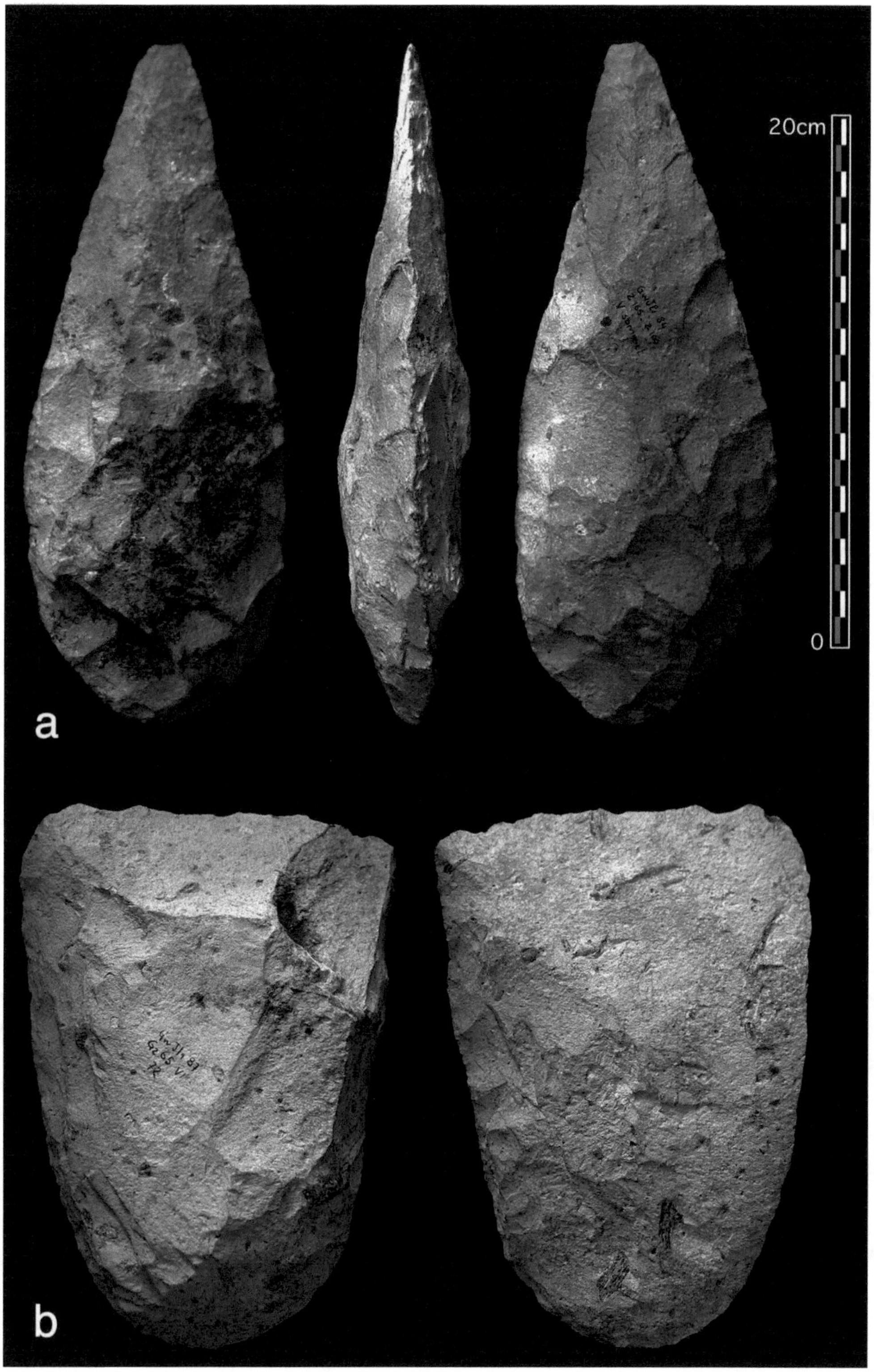

Fig. 3.12 Isenya GwJl1. LCTs in phonolite: lanceolate handaxe (a) and cleaver (b) shaped on flake blanks. Fine-grained phonolite from Kapiti. (Photograph P-JT—MPK/WTAP)

The coming into play of ironwood hammer direct percussion is in itself indicative of a high degree of anticipation from the knappers. One to three generations of thin covering flakes, with concave lower face, were detached. The bifacial balance of the worked piece was permanently established; as for its bilateral balancing, it appeared in the final stages of the shaping. If the worked material allowed it (fine grain), the delineation and the cutting edge of the worked object could be regularized by the carefully controlled removal of small flakes, precisely localized and limited in extent.

At Isenya, the joint production of bifacial pieces and cleavers on flakes was planned up to the choice of the raw material outcrop. The mental images in three dimensions that the knappers had were of great precision. From the outcrop to the finished object, the technical gestures were perfectly controlled, the field of application of the techniques used on these materials was precisely known, and their starting time scrupulously controlled. The shaping of handaxes roughouts by direct percussion with ironwood hammer was a routine operation in GwJl1, while completion by direct percussion with ironwood hammer for 4.1% of the cleavers can be considered as technically over-finished.

Carefully planned knapping operations, a very precise anticipatory vision of the consequences of the technical gesture in three-dimensional space, a perfect mastery of the knapping techniques and their field of application, a well-thought-out scheduling of the technical gestures, and their seriation alternating from one face to another, from one edge to the opposite edge (Fig. 3.13), enabled the knappers of Isenya to realize accurately the mental image of objects made by the hundreds.

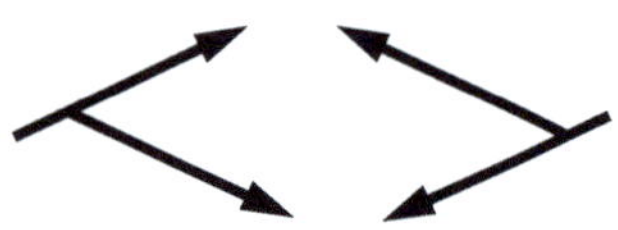

Fig. 3.13 GwJl1: a high level of control is reached removing flakes by mutually organized series in bifacial shaping

3.9 Conclusion

From 3.3 to 1.5 Ma, the species of several types of hominins have rubbed shoulders west of Lake Turkana, in a geographic area corresponding at minimum to that of the sedimentary deposits that have yielded their remains. Since the discovery in Lomekwi 3 of undoubtedly knapped objects, we also have the confirmation that many of these species have practiced hard stone knapping there. These knappers, whose anonymity will probably never be lifted, left behind indisputable evidence, but scattered both in the geographical area concerned and within the time encompassed by the sedimentary deposits involved. However, their technology could only be expressed in a constraining context governed by the laws of solid-state physics. In this context, there were not many alternatives for these early craftsmen, whatever they were, when crossing the technological threshold that could lead to more sophisticated knapping or to the first bifacial shaping. This is what gives cohesion to scattered knapping products and leads us to seek there the regularities that could govern their production mode.

Their examination shows that the slow appropriation of parameters and elementary methods of fracturing rocks suitable for knapping was done according to a recurring mechanism (Fig. 3.14). From LOM3 to KS4, the selection in the near environment of raw materials with specific morphology and module allowed the hominins who were able to develop new knapping concepts to overcome the obstacles that the previously acquired techniques and their field of action at the time did not allow to overcome yet.

At LOM3, the selection of materials and blocks/cobbles with favorable angulation has allowed, despite a still obviously very uncertain control of the technical gestures, to get the very first short series of flake removals ever done by direct or block on block technique.

A million years later, the rigorous and systematic selection of the morphology of blocks of the best materials available on site, made it possible for the LA2C knappers who had already acquired a remarkable precision in the percussion gestures, without particular preparation or reworking of the cores, to produce short but numerous recurring series of flakes.

Six hundred thousand years later, the KS5 knappers left us concrete evidence of their complete control of the sequence of the technical gestures in the three dimensions of space. A decisive step was taken because it is an essential knowledge to the development of a concept such as bifacial shaping and to the early stages of its technological investigation.

Just a few tens of thousands of years apart, in KS4, the descendants or successors of the KS5 knappers did not master yet percussion on anvil and direct hard hammer percussion. They chose large flat cobbles or the splitting on anvil of large thick cobbles to be able to engage into the uncertain sequences of the first bifacial shaping.

During these early and very long stages of industrious humankind that only had at its disposal a few variants of direct hard stone percussion, the basic knapping parameters were gradually integrated. The slow development of simple but efficient débitage methods, then of shaping still at a rudimentary stage enabled them to invest the volume to be knapped in its entirety. Among the available suitable knapping materials, the selection of blocks with specific morphologies reveals the awareness of these basic parameters such as the formalization of new concepts that the

LOM3 (Ken) n=149 *Kenyanthropus platyops* *Australopithecus afarensis*	**3.3 Ma**: superposed and contiguous unidirectional flakes detached but majority of hinge and step fracture terminations; weak understanding of fundamental knapping parameters; percussive motion poorly controlled; battering activities.	Direct percussion, hard stone hammer; percussion on anvil; percussion on passive hammer
La2C (Ken) n=2122 *Homo habilis* *Homo rudolfensis* *Paranthropus boisei*	**2.34 Ma**: production of several recurrent series of 2-5 flakes; one privileged raw material; low striking platform managment but highly selected blocks.	Direct percussion, hard stone hammer
KS5 (Ken) n=1796 *Homo habilis* *Homo rudolfensis* *Paranthropus boisei* *Homo erectus s.l.*	**1.8 Ma**: diversified raw materials, blocks with no specific shape; common striking platform management; production of series of flakes in any direction.	Direct percussion, hard stone hammer
KS4 (Ken) n=192 KGA6-A1 (Eth) *Homo habilis* *Homo rudolfensis* *Paranthropus boisei* *Homo erectus s.l.*	**1.76 Ma**: unifacial or bifacial shaping of large and flat cobbles or of halves of thick splitted cobbles; a privileged raw material; only one series of removals; small number of flake scars; irregular edges; diversified large tools; sequences of removals not mastered	Direct percussion, heavy hard stone hammer (splitting cobbles on anvil) Direct percussion, light hard stone hammer (shaping)
GwJl1 (Ken) n= 10386 *Homo erectus s.l.*	**0.98 Ma**: privileged raw material(s); interwoven production of blanks for handaxes and cleavers; -handaxes: careful bifacial shaping of broad and short flakes; 2-4 well balanced and mastered series of removals; high number of flake scars; edges bifacially and bilaterally regularized. -cleavers: quick but stereotyped shaping of short and broad flakes displaying a laterally predetermined cutting edge.	Direct percussion, heavy stone hammer (blanks production) Direct percussion, light hard stone hammer (handaxe blanks initialization; cleavers shaping) Direct percussion, ironwood hammer (handaxe blanks shaping and finishing, occasional cover-finishing of cleavers)

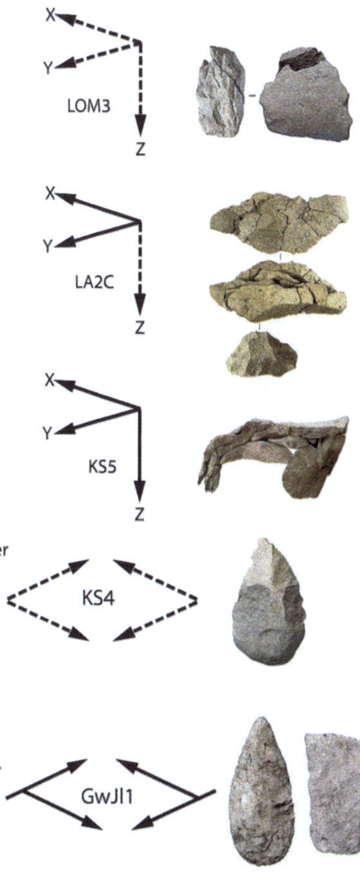

Fig. 3.14 Overview: sites; size of the lithic assemblages; hominins; age and main technological features of the lithic assemblages; knapping techniques; level of technological control of the three directions of space. Dotted arrows: low technological control of the exploitation of the concerned volumes and sub-volumes; plain arrows: full technological control of the concerned volumes or sub-volumes. This last is a fundamental component of the corresponding technical system of lithic production

know-how of the time did not yet allow to formalize in all circumstances.

Circumvention of the technical barrier by selecting appropriate modules and morphologies is the regularity that characterized the first technological advances for 1.5 Ma. It allowed materializing new concepts with a relatively low technological level but it prefigured the control and the rational ordering of the technical gestures in knapping or shaping methods, which occurred much later.

The study of the large cutting tools from the Acheulean site of Isenya vividly shows how, seven hundred thousand years later, what remained was only an attempt, perhaps without future at KS4, turned into a standardized production. The recently established correlations between the ash deposits bracketing the Acheulean bearing layers from Isenya, Kariandusi, and Olorgesailie Formations shed new light on the chronology and flexibility of the Acheulean technologies toward local raw materials in or on the slopes of the meridional section of the Kenyan Rift Valley. For example, comparing assemblages from Olorgesailie (H9A, DE89B…) and Isenya, one can notice the remarkable knowledge that the Acheulean knappers had of the module, fracturing and splitting properties as well as of the strength, hardness and flakability of specific raw materials. At Olorgesailie, the bifacial chaîne opératoire was adapted to the tendency to split of the local trachyte.

Isenya is undoubtedly both the oldest and best-dated Acheulean site, where an in-depth technological study demonstrated the systematic use of ironwood hammer direct percussion when shaping bifacial roughouts and during the finishing touches (Roche and Texier 1996). This technical innovation denotes an excellent knowledge of the mechanical properties and weaknesses of the large worked phonolite blanks. It is also in itself a very strong assessment component of the level of anticipation that some Acheuleans showed, both in the preparation and maintenance of soft hammers and in the predetermination of the blanks to be worked.

With a still approximate knowledge of the properties of the worked materials and with a limited technical range and know-how, the KS4 craftsmen had to compromise with the module and morphology of the available materials to try to realize new projects that went far beyond their technical capabilities of the time.

The technical behavior of knappers like those of Isenya clearly differentiates them from that of their predecessors from Turkana, in the fact that the obstacle to be overcome to realize such a complex project as the production in series and simultaneously of LCTs was no longer circumvented but technically eliminated. The complexity of the project and the technical innovation (direct percussion with ironwood hammer) that accompanied its realization, are two revealing elements of a profound change in the relationship of the knapper to the raw material. The chaîne opératoire analysis of the Isenya's LCTs clearly shows that from 1 Ma in this part of East Africa, the knappers had already conceptually and technically fully subjected some of the materials suitable for knapping in their environment.

The conceptualization and realization of large shaped Acheulean tools are the result of a long technological exploration deeply rooted in the Oldowan. The accessible evidence, scattered in time, shows us how in a constraining technological context the first craftsmen of humankind have been able to appropriate a three-dimensional knapping space where new concepts could take shape.

A more demanding selection of raw materials, an organized layout of the better controlled technical gestures, looking for better balanced shapes and the ability of the knappers to project in a longer time have created a favorable context for the development of a new technique such as direct percussion with organic hammer.

The control and strict delimitation of the field and time of action of the available techniques, a stereotyped execution of the technical gesture proper to each of them, allowed the mass production of large technically standardized cutting tools. What the refits of KS5 prefigured, that the knappers were not as dependent on the morphology of materials suitable to knapping available in their environment, had become a reality.

Around 1.76 Ma, the formulation of a still technically unfeasible new concept momentarily reactivated this dependency. From 1.76 to 1 Ma, the technical background of the knappers expanded considerably. The field, the area of action, and the starting order of the techniques have gradually been defined. The Acheulean knappers then disposed of the technical knowledge and know-how to engage in the mass production of technically standardized tools. Knapped stone technology shifted into another world, which already was or was about to become that of a single species of hominin: *Homo erectus s.l.*

Acknowledgements My grateful thanks to Hélène Roche, head of the Mission Préhistorique au Kenya (MPK), and co-responsible of the West Turkana Archaeological Project (WTAP).

References

Bar-Yosef, O., & Goren-Inbar, N. (Eds.). (1993). *The Lithic Assemblages of 'Ubeidiya', a Lower Paleolithic Site in the Jordan Valley*. Jerusalem: Institute of Archaeology, The Hebrew University of Jerusalem.

Bertouille, H. (1989). *Théories physiques et mathématiques de la taille des outils préhistoriques* (p. 15). Paris: CNRS Editions (Cahiers du Quaternaire).

Beyene, Y., Asfaw, B., Katoh, S., & Suwa, G. (2015). Overview of the Archaeological Research at Konso. In Y. Beyene, B. Afsaw, K. Sano, & G. Suwa (Eds.), *Konso-Gardula Research Project* (pp. 9–24). Tokyo: The University Museum, The University of Tokyo.

Beyene, Y., Katoh, S., Wolde-Gabriel, G., Hart, W. K., Uto, K., Sudo, M., et al. (2013). The characteristics and chronology of the earliest Acheulean at Konso, Ethiopia. *Proceedings of the National Academy of Sciences, 110,* 1563–1564.

Boesch, C., & Tomasello, M. (1998). Chimpanzee and human cultures. *Current Anthropology, 39,* 591–614.

Bouthinon, A. (2002). Apport de l'expérimentation à l'étude des grands outils façonnés du site acheuléen d'Isenya (Kenya). Mémoire de maîtrise, Laboratoire de Préhistoire, Université Paris I – Panthéon-Sorbonne.

Brown, F. H., & Gathogo, P. N. (2002). Stratigraphic Relation between Lokalalei 1A and Lokalalei 2C, Pliocene Archaeological Sites in West Turkana, Kenya. *Journal of Archaeological Science, 29,* 699–702.

Chevrier, B. (2012). Les assemblages à pièces bifaciales au Pléistocène inférieur et moyen ancien en Afrique de l'Est et au Proche-Orient. Ph.D. Dissertation, Université Paris ouest-Nanterre-La Défense.

Cotterell, B., & Kamminga, J. (1979). The mechanics of flaking. In B. Hayden (Ed.), *Lithic Use-Wear Analysis* (pp. 97–112). New York: Academic Press.

de la Torre, I. (2004). Omo revisited. Evaluating the technological skills of Pliocene hominids. *Current Anthropology, 24,* 439–465.

de la Torre, I., & Mora, R. (Eds.). (2005). *Technological Strategies in the Lower Pleistocene at Olduvai Beds I & II*. Liege: ERAUL.

de la Torre, I., Mora, R., & Martinez-Moreno, J. (2008). The early Acheulean in Peninj (Lake Natron, Tanzania). *Journal of Anthropological Archaeology, 27,* 244–264.

Delagnes, A., & Roche, H. (2005). Late Pliocene hominid knapping skills: The case of Lokalalei 2C, West Turkana, Kenya. *Journal of Human Evolution, 48,* 435–472.

Durkee, H., & Brown, F. H. (2014). Correlation of volcanic ash layers between the early Pleistocene Acheulean sites of Isinya, Kariandusi, and Olorgesailie, Kenya. *Journal of Archaeological Science, 49,* 510–517.

Gallotti, R. (2013). An older origin for the Acheulean at Melka Kunture (Upper Awash, Ethiopia): Techno-economic behaviours at Garba IVD. *Journal of Human Evolution, 65,* 594–620.

Gallotti, R., & Mussi, M. (2018). Before, During, and After the Early Acheulean at Melka Kunture (Upper Awash, Ethiopia): A Techno-economic Comparative Analysis. In R. Gallotti & M. Mussi (Eds.), *The Emergence of the Acheulean in East Africa and Beyond. Contributions in Honor of Jean Chavaillon*. Cham: Springer.

Goren-Inbar, N., Feibel, C. S., Verosub, K. L., Melamed, Y., Kislev, M., Tchernov, E., et al. (2000). Pleistocene milestone on the

out-of-Africa corridor at Gesher Benot Ya'aqov, Israel. *Science, 289,* 944–947.
Haile-Selassie, Y., Gibert, L., Mellilo, S. M., Ryan, T. M., Alene, M., Deino, A., et al. (2015). New species from Ethiopia further expands Middle Pliocene hominin diversity. *Nature, 521,* 483–488.
Haileb, B., Brown, F. H., McDougall, I., Gathogo, P. N. (2004). Gombe Group basalts and initiation of Pliocene deposition in the Turkana depression, northern Kenya and southern Ethiopia. *Geological Magazine, 141,* 41–53.
Harmand, S. (2005). Matières premières lithiques et comportements techno-économiques des hominidés plio-pléistocènes du Turkana occidental, Kenya. Ph.D. Dissertation, Université Paris X-Nanterre.
Harmand, S. (2012). Raw materials and techno-economic behaviors at Oldowan and Acheulean sites. In B. Adams & B. S. Blades (Eds.), *Lithic Materials and Paleolithic Societies* (pp. 3–11). New York: Wiley-Blackwell.
Harmand, S., Lewis, J. E., Feibel, C. S., Lepre, C. J., Boës, X., Prat, S., et al. (2015). 3.3 Million-Year-Old Stone Tools from Lomekwi 3, West Turkana, Kenya. *Nature, 521,* 310–316.
Hovers, E. (2009). Learning from mistakes: Flaking accidents and knapping skills in the assemblage of A.L. 894. (Hadar, Ethiopia). In K. Schick & N. Toth (Eds.), *The Cutting Edge: New Approaches to the Archaeology of Human Origins* (pp. 137–150). Gosport, IN: The Stone Age Institute Publication Series 3.
Hovers, E. (2012). Invention, Reinvention and Innovation: The Makings of Oldowan Lithic Technology. In S. Elias (Ed.), *Origins of Human Innovation and Creativity* (pp. 51–68). Elsevier (Development in Quaternary Science, Vol. 16).
Inizan, M.-L., Reduron, M., Roche, H., & Tixier, J. (Eds.). (1995). *Technologie de la pierre taillée*. Meudon: CREP.
Joulian, F. (1996). Comparing chimpanzee and early hominid techniques: some contributions to cultural and cognitive questions. In P. Mellars & K. Gibson (Eds.), *Modelling the early human mind* (pp. 173–189). Cambridge: McDonald Institute Monographs.
Katoh, S., Nagaoka, S., Woldegabriel, G., Renne, P., Snow, M. G., Beyene, Y., et al. (2000). Chronostratigraphy and correlation of the Plio-Pleistocene tephra layers of the Konso formation, southern main Ethiopia rift, Ethiopia. *Quaternary Science Review, 19,* 1305–1317.
Kibunjia, M. (1994). Pliocene archaeological occurrences in the Lake Turkana basin. *Journal of Human Evolution, 27,* 159–171.
Kibunjia, M. (1998). Archaeological investigations of Lokalaleil (GaJh5): A Late Pliocene site, west of Lake Turkana, Kenya. Ph. D. Dissertation, Rutgers University.
Kivell, T. L., Kibii, J. B., Churchill, S. E., Schmid, P., & Berger, L. R. (2011). *Australopithecus sediba* hand demonstrates mosaic evolution of locomotor and manipulative abilities. *Science, 333,* 1411–1417.
Lepre, C. J., Roche, H., Kent, D. V., Harmand, S., Quinn, R. L., Brugal, J.-P., et al. (2011). An earlier origin for the Acheulian. *Nature, 447,* 82–85.
Madsen, B., & Goren-Inbar, N. (2004). Acheulian giant core technology and beyond: An archaeological and experimental case study. *Eurasian Prehistory, 2,* 3–52.
McDougall, I., & Brown, F. H. (2006). Precise $^{40}Ar/^{39}Ar$ geochronology for the upper Koobi Fora formation, Turkana basin, northern Kenya. *Journal of the Geological Society of London, 163,* 205–220.
McDougall, I., Brown, F. H., Vasconcelos, P. M., Cohen, B. E., Thiede, D. S., & Buchanan, M. J. (2012). New single Crystal $^{40}Ar/^{39}Ar$ ages improve time scale for deposition of the Omo group, Omo-Turkana basin, East Africa. *Journal of the Geological Society of London, 169,* 213–226.
McGrew, W. C. (Ed.). (1992). *Chimpanzee Material Culture. Implications for human evolution*. Cambridge: Cambridge University Press.
Mercader, J., Barton, H., Gillespie, J., Harris, J., Kühn, S., Tyler, R., et al. (2007). 4300-year-old chimpanzee sites and the origins of percussive stone technology. *Proceedings of the National Academy of Science, 104,* 3043–3048.
Mercader, J., Panger, M., & Boesch, C. (2002). Excavation of a chimpanzee stone tool site in the African rain forest. *Science, 296,* 1452–1455.
Petraglia, M. D. (2003). The Lower Palaeolithic of the Arabian Peninsula: Occupations, Adaptations and Dispersals. *Journal of World Prehistory, 17,* 131–179.
Petrequin, P., & Petrequin, A.-M. (1993). *Ecologie d'un outil : la hache de pierre en Irian Jaya (Indonésie)* (p. 12). Paris: CNRS Editions (Monographies du CRA).
Pouydebat, E., Gorce, P., Berge, C., & Coppens, Y. (2006). La préhension chez les primates : précision, outils et évolution. *Anthropo, 11,* 159–166.
Prat, S., Brugal, J.-P., Tiercelin, J.-J., Barrat, J.-N., Bohn, M., Delagnes, A., et al. (2005). First occurrence of early *Homo* in the Nachukui Formation (West Turkana, Kenya) at 2.3–2.4 Myr. *Journal of Human Evolution, 49,* 230–240.
Roche, H. (1989). Technological evolution in early Hominids. *Ossa, 14,* 97–98.
Roche, H. (2007). From simple flaking to shaping. Stone knapping evolution among early hominids. In V. Roux & B. Bril (Eds.), *Stone knapping: The necessary conditions for a uniquely hominid behaviour* (pp. 35–48). Cambridge: McDonald Institute.
Roche, H. (2011). Archaeology of human origins: The contribution of West Turkana (Kenya). In J. Sept & D. Pilbeam (Eds.), *Casting the net wide: Papers in honor of Glynn Isaac and his approach to human origins research* (pp. 75–91). Oakville: Oxbow Books.
Roche, H., & Texier, P.-J. (1991). La notion de complexité dans un ensemble lithique. Application aux séries acheuléennes d'Isenya (Kenya). In S. Beyries, L. Meignen, & P.-J. Texier (Eds.), *25 ans d'études technologiques en préhistoire* (pp. 99–108). Juan-les-Pins: APDCA.
Roche, H., & Texier, P.-J. (1996). Evaluation of technical competence of *Homo erectus*. In J. R. F. Bower & S. Sartono (Eds.), *Human evolution in its ecological context. Evolution and Ecology of* Homo erectus (pp. 153–167). Leiden: Pithecanthropus Centennial Foundation.
Roche, H., & Tiercelin, J.-J. (1977). Découverte d'une industrie lithique ancienne in situ dans la formation d'Hadar, Afar Central, Ethiopie. *Comptes rendus de l'Académie des sciences de Paris, 284,* 1871–1874.
Roche, H., Brugal, J.-P., Delagnes, A., Feibel, C., Harmand, S., Kibunjia, M., et al. (2003). Les sites archéologiques plio-pléistocènes de la formation de Nachukui, West Turkana, Kenya : bilan synthétique 1997–2001. *Comptes Rendus Palévol, 2,* 663–673.
Roche, H., Delagnes, A., Brugal, J.-P., Feibel, C., Kibunjia, M., & Texier, P.-J. (1999). Early hominid stone tool production and technical skill 2.34 Myr ago in West Turkana, Kenya. *Nature, 399,* 57–60.
Semaw, S. (2005). Les plus anciens artefacts lithiques (2,6–2,5 millions d'années) des sites archéologiques du Pliocène final de EG-10 et EG-12 à Gona est, Afar, Ethiopie. In M. Sahnouni (Ed.), *Le Paléolithique en Afrique. L'histoire la plus longue* (pp. 211–246). Paris: Artcom / Errance.
Semaw, S., Rogers, M. J., Cáceres, I., Stout D., & Leiss, A. C. (2018). The Early Acheulean ~ 1.6–1.2 Ma from Gona, Ethiopia: Issues on the Emergence of the Acheulean in Africa. In R. Gallotti & M. Mussi (Eds.), *The Emergence of the Acheulean in East Africa and Beyond. Contributions in Honor of Jean Chavaillon.* Cham: Springer.
Semaw, S., Rogers, M. J., & Stout, D. (2010). Insights into late Pliocene lithic assemblage variability: The East Gona and Ounda

Gona south Oldowan archaeology (2.6 million years ago), Afar, Ethiopia. In K. Schick & N. Toth (Eds.), *The Cutting Edge: New Approaches for the Archaeology of Human Origins* (pp. 211–246). Gosport: Stone Age Institute Press.

Stout D., Quade J., Semaw S., Rogers M., Levin N. E. (2005). Raw material selectivity of the earliest stone toomakers at Gona, Afar, Ethiopia. *Journal of Human Evolution*, 48, 365–380.

Stout, D., Semaw, S., Rogers, M. J., & Cauche, D. (2010). Technological variation in the earliest Oldowan from Gona, Afar, Ethiopia. *Journal of Human Evolution, 58,* 474–491.

Texier, P.-J. (1996). L'Acheuléen d'Isenya (Kenya). Approche expérimentale des principales chaînes opératoires lithiques. In *La vie préhistorique* (pp. 58–63). Paris: Société préhistorique française, Faton.

Texier, P.-J. (2012). Les grands singes, les premiers tailleurs et l'outil. In A. Turq, J. Despriée, J. Airvaux, P.-J. Texier, & B. Maureille, (Eds.), *La conquête de l'ouest. Il y a un million d'années en Europe* (pp. 56–74). Les Eyzies: Musée national de préhistoire.

Texier, P.-J., & Roche, H. (1995). The impact of predetermination on the development of some Acheulean chaînes opératoires. In J. M. Bermudez De Castro, J. L. Arsuaga, & E. Carbonell (Eds.), *Evolución humana en Europa y los yacimientos de la Sierra de Atapuerca = Human Evolution in Europe and the Atapuerca Evidence* (pp. 203–420). Valladolid: Junta de Castilla y León.

Texier, P.-J., Roche, H., & Harmand, S. (2006). Kokiselei 5, formation de Nachukui, West Turkana (Kenya) : un témoignage de la variabilité ou de l'évolution des comportements techniques au Pléistocène ancien ? In P. de Maret, E. Cornelissen, & I. Ribot (Eds.), *Actes du XIVe congrès UISPP (Liège, 2001), section 15 : Préhistoire en Afrique / African Prehistory* (pp. 11–22). Liege: Archaeopress, BAR International Series, 1522.

Tixier, J., & de Saint-Blanquat, H. (1992). *Le biface, silex taillé.* Tournai: Casterman.

Tocheri, M. W., Orr, C. M., Jacofsky, M. C., & Marzke, M. W. (2008). The evolutionary history of the hominin hand since the last common ancestor of *Pan* and *Homo*. *Journal of Anatomy, 121,* 544–562.

Tocheri, M. W., Orr, C. M., Larson, S. G., Sutikna, T., Saptomo, E. W., Due, R. A., et al. (2007). The primitive wrist of *Homo floresiensis* and its implications for hominin evolution. *Science, 317,* 1743–1745.

Toth, N., Schick, K. D., Savage-Rumbaugh, E. S., Sevcick, R. A., & Rumbaugh, D. M. (1993). *Pan* the tool-maker: investigations into the stone tool-making and tool-using capabilities of a bonobo (*Pan paniscus*). *Journal of Archaeological Science, 20,* 81–91.

Tsirk, A. (Ed.). (2014). *Fractures in knapping*. Oxford: Archaeopress.

Vayson de Pradenne, A. (1920). La plus ancienne industrie de Saint-Acheul. *L'Anthropologie*, *XXX*, 441–496.

Villmoare, B., Kimbel, W. H., DiMaggio, E. N., Rowan, J., Braun, D. R., Arrowsmith, J. R., et al. (2015). Early *Homo* at 2.8 Ma from Ledi-Geraru, Afar, Ethiopia. *Science, 347,* 1352–1355.

Zarzycki, J. (1991). *Glass and the vitreous state.* Cambridge: Cambridge University Press.

Chapter 4
Before, During, and After the Early Acheulean at Melka Kunture (Upper Awash, Ethiopia): A Techno-economic Comparative Analysis

Rosalia Gallotti and Margherita Mussi

Abstract The emergence of the Acheulean is a major topic, currently debated by archaeologists researching all over East Africa. Despite the ongoing discussion and the increasing amount of available data, the mode(s) of the technological changes leading to this emergence remain(s) largely unexplained. Overall, there is a dearth of continuous stratigraphic sequences recording both the late Oldowan and the early Acheulean at the same site. Accordingly, the technological changes cannot be evaluated taking into account the variability of each microregional context. Besides, the early Acheulean must be defined not only with respect to the Oldowan, but also in comparison with the following middle Acheulean.

At Melka Kunture, on the Ethiopian highlands, the rather continuous record allows a diachronic analysis from ~1.7 to ~0.85 Ma in a single microregion. In this paper we address the emergence and later developments of the Acheulean in the perspective of technical responses to the qualities/limits of raw materials (lithology, dimensions, geometry). A comparative techno-economic perspective makes it possible to investigate the nature of technological change(s) taking into account the role played by lithic resource availability and constraints in the same paleolandscape.

Our results demonstrate that in this area the main novelties leading to the early Acheulean were new concepts in small and large débitage, in addition to the manufacture of large tools. These innovations emerged at Melka Kunture over two hundred thousand years, during a continuous cultural process leading from the late Oldowan to the early Acheulean. On the opposite side, at the end of the Early Pleistocene, the innovations are not a small qualitative step, but rather a gaint leap. We underline the strong techno-economic discontinuity between the early Acheulean and the middle Acheulean.

There is also evidence that *Homo ergaster/erectus* produced both the Oldowan and the early Acheulean at Melka Kunture. Accordingly, the technological changes leading to the emergence of the Acheulean on the Ethiopian highlands are not explained by a newly developing hominin species. Conversely, the middle Acheulean develops while *Homo heidelbergensis*, a new and more encephalized type of hominin, appears on the scene.

Keywords Ethiopian plateau • Melka Kunture • Early Pleistocene • Oldowan • Raw materials • Techno-economic behaviors

4.1 Introduction

Melka Kunture is located 50 km south of Addis Ababa, on the western edge of the Main Ethiopian Rift, in a half-graben depression of the Ethiopian plateau (Fig. 4.1). This cluster of sites preserves one of the longest and most complete prehistoric sequences in East Africa, from the late Oldowan (~1.7 Ma) to the Late Stone Age.

R. Gallotti (✉) · M. Mussi
Dipartimento di Scienze dell'Antichità, Sapienza Università di Roma, Piazzale Aldo Moro 5, 00185 Rome, Italy
e-mail: rosalia.gallotti@uniroma1.it

M. Mussi
e-mail: margherita.mussi@uniroma1.it

R. Gallotti · M. Mussi
Italian Archaeological Mission at Melka Kunture and Balchit

R. Gallotti
Université Bordeaux, UMR 5199, PACEA-PPP, Bâtiment B18 allée Geoffroy Saint-Hilaire, CS 50023, 33615 Pessac Cedex, France

R. Gallotti and M. Mussi (eds.), *The Emergence of the Acheulean in East Africa and Beyond: Contributions in Honor of Jean Chavaillon*, Vertebrate Paleobiology and Paleoanthropology,
https://doi.org/10.1007/978-3-319-75985-2_4

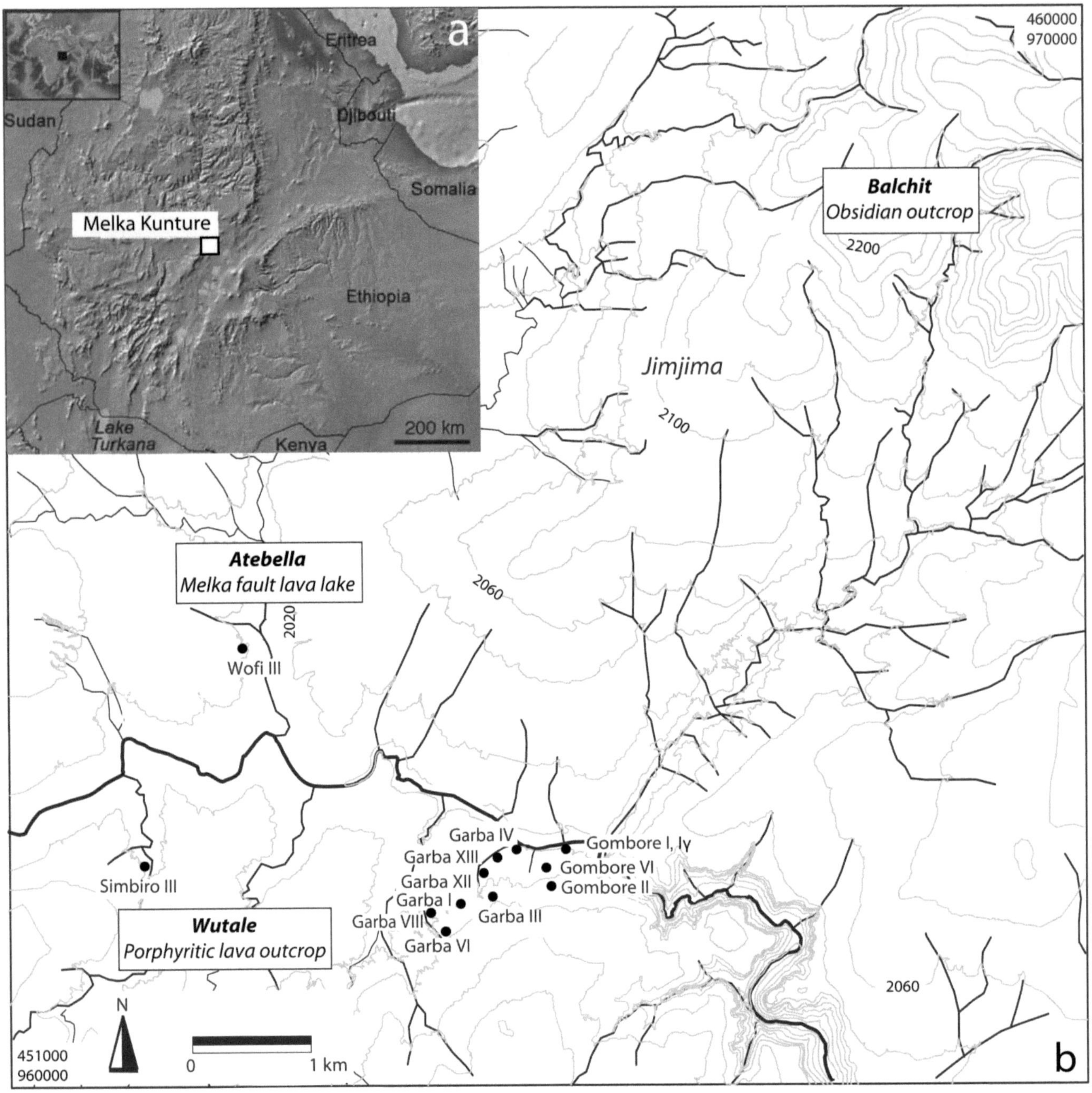

Fig. 4.1 **a** Map showing the location of Melka Kunture on the shoulder of the Main Ethiopian Rift. **b** Location of archaeological sites and primary sources of raw materials in the Melka Kunture area (vector restitution of the 1:50,000 topographic map)

In 1979, Chavaillon et al. recognized locally the first emergence of the Acheulean in level J of Garba XII, a site dated to ~1.0 Ma (Fig. 4.1b; Schmitt et al. 1977; Cressier 1980). They distinguished this "ancient" Acheulean from the Oldowan (Gombore IB; ~1.6 Ma; Fig. 4.1b) and Developed Oldowan (Garba IVD, ~1.6 Ma, and Gombore Iγ, ~1.3 Ma; Fig. 4.1b) on the basis of innovations in lithic productions: the first appearance and consistent manufacture of handaxes, cleavers, and standardized scrapers, and changes in chopper types. In a word, the Acheulean emerged later at Melka Kunture than elsewhere in East Africa (Leakey 1971, 1975). Later, it was also underlined that Melka Kunture Acheulean levels were never discovered above the Oldowan or Developed Oldowan ones in the same stratigraphic sequence (Piperno et al. 2004c). This was contrary to the dichotomy of Developed Oldowan/early Acheulean proposed by M. Leakey (1971, 1975) at Olduvai. On the Ethiopian plateau, there was no evidence of two parallel cultures developing side by side over more than one million years (Chavaillon et al. 1979).

The Acheulean sequence at Melka Kunture was in turn divided into four stages (Chavaillon 1980; Chavaillon and Chavaillon 1980): ancient (Garba XIIJ; ~1.0 Ma), middle (Garba XIIH-D, Simbiro III, Gombore II, Gombore VI, Gotu II, Garba IIIE; 0.8–0.5 Ma), upper (Garba I, Garba VI, Garba VIII; 0.4–0.3 Ma), and final Acheulean (Garba XIIIB, Garba IIIB, Wofi III; 0.25–0.15 Ma; Fig. 4.1b).[1] Chavaillon and Chavaillon (1980) interpreted the cultural change as a gradual local evolution of technology in a unilinear sequence from Oldowan to Acheulean. This scenario was put forward at a time when quantitative and typological features were the basis for any interpretation. The first appearance of certain tools and the variability in the presence of different types of tools were the main factors that made it possible to identify the Oldowan and the Acheulean, while the degree of biface and cleaver refinement was the main parameter defining the evolution of the Acheulean. At the time, this approach was the rule at East African sites (e.g., Kleindienst 1962; Leakey 1971, 1975).

In the last fifteen years, new systematic research at Melka Kunture cast doubt on this technological and stratigraphic characterization of the Oldowan/Developed Oldowan and Acheulean. New excavations and analyses of the Garba IV basal sequence (levels F-E), the review of the Garba IVD lithic assemblage in a techno-economic perspective, the excavations and analysis of new late Early Pleistocene sites (Garba XIIIB and Gombore II OAM), as well as new dates and the stratigraphic construction of the Melka Kunture Formation all pointed to a different scenario (Raynal et al. 2004; Piperno et al. 2009; Gallotti et al. 2010, 2014; Morgan et al. 2012; Gallotti 2013; Gallotti and Mussi 2015, 2017; Mussi et al. 2016; Bonnefille et al. 2018). The emergence of the Acheulean at Melka Kunture is now recognized at ~1.6 Ma (Garba IVD). The attribution to the early Acheulean was made after a detailed comparative analysis with the other early Acheulean sites in East Africa (Gallotti 2013). At the time, technological data from the Oldowan assemblages (Garba IVF-E) were not yet available. They have been published only recently (Gallotti and Mussi 2015).

In this paper, we analyze in a comparative perspective Early Pleistocene techno-economic behaviors at three archaeological sites of Melka Kunture with well-established chronostratigraphies: Garba IVF-D (~1.7–1.6 Ma), Garba XIIIB (~1.0 Ma), and Gombore II OAM (~0.85 Ma). We assess the range of variation and characterize the modes of lithic productions through a descriptive rather than a quantitative approach. Following the chaîne opératoire approach, lithic production, as examined here, is a sequence of technical actions and reductive phases ending in a techno-economic process; that is, we include the technical sequences as well as the technical and cognitive skills involved in tool production (Leroi-Gourhan 1964, 1971; Pelegrin 1985; Geneste 1989, 1991; Perlés 1991; Inizan et al. 1999; Harmand 2009). Additionally, we analyze the location, availability, composition, shape, and dimension of raw materials in order to understand how the knappers responded to lithic resources during the phases of the production process. We discuss the results in a comparative diachronical perspective. Our aim is (1) to identify more precisely the emergence and development of the local Acheulean during the Early Pleistocene, (2) to define the early Acheulean with respect both to the late Oldowan and to the middle Acheulean, and (3) to evaluate the role played by raw material availability and constraints on techno-economic change.

4.2 Melka Kunture: The Geological and Archaeological Setting

The Upper Awash River and its tributaries drain the area of Melka Kunture, and Pliocene volcanoes surround it (Mohr 1999). The major volcanic events started 5–4 million years ago, but later eruptions also modified the environment when hominin groups were already present in the area. The Awash re-established its course after each volcanic event. As described in the literature, the currents in the main river and its tributaries reworked and transported loads of sediments, including volcanic material, that buried and preserved archaeological sites (Kieffer et al. 2002, 2004; Bardin et al. 2004; Raynal and Kieffer 2004).

The archaeological sites are clustered over some 100 km^2 (Fig. 4.1b). During the last 50 years they have been intensely researched and excavated over some 20 km^2. They are located on the edges of gullies and valleys which cut through alluvial sediments from the Early and Middle Pleistocene ages. Upper Pleistocene deposits are less extensively preserved. To date around 30 of the over 70 archaeological levels located on both banks of the river have been tested or extensively excavated (Chavaillon and Piperno 2004). The levels are named after the gully where they are located (e.g., Simbiro, Garba, Gombore), followed by a Roman numeral (e.g., Atebella II). If a level contains multiple sub-levels, a capital letter is added for each one (e.g., Garba IVE).

[1]No systematic excavations were carried out at Gombore VI, Gotu II, Garba VI and Garba VIII. The lithic artifacts were collected along exposed sections. The location of Gotu II is unknown, hence it is not shown in Fig. 4.1b. The Garba IIIB industry, heretofore attributed to the final Acheulean, was recently reanalyzed and assigned to the early Middle Stone Age (Mussi et al. 2014).

4.3 The Analyzed Early Pleistocene Sites

Hominins settled again and again in this part of the Upper Awash valley from the end of the Olduvai Polarity Subzone to later than the Brunhes/Matuyama Reversal, as evidenced by dated chrono-stratigraphic successions (Schmitt et al. 1977; Cressier 1980; Morgan et al. 2012; Tamrat et al. 2014). All the Early Pleistocene sites discovered along the Garba (Garba IV, Garba XII and Garba XIII) and Gombore gullies (Gombore I, Gombore Iγ, and Gombore II) are included in the Melka Kunture Formation (hereinafter MKF; Fig. 4.2a, b; Raynal et al. 2004). However, while the overall chronostratigraphy is well understood, a detailed reassessment is under way for each site. Accordingly, in this paper we shall discuss Garba IVD-F, Garba XIIIB, and Gombore II OAM, which have already been reviewed.

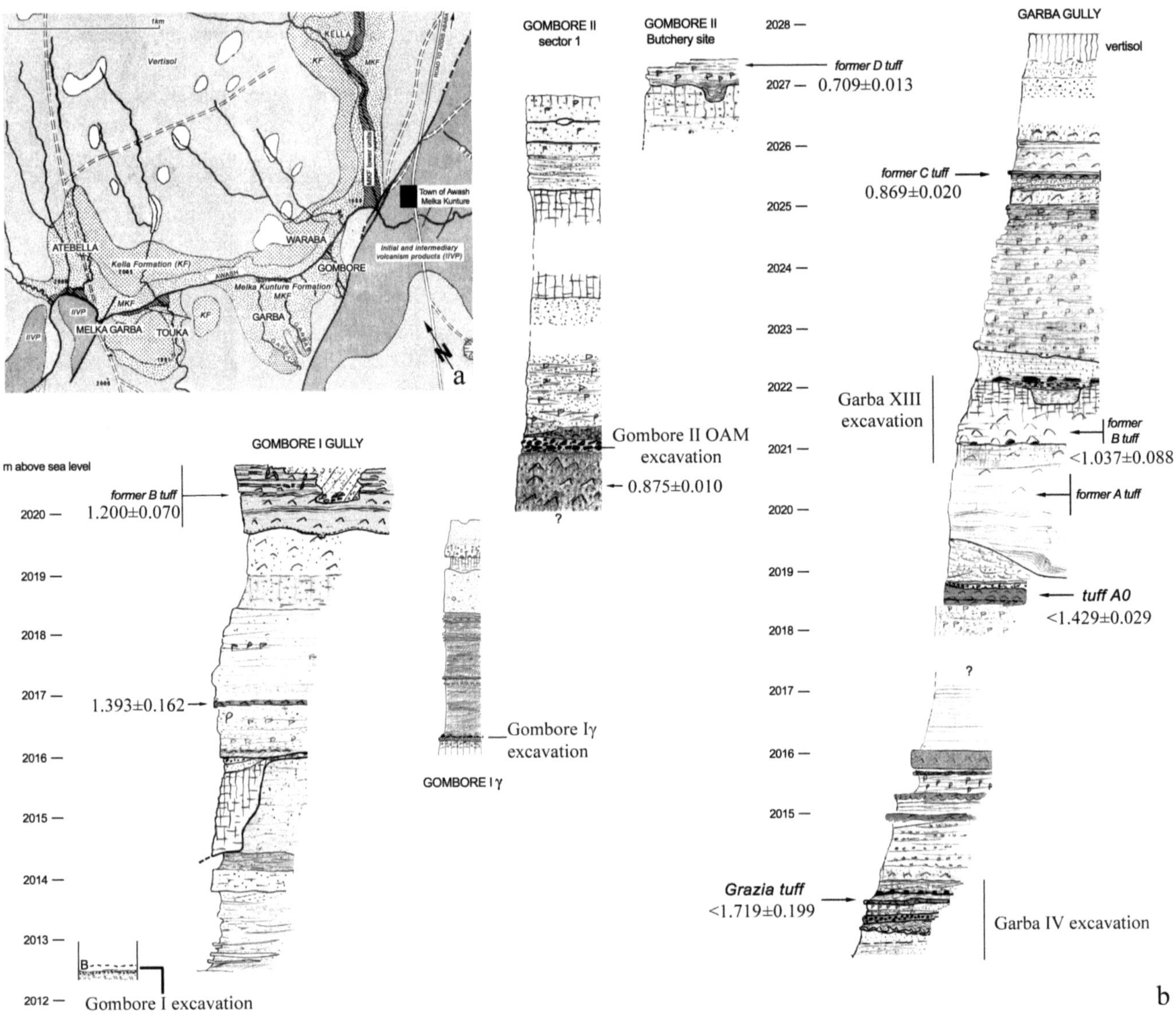

Fig. 4.2 **a** Geological sketch map of the Melka Kunture area (after Taieb 1974, revised). **b** The Melka Kunture Formation (after Raynal et al. 2004, revised; radiometric dates as in Morgan et al. 2012)

4.3.1 Garba IV (~1.7–1.6 Ma)

Garba IV is located on the right bank of the Awash at the confluence of the Garba creek (Fig. 4.1b). It was discovered in 1972 by Jean Chavaillon, who excavated it from 1973 to 1982 (Chavaillon and Piperno 1975; Piperno and Bulgarelli-Piperno 1975; Piperno and Bulgarelli 2004). The deposit, which belongs to the lowest parts of the MKF, consists of a stratigraphy that is approximately 3 m high. Three stratigraphic units were recognized in fluvial sedimentary series, and several archaeological horizons were discovered (Raynal et al. 2004). The sequence lies below tuff A0, dated to <1.429 ± 0.029 Ma (Morgan et al. 2012), which caps levels C and D. The lithic assemblage found in level D documents the emergence of the Acheulean at Melka Kunture at approximately 1.6 Ma (Gallotti 2013). The Grazia tuff sandwiched between level D and the underlying level E is dated to <1.719 ± 0.199 Ma (Morgan et al. 2012) (Figs. 4.2b, 4.3a–c). Levels E and F, below the Grazia Tuff, are included by Tamrat et al. (2014) in the normal polarity interval (N1), which they assign to the end of the Olduvai subchron. The Garba IVE-F lithic assemblages have been attributed to the late Oldowan (Gallotti and Mussi 2015).

4.3.1.1 Levels E-F

In 1982, level E was tested over 4 m^2 (Piperno et al. 2004d, 2009). The fragmented mandible of a two- or three-year-old *Homo erectus s.l.* child was discovered here (Condemi 2004; Zilberman et al. 2004a, b) together with lithics and faunal remains. In 2005, 2008 and 2009, level E and the underlying level F were expanded over approximately 34 m^2 and 12 m^2, respectively (Fig. 4.4). Every single item was recovered, including unworked lithic items. Level E yielded 504 unworked lithic objects, 718 artifacts, and 774 faunal remains; level F yielded 80 unworked lithic objects, 113 artifacts and 110 faunal remains. The spatial data of each object >1 cm were recorded in three dimensions. Hundreds of small flakes and unidentifiable fragments less than one cm long were also collected by systematically sieving the sediment from each level in half-square-meter sections (Gallotti and Mussi 2015). According to Raynal et al. (2004), the assemblages from levels E and F (Table 4.1) had deposited within a relatively short period of time.

4.3.1.2 Level D

The Garba IVD site displays a high-density distribution of artifacts and faunal remains. More than 100 m^2 were excavated here between 1972 and 1982. The Awash destroyed an unknown portion of the northern part of the site, and the Garba creek washed away the central part of this level (Gallotti and Piperno 2004; Piperno and Bulgarelli 2004; Fig. 4.5a).

A total of 19,055 finds (9,821 lithic artifacts, 6,654 unworked lithic objects, and 2,580 faunal remains) were systematically recorded (Gallotti and Piperno 2003, 2004; Piperno et al. 2004a, b; Gallotti 2013). The unworked lithic objects found during the excavations are no longer available for study. They had been neither catalogued nor stored, but only drawn on two-dimensional maps (D'Andrea et al. 2000, 2002; D'Andrea and Gallotti 2004). When we re-examined 9,028 of the 9,821 items originally described as "artifacts", we found that 6,986 had indeed been knapped, while 2,042 were unworked objects (Gallotti 2013; Table 4.1). The 793 items we did not re-examine included 499 obsidian pieces that were completely broken and 294 items that were missing.

Unit D is made of tightly packed lithic artifacts and faunal remains embedded in sands and gravels (Fig. 4.3d). The gravels are fine, but also include numerous unworked pebbles, cobbles, and blocks. The lithic objects (both knapped and unworked) display different degrees of abrasion. Fresh surfaces and edges sometimes coexist on the same object with abraded areas (Gallotti 2013). Raynal et al. (2004) suggest two alternative hypotheses regarding the deposition processes that formed this unit. According to the first hypothesis, a flood simultaneously transported and redeposited archaeological remains and unworked materials, and hominins had no impact on this deposit thereafter. According to the second hypothesis, when the water receded hominins did settle on the lag deposit. They knapped lithic tools using raw materials available on the spot. Later on, this new horizon was buried and partially reworked when a low-energy flow deposited sands on top of it.

4.3.2 Garba XIIIB (~1.0 Ma)

Garba XIII is likewise located along the Garba creek, not far from Garba IV and some meters higher up in the stratigraphic sequence (Fig. 4.1b). A stratigraphic section of ~2 m has been fully documented (Raynal et al. 2004; Gallotti et al. 2014). The main archaeological feature is level B, which has been excavated over ~15 m^2. It lies stratigraphically below a tuff unit dated to 0.869 ± 0.020 Ma (former C tuff) and immediately above a tuff unit dated to <1.037 ± 0.088 Ma (former B tuff, Figs. 4.2b, 4.6a; Morgan et al. 2012). Both tuffs are of reverse polarity (Westphal et al. 1979; Cressier 1980).

The lithic assemblage from level B (176 artifacts and 295 unworked objects) was analyzed in its entirety (Table 4.1). The artifacts are very fresh, whereas the unworked materials are very abraded. Accordingly, the hominins left artifacts on a lag deposit winnowed by fluvial agents (Gallotti et al. 2014; Fig. 4.6b–d).

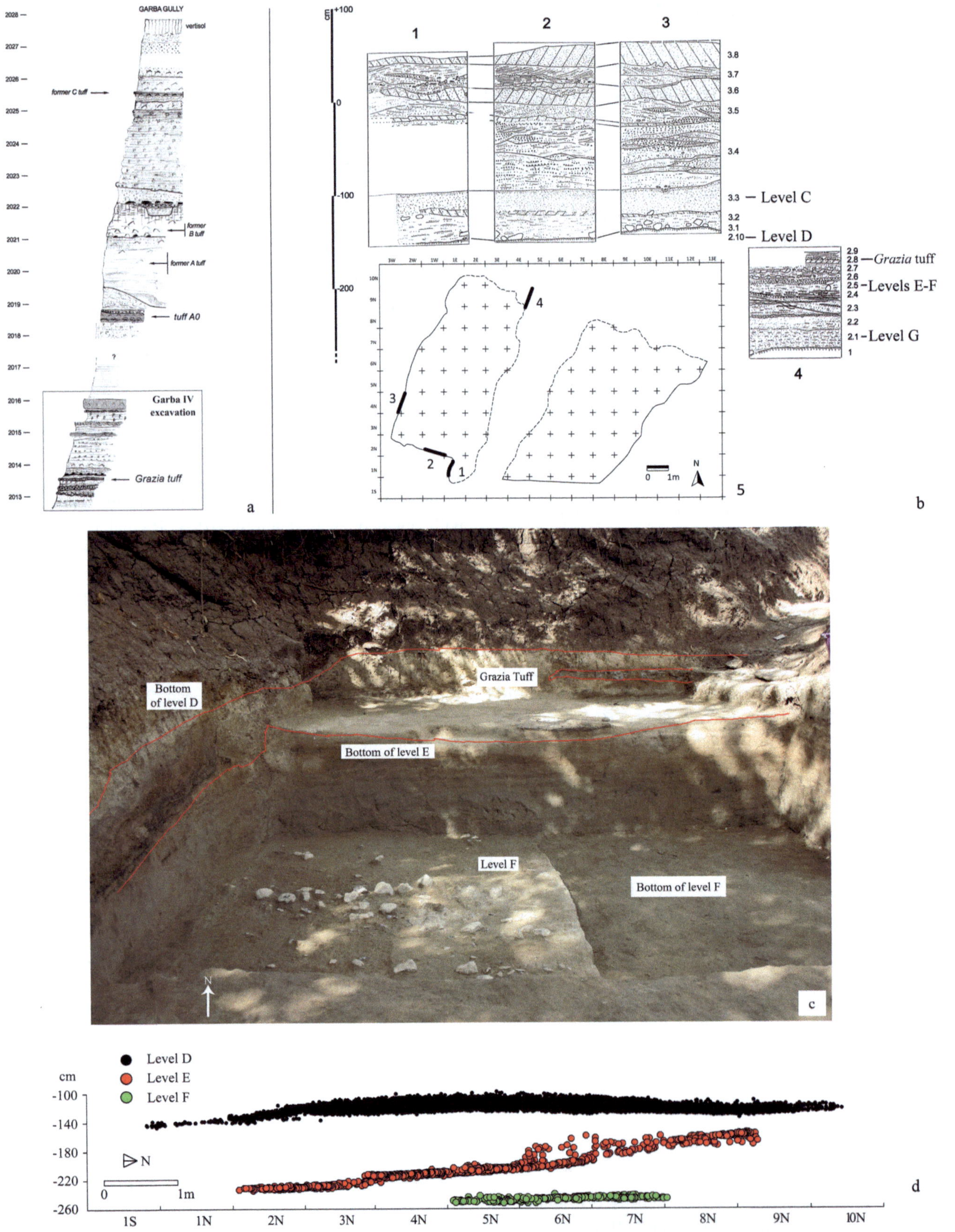

Fig. 4.3 **a** Stratigraphic position of Garba IV at the bottom of the Garba gully. **b** Lithostratigraphy of the Garba IV site. **c** Garba IV during the 2009 excavations. **d** S-N projections of lithic artifacts and faunal remains

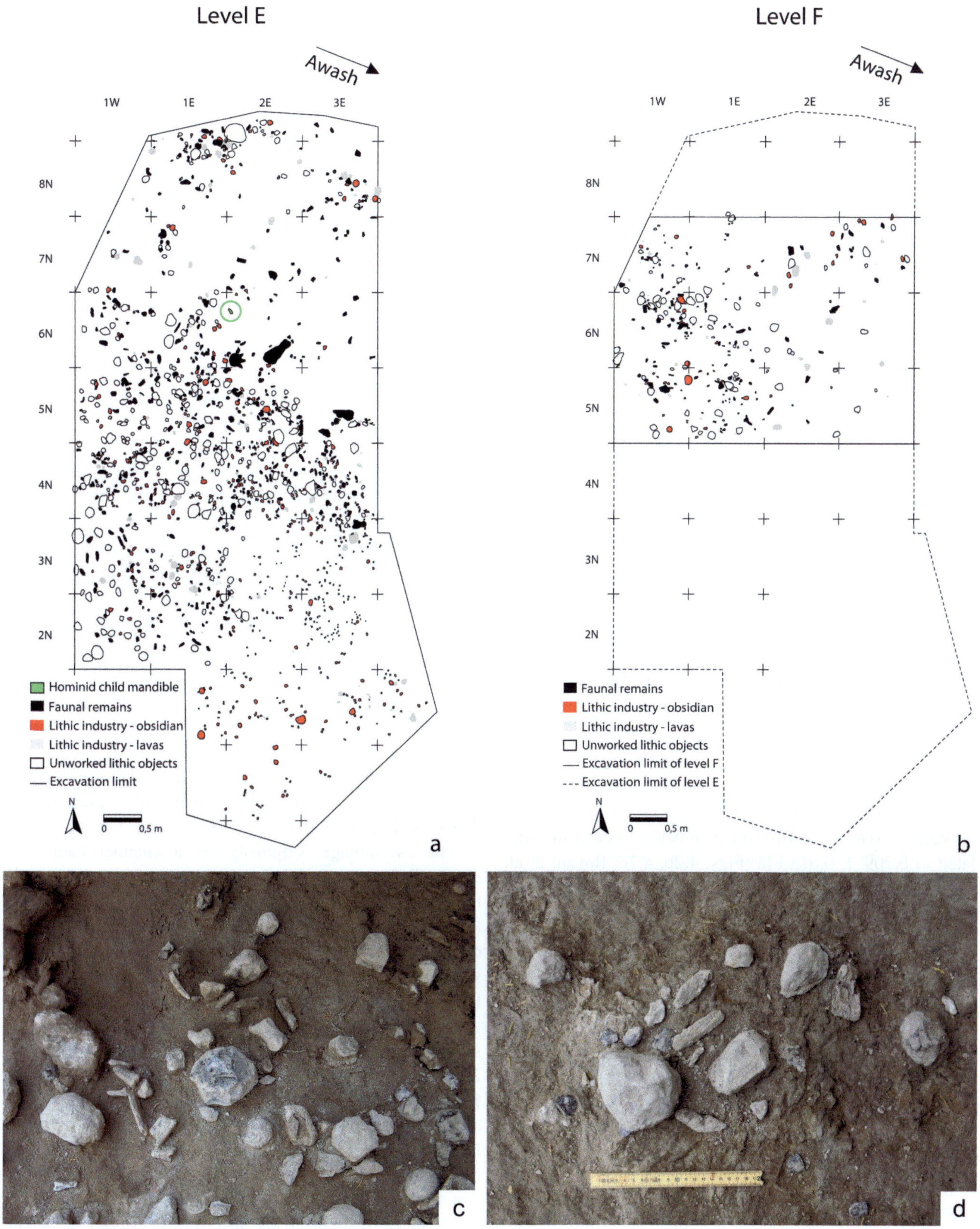

Fig. 4.4 Garba IVE-F. Horizontal maps and details of the excavation of layers E (**a**, **c**) and F (**b**, **d**)

Table 4.1 Components of the worked and unworked assemblages from Garba IVE-D, Garba XIIIB, and Gombore II OAM

Components		Garba IVE-F		Garba IVD		Garba XIIIB		Gombore II OAM	
		N	%	N	%	N	%	N	%
Cores		92	*8.2*	1816	*26.0*	32	*18.2*	35	*14.5*
Core fragments		9	*0.8*	107	*1.5*	0	*0*	0	*0*
Small flakes (<1 cm for Garba IVE-F, <2 cm for the others)		294	*26.1*	188	*2.7*	12	*6.8*	0	*0*
Broken flakes		158	*14*	874	*12.5*	0	*0*	0	*0*
Flakes		295	*26.3*	2508	*35.9*	79	*44.9*	109	*45.0*
Retouched flakes		74	*6.6*	193	*2.8*	1	*0.6*	5	*2.1*
Large flakes (>10 cm)		0	*0*	41	*0.6*	0	*0*	2	*0.8*
LCTs	*Bifaces*	0	*0*	0	*0*	18	*10.2*	58	*24.0*
	Cleavers	0	*0*	2	*0.03*	8	*4.5*	7	*2.9*
	Massive scrapers	0	*0*	21	*0.27*	0	*0*	0	*0.0*
Twisted bifaces		0	*0*	0	*0*	0	*0*	24	*9.9*
Indeterminable fragments		203	*18*	1151	*16.5*	26	*14.8*	0	*0*
Percussion elements		0	*0*	85	*1.2*	0	*0*	2	*0.8*
Total of the worked material		**1125**	***100***	**6986**	***100***	**176**	***100***	**242**	***100***
Angular items	*Small elements*	133	*22.8*	289	*14.2*	86	*29.2*	36	*24.0*
	Angular elements	71	*12.2*	400	*19.6*	94	*31.9*	17	*11.3*
	Small blocks	5	*0.8*	58	*2.8*	11	*3.7*	1	*0.7*
	Blocks	7	*1.2*	10	*0.5*	0	*0.0*	2	*1.3*
Rounded items	*Pebbles*	74	*12.7*	82	*4.0*	28	*9.5*	24	*16.0*
	Cobbles	279	*47.7*	1150	*56.3*	74	*25.1*	67	*44.7*
	Large cobbles	15	*2.6*	53	*2.6*	2	*0.6*	3	*2.0*
Total of the unworked material		**584**	***100***	**2042**	***100***	**295**	***100***	**150**	***100***

4.3.3 Gombore II OAM (~0.85 Ma)

Gombore II OAM, located in the Gombore gully at a distance of 400 m from the sites discussed above, is one of several sectors that were excavated in the same archaeostratigraphic unit (Figs. 4.1b, 4.7a). It lies above a tuff dated to 0.875 ± 0.010 Ma, 5 m below another tuff unit dated to 0.709 ± 0.013 Ma (Figs. 4.2b, 4.7c; Raynal et al. 2004; Morgan et al. 2012; Mussi et al. 2016) and belongs to the upper part of the Matuyama chron (Tamrat et al. 2014). This sector was chosen to become an open-air exhibit where visitors can see 35 exposed square meters of the archaeological surface, with its various materials still in place (Chavaillon and Piperno 2004; Gallotti et al. 2010).

A 2-m-high section located at the northwest end of the Gombore II OAM excavation has been described in detail (Raynal et al. 2004; Gallotti et al. 2010). The archaeological level is a thin clast-supported pebble bed, 0.10 m thick, which was once part of a paleochannel. In the section it looks like a row of stones containing many artifacts and bones. Where the surface of this unit is exposed, we see that the components are imbricated and graded by kinetic sieving (e.g., smaller items such as obsidian bifaces stand upright in gaps between larger items). In mid-channel, items from the archaeological level float in bedded sands, thereby demonstrating the mixed history of the archaeological surface. Large flat artifacts and long bones are clearly oriented, indicating that the current mainly flowed ENE.

The archaeological level contains unworked and knapped lithic objects as well as faunal remains, all present in high densities (Fig. 4.7b, d, e).

The assemblage, originally on a channel bank, was eventually partly displaced, concentrated and reoriented by the stream flow, a process that was probably repeated several times. This layer can be described as an ancient alluvial deposit. The lithic assemblage extracted from the exposed surface consists of 242 artifacts and 150 unworked pieces (Table 4.1; Gallotti et al. 2010; Gallotti and Mussi 2017).

The nearby Gombore II1 exposure, which belongs to the same archaeostratigraphic unit (Fig. 4.7a; Gallotti et al. 2010), yielded two human fossils. A left parietal bone fragment (GOM II1-6169) was discovered in situ in 1973, and a frontal bone fragment (GOM II1-576) was recovered in 1975 from the section dug in a small stream that flows through the excavation area. Profico et al. (2016) conclude that the hominins from whom these specimens came are likely to be recognized as ancestors of *Homo heidelbergensis* in sub-Saharan Africa.

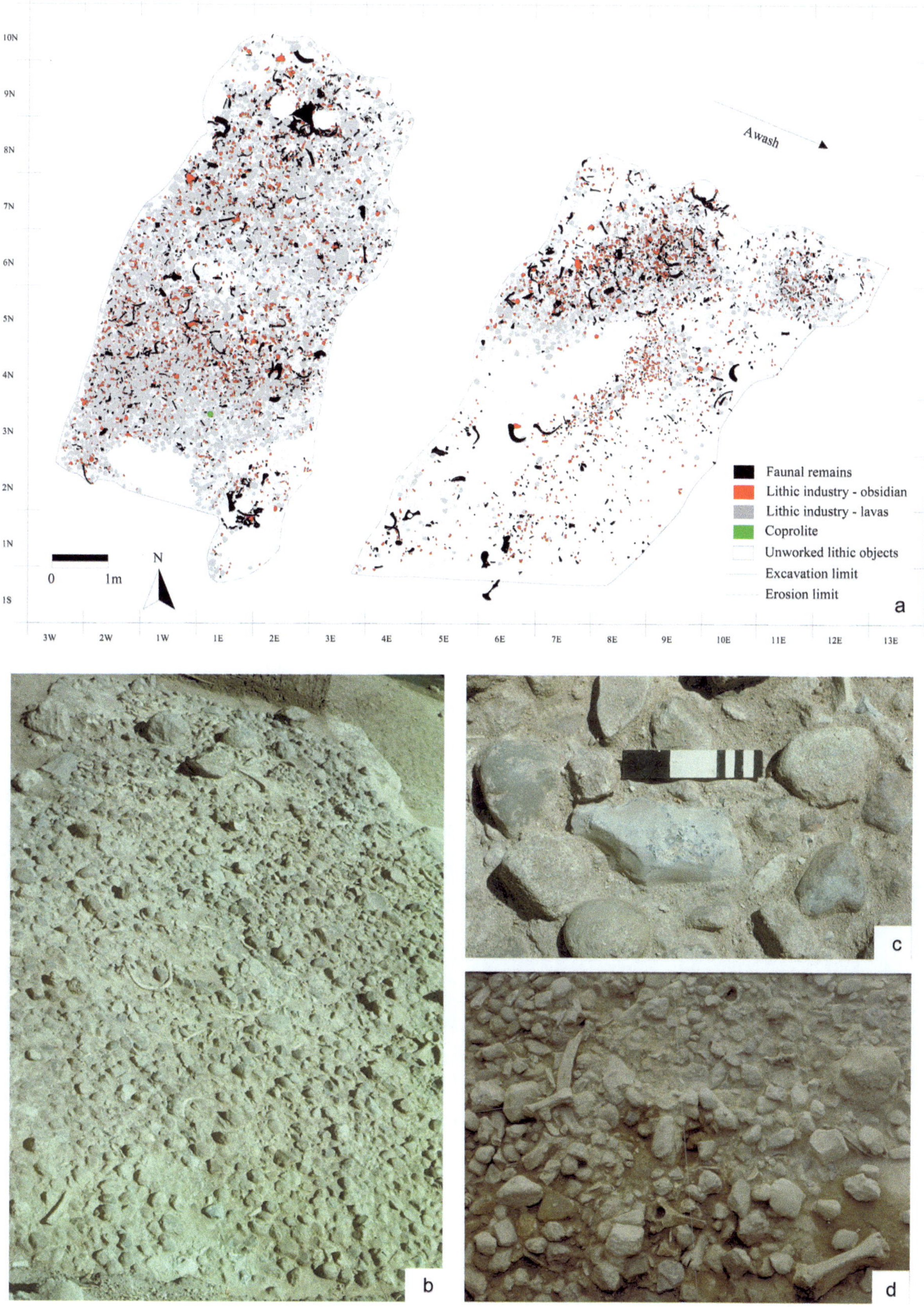

Fig. 4.5 Garba IVD. Horizontal map (**a**) and details of the excavation (**b–d**) of layer D. The middle part of the deposit is eroded

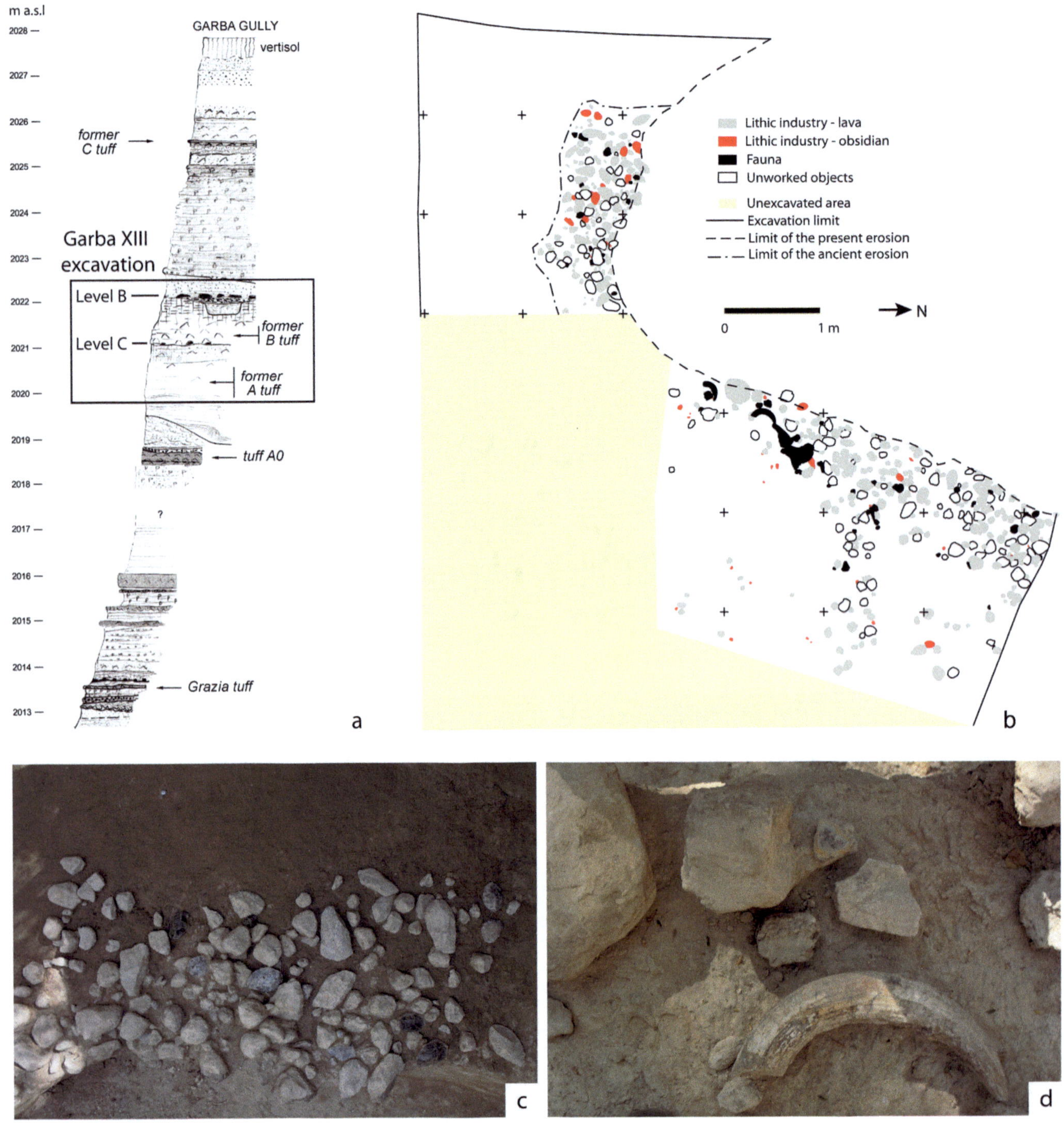

Fig. 4.6 Garba XIIIB. **a** Stratigraphic position of Garba XIII along the Garba gully. **b** Horizontal map of layer B. **c–e** Details of the excavation of layer B

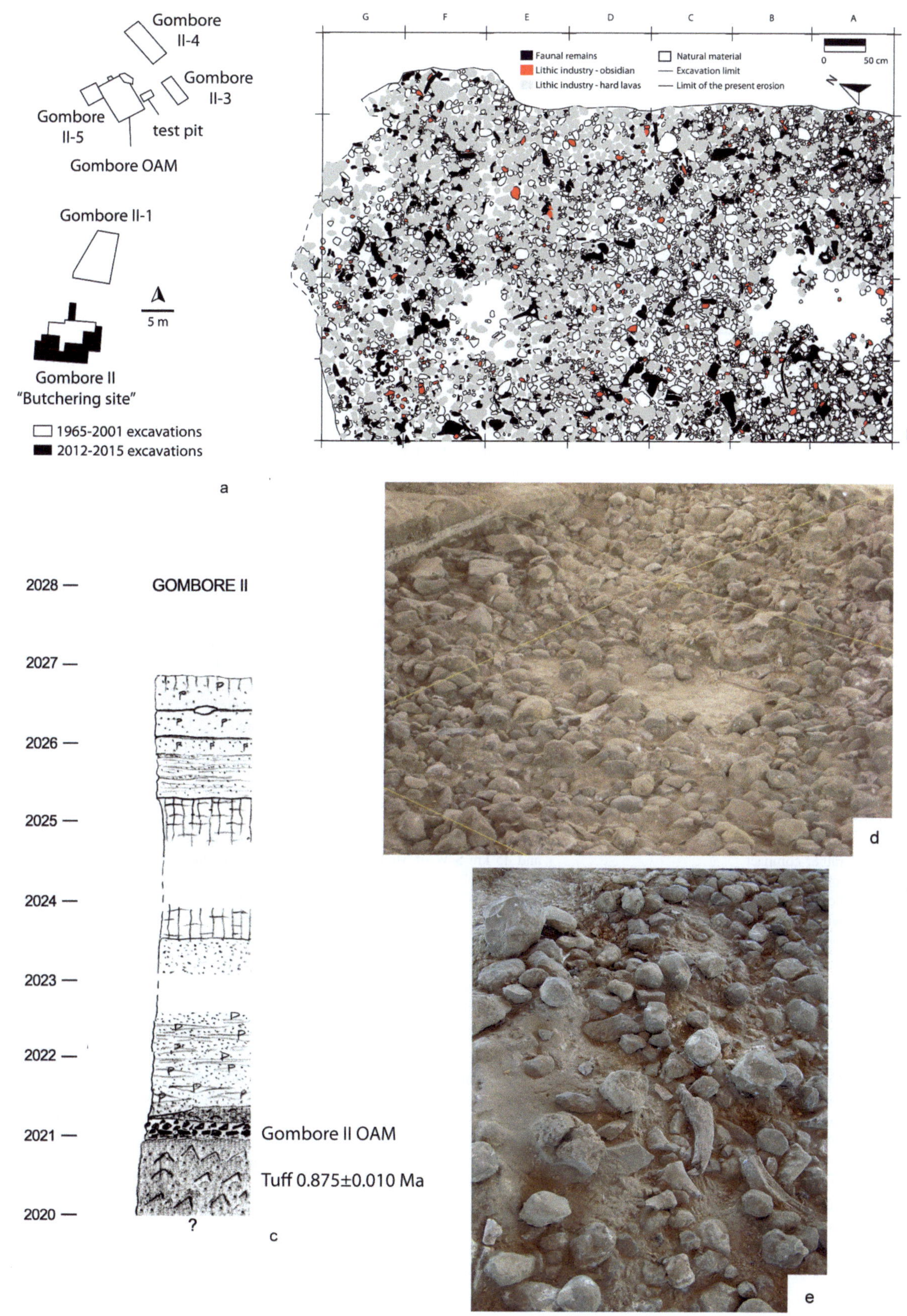

Fig. 4.7 Gombore II. **a** Excavation sectors. **b** Horizontal map of Gombore II OAM. **c** Stratigraphic position of Gombore II OAM along the Gombore II gully. **d–e** Details of the excavation of Gombore II OAM

4.4 Availability of Raw Materials

The raw materials available in the Melka Kunture region are of volcanic origin. Some aphyric, porphyritic, and micro-doleritic rocks were identified in the alluvial deposits of the Awash River and its tributaries, together with Melka Fault lava, different kinds of welded and non-welded ignimbrites and obsidian (Table 4.2; Kieffer et al. 2002, 2004; Gallotti and Mussi 2017). The lithological and morphometric composition of the Early Pleistocene alluvial deposits is very similar to that of the archaeological sites (Gallotti and Mussi 2017).

Table 4.2 Lithotypes identified in old alluvial deposits of the Melka Kunture region

Lithotype	Grain size and homogeneity	Knapping suitability		Primary sources	Secondary sources	Description
		Hard hammer	Soft hammer			
Obsidian	Very fine-grained, compact, and homogeneous	****	*****	Melka Kunture region Angular elements and fragments widely distributed across the paleolandscapes as products of erosion and redeposition from the primary source	Pebbles and cobbles in low frequencies	The obsidian color is dominantly black, but blue, green, red and beige colors have been observed. The unweathered lava is massive, black, very finely banded and breaks easily with conchoidal fracture, giving more or less translucent flakes with excellent cutting edges. Obsidian outcrops at Balchit, 7 km north from Melka Kunture sites
Aphyric to subaphyric basalts	Fine-grained, compact and homogeneous	****	**	Known source at 26 km NE and 45 km SW of Melka Kunture	Pebbles, small elements, angular elements and cobbles in high frequencies Large cobbles and blocks are exceptional	Aphiric basalts do not exhibit visible crystals. They are characterized by a compact texture formed by very fine minerals (microlites), plagioclase, augite and olivine, and by glass giving them a dark gray-blue color. They seldom have more than 50% of SiO_2, and accordingly are fairly fluid at eruption temperature
Differentiated aphyric to subaphyric lavas	Fine-grained, rather compact and homogeneous with few small crystals	***	**	Known source at 26 km N-E and 45 km S-W of Melka Kunture	Pebbles, small elements, angular elements and cobbles in medium frequencies	Aphyric and subaphiric differentiated lavas are trachytes or rhyolites (60/70% of SiO_2) with a compact gray, green or yellow fine-grained texture, more or less bright, and variable porosity sometimes including a few small crystals. At the time of eruption they were fairly viscous
Melka Fault lava	Compact facies: fine-grained, rather compact and homogeneous	***	**	Melka Kunture region	Angular elements and cobbles in high frequencies Large cobbles and blocks are not frequent	Aphyric fluidal lava, related to the limit-fault (south/southeast) of Melka Kunture Basin, is particularly abundant. This gray to blue facies resembles the benmoreite (similar to alkaline trachytes, with 62/65% of SiO_2). Notwithstanding the peculiar fluidal texture, it can be very compact, but it often displays vesicles oriented following the sense of fluidality
	Vesciculated facies: inhomogeneous and even compact	*	*			

(continued)

Table 4.2 (continued)

Lithotype	Grain size and homogeneity	Knapping suitability		Primary sources	Secondary sources	Description
		Hard hammer	Soft hammer			
Porphyritic basalts	Fine-grained, rather compact and homogeneous with large crystals	***	**	Melka Kunture region	Angular elements, cobbles and large cobbles in low frequencies	Porphyritic basalts have large crystals of up to 1 cm (phenocrystals), within a microlithic or vitreous groundmass. Poorly porphyritic to semiporphyritic basalts have some visible (1 to 3 mm) crystals (olivine and augite) within a microlithic or vitreous groundmass
Differentiated porphyritic lavas	Fine-grained, rather compact and homogeneous with large crystals	***	**	Melka Kunture region	Angular elements and cobbles in low frequencies Large cobbles and blocks exceptional	Porphyritic differentiated lavas are generally brighter and less compact. They usually have either some phenocrystals of alkaline feldspars (sanidine), or of augite and hornblende or quartz. They are usually related to trachytes or to rhyolites, and rarely to phonolites. Very viscous
Microdoleritic basalts	Very fine-grained, very compact, and homogeneous with microcrystals	****	****	Known sources 15–20 km north and south from Melka Kunture	Pebbles, small elements, angular elements and cobbles in low frequencies	Microdoleritic basalts display a fine-grained and gray texture rich in plagioclase, including augite and olivine microlites. Some basalts with a larger doleritic texture, and interstitial wides, outcrop around Melka Kunture. Their flows are easily recognizable because of the erosion which shaped them into large blocks (e.g., some kilometers northeast from Awash village)
Trachybasalt	Fine-grained, compact, and rather homogeneous with crystals	***	**	Unknown	Pebbles, small elements, angular elements and cobbles in low frequencies or absent	Trachybasalts have rare olivine crystals and they include feldspars phenocrystals (plagioclase or alkaline feldspar as anorthose) within a texture of basaltic composition, whose compactness determines the rock hardness
Trachyandesite	Fine-grained, tender with large crystals	**	*	Unknown	Pebbles, small elements, angular elements and cobbles in very low frequencies or absent	Mesocratic lava, lighter than the basalts, usually with numerous large phenocrystals of alkaline feldspaths
Welded ignimbrite n.1	Coarse-grained, inhomogeneous, and even compact	*	*	Melka Kunture region	Pebbles, small elements, angular elements and cobbles in high frequencies Few large cobbles and blocks	Welded ignimbrite, which probably originally extended over hundreds of km^2, is the more remarkable formation of this type in the area of Melka Kunture. It was produced during a major peralkaline rhyolitic eruption, and moved like an aerosol before releasing lava elements which were welded when gas energy diminished

(continued)

Table 4.2 (continued)

Lithotype	Grain size and homogeneity	Knapping suitability		Primary sources	Secondary sources	Description
		Hard hammer	Soft hammer			
Other types of ignimbrites and ignimbritic tuff	Coarse-/fine-grained, inhomogeneous and even compact	*	*	Melka Kunture region	Pebbles, small elements, angular elements and cobbles in very low frequencies or absent	Thirteen types of welded ignimbrites with various facies and lithological characters have been identified in the ancient alluviums of Simbiro along with some ignimbritic tuff. They all refer to welded or non-welded pyroclastic flows emitted during peralkaline rhyolitic (pantelleritic) syn-rift eruptions
Obsidian lavas	Fine-grained, homogeneous and more or less compact	***	***	Melka Kunture region	Pebbles, small elements, angular elements and cobbles in low frequencies or absent Very few blocks	Distinct from real obsidian, as Balchit obsidian. They were produced by quick cooling, as at the base of some welded ignimbrites. Vitrification is more or less developed, and accordingly the rock is more or less compact, following local features, notably the thickness of cooling and the presence/absence of water
Syenitic inclusions	Very fine-grained, compact and homogeneous	*****	*****	Unknown	Exceptional small elements and pebbles	They were produced in a magma chamber where crystallized minerals had accumulated before being brought up, dislocated and thrown out of the volcano during a major eruption which also expelled the magma. They are mainly constituted by alkaline feldspars and they document the trachytic, phonolithic or rhyolitic composition of the magma
Amorphic silica	Very fine-grained, compact and homogeneous	*****	*****	Unknown	Exceptional small elements and pebbles	Amorphic silica material (flint and opale), usually classified as sedimentary rocks, in the Melka Kunture Basin is the outcome of hydrothermal circulations and of amorphic silica precipitation (silcrete)

The closest known primary source of obsidian is Balchit, 7 km north of the Awash (Fig. 4.1b). Balchit—the name means "obsidian" in Amharic—is a flat dome flow aged 4.37 ± 0.07 Ma (Chernet et al. 1998), that outcrops today over an area of about 4 km^2 (Salvi et al. 2011). This obsidian flow can be seen in situ on the Jimjima plateau, and near a village that is likewise named Balchit (Fig. 4.8a). Pure and massive obsidian amygdales up to 1 m long are scattered among the weathered rocks (Fig. 4.8b, c). This obsidian is mainly black. It breaks easily (the fracture is conchoidal), producing more or less translucent flakes with excellent cutting edges. The analysis of 12 obsidian samples from Balchit, two obsidian samples from alluvial deposits (Poupeau et al. 2004; Le Bourdonnec 2007) and 10 obsidian artifacts from Gombore I, Gombore II and Garba IV (Negash et al. 2006) showed that they all had the same chemical composition. Angular pieces of obsidian eroded from the primary source lie on the ground near the outcrop (Fig. 4.9a, b). Rounded pieces, i.e., pebbles and cobbles, have been found in small percentages in paleochannels (Fig. 4.9c); blocks are rare (Fig. 4.9d; Piperno et al. 2009; Salvi et al. 2011).

Obsidian lava, which differs from true obsidian, was produced by quick cooling, as at the base of some welded ignimbrites. Vitrification is more or less developed, and accordingly the rock is fine-grained and more or less compact. Obsidian lava is sparsely represented in ancient alluviums; the most frequent forms are small to large cobbles (Gallotti and Mussi 2017).

Welded ignimbrite cobbles, angular pieces, and blocks are widely distributed in the ancient alluvial deposits. This type of rock is easily recognizable all over the area. It was produced during a major eruption and probably originally extended over hundreds of km^2 (Fig. 4.8d–e; Kieffer et al. 2002, 2004). It is not suitable for knapping and was never used by hominins in lithic percussion (Gallotti and Mussi 2017).

Aphyric rocks—fine-grained, homogeneous, and compact, with few or no small crystals—are abundantly available. They can be found in alluvial deposits, mainly in the form of medium-sized angular, elongated and spherical cobbles; large forms are rare. Though they are abundant in secondary sources, no primary sources have yet been discovered close to the archaeological sites (Kieffer et al. 2002, 2004; Gallotti and Mussi 2017). The closest primary sources are in the many pre-rift flows of the Addis Ababa and Guraghe–Anchar basalts, respectively, 26 km NE and 45 km SW of Melka Kunture (Abebe et al. 2005).

Conversely, Melka Fault lava is related to the limit-fault (SSE) of the Melka Kunture Basin. A so-called lava lake structure outcrops 500 m upstream from the places where the Atebella and Balchit creeks flow into the Awash (Figs. 4.1b, 4.8f). Accordingly, it was abundant in the paleolandscape, mainly in a vesciculated facies, less frequently in a compact facies (Kieffer et al. 2002, 2004). Despite its typical fluidic texture, this lava is very compact, but it often contains oriented vesicles (vesciculated facies). In the paleochannels, the compact facies is less frequent than the vesciculated one, but it is the only one that was used by knappers. It is available mainly as angular cobbles; large forms (mainly blocks) are more abundant than is the case with other aphyric rocks (Gallotti and Mussi 2017).

Porphyritic rocks—fine-grained, rather compact and homogeneous, but containing large crystals—are found in low percentages in the alluvial deposits, mainly in the form of medium-sized and large cobbles. In the Wutale area, some 2.5 km southwest of the Garba gully (Fig. 4.1b), porphyritic and vesicular lavas with large plagioclase crystals have been observed in primary position (Raynal and Kieffer 2004). Porphyritic lava also outcrops in the Boti and Guraghe volcanic areas west and south of Melka Kunture.

Microdoleritic basalts are very fine-grained, very compact and homogeneous rocks containing microcrystals. The outcrops are 15–20 km north and south of Melka Kunture, where they are fragmented by erosion into very large blocks and are easily recognizable (Fig. 4.8g). Microdoleritic basalts are found in very low numbers in secondary sources, mainly in small or medium-sized rounded forms (Kieffer et al. 2002, 2004; Gallotti et al. 2010; Gallotti and Mussi 2017). The other rocks listed in Table 4.2 are only rarely discovered.

4.5 Techno-economic Behaviors

4.5.1 *Production of Small- and Medium-Sized Flakes at Garba IVE-F, ~1.7 Ma*

Knapping activities documented in levels E-F at Garba IV focused solely on the production of small or medium-sized flakes (Table 4.1). All stages of the chaînes opératoires are present in both levels. Our comparative analysis of the Garba IVE and Garba IVF series showed very similar technical patterns. Accordingly, the two series are analyzed jointly with no stratigraphic subdivision (Gallotti and Mussi 2015).

Obsidian, present in low percentages in unworked material, accounts for 79.4% of the artifacts in level E and 86.7% in level F (Fig. 4.10a); this indicates that knappers selected it intentionally. Lavas suitable for knapping are abundant in the unworked assemblage (Fig. 4.11), but obsidian pebbles and cobbles are the most frequently exploited lithic resources in both levels E and F.

The cores exploited for small- to medium-sized flake production can be grouped in two technologically significant sets. The first set corresponds to a type of débitage described by Delagnes and Roche (2005) as "simple"; it consists of items that have few unorganized flake scars. The second set, which is quantitatively and qualitatively more significant, comprises cores that show organized and structured exploitation patterns (Delagnes and Roche 2005). Omitting core fragments, the flake-to-core ratio[2] is 5.7:1, which is not consistent with the average number of removal scars observed on the cores. Moreover, most of these cores were intensively exploited, and most of the flakes have a large number of removal scars on their dorsal face, which suggests that the count of detached pieces does not reflect the number of flakes that were originally obtained from the cores.

Five organized débitage methods have been identified in this level (Gallotti and Mussi 2015; Fig. 4.10b):

[2]Flakes include whole, broken, and retouched flakes of small-medium dimensions, i.e., with length or width <10 cm.

Fig. 4.8 **a** Obsidian outcrop at Balchit. **b–c** Obsidian blocks among weathered rocks. **d** Welded ignimbrite at Atebella. **e** Non-welded ignimbrite overlying welded ignimbrite with large columnar jointing. **f** Lava lake structure 500 m downstream from the confluence of the Atebella and Balchit creeks. **g** Large blocks of microdoleritic basalt scattered on the ground 15 km north of the Melka Kunture area

- Unifacial unidirectional. The flaked surface is the convex face of the cobble and is located either on the transversal or longitudinal plane of elongated and spherical cobbles, or on the sagittal plane of flat cobbles. A flaked surface on the sagittal/longitudinal plane (i.e., on the blank's longest natural convex face) strongly suggests an intention to produce elongated products. The striking platform is a flat natural surface or a platform rectified by removals to create a suitable angle (Fig. 4.12: A–E). The corresponding flakes show a unidirectional removal-scar pattern. These flakes are more elongated than the other whole flakes and show a different

Fig. 4.9 **a–b** Angular pieces of obsidian lying on the ground near the outcrop. **c** Obsidian pebbles and cobbles in an ancient alluvial deposit. **d** Obsidian block in an ancient alluvial deposit

flake-scar pattern on the dorsal face. Although several flakes show a backed lateral and/or distal edge, long portions of the edges are sharp and suitable for cutting (Fig. 4.12: F–J).

- Peripheral unidirectional. A peripheral flaking surface exploits the whole or nearly whole thickness of the cobbles by detaching several generations of flakes, resulting in short and wide/thick cores. The natural striking platform corresponds to the flat surface of plano-convex cobbles or to a surface rectified by few removals (Fig. 4.12: K). The angle of interaction between the striking and flaking surfaces is abrupt. Flakes produced from these cores show technical patterns similar to those of flakes produced from unifacial unidirectional cores.
- Partial bifacial. This method exploits the core by unidirectional removals on two adjacent surfaces. Each removal scar is used alternatively as a striking platform to flake the adjacent plane. The resulting flakes are short and wide and have plain or dihedral butts (Fig. 4.12: L).
- Centripetal. Centripetal/tangential exploitation is usually performed on one flaking surface from a cortical peripheral platform (Fig. 4.12: O) or from a striking platform rectified by a few removals (Fig. 4.12: M). There is no management of volume or convexity configurations, no recurrence or preparation, and no hierarchy. This method seeks to achieve the best solution for exploiting (sub)spherical cobbles. Only one core shows bifacial centripetal exploitation of a blank consisting of a thick flake (Fig. 4.12: N). The flakes produced by this method are circular, triangular or sub-quadrangular with centripetal or tangential removals on the dorsal face. The butts are natural, plain, dihedral or rarely facetted, wide and thick, and the flaking angle is generally obtuse (Fig. 4.12: P–Q).

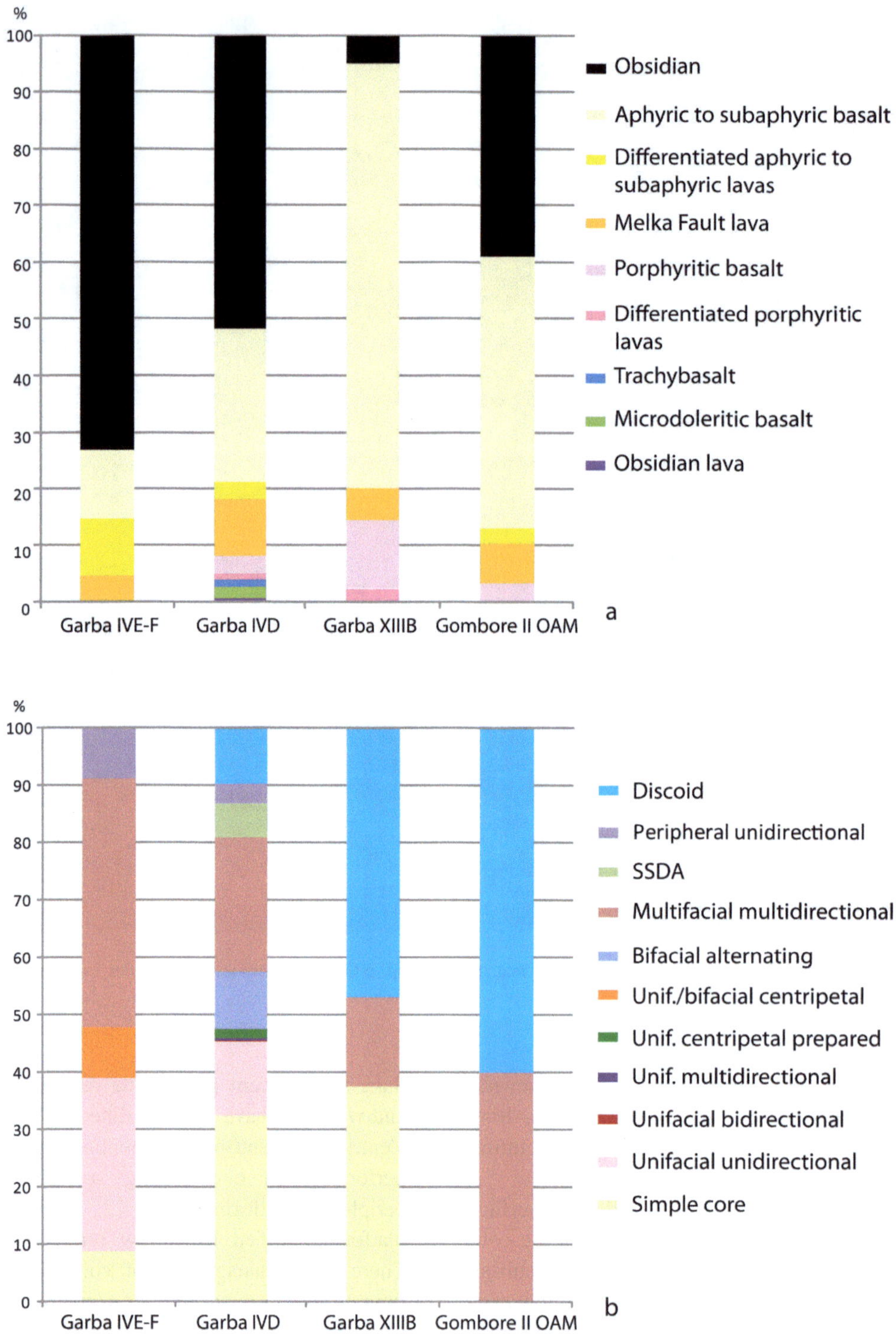

Fig. 4.10 **a** Lithological composition of the small- and medium-sized flake tools. **b** Frequency of the small débitage methods identified in the assemblages studied

- Multifacial multidirectional irregular. This method was used on most of the cores. The core surfaces were alternately flaked with multidirectional removals with no clear organization of the reduction process (Fig. 4.12: R–S). Nevertheless, some cores attest to the knapper's attempt to maintain an orthogonal organization during flaking (Fig. 4.12: U). Moreover, a few cores have long unidirectional flake scars on their major flaked surfaces, which were exploited at the beginning of the reduction process, and multidirectional removals on their other faces, which correspond to the final core reduction phase, when the main flaking surfaces were probably no longer functional (Fig. 4.12: T). No specific platform preparation was carried out, because each removal scar served as

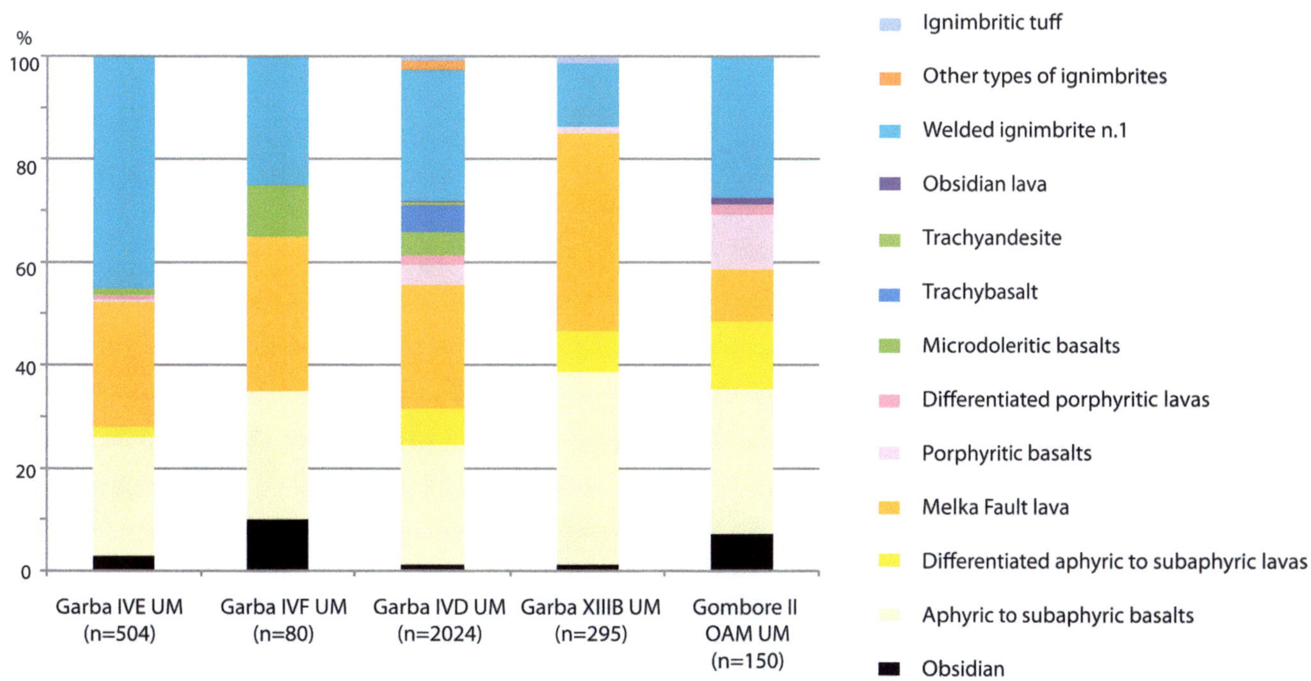

Fig. 4.11 Lithological composition of the unworked materials (UM) found at the analyzed sites

a striking platform for the next removal on a secant face. The flakes thus produced have various shapes. They are frequently short, with thick, asymmetrical cross-sections. The multiple removal scars on the dorsal face have irregular or perpendicular directions (Fig. 4.12: V–X, b–d). A few flakes present one or two series of unidirectional flake scars together with multidirectional removal scars, which are correlated with the multifacial multidirectional cores with one preferred unidirectional flaking surface (Fig. 4.12: e, Y). The percentage of core-edge flakes is high, which implies continuous rotation of the flaking surfaces (Fig. 4.12: Z–a, f).

A relatively large percentage of the whole obsidian flakes are retouched (31%). They can be grouped in two sets. The first one, identified only in level E, consists of flakes (n = 32) whose edges were modified by continuous retouch, ranging from marginal to invasive, that modified one or two edges. The resulting tools display considerable dimensional and morphological variability (Gallotti and Mussi 2015). Most blanks are first flakes (opening flakes), or flakes with natural residual parts on the dorsal face (Fig. 4.13: L, N), or core-edge flakes (Fig. 4.13: M).

The second set (41 tools) displays a retouch process aimed specifically at producing a small point by modifying the distal part of the blank. The percussion axis may or may not correspond to the morphological axis. The point is produced in different ways: (a) by two or more notches on two convergent edges (Fig. 4.13: B, C, F, I, J); (b) by one or more notches opposite a retouched edge (Fig. 4.13: A, K); (c) by one or more notches opposite a (natural) back (Fig. 4.13: G, H); (d) by a retouched edge opposite a back (Fig. 4.13: E); or (e) by a convergent side-scraper (Fig. 4.13: D).

Accordingly, what we are seeing here is some degree of standardization. By the term "standard product" we mean a product that conforms to specifications resulting from the same technical requirements (Daniel and Lapedes 1978; Gallotti and Mussi 2015). In the small pointed tools found at Garba IVE-F, standardization is expressed by (a) the repetitive aim to shape the distal portion of the flake into a tip; (b) the repetitive aim to create a convergence, either by modifying both edges or by modifying one edge and using the technical properties of the other edge; and (c) recurrent efforts on the knappers' part to achieve a small and homogeneous size.

Intentional behavior is further proved by (a) hundreds of whole small flakes (<1 cm) found in the same levels but clearly distinct from natural or knapped fragments of the same size; (b) the lack of any bipolar technique on anvil, which is often responsible for pseudo-retouching (de la Torre and Mora 2005; Zaidner 2013); (c) the similar technical marks left by retouch on both lava and obsidian flakes in more recent Melka Kunture assemblages (Gallotti 2013); (d) the lack of any evidence of bioturbation or trampling in the deposit (Raynal et al. 2004).

In brief, the unusually high number of retouched and often pointed tools found at Garba IVE-F is not the outcome of any natural process that altered their edges (Gallotti and Mussi 2015).

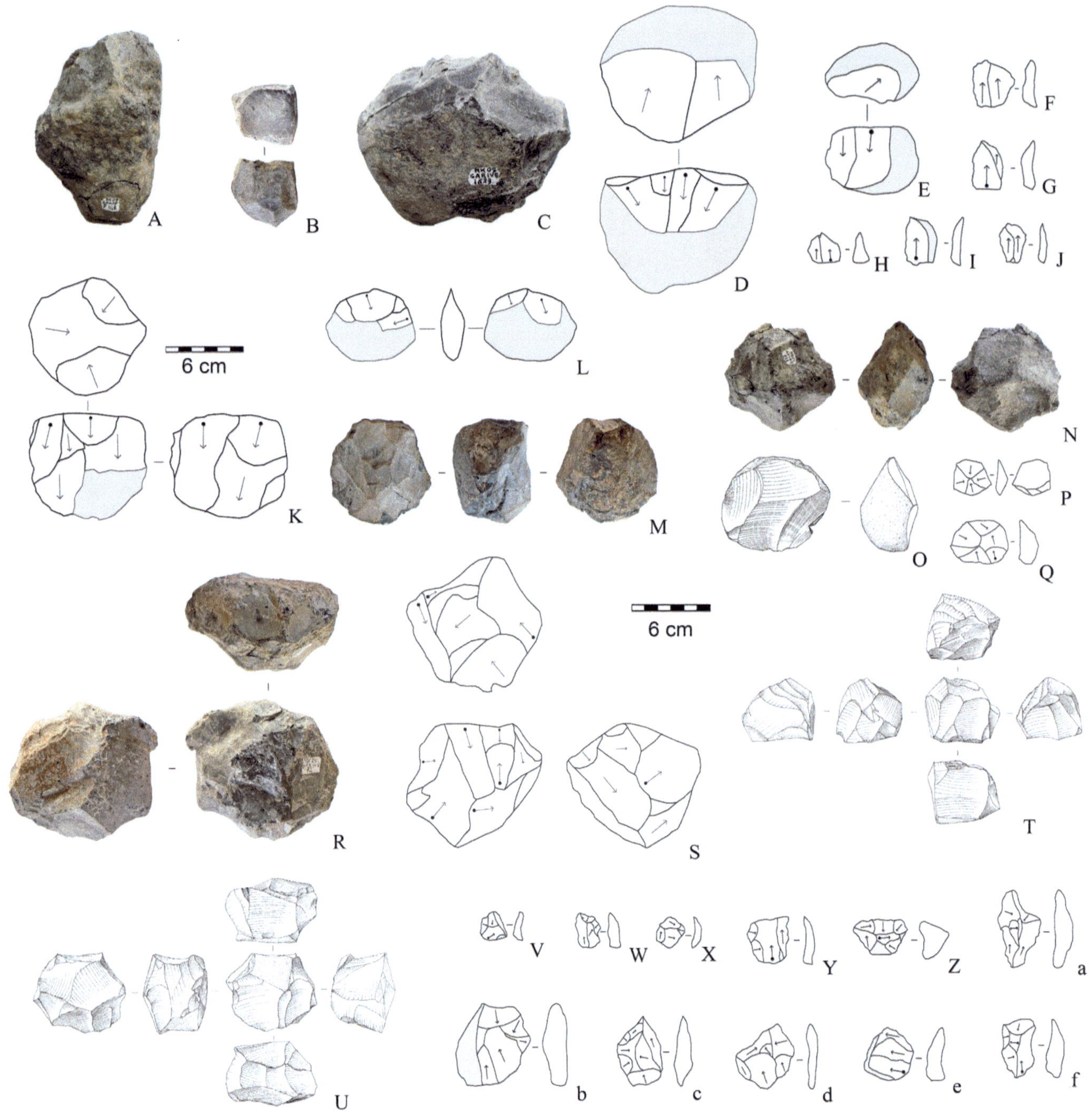

Fig. 4.12 Garba IVE-F. **A–E**: unifacial unidirectional cores (**A–C**, **E**: OBS; **D**: MFL). **F–J**: flakes with unidirectional removal scars on the dorsal face (OBS). **K**: peripheral unidirectional core (OBS). **L**: partial bifacial core (MFL). **M–O**: centripetal/tangential cores (OBS). **P**, **Q**: flakes with centripetal/tangential removal scars on the dorsal face (OBS). **R–S**: irregular multifacial multidirectional cores (OBS). **T**: multifacial multidirectional cores with major unidirectional flaking surfaces (OBS). **U**: orthogonal multifacial multidirectional core (OBS). **V**, **W**, **b**, **c**: flakes with irregular multidirectional removal scars on the dorsal face (**b**: ASB; **V**, **W**, **c**: OBS). **X–Y**, **d–e**: flakes with orthogonal removal scars on the dorsal face (OBS). **Z–a**, **f**: core-edge flakes with multidirectional removals (OBS). ASB: aphyric to subaphyric basalt. MFL: Melka Fault lava. OBS: obsidian. **O**, **T–U**: drawings by M. Pennacchioni

4.5.2 Production of Small- and Medium-Sized Flakes at Garba IVD, ~1.6 Ma

All the steps in the small débitage chaînes opératoires, representing the main knapping activities, are documented at Garba IVD. With a flake-to-core ratio of 2:1, the number of flakes is very low compared to the number of negative scars on both cores and flakes. As in the case of Garba IVE-F, the flakes correspond to all the flaking stages and methods identified in core analysis. Accordingly, the flake deficit is the outcome of post-depositional processes (Figs. 4.14, 4.15; Raynal et al. 2004; Gallotti 2013).

The most exploited raw materials are obsidian and aphyric rocks (Fig. 4.10a). Aphyric rocks are also abundant in the unworked assemblage, while obsidian, as noted above, is very scarce (Fig. 4.11). Medium-sized angular forms are the most frequently exploited blanks. This is clearly shown by a certain number of first flakes whose thick triangular cross-section is produced by two flat faces of the natural blank (Fig. 4.15: A). Medium-sized elongated and spherical obsidian cobbles were also used frequently. Conversely, in the available unworked assemblage from the site, obsidian cobbles are present only in low percentages, which shows that the available obsidian was exploited intensely. Knappers probably also collected this raw material in the nearby alluvial deposits. They also flaked angular pieces of obsidian, which are not found in the alluvial deposits but are abundantly scattered on the ground near the Balchit outcrop (Gallotti and Mussi 2015; Fig. 4.1b). Therefore, the knappers must have collected angular obsidian rocks at a distance of 5–7 km from the site, or maybe from closer by if the primary source was more extended ~1.6 Ma than it is today, and/or was less covered by the subsequent accumulation of fluviolacustrine and volcanic deposits (Gallotti 2013).

Small débitage is the outcome of several exploitation methods. The ones identified at Garba IVD are the following: simple, unifacial uni/bi/multidirectional, peripheral unidirectional, alternating bifacial, multifacial multidirectional, alternating débitage surface system, or SSDA (Système par Surface de Débitage Alternée; Forestier 1992, 1993), prepared unifacial centripetal and discoid[3] (Figures 4.10b, 4.14; Gallotti 2013).

The choice of a specific method was influenced by the geometry of the raw materials and by the knappers' technical purposes. The peripheral unidirectional, SSDA, and multifacial multidirectional irregular exploitation methods all suggest that the knappers found solutions adapted to the cobbles' natural morphology, in order to maximize the production of flakes without pursuing any preferred exploitation pattern.

A peripheral flaking surface makes it possible to exploit nearly the whole thickness of plano-convex cobbles by means of unidirectional removals, thereby producing short and wide/thick cores. A flat surface on the cobble is used as striking platform (Fig. 4.14: J). The angle of interaction between the striking and flaking surfaces is abrupt. The number of detached flakes is high (average: 17), because knappers exploited the same surface again and again.

The SSDA cores found here, all of which are of lava rocks, have three to four flaked surfaces. They are produced by unidirectional removals from angular cobbles that have both flat and convex surfaces. During flaking, the cobble is rotated continuously, in order to follow the orthogonal angles from one face to the next. The first striking platform is a natural face; thereafter, each flaked surface becomes the striking platform of the adjacent face (Fig. 4.14: M). The flakes display unidirectional removals on adjacent orthogonal faces that are, respectively, the dorsal face and the distal face (Fig. 4.15: K), or lateral edges.

In several cases, edge-core flakes indicate that the knapper was seeking new angles so as be able to continue to exploit exhausted irregular multifacial multidirectional SSDA cores (Fig. 4.15: I, L).

Most cores are irregular, multifacial and multidirectional. The flakes produced from them vary in shape; they are frequently short with thick, asymmetrical cross-sections (Fig. 4.14: N, P; Fig. 4.15: D, F–G). Many are core-edge flakes produced with continuous rotation of the core. The removals sometimes follow orthogonal directions (Fig. 4.15: J). The cores are typically small and overexploited. The largest ones (n = 14) have a major flaked surface, displaying long unidirectional flake scars, that was exploited at the beginning of the reduction process. The other core faces bear evidence of the multidirectional removals made in the final reduction phase (Fig. 4.14: O). On the dorsal face of the resulting flakes there are one or two series of unidirectional flake scars and multidirectional removal scars (Fig. 4.15: E, H).

Cores that show evidence of unifacial unidirectional exploitation are mainly on elongated cobbles. In a few cases, the removals are bi- and multidirectional (Fig. 4.14: H). The knappers chose the longest available convex face to produce elongated flakes (Fig. 4.14: A–F; Fig. 4.15: B–C). In elongated cobbles, the striking platform had to be rectified in order to create a suitable angle and begin the débitage. With angular cobbles, the flaked surface is the convex face of the cobble, and a flat natural surface is used as striking platform, with no rectification. Sometimes such cores were also retouched, and in these rare cases a continuous uni- or bifacial retouch is located on the transversal edge of a plano-convex cobble at an angle close to 75°. This

[3]The term discoid corresponds to the discoid concept, as defined by Boëda 1993.

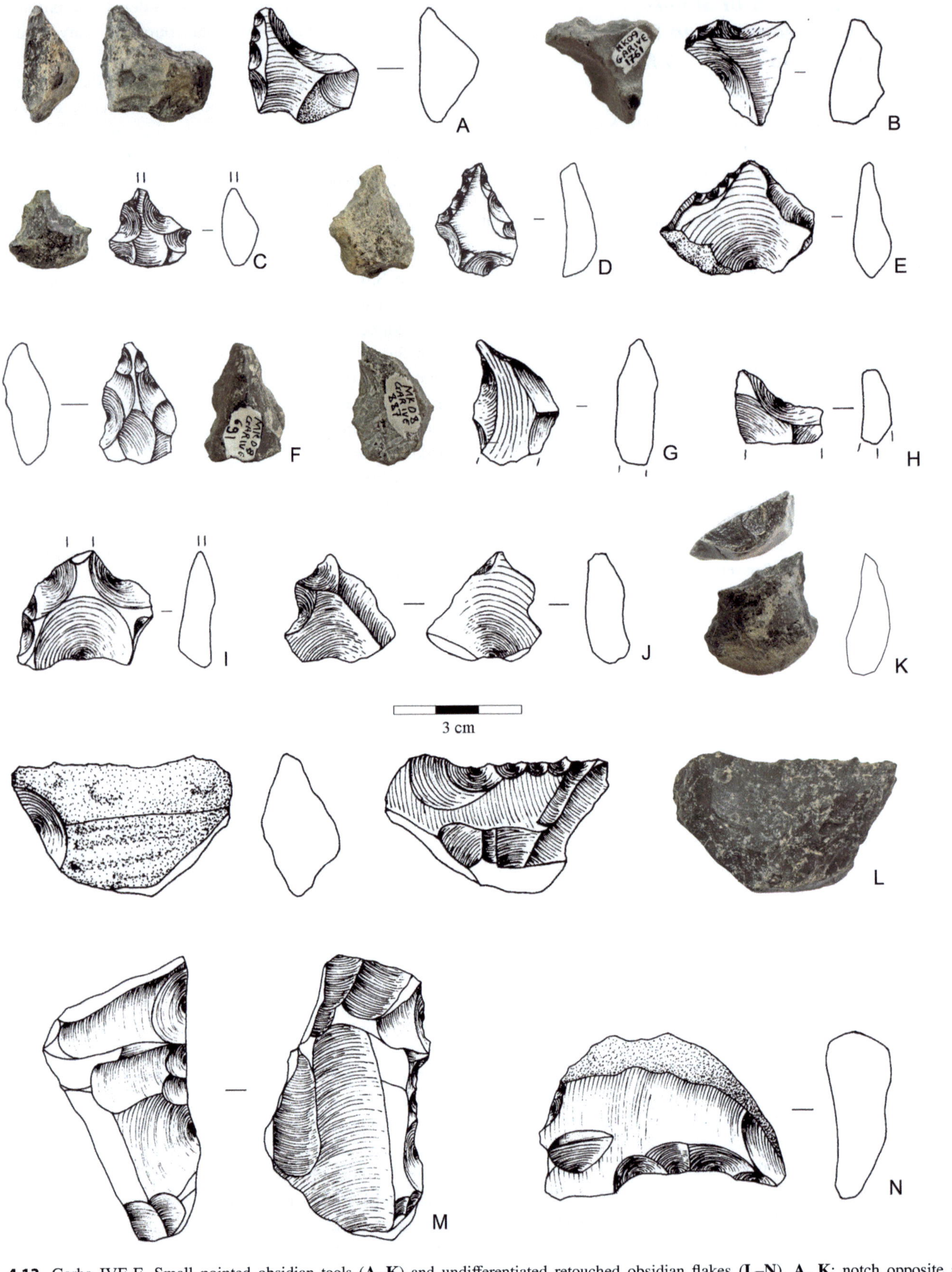

Fig. 4.13 Garba IVE-F. Small pointed obsidian tools (**A–K**) and undifferentiated retouched obsidian flakes (**L–N**). **A**, **K**: notch opposite a retouched edge. **B**, **C**, **F**, **I**, **J**: two or more notches on two convergent edges. **D**: convergent side-scraper. **E**: retouched edge opposite a back. **G**, **H**: notch opposite a back. **L**: transverse side-scraper. **M**: lateral side-scraper on core-edge flake. **N**: retouched proximal notch. Drawings by N. Tomei

corresponds to the last phase of reduction, as evidenced by scars that are altogether different from the ones produced in the débitage process (Fig. 4.14: G).

Biconvex and flat cobbles were used for partial bifacial exploitation, and exhibit unidirectional removals on two adjacent surfaces. Each removal scar was used as a striking platform to flake the adjacent plane (Fig. 4.14: I).

Unifacial prepared centripetal method required the selection of round cobbles or of angular cobbles that have one convex surface (Fig. 4.14: K, L). This débitage method was used only to exploit lava cores. The volume of these cores is divided into two asymmetrical hierarchical surfaces. One is the flaking surface, which was exploited by repeated centripetal or tangential removals. The other one is a peripheral or semi-peripheral striking platform perpendicular to the flaking surface; it was prepared by unidirectional removals. Angular flat surfaces were chosen as striking platforms, as appears clearly when parts of the cobbles' natural surfaces are preserved. The flaking surface is generally secant to the plane of intersection of both surfaces. Conversely, the flaking surface is parallel and flat when the centripetal exploitation continues, detaching second-generation flakes. The role of the surfaces is maintained throughout the reduction sequence.

Discoid cores too are mainly produced on round cobbles, which are ideal blanks for meeting the technical requirements of the discoid concept. Discoid cores have two convex surfaces that are usually asymmetrical, i.e., the flaking surface and the striking platform, whether hierarchized or not. Flakes are detached according to a plane that is secant to the plane of intersection of the two surfaces, or sub-parallel when the core becomes overexploited. The striking platform may be a natural surface, or the convex dorsal face of a flake, or a prepared peripheral surface, or two fairly similar symmetrical surfaces. The latter are created by removals and can be used either simultaneously or, during alternating series of removals, one as a striking platform and the other as a flaking surface (Boëda 1993; Jaubert and Mourre 1996; Mourre 2003; Terradas 2003). For the most part, peripheral convexity is created by centripetal flaking and tangential removals (Fig. 4.14: Q–V). The flakes are mostly short and wide with centripetal/tangential removals on the dorsal face, with wide and thick plain, dihedral, or facetted butts, with at least one cutting edge and frequently with a pointed distal-lateral end (Fig. 4.15: M–R).

However, if the geometry of the raw material was not suitable for discoid exploitation, as in angular obsidian elements, the knappers prepared an initial configuration that modified the original geometry of the natural blanks (Gallotti 2013).

Five percent of the flakes were retouched as side-scrapers, denticulates, and notches. No specific correlation is observed between flake blank types and retouch systems on the one hand, and the choice of a specific flaking method on the other. Retouch was usually applied to only one edge, either lateral (Fig. 4.16: A–C) or transversal (Fig. 4.16: D–H), and only rarely on two edges (Fig. 4.16: I–M). It is sometimes convergent (Fig. 4.16: N–Q) or skewed (Fig. 4.16: R–U), so as to produce pointed shapes. Retouch scars are marginal, abrupt, denticulate or invasive, producing a linear, convex or concave cutting edge. Notches are frequent; sometimes they are retouched (Fig. 4.16: X) and sometimes they are not (Fig. 4.16: B, V–W, Y–Z). They are often associated with a retouched edge (Fig. 4.16: B, W, Y). However, the retouch never modifies the shape of the blank. No standardization develops.

4.5.3 Production of Small- and Medium-Sized Flakes at Garba XIIIB, ~1.0 Ma

Small and medium-sized flakes were produced at Garba XIIIB mainly on cobbles of aphyric basalt which is found in high percentages in the unworked assemblage (Figs. 4.10a, 4.11).

The core-to-flake ratio is 1.8 flakes per core, which does not match the average number of removal scars on the cores (average: 22). Furthermore, the flakes do not represent all the flaking stages identified by core analysis. Notably, the first flaking stage is represented by only two first flakes, and a few more that preserve small portions of the natural surfaces.

Twelve cores are characterized by a low count of removal scars (distributed at random on distinct faces), which indicates simple débitage. The other cores were exploited through two débitage methods: discoid and multifacial multidirectional irregular (Fig. 4.10b).

Discoid technology is the prevailing débitage method (Fig. 4.10b). The exploitation is bifacial (Fig. 4.17: G): two fairly similar symmetrical surfaces created by removals were used as striking platforms and flaking surfaces, either simultaneously or by alternate series of removals. Both cobbles and flakes were used as blanks. The natural surfaces of the original blank, when preserved, show that the knappers either took advantage of any significant convexity to perform removals (Defleur and Crégut-Bonnoure 1995; Moncel 1998; Pasty 2000; Terradas 2003), or that they configured the core's initial geometry by removing its natural angles. The flakes produced by this method are either (a) circular, triangular or sub-quadrangular, with centripetal removals (Fig. 4.17: F), or (b) triangular or irregular with tangential removals (Fig. 4.17: D–E, H), i.e., flakes with a pointed distal-lateral end and pseudo-Levallois points. The butts are mostly plain, dihedral or facetted, wide and thick. The flaking angle is generally obtuse. Few flakes have a cutting edge opposed to a prepared or natural back.

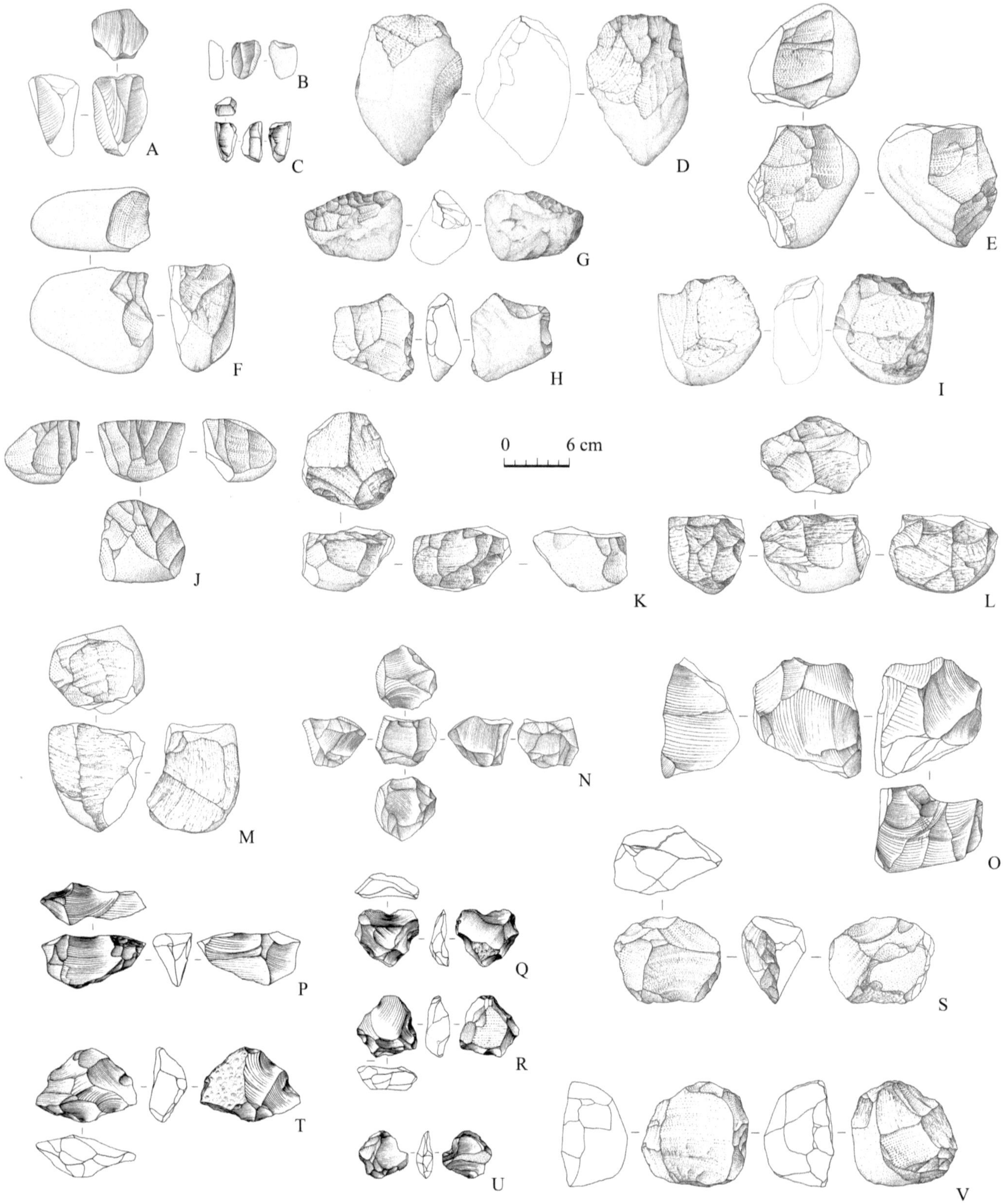

Fig. 4.14 Garba IVD. **A–F**: unifacial unidirectional cores. Core **E** is one of the few specimens that was exploited by bipolar technique (**A–C**: OBS; **D**: DPL; **E–F**: ASB). **G**: retouched unifacial unidirectional core (ASB). **H**: retouched unifacial bidirectional core (ASB). **I**: bifacial partial core (DASL). **J**: peripheral unidirectional core (OL). **K, L**: prepared unifacial centripetal cores (ASB). **M**: SSDA core (ASB). **N**: irregular multifacial multidirectional core (OBS). **O, P**: irregular multifacial multidirectional cores that have a major flaked surface showing unidirectional removals (OBS). **Q–V**: discoid cores (**Q–R**, **T–U**: OBS; **S**, **V**: ASB). ASB: aphyric to subaphyric basalts. DASL: differentiated aphyric to subaphyric lavas. DPL: differentiated porphyritic lavas. OBS: obsidian. OL: obsidian lava (after Gallotti 2013, modified). Drawings by M. Pennacchioni

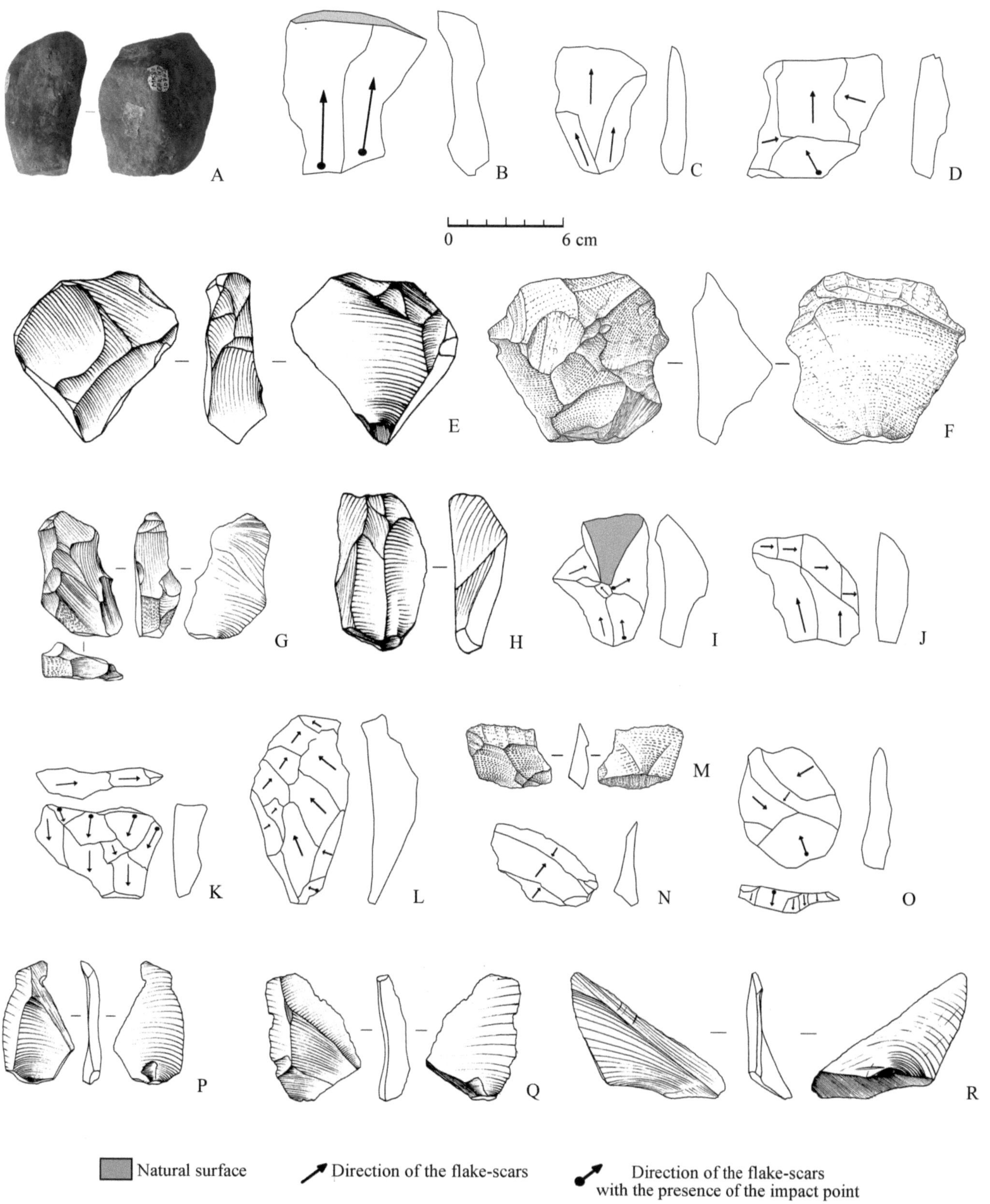

Fig. 4.15 Garba IVD. **A**: first flake (MFL). **B**, **C**: flakes with unidirectional flake scars on the dorsal face (**B**: PB; **C**: DASL). **D–H**: flakes with multidirectional removal scars on the dorsal face (**D**, **F**: ASB; **E**, **G–H**: OBS). **I**, **K**, **L**: flakes with unidirectional removals on orthogonal surfaces (**I**: ASB; **K**: MFL; **L**: PB). **J**: flake with orthogonal removal scars on the dorsal face (PB). **M–R**: flakes with centripetal or tangential flake scars on the dorsal face (**M**, **O**: MFL; **N**, **P–R**: OBS). ASB: aphyric to subaphyric basalts. DASL: differentiated aphyric to subaphyric lavas. MFL: Melka Fault Lava. OBS: obsidian. PB: porphyritic basalt (after Gallotti 2013). **E–H**, **M**, **P–R**: drawings by M. Pennacchioni

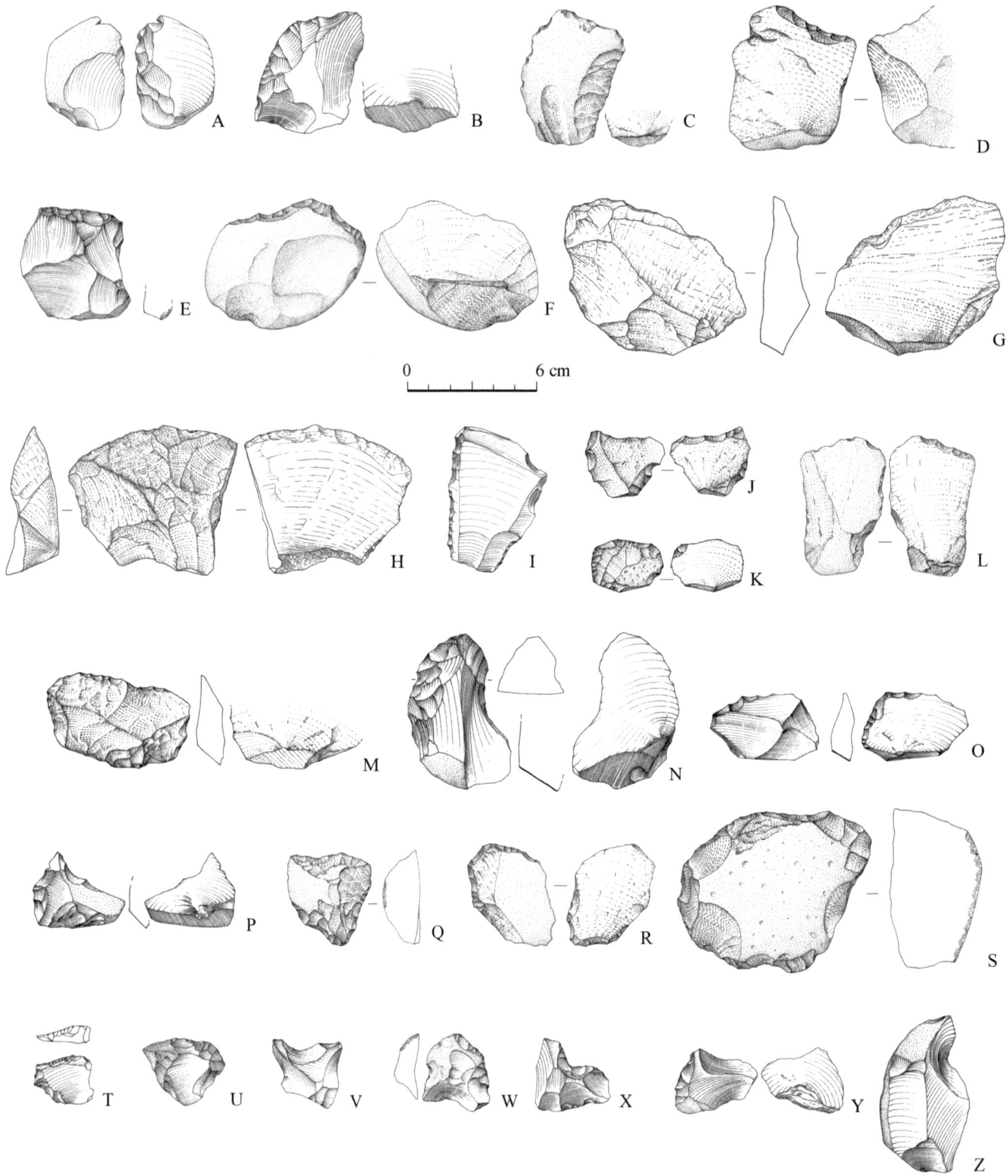

Fig. 4.16 Garba IVD. Retouched flakes. **A**, **C**: lateral side-scrapers (**A**: OBS; **C**: DASL). **B**, **W**, **Y**: lateral side-scrapers and notches (OBS). **D–H** transverse side-scraper (**D**: MFL; **E**: OBS; **F**: PB; **G–H**: ASB). **I**: bilateral side-scraper (OBS). **J–M**: lateral-transverse side-scrapers (**J**, **L**: DASL; **K**, **M**: ASB). **N–Q**: convergent side-scrapers (**N–P**: OBS; **Q**: ASB). **R–U**: skewed side-scrapers (**R**: DASL; **S**: ASB; **T–U**: OBS). **V**, **Z**: notches (OBS). **Y**: retouched notch (OBS) (after Piperno et al. 2004b, modified). ASB: aphyric to subaphyric basalts. DASL: differentiated aphyric to subaphyric lavas. MFL: Melka Fault Lava. OBS: obsidian. PB: porphyritic basalt (after Gallotti 2013)

Multifacial multidirectional irregular cores were alternately flaked through multidirectional removals without any recognizable organization of the reduction process and without any specific platform preparation (Fig. 4.17: A). The flakes thereby produced display three to seven negative scars, thus confirming that the multifacial multidirectional cores were overexploited. Flake shapes vary; they are often sub-quadrangular with a thick and asymmetrical cross-section. Core-edge flakes (Fig. 4.17: B–C) indicate continuous rotation of flaking surfaces. Nonetheless, these flakes are apparently the outcome of the search for new angles, rather than of a rejuvenation phase aimed at rearranging the flaking and striking surfaces (Gallotti et al. 2014).

4.5.4 Production of Small- and Medium-Sized Flakes at Gombore II OAM, ~0.85 Ma

As is the case at Garba XIIIB, discoid technology is the dominant débitage concept at Gombore II OAM. The second most prevalent technology is multifacial multidirectional exploitation. Aphyric lavas are the most frequently exploited raw materials, followed by obsidian and porphyric basalt (Fig. 4.10a, b), which are also well documented in the unworked assemblage (Fig. 4.11). Because of winnowing by natural agents, most of the flakes are missing (flake-to-core ratio 3.2:1) (Raynal et al. 2004; Gallotti and Mussi 2017).

Discoid exploitation is systematically bifacial (Fig. 4.17: J, K). The natural surfaces of the original blank, when preserved, show that the knappers used round or angular cobbles and stones as blanks, adapting or modifying their natural geometry in order to meet the method's requirements. Flakes too were used as core blanks. The average removal count of these discoid cores is 22 flakes per core. This result was made possible by maintaining peripheral convexity. In fact, despite the overexploited appearance of the cores, negative removal scars do not overlap the prominent central part of the flaking surface, which remains convex until its final state. The many flakes that show a tangential flaking direction on their dorsal face, and a cutting edge opposite a back, also suggest that convexity was maintained intentionally. These flakes usually display a deviation of the flaking axis from the morphological axis (Fig. 4.17: N, P–Q). The other flakes were usually obtained by centripetal flaking, in which case the morphological and flaking axes coincide (Fig. 4.17: M).

Multifacial multidirectional exploitation is irregular. Angular cobbles were intensely exploited (with an average of 39 removal scars per core); the cores were rotated and removals were made in multiple directions, as documented by many edge-core flakes (Fig. 4.17: I, L, O).

There are very few retouched flakes. There are some denticulates (n = 3) and notches (n = 2). Retouch did not modify the shape of the blank, only the edges (Gallotti and Mussi 2017).

4.5.5 LCT[4] Production at Garba IVD, ~1.6 Ma

The LCT chaînes opératoires appear for the first time in the Melka Kunture region in level D of Garba IV, ~1.6 Ma, where they constitute a minor aspect of the technical activities (1.5% of the artifacts). A specific knapping process was adopted to produce large flakes that were to be turned into LCTs. There is also a notable difference in flake dimensions (Gallotti 2013).

All the LCTs are on large flakes except one obsidian discoid core that was eventually turned into a massive scraper. The raw materials most exploited are porphyritic rocks, Melka Fault lava and obsidian (Fig. 4.18). This circumstance is noteworthy because the percentages of these three materials in the unworked assemblage are low in comparison with aphyric basalts. Overall, in the unworked assemblage, as in the other ancient alluvial deposits, large cobbles and blocks—i.e., blanks large enough to allow at least one large flake to be detached from them—are usually either porphyritic rocks or Melka Fault lava (Gallotti and Mussi 2017).

Lavas and obsidian were exploited in different ways for the production of LCT blanks. Three obsidian cores were treated by simple débitage; i.e., they are small blocks displaying just one or two removals, either longitudinal or transversal. The corresponding flakes were turned into massive transverse side-scrapers by a unifacial, non-invasive continuous retouch on the dorsal face (Fig. 4.19: E).

[4]LCTs are intended here as shaped or retouched tools with a length or width > 10 cm. They include massive scrapers, bifaces, and cleavers. Massive scrapers are large flake blanks with retouched edge(s). Bifaces are LCTs whose morphology «résulte de l'aménagement simultané de deux convexités, de manière à ce que l'une soit à l'image de l'autre en fonction d'un *plan d'équilibre bifacial* … De l'intersection de ces deux convexités naît une silhouette « lissée » par retouche, qui se distribue par rapport à un *plan d'équilibre bilatéral* » (Roche and Texier 1991: 102). Cleavers (sensu Tixier 1956) are intended here as LCTs obtained either by débitage only, or by débitage followed by shaping. The cutting edge must be left unretouched, i.e. it is the outcome of the débitage of the blank. Bifacial pieces with a bit achieved by shaping or by lateral tranchet blow technique are not cleavers but handaxes with a transverse (or terminal) cutting edge (Inizan et al. 1999).

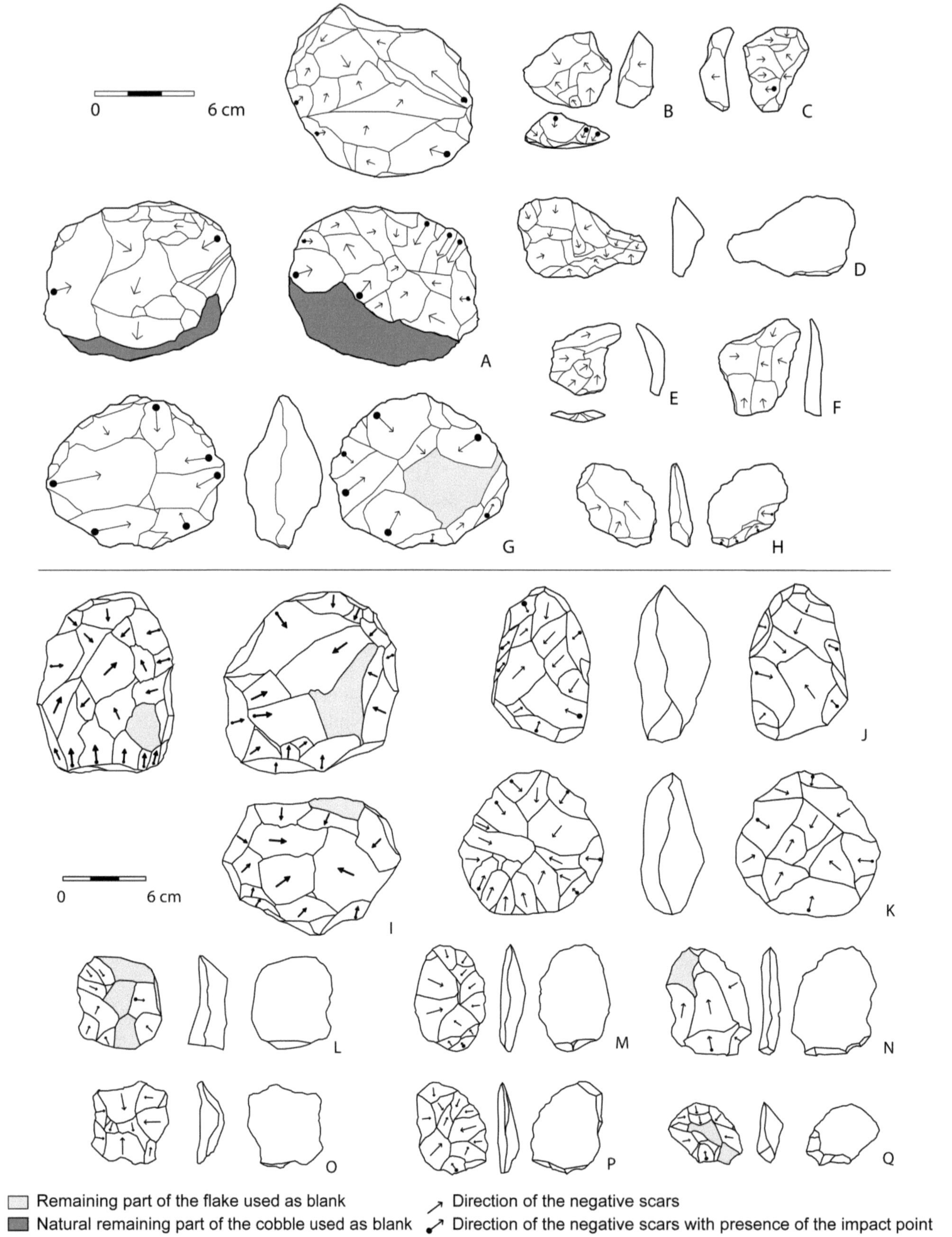

Fig. 4.17 Small débitage cores and flakes from Garba XIIIB (**A–H**) and Gombore II OAM (**I–Q**), **A**: irregular multifacial multidirectional core (ABS). **B**, **C**: flakes obtained by multifacial exploitation (ASB). **D–F**, **H**: flakes obtained by discoid exploitation (ABS). **G**: bifacial discoid core on flake (ASB). **I**: multifacial multidirectional core (ASB). **J**, **K**: bifacial discoid cores (**J**: OBS; **K**: ASB). **L**, **O**: flakes obtained by multifacial exploitation (**L**: ASB; **O**: OBS). **M**, **N**, **P**, **Q**: flakes obtained by discoid exploitation (**M**, **N**: ASB; **P**, **Q**: OBS). ASB: aphyric to subaphyric basalts. OBS: obsidian (after Gallotti et al. 2010, 2014, modified)

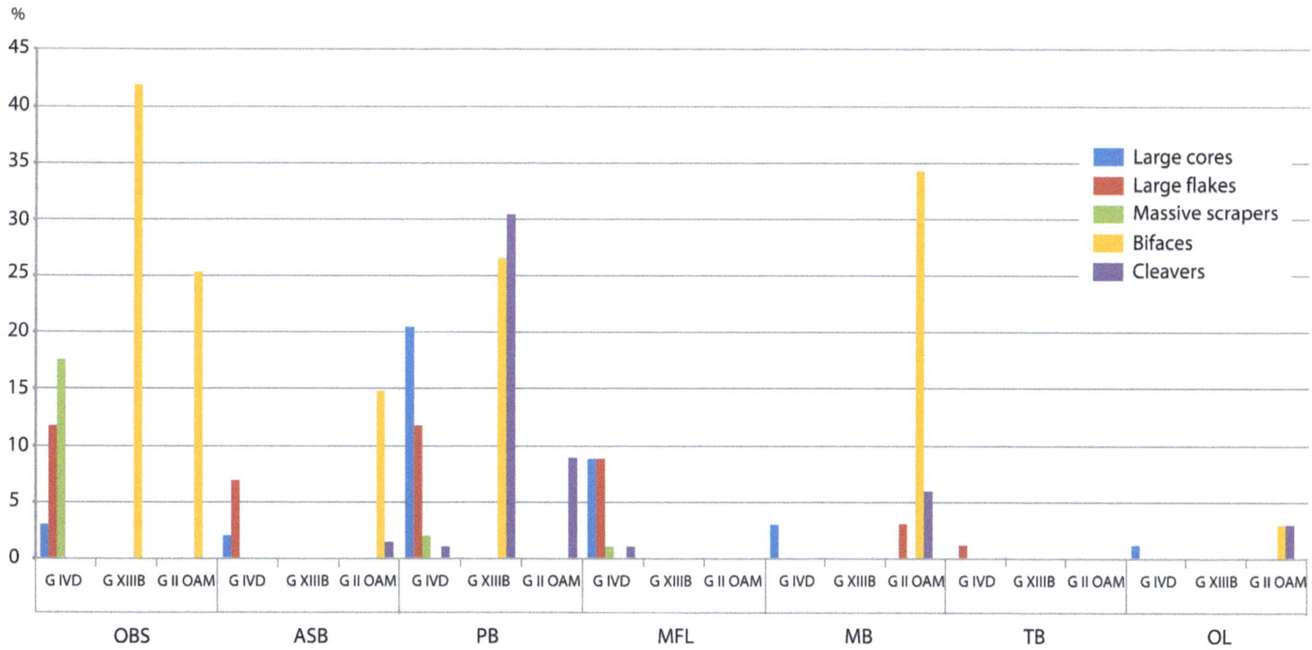

Fig. 4.18 Lithological composition and technological components of the LCT productions at Garba IVD (G IVD), Garba XIIIB (G XIIIB) and Gombore II OAM (G II OAM). ASB: aphyric to subaphyric basalts. MB: microdoleritic basalt. MFL: Melka Fault Lava. OBS: obsidian. OL: obsidian lava. PB: porphyritic basalt. TB: trachybasalt. Twisted bifaces are not included in this figure because they have not been found anywhere else at Melka Kunture nor in East Africa

Most of the LCTs (n = 10) were produced from flakes resulting from discoid exploitation. The obsidian discoid core mentioned above is long and narrow. It has a thick pyramidal cross-section created by two flaking surfaces created by means of centripetal/tangential exploitation. One flaking surface is marked by the large removal scar of a short and wide flake that could have been used as LCT blank. After this large flake was detached, the core was likewise turned into an LCT by marginal non-continuous retouch (Fig. 4.19: A). In all other cases, wide and thick flakes obtained by discoid exploitation were turned into massive scrapers through unifacial or bifacial retouch on the transverse or lateral edges (Fig. 4.19: F, H). Three massive scrapers present marginal, denticulate and non-continuous retouch along two convergent and skewed edges. On some of them, an invasive retouch on the dorsal face, or on both faces, thinned the butt-bulb part (Fig. 4.19: G–H). There is only one massive transverse scraper; it was produced by retouching a wide and thick flake with unidirectional removals on the dorsal face (Gallotti 2013; Fig. 4.19: D).

One débitage method is documented by the lava-core analysis, which points to a prepared unifacial centripetal method similar to that used for small and medium-sized flake production (Fig. 4.19: B–C). The flaking surface is a horizontal plane on a large cobble or block, whereas a peripheral or semi-peripheral plane makes it possible to prepare the striking platform by unidirectional removals. On most cores, evidence of large flake production is limited to that provided by a single scar. On a few cores, a few centripetal scars precede this final large scar, suggesting the preparation of peripheral convexity, and predetermination. After the flake was detached by direct freehand percussion, the core was discarded.

The resulting large flakes display remains of the natural surface in the middle of the dorsal face, peripheral centripetal removals and wide, thick, facetted butts. The latter are often adjacent to one of the backed side edges. Some flakes look like large éclats débordants. However, three flakes have on their dorsal face invasive negative centripetal and chordal scars, thereby documenting lengthy preparation of the convexity of the flaking surface. Oddly, only three such flakes were retouched: one by large notches on the dorsal face, one by a bifacial, rather invasive and non-intensive retouch that created a convex cutting edge (Fig. 4.19: J), and the third by a fairly invasive and continuous retouch on one edge of the ventral face, which thinned and removed the butt-bulb part. Small isolated and non-continuous removals were made on thin portions of the edges (Fig. 4.19: I).

The edge morphology of the two cleavers (sensu Tixier 1956) was determined prior to the blank's detachment. The flaking axis is either perpendicular (Fig. 4.19: L) or oblique (Fig. 4.19: K) to the cleaver's axis. However, we have not found any large core related to the production of flake blanks that have a predetermined cutting edge. Accordingly, the débitage method remains unknown. One of the

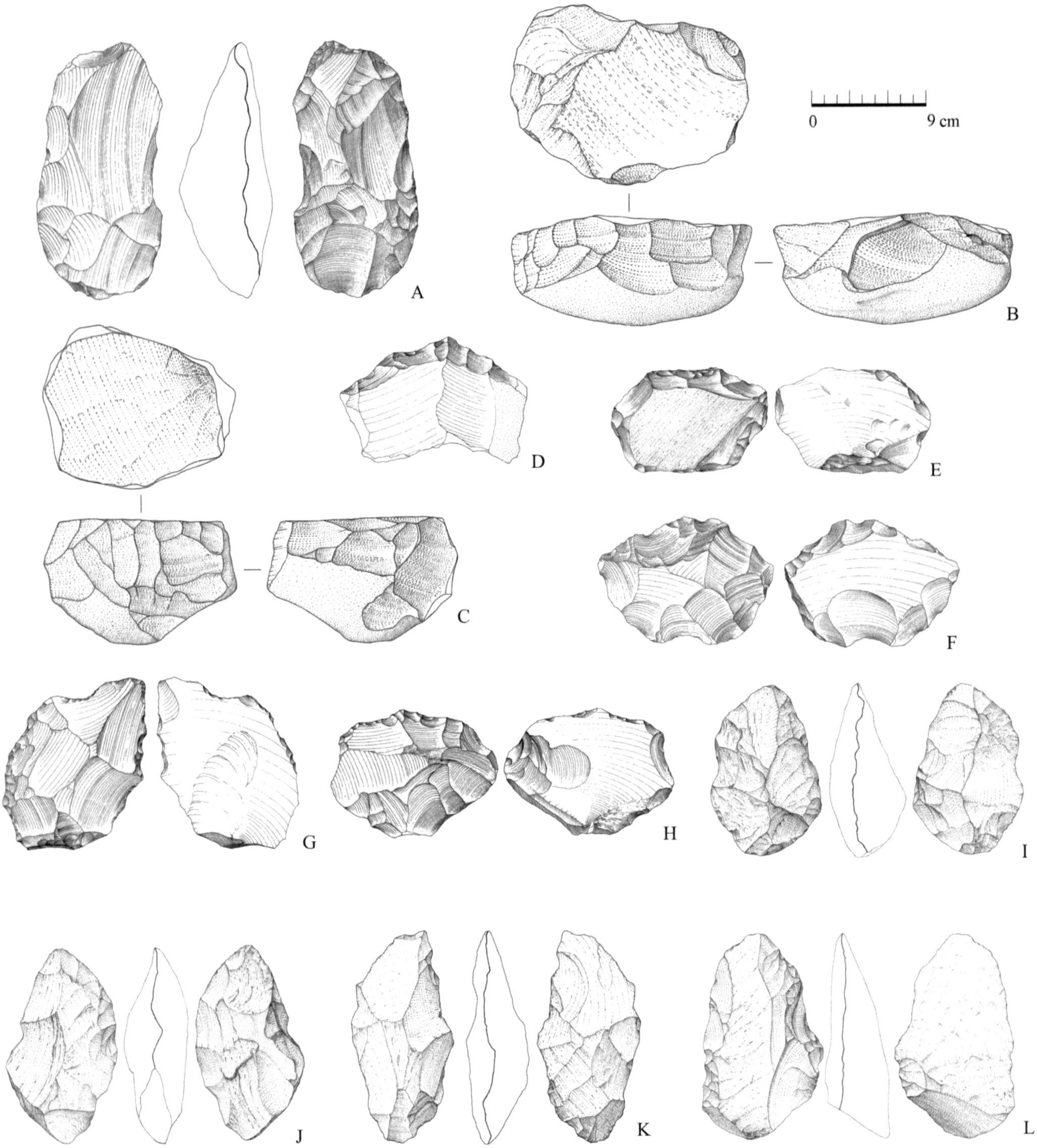

Fig. 4.19 Garba IVD. **A**: large retouched core (OBS). **B–C**: large cores (MFL). **D–J**: massive scrapers (**D–H**: OBS; **I–J**: PB). **K–L**: cleavers (**K**: MFL; **L**: PB). MFL: Melka Fault Lava. OBS: obsidian. PB: porphyritic basalt (after Gallotti 2013, modified). Drawings by M. Pennacchioni

cleavers was shaped by reducing the irregular convexity of the dorsal face and thinning the butt-bulb part (Fig. 4.19: K). The retouch did not regularize the edges; they retain a rather denticulate and irregular outline. The second cleaver was not shaped, just retouched on its dorsal face (Gallotti 2013; Fig. 4.19: L).

4.5.6 LCT Production at Garba XIIIB, ~1.0 Ma

At Garba XIIIB, the LCT chaînes opératoires are represented by bifaces and cleavers. Neither large cores nor unmodified large flakes were discovered (Table 4.1). Bifaces are in obsidian (n = 11) or porphyritic rocks (n = 7). All the cleavers are made of porphyritic basalt (Fig. 4.18). The Kombewa method was the only débitage mode used on large flake blanks. It made it possible to produce flakes whose technical, morphological and dimensional features were extremely homogeneous. In other words, the Kombewa flakes point to a high degree of predetermination (Texier and Roche 1995; Inizan et al. 1999).

As to bifaces, the shaping of Kombewa flakes is either intense on both faces (Fig. 4.20: B), or intense on the upper face and limited on the lower one (Fig. 4.20: A, C). The shapes thus produced are rather pointed, standardized in size, and bifacially symmetrical. The flakes were shaped in two phases: first, alternate invasive removals involving all or part of both faces; second, thinning the bulb, removing the butt and producing a biconvex section.

The bits of the cleavers are either convex (Fig. 4.20: E) or oblique (Fig. 4.20: D, F) and are predetermined by the morphology of the Kombewa flake blanks. The cleavers' size was predetermined by the unmodified edge and butt; their bifacial symmetry was partly predetermined by the Kombewa flake blank and partly achieved through the shaping process.

The cleavers have standardized intra-shape dimensions and the bifaces have standardized intra-type dimensions, which were obtained partly by means of strict predetermination of the blank type and partly by means of standardized shaping procedures (Gallotti et al. 2014).

4.5.7 LCT Production at Gombore II OAM, ~0.85 Ma

As was the case at Garba XIIIB, only the final stage of LCT production is documented at Gombore II OAM, i.e., bifaces and cleavers. A specific aspect of the technical activities is the manufacture of obsidian twisted bifaces, which have not been found anywhere else in the Melka Kunture archaeological sequence, nor at any other prehistoric African site documented in stratigraphic context (Fig. 4.21; Gallotti et al. 2010). Their length (between 6 and 10 cm) is smaller than is generally recognized as diagnostic for LCTs, i.e., length or width >10 cm (cf. Sharon 2010; de la Torre 2011; Beyene et al. 2013; Gallotti 2013). Accordingly, they are omitted from the lithotype counts discussed here.

Bifaces are mainly made of microdoleritic basalt, porphyritic basalt or obsidian. Most of the cleavers are made of microdoleritic basalt (Fig. 4.18). Biface blanks, if recognizable, are Kombewa flakes or large skewed flakes. Given the absence of large cores and the near absence of residual natural surfaces, the débitage method cannot be determined. The shaping of skewed flake blanks is intense on both faces, thereby producing more or less pointed ovoid or limande bifaces characterized by intra-shape standard sizes, bifacial asymmetry and poorly delineated edges (Fig. 4.22: E, F). Kombewa flakes are solely made of obsidian or microdoleritic basalt. Predetermined bifacial symmetry leads to more limited shaping procedures, which generally require one or two series of invasive removals and marginal retouch. Symmetry is fully achieved, the sections are biconvex and the edges are well delineated and continuous (Gallotti et al. 2010; Gallotti and Mussi 2017; Fig. 4.22: A–D).

Cleaver flake blanks were detached by means of the Kombewa method and a method involving bidirectional removals from opposite core platforms. Due to the lack of cores, we cannot determine whether only bidirectional débitage with opposed striking platforms was performed, or if a more complex predetermined flaking system involving the whole periphery of the core was also applied. The systematic and frequent use of Kombewa flakes and of prepared striking platforms on some cleavers (Fig. 4.22: H) both suggest a predetermined Kombewa strategy. In general, both débitage methods produce flakes that have similar technical aspects, morphology and dimensions, i.e., highly predetermined flakes. The intensity of shaping depended on the degree of predetermined bifacial symmetry of the flake blank. Finishing consisted of edge delineation (Gallotti et al. 2010; Gallotti and Mussi 2017; Fig. 4.22: G, H).

4.6 Discussion: The Origin of the Acheulean at Melka Kunture

The Early Pleistocene sequence at Melka Kunture offers a rare opportunity to analyze technological changes related to the emergence and development of the Acheulean in a microregion.

This sequence starts with the basal levels (E–F) of Garba IV, dated to ~1.7 Ma (Raynal et al. 2004; Morgan et al. 2012; Gallotti and Mussi 2015). The artifact assemblage is made up of all the components that belong to small débitage, and no others. The unworked assemblage consists mainly of aphyric lavas that are quite suitable for knapping but were not exploited to any great degree. Conversely, obsidian was very clearly chosen as the preferred raw material. Procurement was evidently local, i.e., the necessary raw materials

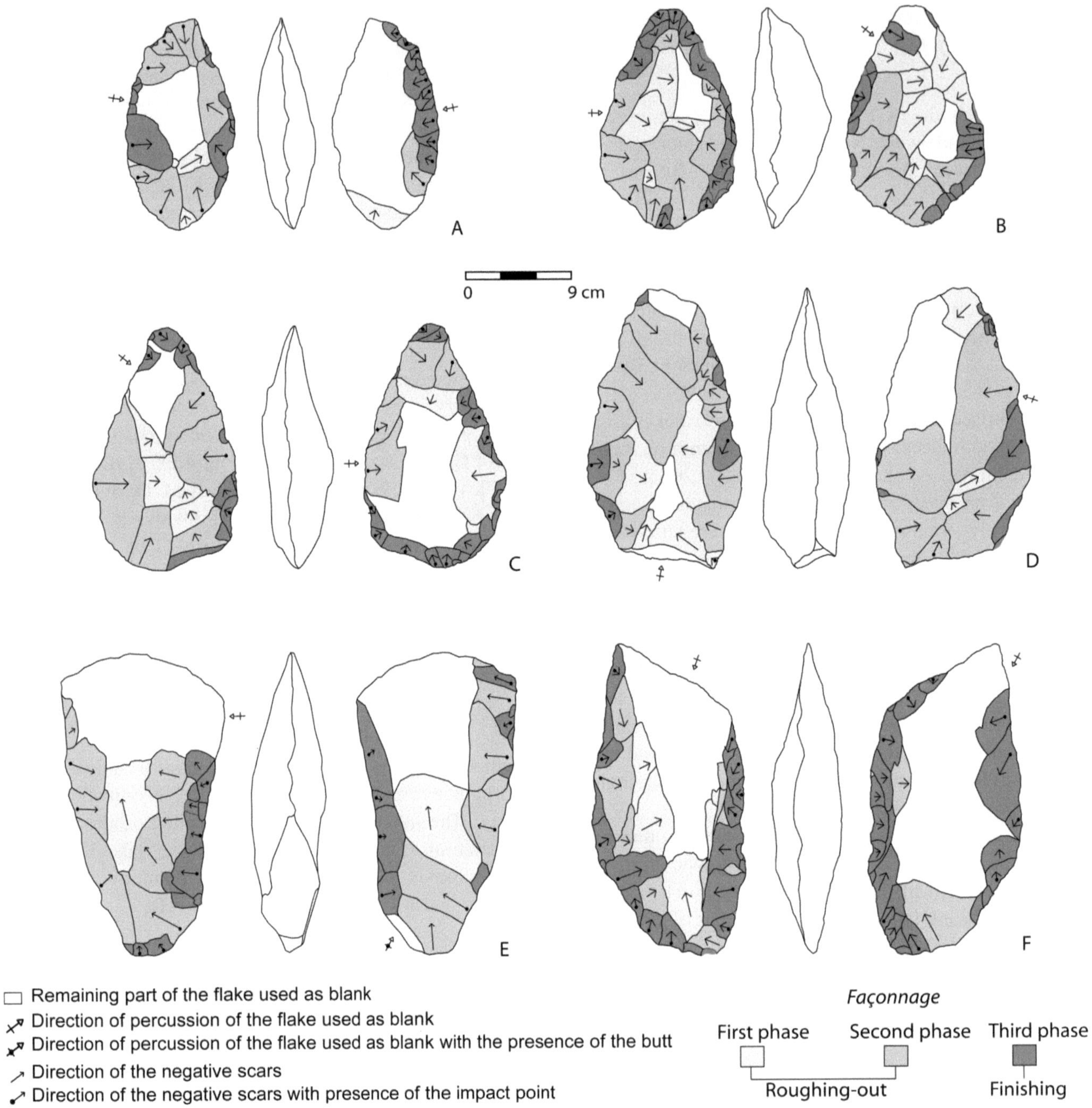

Fig. 4.20 Garba XIIIB. **A**–**C**: bifaces (**A**, **C**: PB; **B**: OBS); **D**–**F** cleavers (PB). OBS: obsidian. PB: porphyritic basalt (after Gallotti et al. 2014)

were collected in alluvial deposits near the sites. This conclusion is supported by the systematic use of pebbles and cobbles as blanks, i.e., rounded stones available in the paleochannels, where there were no angular natural blanks. Conversely, the latter are widely distributed in the vicinity of the primary source, within a radius of one or two km from Balchit (Gallotti and Mussi 2015; Fig. 4.1b).

Small and medium-sized flakes were produced by several different débitage methods that developed as the best ways to exploit cobble geometries. Technical patterns document the efficient use of any available angle through continuous core rotation, while the surfaces were rectified in order to create suitable striking platforms and angles that made it possible to detach flakes. However, there was no systematic preparation of striking platforms, no recurrence, no volume/convexity management or hierarchy among surfaces and no modification of the natural blank geometry that would have made it possible to use a particular flaking method. The débitage

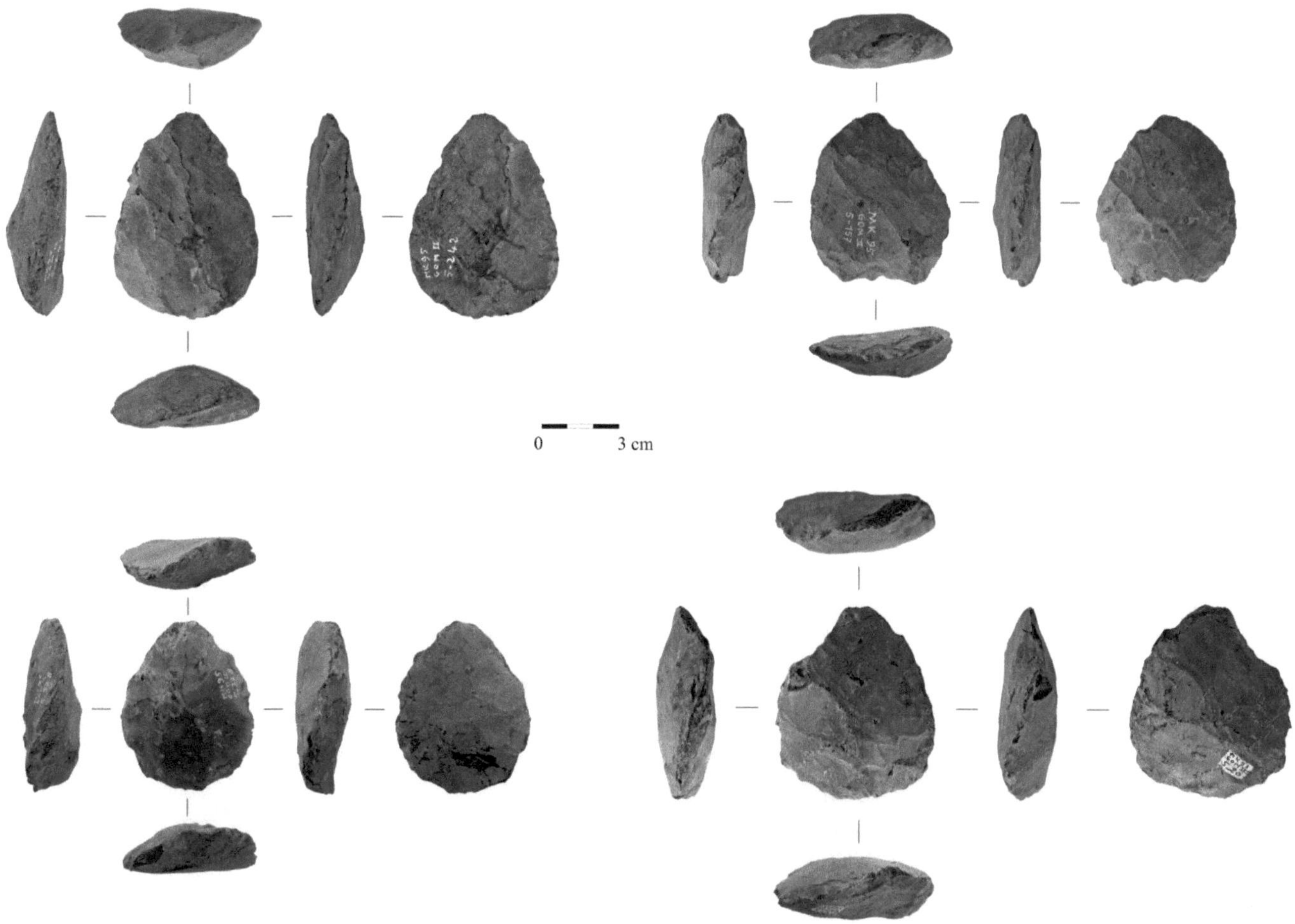

Fig. 4.21 Gombore II5. Twisted obsidian bifaces

methods were the outcome of technical structures, skills and cognitive abilities similar to those identified at the other Oldowan sites in East Africa (e.g., de la Torre 2004; Delagnes and Roche 2005; de la Torre and Mora 2005; Semaw 2006; Braun et al. 2009; Gallotti 2018). The intra-site variability of débitage modalities is also observed in late Oldowan complexes (Gallotti 2018). This was happening at a time when the early Acheulean had already appeared elsewhere in East Africa (Quade et al. 2004, 2008; Lepre et al. 2011; Beyene et al. 2013; Díez-Martin et al. 2015; Semaw et al. 2018; Texier 2018).

At Melka Kunture there is also evidence of a specific technical process that has never been recorded elsewhere in the Oldowan, i.e., a systematic search for small pointed forms. This process is closely linked to obsidian exploitation, and most probably aimed to serve a specific techno-functional purpose (Gallotti and Mussi 2015; Gallotti 2018).

The variability of small débitage methods increases at Garba IVD, ~1.6 Ma. The knappers used as core blanks all the clasts available in the alluvial deposits, which were mainly obsidian and aphyric cobbles (Gallotti 2013; Gallotti and Mussi 2017). However, the knappers also used angular pieces of obsidian, produced by fracturing of the primary source due to weathering. Ongoing site formation analysis further confirms that angular pieces are not natural components of alluvial deposits. They are exogenous and were brought to the site as manuports by the knappers. Besides, they are not found in other penecontemporaneous alluvial deposits (Gallotti and Mussi 2017). This may also confirm that hominins did use a lag deposit (Raynal et al. 2004). They used local raw materials and angular pieces of obsidian that they found scattered near the obsidian outcrop; today such stones can be collected at distances ranging from 5 to 7 km. This means that during the procurement phase, the chaînes opératoires correlated with the exploitation of these specific forms were fragmented. However, we do not know how extensive the primary obsidian source was ~1.6 million years ago. If at that time some obsidian outcropped closer to the site, or if the pattern of the dispersion of angular stones was different from what it is today, the chaînes opératoires would not have been fragmented.

In any case, the use of such stones for discoid exploitation reflects a new technical skill: the ability to produce an initial

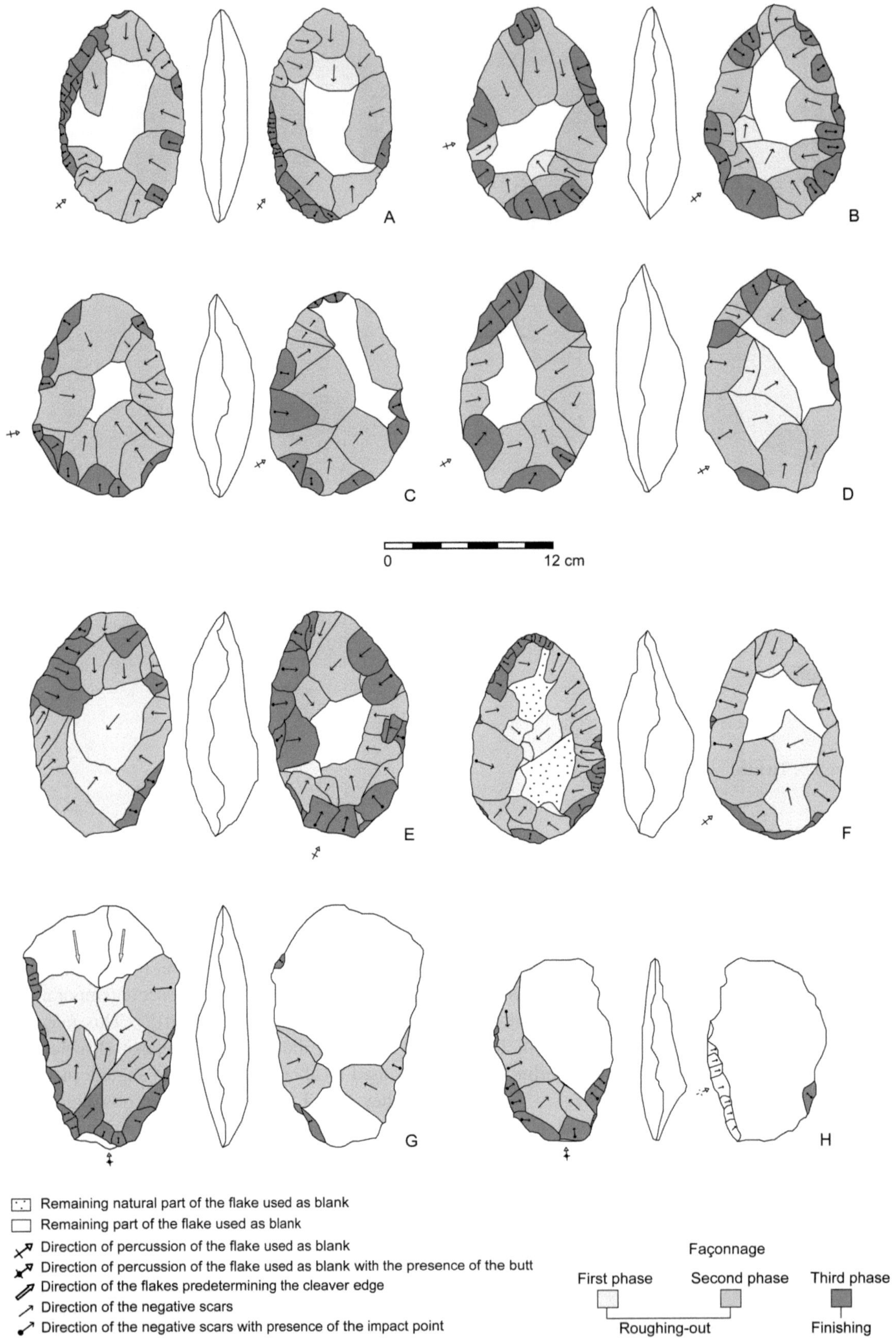

Fig. 4.22 Gombore II OAM. **A–D**: bifaces on Kombewa flakes (**A**, **B**, **D**: MB; **C**: OBS). **E–F**: bifaces on skewed flakes (**E**: MB; **F**: ASB). **G**: cleaver on flake with bidirectional removals (OL). **H**: cleaver on Kombewa flake (MB). ASB: aphyric to subaphyric basalts. MB: microdoleritic basalt. OBS: obsidian. OL: obsidian lava (after Gallotti and Mussi 2017)

configuration of the core that modified the blank's natural geometry. Discoid and unifacial centripetal exploitations reflect other technical innovations as well: preparation of the striking platform, recurrence, volume/convexity management and hierarchy among surfaces. The variability of the operational scheme of the discoid concept (Peresani 2003) is fully observed at Garba IVD. The other débitage methods are the outcome of geometrical constraints imposed by raw materials (Gallotti 2013). The technical parameters are the same as those identified in the older Oldowan levels (E–F).

In later Acheulean sites the variability of the small débitage processes unquestionably decreases. At Garba XIIIB and Gombore II OAM ($\sim$1.0–0.85 Ma), the only débitage methods are the discoid and the multifacial multidirectional ones. The knappers' very effective volume control and maintenance allowed greater recurrence and higher productivity. For the first time, both spherical cobbles and angular lava cobbles/elements were used for discoid exploitation. The knappers succeeded in overcoming the lithological and geometric constraints. Conversely, in the early Acheulean at Garba IVD, it was precisely the suitability of obsidian for knapping that allowed a limited degree of independence from the original shape of the raw material. Angular lava cobbles with one convex surface were chosen because they were easy to turn into cores for unifacial centripetal exploitation. The convex surface was exploited by using the existing angular planes as striking platforms. The angle between the two surfaces remained the original one, i.e., $\sim$90°. At Garba XIIIB and Gombore II OAM, discoid exploitation of angular lava cobbles replaced this method.

At Garba IVD, just as at Garba XIIIB and Gombore II OAM, small débitage was used to produce small- and medium-sized flakes that were rarely modified by retouch. The fairly standardized small tools at Garba IVE-F are no longer found in later periods. This suggests that it was only an occasional technological development, possibly driven by practical needs and facilitated by obsidian's high suitability for knapping. The absence of such tools at later sites is at odds with hypotheses such as those advanced by Barsky et al. (2014), which suggests that the growing numbers of retouched flakes and the emergence of standardization in toolkits are related to improvements in technical skills, hence they reflect a major step in cultural evolution. However, the question of why small-tool production was not actually part of the emerging Acheulean is still open.

The other major technical innovation first seen at Garba IVD was the production of large flakes to be turned into LCTs. LCT production is linked to the exploitation of raw materials—namely obsidian, porphyritic basalt and Melka Fault lava—found as large blanks in alluvial deposits. However, these large forms are scarce in secondary sources, which may explain why LCT production was very limited. In this region, only later on there is any evidence of procurement at primary sources and of the fragmentation of the related chaîne opératoire, both of which are considered typical of Acheulean techno-complexes (e.g., de la Torre and Mora 2005; Goren-Inbar and Sharon 2006; de la Torre et al. 2008; de la Torre 2011). Systematic procurement at a primary source—i.e., at the porphyritic lava and obsidian outcrops located 2.5 to 7 km away—enabled large-scale manufacturing of LCTs at Garba XIIIB and Gombore II OAM. At the latter site, there is evidence that hominins ranged far afield in their search for microdoleritic basalt, a very fine-grained rock. This reflects broadening of landscape cognition over a radius of at least 15–20 km. As a consequence of procurement right at the primary sources, the LCT chaînes opératoires were fragmented (Gallotti and Mussi 2017).

At Garba IVD, the LCTs were always made on large flakes. The two débitage methods used to produce such flakes were the discoid and the prepared unifacial centripetal. Both are related to innovations in small débitage: systematic preparation of the striking platform; recurrence in exploitation, which made it possible to obtain flakes with longer suitable edges or large flakes that were wider than they were long; management and maintenance of volume and convexities; and hierarchy among surfaces. To achieve their purposes, the knappers systematically selected geometrically suitable large blanks. At the end of the Early Pleistocene, new large flake débitage methods emerged for the production of highly standardized flake blanks. Volume management eventually made it possible to produce highly predetermined flakes that resembled each other in their technical and morphological features and in their dimensions.

Accordingly, the ability to extract large flakes emerged well before the ability (or maybe the need) to manage the volume of objects by means of bifacial and bilateral equilibrium and two convergent edges. At $\sim$1.6 Ma, LCT manufacturing was limited to edge retouch and produced generic tool types, better described as massive scrapers. One of the criteria followed was to use thick, wider-than-long flakes. Another repetitive pattern was the thinning of percussion platforms and bulbs (Gallotti 2013).

At $\sim$1.0–0.85 Ma, there was a surge in the production of bifaces and cleavers. Bifaces were produced by balancing the bifacial and bilateral planes and creating systematic convergence of the two edges. The intensity of shaping was directly linked to the degree of predetermination of bifacial and bilateral equilibrium in the flake blanks. The shaping process was aimed at refining this equilibrium, at creating convergent edges in order to manufacture bifaces, and at producing specific tool types. The Kombewa method made it possible to produce highly predetermined flakes that resembled each other in their technical and morphological features and in their dimensions (Gallotti et al. 2010, 2014; Gallotti and Mussi 2017). It was the only débitage method in use at Garba XIIIB and it was widely used at Gombore II OAM.

4.7 Conclusions

Melka Kunture preserves one of the longest and most complete sequences that document the transition from the late Oldowan to the early Acheulean in East Africa, as well as the development of the Acheulean Industrial Complex up to the transition to the Middle Stone Age (Chavaillon et al. 1979; Chavaillon and Piperno 2004; Mussi et al. 2014; Gallotti and Mussi 2015, 2017). Furthermore, East African sites where both the late Oldowan and the early Acheulean have been recovered are very few (Gallotti 2018). Despite the great importance of the Oldowan–Acheulean transition for understanding human evolution, the biological and cultural mechanisms underlying this process are still poorly understood (e.g., de la Torre et al. 2011; Gallotti 2013; de la Torre 2016).

Until now, the early Acheulean at Melka Kunture has been characterized through techno-economic comparison with other East African sites (Gallotti 2013). We are now able to add a detailed comparative analysis of Early Pleistocene lithic assemblages dated between ~1.7 Ma and ~0.85 Ma, highlighting the changes which are specific to this region.

If we consider the diachronic variations in a techno-economic perspective, the early Acheulean, as identified at Garba IVD at ~1.6 Ma, has the following ontological characteristics:

- it includes two main lithic activities, i.e., production of small and medium-sized flakes and LCTs. The production of LCTs and of large flakes do not appear in the late Oldowan at Garba IVE–F;
- both productions are based on a local procurement system that exploits nearby alluvial deposits;
- the selection of lithologies and natural morphometrical types for small débitage is very similar in both the late Oldowan and the early Acheulean assemblages. Aphyric lavas and obsidian are the most exploited lithotypes in both assemblages. Nevertheless, the late Oldowan small débitage is more focused on the selection of the rock type easiest to knap, i.e., on obsidian. The early Acheulean small débitage is apparently less constrained by raw material lithology;
- small débitage chaînes opératoires were not fragmented in the early Acheulean, just as they were not in the late Oldowan. The incipient fragmentation that may have emerged in the early Acheulean between the procurement and production phases would have been connected to the exploitation of a specific lithotype, i.e., obsidian;
- the small débitage methods used in the late Oldowan also persist in the early Acheulean, as they were the best ways to exploit the natural geometries of rocks available in alluvial deposits. However, these technical solutions were more variable in the early Acheulean, when knappers started to exploit a wider variety of natural blanks;
- although there were strong similarities with the late Oldowan, new technical criteria emerge in small débitage, notably preparation of the striking platform, recurrence, volume/convexity management, hierarchy among surfaces and modification of the natural geometry of the blanks. The adoption of these technical criteria is closely linked to new débitage methods and concepts—i.e., discoid and prepared unifacial centripetal exploitations—which made it possible to produce flakes with long serviceable edges. However, the incipient ability to configure for discoid exploitation natural clasts that were not ideal as blanks for cores depended strongly on raw material lithology. This constraint was systematically overcome only at the end of the Early Pleistocene at Garba XIII and Gombore II;
- the same innovative criteria adopted in small débitage are also seen in large flake débitage. Discoid and prepared unifacial centripetal exploitation were used to produce large flakes and blanks to be turned into LCTs. The knappers' main purpose was to detach fairly thick flakes that would be wider than they were long. Predetermination of the shapes, sizes and technical features of large blanks did not appear at Melka Kunture until later on, i.e., at ~1.0 Ma;
- LCT production is not fragmented and is quantitatively limited by the scarcity of suitable large forms in the paleochannels. This may mean that (a) the knappers did not organize systematic production of large flakes because they were unable to conceive of a wider-ranging provisioning system in which raw materials would be obtained directly at their primary sources, and the production phase as well as the procurement phase would be fragmented; or that (b) the actual functional needs for which the LCTs were meant were not yet strong enough to drive the knappers to change their landscape perception and temporal sequencing;
- after ~1.0 Ma, provisioning at primary sources, wider knowledge of the landscape and chaîne opératoire fragmentation co-occurred systematically. This happened in parallel with the frequent manufacturing of bifaces and cleavers: specific tool types that were technologically (and most probably functionally) different from small tools. With the exception of two cleavers, at this site

large tool types are not found in the early Acheulean, and LCT retouching was limited to edge retouching that did not modify the shape of the blank, just as was the case with small tools.

Summing up, the transition from the late Oldowan to the early Acheulean at Melka Kunture was a cultural continuum, during which new concepts and behaviors gradually emerged. The production of large tools is a diagnostic aspect of the early Acheulean. Nevertheless, these large tools are the outcome of retouch modifying only edges, not the shape of the blank, and shaping is absent. Bifaces are missing and there are only two cleavers. Actually, the main conceptual innovations are just the introduction of some technical criteria into débitage for the extraction of small, medium and large-sized flakes. For the knappers, these innovations were a first step in gaining independence from constraints imposed by nature, i.e., by lithic resources. This local evolution was accomplished by *Homo erectus* sensu lato, whose remains have been recorded both in Oldowan levels (Garba IVE) and in early Acheulean ones (Gombore IB; Di Vincenzo et al. 2015).[5]

Based on a techno-economic approach, the transition from the late Oldowan to the early Acheulean at Melka Kunture was completed within 200 kyr- not, as Chavaillon et al. (1979) had suggested, in 500 kyr. At Melka Kunture, biface and cleaver production does not characterize the emergence of the Acheulean; it pertains to the Acheulean of the late Early Pleistocene, which is distinguished from the early Acheulean by a sharp temporal, techno-economic and human paleontological discontinuity.

Acknowledgements We thank the Authority for Research & Conservation of the Cultural Heritage of Ethiopia's Ministry of Culture & Tourism, the National Museum of Addis Ababa, and the Oromia Culture and Tourism Bureau for fieldwork permits and access to the lithic collections. The research was supported by grants from "La Sapienza" University of Rome ("Grandi scavi archeologici") and from the Italian Foreign Ministry, awarded to MM. The study of the Garba IVD lithic artifacts was made possible by a Post-Ph.D research grant from the Wenner-Gren Foundation (number 7715 "Technical Behaviors During the Oldowan at Garba IVD, Melka Kunture, Ethiopia") awarded to RG. We also wish to thank Guy Kieffer and Jean-Paul Raynal for lithological analysis of raw materials, Massimo Pennacchioni and Noemi Tomei for their drawings of artifacts, and Jean-Paul Raynal for the photos in Fig. 4.8e, f. We express deep thanks to the anonymous reviewers as well as to the VERT Series editors, who all helped us to improve this paper.

RG studied the lithic collections. MM, director of the Italian Archeological Mission at Melka Kunture and Balchit, coordinates research and designed and organized this project. RG wrote the paper and MM contributed to the draft.

[5]Cf. Gallotti and Mussi (2017) for a redefinition of the cultural attribution of level Gombore IB, previously thought to be related to the Oldowan.

References

Abebe, T., Manetti, P., Bonini, M., Corti, G., Innocenti, F., & Mazzarini, F. (2005). *Geological Map of the Northern Main Ethiopian Rift*. Map and chart series MCH094, The Geological Society of America. USA: Inc. Boulder.

Bardin, G., Raynal, J.-P., & Kieffer, G. (2004). Drainage patterns and regional morphostructure at Melka Kunture (Upper Awash, Ethiopia). In J. Chavaillon & M. Piperno (Eds.), *Studies on the Early Paleolithic Site of Melka Kunture, Ethiopia* (pp. 83–92). Florence: Origines.

Barsky, D., Garcia, J., Martínez, K., Sala, R., Zaidner, Y., Carbonell, E., et al. (2014). Flake modification in European Early and Early-Middle Pleistocene stone tool assemblages. *Quaternary International, 316,* 140–154.

Beyene, Y., Katoh, S., WoldeGabriel, G., Hart, W. K., Uto, K., Sudo, M., et al. (2013). The characteristics and chronology of the earliest Acheulean at Konso, Ethiopia. *Proceedings of the National Academy of Sciences, 110,* 1584–1591.

Boëda, E. (1993). Le débitage discoïde et le débitage Levallois récurrent centripète. *Bulletin de la Société Préhistorique Française, 90–96,* 392–404.

Bonnefille, R., Melis, R. T., & Mussi, M. (2018). Variability in the Mountain Environment at Melka Kunture Archaeological Site, Ethiopia, During the Early Pleistocene (~1.7 Ma) and the Mid-Pleistocene Transition (0.9–0.6 Ma). In R. Gallotti & M. Mussi (Eds.), *The Emergence of the Acheulean in East Africa and Beyond. Contributions in Honor of Jean Chavaillon.* Cham: Springer.

Braun, D. R., Plummer, T., Ferraro, J. V., Ditchfield, P., & Bishop, L. C. (2009). Raw material quality and Oldowan hominin toolstone preferences: Evidence from Kanjera South, Kenya. *Journal of Archeological Science, 36,* 1605–1614.

Chernet, T., Hart, W. K., Aronson, J. L., & Walter, R. C. (1998). New age constraints on the timing of volcanism and tectonism in the northern Main Ethiopian Rift and the southern Afar transition zone (Ethiopia). *Journal of Volcanology and Geothermal Research, 80,* 267–280.

Chavaillon, J. (1980). Chronologie archéologique de Melka-Kunturé (Ethiopie). In R. E. F. Leakey & B. A. Ogot (Eds.), *Proceedings VIII Panafrican Congress of Prehistory and Quaternary Studies, Nairobi 1977* (pp. 200–201). Nairobi: The International Louis Leakey Memorial Institute for African Prehistory.

Chavaillon, J., & Chavaillon, N. (1980). Évolution de l'Acheuléen à Melka-Kunturé (Éthiopie). *Anthropologie, XVIII*(2/3), 153–159.

Chavaillon, J., Chavaillon, N., Hours, F., & Piperno, M. (1979). From the Oldowan to the Middle Stone Age at Melka-Kunture (Ethiopia). Understanding cultural changes. *Quaternaria, XXI,* 1–26.

Chavaillon, J., & Piperno, M. (1975). Garba IV, site paléolithique ancien de Melka-Kunturé (Éthiopie). *Bulletin de la Société Préhistorique Française, 72,* 134–138.

Chavaillon, J., & Piperno, M. (2004). *Studies on the Early Paleolithic site of Melka Kunture, Ethiopia.* Florence: Origines.

Condemi, S. (2004). The Garba IVE mandible. In J. Chavaillon & M. Piperno (Eds.), *Studies on the Early Paleolithic site of Melka Kunture, Ethiopia* (pp. 687–701). Florence: Origines.

Cressier, P. (1980). Magnétostratigraphie du gisement pléistocène de Melka-Kunturé (Éthiopie). Datation des niveaux oldowayens et acheuléens. Ph.D. Dissertation, Université Louis Pasteur.

D'Andrea, A., & Gallotti, R. (2004). GIS and intra-site spatial analysis. In J. Chavaillon & M. Piperno (Eds.), *Studies on the Early Paleolithic site of Melka Kunture, Ethiopia* (pp. 589–597). Florence: Origines.

D'Andrea, A., Gallotti, R., & Piperno, M. (2000). Applicazione di un GIS intra-site al giacimento paleolitico di Garba IV (Melka Kunture, Etiopia). *Archeologia e Calcolatori, 11,* 319–338.

D'Andrea, A., Gallotti, R., & Piperno, M. (2002). Taphonomic interpretation of the Developed Oldowan site of Garba IV (Melka Kunture, Ethiopia) through a GIS application. *Antiquity, 76,* 991–1001.

Daniel, N., & Lapedes, D. N. (1978). *McGraw-Hill Dictionary of Scientific and Technical Terms* (2nd ed.). New York: McGraw-Hill Inc.

de la Torre, I. (2004). Omo Revisited: Evaluating the Technological Skills of Pliocene Hominids. *Current Anthropology, 45,* 439–465.

de la Torre, I. (2011). The Early Stone Age lithic assemblages of Gadeb (Ethiopia) and the developed Oldowan/early Acheulean in East Africa. *Journal of Human Evolution, 60,* 768–812.

de la Torre, I. (2016). The origins of the Acheulean: Past and present perspectives on a major transition in human evolution. *Philosophical Transactions of the Royal Society of London. Series B, 37,* 20150245.

de la Torre, I., & Mora, R. (2005). *Technological Strategies in the Lower Pleistocene at Olduvai Beds I & II.* Liege: ERAUL.

de la Torre, I., Mora, R., & Martínez-Moreno, J. (2008). The early Acheulean in Peninj (Lake Natron, Tanzania). *Journal of Anthropological Archeology, 27,* 244–264.

Defleur, A., & Crégut-Bonnoure, E. (1995). Le gisement paléolithique moyen de la grotte des Cèdres (Le Plan-d'Aups, Var). Paris: *Documents d'Archéologie Française, 49,* CNRS.

Delagnes, A., & Roche, H. (2005). Late Pliocene knapping skills: The case of Lokalalei 2C, West Turkana, Kenya. *Journal of Human Evolution, 48,* 435–472.

Diez-Martín, F., Sánchez-Yustos, P., Uribelarrea, D., Baquedano, E., Mark, D. F., Mabulla, A., et al. (2015). The Origin of the Acheulean: The 1.7 Million-Year-Old Site of FLK West, Olduvai Gorge (Tanzania). *Scientific Reports, 5,* 17839.

Di Vincenzo, F., Rodriguez, L., Carretero, J. M., Collina, C., Geraads, D., Piperno, M., et al. (2015). The massive fossil humerus from the Oldowan horizon of Gombore I, Melka Kunture (Ethiopia, >1.39 Ma). *Quaternary Science Review, 122,* 207–221.

Forestier, H. (1992). Approche technologique de quelques séries dites clactoniennes du nord-ouest de la France et du sud-est de l'Angleterre. Mémoire de Maîtrise, Université Paris X Nanterre.

Forestier, H. (1993). Le Clactonien : mise en application d'une nouvelle méthode de débitage s'inscrivant dans la variabilité des systèmes de production lithique du paléolithique ancien. *Paléo, 5,* 53–82.

Gallotti, R. (2013). An older origin for the Acheulean at Melka Kunture (Upper Awash, Ethiopia). Techno-economic behaviors at Garba IVD. *Journal of Human Evolution, 65,* 594–620.

Gallotti, R. (2018). Before the Acheulean in East Africa: An Overview of the Oldowan Lithic Assemblages. In R. Gallotti & M. Mussi (Eds.), *The Emergence of the Acheulean in East Africa and Beyond. Contributions in Honor of Jean Chavaillon.* Cham: Springer.

Gallotti, R., & Mussi, M. (2015). The Unknown Oldowan: ~1.7-Million-Year-Old Standardized Obsidian Small Tools from Garba IV, Melka Kunture, Ethiopia. *PLoS ONE, 10,* e0145101.

Gallotti, R., & Mussi, M. (2017). Two Acheuleans, two humankinds: From 1.5 to 0.85 Ma at Melka Kunture (Upper Awash, Ethiopian highlands). *Journal of Anthropological Sciences, 95,* 1–46.

Gallotti, R., & Piperno, M. (2003). Recent activities of the Italian Archeological Mission at Melka Kunture: The Open Air Museum Project and the GIS application to the study of the Oldowan sites. In J. M. Moreno, R. M. Torcal, & I. de la Torre Sainz (Eds.), *Oldowan: Rather more than smashing tools, First Hominid Technology Workshop* (pp. 37–75). Bellaterra: Treballs d'Arqueologia, 9.

Gallotti, R., & Piperno, M. (2004). The site of Garba IV. Spatial analysis of the lithic material from Level D. In J. Chavaillon & M. Piperno (Eds.), *Studies on the Early Paleolithic site of Melka Kunture, Ethiopia* (pp. 599–635). Florence: Origines.

Gallotti, R., Collina, C., Raynal, J.-P., Kieffer, G., Geraads, D., & Piperno, M. (2010). The Early Middle Pleistocene Site of Gombore II (Melka Kunture, Upper Awash, Ethiopia) and the Issue of Acheulean Bifacial Shaping Strategies. *African Archeological Review, 27,* 291–322.

Gallotti, R., Raynal, J.-P., Geraads, D., & Mussi, M. (2014). Garba XIII (Melka Kunture, Upper Awash, Ethiopia): A new Acheulean site of the late Lower Pleistocene. *Quaternary International, 343,* 17–27.

Geneste, J. M. (1989). Économie des ressources lithiques dans le Moustérien du sud ouest de la France. In M. Otte (Ed.), *L'Homme de Néanderthal, La Subsistance vol. 6* (pp. 75–97). Liege: ERAUL.

Geneste, J. M. (1991). Systèmes techniques de production lithique : variations technoéconomiques dans les processus de réalisation des outillages paléolithiques. *Technique et Culture, 17*(18), 1–35.

Goren-Inbar, N., & Sharon, G. (2006). Invisible handaxes and visible Acheulan biface technology at Gesher Benot Ya'aqov, Israel. In N. Goren-Inbar & G. Sharon (Eds.). *Axe Age. Acheulean Toolmaking from Quarry to Discard* (pp. 111–135). London: Equinox.

Harmand, S. (2009). Raw material and techno-economic behaviours at Oldowan and Acheulean sites in the West Turkana region, Kenya. In B. Adams & B. Blades (Eds.), *Lithic Materials and Paleolithic Societies* (pp. 3–14). Oxford: Blackwell.

Inizan, M. L., Reduron-Ballinger, M., Roche, H., & Tixier, J. (1999). *Technology and Terminology of Knapped Stone (Préhistoire de la Pierre taillée 5).* Nanterre: CREP.

Jaubert, J., & Mourre, V. (1996). Coudoulous, le Rescundudou, Mauran : diversité des matières premières et variabilité des schemas de production d'éclats. In A. Bietti & S. Grimaldi (Eds.), *Reduction Processes* (chaînes opératoires) *in the European Mousterian* (pp. 313–341). Rome: Quaternaria Nova 6, Istituto Italiano di Paleontologia Humana.

Kieffer, G., Raynal J.-P., & Bardin, G. (2002). Cadre structural et volcanologiques des sites du Paléolithique ancien de Melka Kunture (Awash, Ethiopie) : premiers résultats. In J.-P. Raynal, C. Albore-Livadie, & M. Piperno (Eds.), *Hommes et Volcans. De l'éruption à l'objet* (pp. 77–92). Les Dossiers de l'Archéo-Logis 2, Proceedings XIV UISPP Congress, Symposium 15.2, Liege, 2001.

Kieffer, G., Raynal, J.-P., & Bardin, G. (2004). Volcanic markers in coarse alluvium at Melka Kunture (Upper Awash, Ethiopia). In J. Chavaillon & M. Piperno (Eds.), *Studies on the Early Paleolithic site of Melka Kunture, Ethiopia* (pp. 93–101). Florence: Origines.

Kleindienst, M. R. (1962). Components of the East African Acheulian assemblages: An analytic approach. In G. Mortelmans & J. Nenquin (Eds.), *Actes du IV° Congrès Panafricain de Préhistoire et d'Études du Quaternaire, Leopoldville, 1959* (pp. 81–108). Tervuren: Belgie Annalen, Musée Royal de l'Afrique Centrale.

Leakey, M. D. (1971). *Olduvai Gorge. Excavations in Bed I and II, 1960–1963* (Vol. 3). Cambridge: Cambridge University Press.

Leakey, M. D. (1975). Cultural patterns in the Olduvai sequence. In K. W. Butzer & G. L. Isaac (Eds.), *After the Australopithecines. Stratigraphy, Ecology, and Cultural Change in the Middle Pleistocene* (pp. 477–493). Chicago: Mouton.

Le Bourdonnec, F.-X. (2007). Aspects archéométriques de la circulation de l'obsidienne préhistorique. Développements analytiques et applications en Corse, Sardegne, Éthiopie. Ph.D. Dissertation, Université Michel de Montaigne Bordeaux 3, Pessac.

Leroi-Gourhan, A. (1964). *Le geste et la parole. Technique et langage.* Paris: Albin Michel.

Leroi-Gourhan, A. (1971). *Évolution et Technique. L'Homme et la Matière.* Paris: Albin Michel.

Lepre, C. J., Roche, H., Kent, D. V., Harmand, S., Quinn, R. L., Brugal, J.-P., et al. (2011). An earlier origin for the Acheulian. *Nature, 477,* 82–85.

Mohr, P. (1999). Le système des rifts Africains. Environnement géologique et géographique. In A. Gallay (Ed.), *Comment l'Homme ? A la découverte des premiers hominidés d'Afrique de l'Est* (pp. 231–288). Paris: Editions Errance.

Moncel, M. H. (1998). Les niveaux moustériens de la grotte de Saint-Marcel (Ardèche). Fouilles René Gilles. Reconnaissance de niveaux à débitage discoïde dans la vallée du Rhône. *Bulletin de la Société Préhistorique Française, 95,* 141–170.

Morgan, L. E., Renne, P. R., Kieffer, G., Piperno, M., Gallotti, R., & Raynal, J.-P. (2012). A chronological framework for a long and persistent archeological record: Melka Kunture, Ethiopia. *Journal of Human Evolution, 62,* 104–115.

Mourre, V. (2003). Discoïde ou pas Discoïde ? Réflexions sur la pertinence des critères techniques définissant le débitage Discoïde. In M. Peresani (Ed.), *Discoid Lithic Technology—Advances and implications* (pp. 1–18). Oxford: BAR International Series 1120.

Mussi, M., Altamura, F., Bonnefille, R., De Rita, D., & Melis, R. T. (2016). The environment of the Ethiopian highlands at the Mid Pleistocene Transition: Fauna, flora and hominins in the 850–700 ka sequence of Gombore II (Melka Kunture). *Quaternary Science Review, 149,* 259–268.

Mussi, M., Altamura, F., Macchiarelli, R., Melis, R. T., & Spinapolice, E. E. (2014). Garba III (Melka Kunture, Ethiopia): A MSA site with archaic *Homo sapiens* remains revisited. *Quaternary International, 343,* 28–39.

Negash, A., Shackley, M. S., & Alene, M. (2006). Source provenance of obsidian artifacts from the Early Stone Age (ESA) site of Melka Kunture, Ethiopia. *Journal of Archeological Science, 33,* 1647–1650.

Pasty, J.-F. (2000). Le gisement paléolithique moyen des Meillers (Allier) : un exemple de la variabilité du débitage discoïde. *Bulletin de la Société Préhistorique Française, 97,* 165–190.

Pelegrin, J. (1985). Réflexions sur le comportement technique. In M. Otte (Ed.), *La signification culturelle des industries lithiques* (pp. 72–91). Liege: Studia Praehistorica Belgica 4, BAR International Series 239.

Peresani, M. (Ed.). (2003). *Discoid Lithic Technology—Advances and implications.* Oxford: BAR International Series 1120.

Perlès, C. (1991). Économie des matières premières et économie du débitage : deux conceptions opposées ? In S. Beyries, L. Meignen, & P. J. Texier (Eds.), *25 ans d'études technologiques en Préhistoire : bilan et perspectives. Proceedings XIème rencontres internationales d'Archéologie et d'Histoire d'Antibes* (pp. 35–46). Juan-les-Pins: APDCA.

Piperno, M., & Bulgarelli, G. M. (2004). The site of Garba IV. Excavation 1973–1982. In J. Chavaillon & M. Piperno (Eds.), *Studies on the Early Paleolithic site of Melka Kunture, Ethiopia* (pp. 449–458). Florence: Origines.

Piperno, M., & Bulgarelli-Piperno, G. M. (1975). First approach to the ecological and cultural significance of the early Palaeolithic occupation site of Garba IV at Melka-Kunture (Ethiopia). *Quaternaria, XVIII,* 347–382.

Piperno, M., Bulgarelli, G. M., & Gallotti, R. (2004a). The site of Garba IV. The lithic industry of Level C. Typological and technological study. In J. Chavaillon & M. Piperno (Eds.), *Studies on the Early Paleolithic site of Melka Kunture, Ethiopia* (pp. 459–468). Florence: Origines.

Piperno, M., Bulgarelli, G. M., & Gallotti, R. (2004b). The site of Garba IV. The lithic industry of Level D. Débitage and tools on flake. In J. Chavaillon & M. Piperno (Eds.), *Studies on the Early Paleolithic site of Melka Kunture, Ethiopia* (pp. 469–544). Florence: Origines.

Piperno, M., Bulgarelli, G. M., & Gallotti, R. (2004c). The site of Garba IV. The lithic industry of Level D. Tools on pebble and percussion material. In J. Chavaillon & M. Piperno (Eds.), *Studies on the Early Paleolithic site of Melka Kunture, Ethiopia* (pp. 545–580). Florence: Origines.

Piperno, M., Bulgarelli, G. M., & Gallotti, R. (2004d). The site of Garba IV. The test trenches A and B and the sounding in Level E. In J. Chavaillon & M. Piperno (Eds.), *Studies on the Early Paleolithic site of Melka Kunture, Ethiopia* (pp. 581–587). Florence: Origines.

Piperno, M., Collina, C., Gallotti, R., Raynal, J.-P., Kieffer, G., Le Bourdonnec, F.-X., et al. (2009). Obsidian exploitation and utilization during the Oldowan at Melka Kunture (Ethiopia). In E. Hovers & D. R. Braun (Eds.), *Interdisciplinary Approaches to the Oldowan* (pp. 111–128). Dordrecht: Springer.

Poupeau, G., Kieffer, G., Raynal, J. P., Milton, A., & Delerue, S. (2004). Trace element geochemistry in Balchit obsidian (Upper Awash, Ethiopia). In J. Chavaillon & M. Piperno (Eds.), *Studies on the Early Paleolithic site of Melka Kunture, Ethiopia* (pp. 103–110). Florence: Origines.

Profico, A., Di Vincenzo, F., Gagliardi, L., Piperno, M., & Manzi, G. (2016). Filling the gap. Human cranial remains from Gombore II (Melka Kunture, Ethiopia; ca. 850 ka) and the origin of *Homo heidelbergensis*. *Journal of Anthropological Sciences, 94,* 1–24.

Quade, J., Levin, N., Semaw, S., Stout, D., Renne, P., Rogers, M. J., et al. (2004). Paleoenvironments of the earliest stone toolmakers, Gona, Ethiopia. *Geological Society of America Bulletin, 116,* 1529–1544.

Quade, J., Levin, N., Simpson, S., Butler, R., McIntosh, W., Semaw, S., et al. (2008). The Geology of Gona, Afar, Ethiopia. *Geological Society of America Bulletin, Special Paper, 446,* 1–31.

Raynal, J.-P., & Kieffer, G. (2004). Lithology, dynamism and volcanic successions at Melka Kunture (Upper Awash, Ethiopia). In J. Chavaillon & M. Piperno (Eds.), *Studies on the Early Paleolithic site of Melka Kunture, Ethiopia* (pp. 111–135). Florence: Origines.

Raynal, J.-P., Kieffer, G., & Bardin, G. (2004). Garba IV and the Melka Kunture Formation. A preliminary lithostratigraphic approach. In J. Chavaillon & M. Piperno (Eds.), *Studies on the Early Paleolithic site of Melka Kunture, Ethiopia* (pp. 136–166). Florence: Origines.

Roche, H., & Texier, P.-J. (1991). La notion de complexité dans un ensemble lithique. Application aux séries acheuléennes d'Isenya (Kenya). In S. Beyries, L. Meignen, & P.-J. Texier (Eds.), *25 ans d'études technologiques en Préhistoire : bilan et perspectives. Proceedings XIème rencontres internationales d'Archéologie et d'Histoire d'Antibes* (pp. 99–108). Juan-les-Pins: APDCA.

Salvi, M. C., Salvini, R., Cartocci, A., Kozciak, S., Gallotti, R., & Piperno, M. (2011). Multitemporal analysis for preservation of obsidian sources from Melka Kunture (Ethiopia): Integration of fieldwork activities, digital aerial photogrammetry and multispectral stereo-IKONOS II analysis. *Journal of Archeological Science, 38,* 2017–2023.

Schmitt, J.-J., Wempler, J.-M., Chavaillon, J., & Andrews, M. C. (1977). Initial K/Ar and paleomagnetic results of the Melka-Kunturé early-man sites, Ethiopia. In *Proceedings VIII Panafrican Congress of Prehistory and Quaternary Studies, Nairobi.*

Semaw, S. (2006). The Oldest Stone Artifacts from Gona (2.6–2.5 Ma), Afar, Ethiopia: Implications for understanding the earliest stages of stone knapping. In N. Toth & K. Schick (Eds.), *The Oldowan: Case Studies into the Earliest Stone Age* (pp. 43–75). Gosport, Indiana: Stone Age Institute Press.

Semaw, S., Rogers, M. J., Cáceres, I., Stout D., & Leiss, A. C. (2018). The Early Acheulean ~1.6–1.2 Ma from Gona, Ethiopia: Issues on the Emergence of the Acheulean in Africa. In R. Gallotti & M. Mussi (Eds.), *The Emergence of the Acheulean in East Africa and Beyond. Contributions in Honor of Jean Chavaillon.* Cham: Springer.

Sharon, G. (2010). Large flake Acheulan. *Quaternary International, 223–224,* 226–233.

Tamrat, E., Thouveny, N., Taieb, M., & Brugal, J. P. (2014). Magnetostratigraphic study of the Melka Kunture archeological site (Ethiopia) and its chronological implications. *Quaternary International, 343,* 5–16.

Taieb, M. (1974). Evolution quaternaire du bassin de l'Awash. Ph.D. Dissertation, University of Paris VI.

Terradas, X. (2003). Discoid flaking method: Conception and technological variability. In M. Peresani (Ed.), *Discoid Lithic Technology—Advances and implications* (pp. 19–32). Oxford: BAR International Series 1120.

Texier, P.-J. (2018). Technological Assets for the Emergence of the Acheulean? Reflections on the Kokiselei 4 Lithic Assemblage and Its Place in the Archaeological Context of West Turkana, Kenya. In M. Mussi & R. Gallotti (Eds.), *The Emergence of the Acheulean in East Africa and Beyond. Contributions in Honor of Jean Chavaillon*. Cham: Springer.

Texier, P.-J., & Roche, H. (1995). The impact of predetermination on the development of some Acheulan chaînes opératoires. In J. M. Bermudez, J. L. Arsuaga, & E. Carbonell (Eds.), *Human Evolution in Europe and the Atapuerca Evidence* (pp. 403–420). Valladolid: Regional Junta de Castilla y León.

Tixier, J. (1956). Le hachereau dans l'Acheuléen nord-africain. Notes typologiques. In *Proceedings of the XV Session of the Congrès préhistorique de France, July 15–22, 1956, Poitiers-Angoulême* (pp. 914–923).

Westphal, M., Chavaillon, J., & Jaeger, J. J. (1979). Magnétostratigraphie des dépôts Pléistocènes de Melka-Kunturé (Éthiopie) : premières données. *Bulletin de la Société Géologique Française, 7,* 237–241.

Zaidner, Y. (2013). Adaptive flexibility of Oldowan hominins: Secondary use of flakes at Bizat Ruhama, Israel. *PlosOne, 8,* e66851.

Zilberman, U., Smith, P., & Condemi, S. (2004a). Evidence for a genetic disorder affecting tooth formation in the Garba IV child. In J. Chavaillon & M. Piperno (Eds.), *Studies on the Early Paleolithic site of Melka Kunture, Ethiopia* (pp. 703–713). Florence: Origines.

Zilberman, U., Smith, P., Piperno, N., & Condemi, S. (2004b). Evidence of amelogenesis imperfecta in an early African *Homo erectus*. *Journal of Human Evolution, 46,* 647–653.

Chapter 5
Variability in the Mountain Environment at Melka Kunture Archaeological Site, Ethiopia, During the Early Pleistocene (~1.7 Ma) and the Mid-Pleistocene Transition (0.9–0.6 Ma)

Raymonde Bonnefille, Rita Teresa Melis, and Margherita Mussi

Abstract In this paper, we present and discuss pollen data from the Early Pleistocene (1.8 to 1.6 Ma) – we use the revised timescale approved by IUGS, in which the base of the Pleistocene is defined by the GSSP of the Gelasian Stage at 2.588 (2.6) Ma (Gibbard et al. 2010) – and from the Mid-Pleistocene Transition (0.9 to 0.6 Ma) at Melka Kunture (Upper Awash, Ethiopia). At 2000 m asl in the Ethiopian highlands, these deposits yield many rich and successive archaeological sites, notably documenting the late Oldowan, the emergence of the Acheulean and the middle Acheulean. The stratigraphic position of the fifteen pollen samples is checked by $^{40}Ar/^{39}Ar$ dating and by geological investigation. Furthermore, they are now correlated to archaeological layers whose excavated lithic industries have been reinterpreted. Our study shows that mountain forest trees belonging to the present-day Afromontane complex were already established in Ethiopia at ~1.8 Ma and that the knappers of the Oldowan and early Acheulean could cope with mountain climatic conditions that had a large diurnal temperature range. Moreover, the new interpretation of pollen results emphasizes changes that occurred in the vegetation cover at 200- or 300-thousand-year snapshot intervals, one during the Early Pleistocene and another one later on, during the Mid-Pleistocene Transition. These changes concerned plant species and their respective abundance and appear to have been related to rainfall and temperature variability. The proportion of forest trees increased during wet episodes, whereas the influence of Afroalpine grassland indicators increased during cool and dry episodes. Variations in Early Pleistocene pollen data from Melka Kunture at ~1.8–1.6 Ma are consistent with isotopic evidence of precession variability as recorded at Olduvai and Turkana archaeological sites at ~2–1.8 Ma. For the Mid-Pleistocene Transition, variations in pollen data seem to match the climatic variability of isotopic and long pollen records from the Mediterranean region, notably upon the onset of dominant 100 ka-long glacial/interglacial cycles.

Keywords Melka Kunture • Early Pleistocene • Mid-Pleistocene Transition • Pollen • Mountain forest history • Hominin adaptation

5.1 Introduction

Melka Kunture is known for its extensive archaeological and paleontological record, spanning over most of the Early, Middle, and Upper Pleistocene (Chavaillon and Piperno 2004). Located in Ethiopia 50 km southwest of Addis Ababa, at 2000 m asl, in the Upper Awash valley and on the border of the Northern Ethiopian plateau, this cluster of archaeological sites is one of the few known mountain habitats of early hominins (Fig. 5.1). It is situated at a slightly lower altitude than the Gadeb site in the Bale Mountains (Southern Ethiopia), unfortunately inaccessible since it was flooded after the construction of a dam (Clark and Kurashina 1979; de la Torre 2011). The Melka Kunture sites yield evidence of Oldowan assemblages at ~1.7 Ma (Gallotti and Mussi 2015) and record the development of the

R. Bonnefille
CNRS, Université Aix-Marseille, UMR 7330, Cerege BP 80, 13545 Aix-en-Provence, France
e-mail: rbonnefille@orange.fr

R.T. Melis
Dipartimento di Scienze chimiche e geologiche, Università di Cagliari, Via Trentino 51, 09127 Cagliari, Italy
e-mail: rtmelis@unica.it

R.T. Melis · M. Mussi (✉)
Italian Archaeological Mission at Melka Kunture and Balchit
e-mail: margherita.mussi@uniroma1.it

M. Mussi
Dipartimento di Scienze dell'Antichità, Sapienza Università di Roma, Piazzale A. Moro 5, 00185 Rome, Italy

R. Gallotti and M. Mussi (eds.), *The Emergence of the Acheulean in East Africa and Beyond: Contributions in Honor of Jean Chavaillon*, Vertebrate Paleobiology and Paleoanthropology,
https://doi.org/10.1007/978-3-319-75985-2_5

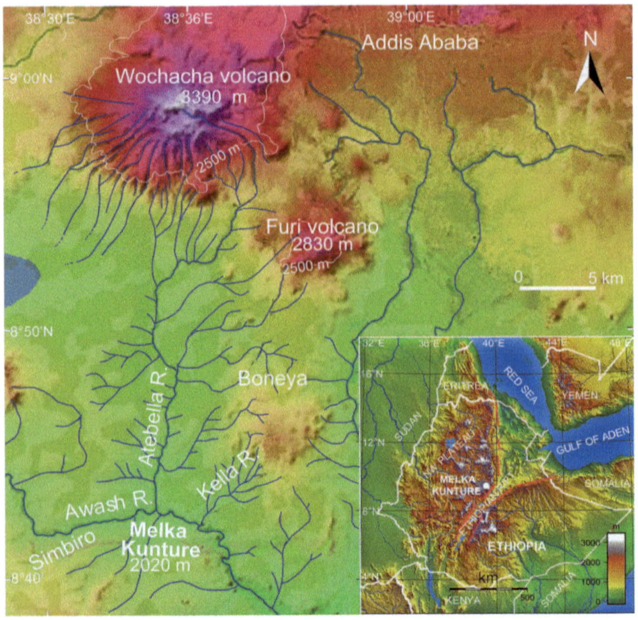

Fig. 5.1 Location map of Melka Kunture archaeological site in the Upper Awash valley, on the border of the Rift Escarpment, Ethiopia

Acheulean starting from its emergence at ~1.6 Ma (Gallotti 2013; Gallotti and Mussi 2018). Its location in the Ethiopian highlands is unlike that of other Acheulean sites in East Africa: those in the Ethiopian Rift, at 500 m asl (de Heinzelin et al. 2002); around Lake Turkana in Kenya (Isaac and Isaac 1997; Roche et al. 2003; de la Torre 2004; Lepre et al. 2011); and around Lake Natron as Peninj in Tanzania (de la Torre et al. 2008). Other Acheulean sites such as Olduvai in Tanzania (Leakey 1971) and Konso in southwestern Ethiopia (Beyene et al. 2013) are situated at medium altitude, i.e. at 1500 m asl.

Past vegetation is one of the environmental features that may help us understand the emergence of technological changes in lithic productions (Gallotti et al. 2010) and the long-lasting occupation of the Ethiopian highlands. Isotopic analysis and Carbon δC^{13} measurements on soil carbonates or organic matter provide information on the proportion of C_3 versus C_4 plants in the vegetal cover, a ratio that can be interpreted as an estimate of tree cover density in past vegetation (Cerling et al. 2011; Magill et al. 2013). However, at high altitudes the C_3 signal may also reflect a certain proportion of C_3 grass. Moreover isotopic analyses do not yield knowledge regarding the botanical identification of the plant themselves, a critical information that is necessary for assessing vegetation type, hominin diet (Teaford 2000; Goren-Inbar et al. 2007, 2014) or habitat. Pollen grains are distinctive of the plants that produce them and provide such botanical precision. In this paper, we present and discuss fossil pollen data linked to the discovery of two hominin fossils: a massive humerus of *Homo* sp. found at ~1.6 Ma in level Gombore IB with early Acheulean lithic productions (Chavaillon et al. 1977; Di Vincenzo et al. 2015; Gallotti and Mussi 2018) and cranial fragments of *Homo* cf. *heidelbergensis* associated with the middle Acheulean of Gombore II-1, which is slightly younger than 0.875 Ma (Chavaillon et al. 1974; Chavaillon and Coppens 1975, 1986; Profico et al. 2016). Within this time interval, fossil pollen data were obtained from 15 distinct stratigraphic layers, that are partly associated with Oldowan and Acheulean lithic implements. They are centered around two critical periods: 1.8–1.6 Ma and 0.9–0.6 Ma, and there is additional information at ~1.4 Ma.

The sites are named after the gully or valley where they are located, for example Garba (or Gombore or Kella), followed by a Roman numeral that refers to the order in which they were discovered (e.g. Gombore II was discovered after Gombore I). At Gombore II, two distinct archaeological horizons mentioned in this paper are named Gombore II-1 and Gombore II-2, respectively. KII and KIII refer to geological sections in the Kella valley, as reported in Taieb (1974).

5.2 General Background

Melka Kunture (8°42 N; 38°36 E) is located in the tropical zone, but at high altitude, a geographical environment which modulates climatic conditions. At 2000 m asl, the maximum temperature averages 20° and the minimum one 8°C. Large diurnal temperature variations and nighttime frost may occur. The mean annual rainfall is about 860 mm at Boneja, the nearest meteorological station, 20 km north of Melka Kunture. There is great annual variability in the 1974–2010 record, with a minimum value of 388 mm, and a peak of 1419 mm. Rainfall distribution is usually bimodal, with a short rainy season in spring and long summer rains from June to September. Monsoon rainfall is primarily controlled by the seasonal migration of the Inter Tropical Convergence Zone (ITCZ) which lies north of Ethiopia at that time. The dry season is from October to February when the ITCZ lies south of Ethiopia. The northeasterly winds that blow in winter carry little moisture from the Red Sea and are dry when they reach central Ethiopia. The short rainy season is from March to May, when the ITCZ moves from south to north.

5.2.1 Vegetation

In the upper Awash valley of central Ethiopia, the archaeological area is surrounded by pastoral and intensely cultivated land that leaves very small patches of natural vegetation. In this region, agriculture developed thousands of years ago (Terwilliger et al. 2011). After a pastoral Oromo migration, in the second half of the sixteenth century, cereal cultivation was replaced by pastoralism and abandoned areas turned into bushland (Zerihun and Backéus 1991). However, the "potential vegetation" is mapped by botanical studies as a unit of the "Dry evergreen Afromontane forest and grassland complex" (DAF) (Friis et al. 2010). DAF vegetation, widespread on the Ethiopian plateaux between 1800 and 3000 meters, includes mountain forests, woodlands, wooded grasslands and grasslands. It is a complex unit where the various vegetation types are characterized by heterogeneous distribution of trees or shrubs and herbaceous plants, but include the same botanical components. The Afromontane vegetation type is clearly distinguished from that of lowland savanna or woodland by the botanical attribution of characteristic plants, notably trees. It is defined by the presence of tall tree species such as *Juniperus procera* (Cupressaceae), *Podocarpus falcatus* (Podocarpaceae), *Olea europea* (Oleaceae), *Croton macrostachys* (Euphorbiaceae) associated with smaller trees such as *Celtis africana* (Ulmaceae) and with shrubs among which *Myrsine africana* (Myrsinaceae) frequently occurs. Mountain woodland, wooded grassland, and grassland are DAF subtypes and include many species of *Acacia* (Fabaceae, subfamily Mimosoideae) together with many *Maytenus obscura* (Celastraceae) and several species of endemic *Echinops* (Asteraceae). At its lower altitudinal limit, a drier type of vegetation marks the transition between Afromontane vegetation and the lowland vegetation described as *Acacia-Commiphora* bushland. Scattered *Juniperus procera* may occur, sometimes associated with *Dracaena*, *Barbeya oleoides* (Barbeyaceae), *Tarchonanthus camphoratus* (Asteraceae), or *Dodonaea angustifolia* (new name for *Dodonaea viscosa*) (Sapindaceae) and with various species of *Euclea* (Ebenaceae). The vegetation zone located at 1600 m asl, like the one around the lakes in the Rift, is occupied by *Combretum-Terminalia* (Combretaceae) woodland (CTW), and by wooded grassland (Friis et al. 2010). Above the Afromontane forest, at ~3000 m asl, the vegetation consists of *Hagenia abyssinica-Hypericum revolutum* woodland, overlapping in part with an Ericaceous bushland dominated by *Erica arborea*; Asteraceae and *Artemisia* abound in the Afroalpine grassland up to 4000 m asl (Friis 1992; Friis and Sebsebe 2001). This altitudinal distribution of modern vegetation is observed along the slopes of the Wochacha volcano, some 20 km away from Melka Kunture (Fig. 5.1). The succession of woodland, mixed forest and juniper-dominated forest, followed by remnants of the Afroalpine bushland on the top, reflects variations in ecological conditions along the altitudinal gradient. Between 2200 and 3200 m the mean annual temperature decreases by about 5 °C to 6 °C, whereas rainfall increases up to 3000 m (upper cloud limit) and then decreases above the cloud forest in the Afroalpine vegetation. Pollen analysis of present-day surface soil samples, collected under different types of vegetation still preserved on the slopes of the Wochacha and Zuqualla volcanoes, clearly indicates several pollen markers

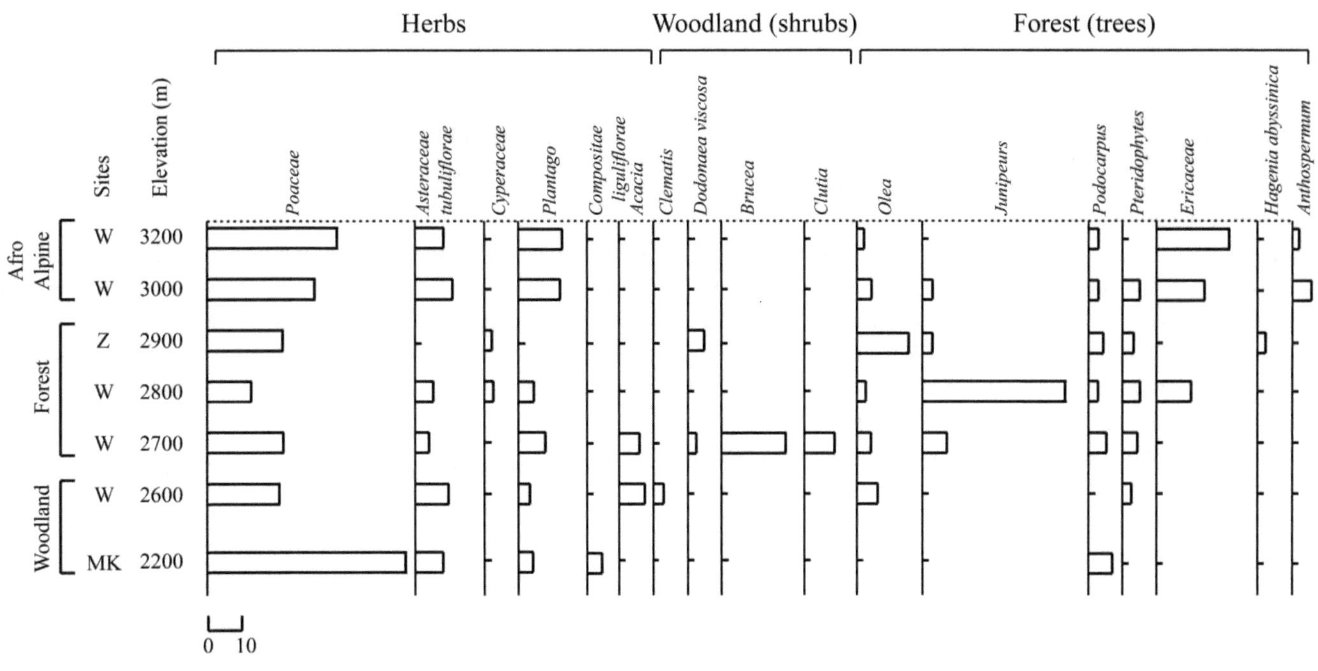

Fig. 5.2 Distribution of selected pollen markers in pollen analysis of soil samples collected in modern vegetation zones encountered along an altitudinal gradient on slopes of the Wochacha and Zuqualla volcanoes. New name for *Dodonaea viscosa* is *Dodonaea angustifolia*

of altitudinal vegetation zones (Fig. 5.2). Grass pollen dominates in the lowermost woodland. *Dodonaea* shrubs and *Acacia* indicate the transition zone, while the forest is indicated by abundant *Olea* and *Juniperus* pollen. Ericaceae and *Plantago* are good markers of the Afroalpine vegetation.

5.2.2 Geology, Chronology, and Archaeology

The Melka Kunture sedimentary basin lies on the shoulder of the northern Ethiopian plateau, near the western margin of the active Wonji fault belt that marks the axis of the Ethiopian Rift (Mohr 1971). Outcrops of Pleistocene deposits exposed on both banks of the Awash River hardly amount to a total combined thickness of 40 meters. Silt, sand, clay, and coarse layers inter-bedded with tephra falls and tuffaceous clay were deposited by the Awash River and its tributaries. Fluvial sedimentation proceeded mainly in an alluvial braided context influenced by the deposition of volcanic ashes and lavas. The fluvial-volcanic succession accumulated on top of the "basal clays", deposited when a shallow body of water had developed in the basin. The "basal clays" were initially measured by geophysical echosounding and found a few meters thick near the mouth of the Garba gully (Taieb 1974). A normal paleomagnetic signal places their deposition at the end of the Olduvai subchron, with an upper limit at 1.778 Ma (Tamrat et al. 2014). Recent fieldwork enabled additional observation of the basal clays in the Kella tributary valley. Evidence of erosion at the top of the clays was mentioned in Jean Chavaillon's archives and observed again recently, when water level was low, along the Kella tributary close to its confluence in the Awash River. The earliest volcanic deposit (tuff sample 27-17) capping the basalt clays was dated at 1.874 ± 0.012 Ma (Morgan et al. 2012). Accordingly, we consider that the extant, eroded basal clays are related to the lower part of the Olduvai paleomagnetic event. The earliest dated tuff provides at ~1.9 Ma the lower limit for the archaeological and paleontological record of Melka Kunture.

Soon after the area's archaeological potential was recognized (Bailloud 1965), exploration surveys and extensive excavations unearthed abundant Oldowan and Acheulean artifacts, as well as later lithic productions (Chavaillon 1971, 1979), whereas geological studies (Taieb 1974) were undertaken together with pollen sampling (Bonnefille 1972). Discontinuities in the sedimentary record, some indicated by erosion surfaces, as well as the lack of easily recognizable marker beds, made it difficult at first to establish stratigrafic correlations between exposures outcropping in different gullies, partly covered by abundant bushy vegetation. Initially, the succession of Oldowan, Acheulean, Middle Stone Age, and Late Stone Age lithic productions helped place the local stratigraphy within the Pleistocene through typological comparison with dated industries from Olduvai (Chavaillon 1973, 1980; Chavaillon and Piperno 2004). An estimated time range from at least 1.5 Ma to the last ten thousand years was supported by early paleomagnetic measurements (Westphal et al. 1979; Cressier 1980) until $^{40}Ar/^{39}Ar$ dating provided a more precise chronological framework (Morgan et al. 2012). With the help of new fieldwork and recent archaeological investigation (Gallotti et al. 2014; Mussi et al. 2014, 2016), formerly collected pollen samples are now securely placed in this new chronology.

Initial sampling for pollen was done in the sedimentary sequences along three gullies: Garba (G) and Gombore (Gomb) which are tributaries of the Awash River on the right bank, and mostly dry all year long, and Kella (K) on the left bank (Fig. 5.3). A few more samples were collected directly on excavated archaeological layers. Accordingly, they are numbered from top to bottom as it is usual in archaeological practice. The excavated archaeological localities are Gombore I (level C and overlying level B) and Garba IV (levels E and F and overlying level D). The pollen data discussed in this paper are constrained by $^{40}Ar/^{39}Ar$ dating and tephrochronology in a time interval starting at ~1.8 Ma and ending before 0.709 ± 0.013 Ma (Morgan et al. 2012), except for one sample which is more recent. Although initially collected at more or less regular depth interval throughout the sedimentary sequences, in light of current fieldwork the stratigraphic positions of the pollen samples appear to be grouped in two time periods. In the Early Pleistocene sequence, the pollen data correspond to the time of the late Oldowan and emerging Acheulean (1.8–1.6 Ma); during the Mid-Pleistocene Transition (0.9–0.6 Ma), they document past vegetation of the middle to final Acheulean.

5.3 Pollen Data from the Early Pleistocene, ~1.8–1.6 Ma

Past vegetation in the 1.8–1.6 Ma period is documented by eight pollen samples collected in deposits at the base of stratigraphic successions in the Kella, Garba, and Gombore gullies. Intercorrelation between the three sections relies on current research that focuses on volcanic material.

5.3.1 Chronostratigraphic Placement of the Pollen Samples

Four samples come from two sections, Kella II and Kella III, which are exposed 500 m apart from each other in the Kella valley (Fig. 5.3). Three of them were collected in a three-meters sedimentary sequence outcropping on a cliff at Kella III. From bottom to top, sample K263 was extracted from

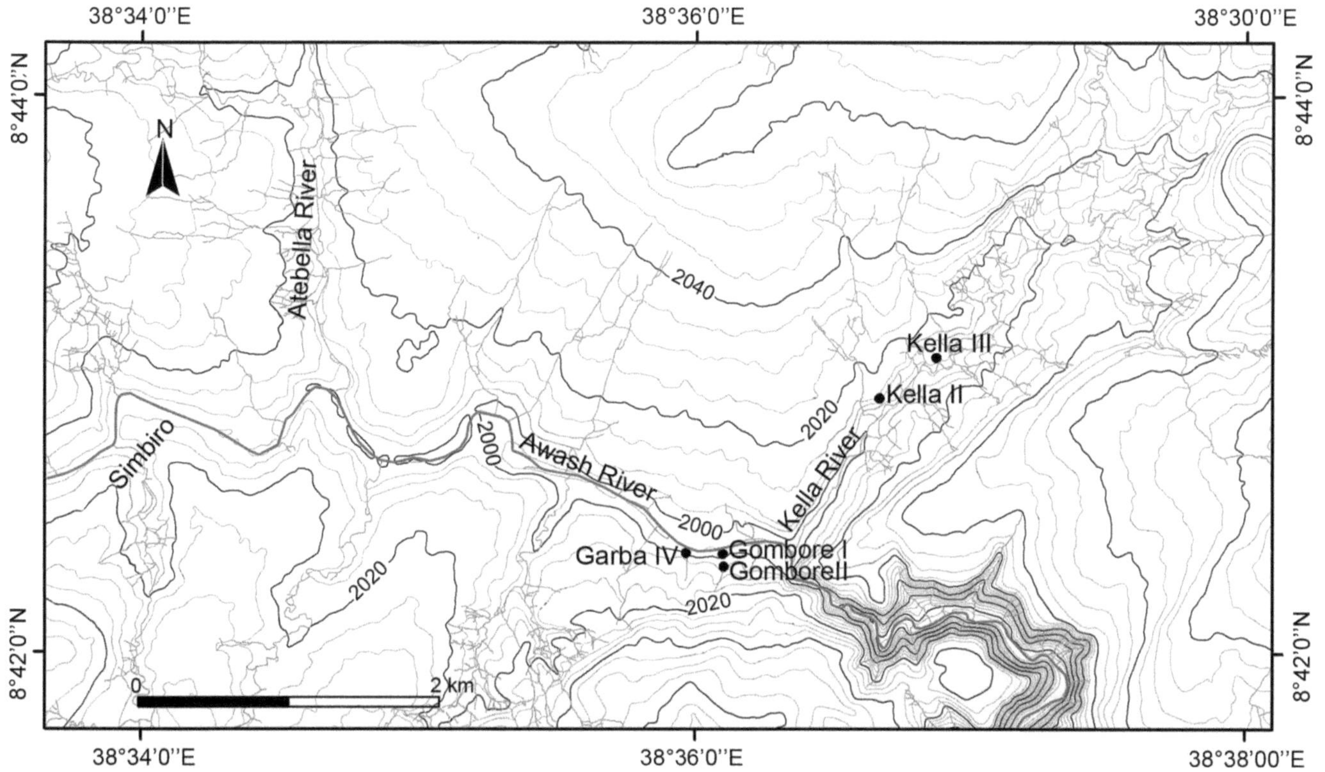

Fig. 5.3 Melka Kunture, location map of the litho-stratigraphic sections and archaeological localities mentioned in the text

clayey sands with obsidian gravels, K265 from sands and K269 from cross-bedded sands (Fig. 5.4). The Kella III sequence is overlaid, 2 m above K269, by a tuff which, based on stratigraphic correlation of the outcrops along the Kella valley, corresponds to tuff layer MK27-18, dated 1.666 ± 0.009 Ma, which lies a few hundred meters downstream (Morgan et al. 2012). At Kella II, also called *Butte Kella*, pollen sample K242 comes from the infill of a bone extracted from a sand layer capping this same dated volcanic ash (Bonnefille 1968; Taieb 1974). Therefore K242 is slightly younger than 1.666 Ma whereas K263, K265, and K269 are older (Table 5.1). The stratigraphic position of these samples is well constrained by the 1.666 ± 0.009 Ma dated tuff intercalated between samples K269 and K242.

Further chronological constraint is provided by Early Pleistocene deposits that are likewise exposed above the basal clays in the Garba and Gombore gullies near the Awash River, two km downstream (Fig. 5.3).

Two pollen samples from the excavated archaeological horizons Gombore IC and Gombore IB are separated from each other by an interval of few dozen centimeters (Chavaillon and Berthelet 2004). Archaeological layer Gombore IC lies 10–15 cm below a tuff consisting of ryolithic ashfall material that showed an elementary composition “quasi-identical” to the “Grazia Tuff” (Raynal and Kieffer 2004, p. 152), dated at <1.719 ± 0.199 Ma at the Garba IV archaeological site (sample 27-23 in Morgan et al. 2012) (Fig. 5.5). Layer Gombore IB, 20 cm above the local ryolithic tuff, is considered to be ~1.6 Ma (Gallotti and Mussi 2018).

In the Garba gully, the lowest and earliest Oldowan horizons, i.e., Garba IVE and Garba IVF, are older than the <1.719 ± 0.199 Ma “Grazia tuff” (Piperno et al. 2009; Morgan et al. 2012; Gallotti and Mussi 2015) (Fig. 5.5). Pollen samples were not collected either from layer Garba IVF, or from layer Garba IVE. This is unfortunate since the latter layer yielded a hominin child mandible, tentatively attributed to *Homo erectus s.l.* (Condemi 2004; Zilberman et al. 2004). We have no palynological samples from archaeological horizon Garba IVD, which caps the <1.719 ± 0.199 Ma tuff and contains implements previously attributed to the Oldowan (Chavaillon and Piperno 2004) and now to the earliest Acheulean of Melka Kunture (Gallotti 2013). However two pollen samples had been collected (Bonnefille 1972) from an exposed section later described by Kieffer et al. (2002, Fig. 21, p. 89). Pollen sample G393 extracted from a manganese clay layer with embedded obsidian artifacts corresponds well to archaeological horizon Garba IVD. Sample G133 from a fine cross-bedded sand with bones 80 cm above this horizon is equivalent to subunits 3.4 or 3.5 in Kieffer et al. (2002, Fig. 21, p. 89). This sample is much older than tuff 27-19 dated 1.429 ± 0.029 Ma and located 4 m higher up in the stratigraphy (Morgan et al. 2012).

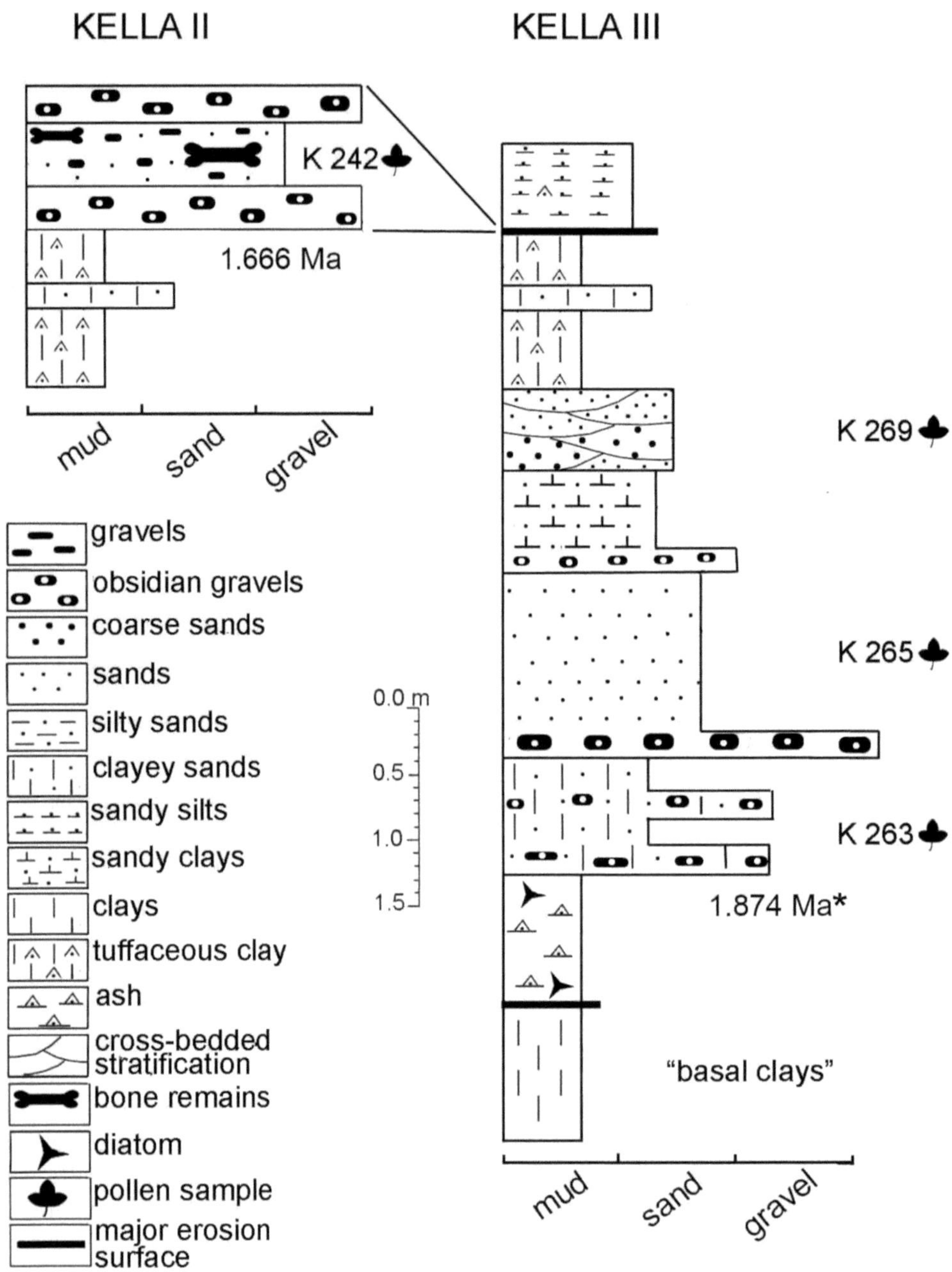

Fig. 5.4 Stratigraphic position and correlation between the Early Pleistocene (1.8–1.6 Ma) pollen samples collected at Kella, with respect to dated volcanic deposits. *The dated sample was collected 80 m downstream of Kella III

The <1.719 ± 0.199 Ma "Grazia tuff" at Garba IV and the undated ash layer from Gombore I, which are related to the same volcanic activity, are used as a marker bed for correlating the deposits in the two sections. The ~1.66 Ma tuff at Kella III, in turn, is correlated to this marker bed by geostratigraphy. All the samples detailed above were collected in close stratigraphic proximity to each other and are considered to date from ~1.7 Ma. Although the time span cannot be estimated exactly, no more than about two hundred thousands years separate the oldest K263 from the youngest G133 (Table 5.1).

5.3.2 Pollen Results from the Early Pleistocene, ~1.8–1.6 Ma

The eight samples listed in Table 5.1 provided abundant pollen that document past vegetation around Melka Kunture during the Early Pleistocene. A total of 58 pollen taxa have been identified; 32 of them are trees and 8 to shrubs and climbers. The others are from herbs, aquatic plants, and ferns (Table 5.2). The great number of tree species indicates a highly diversified woody vegetation, although the high

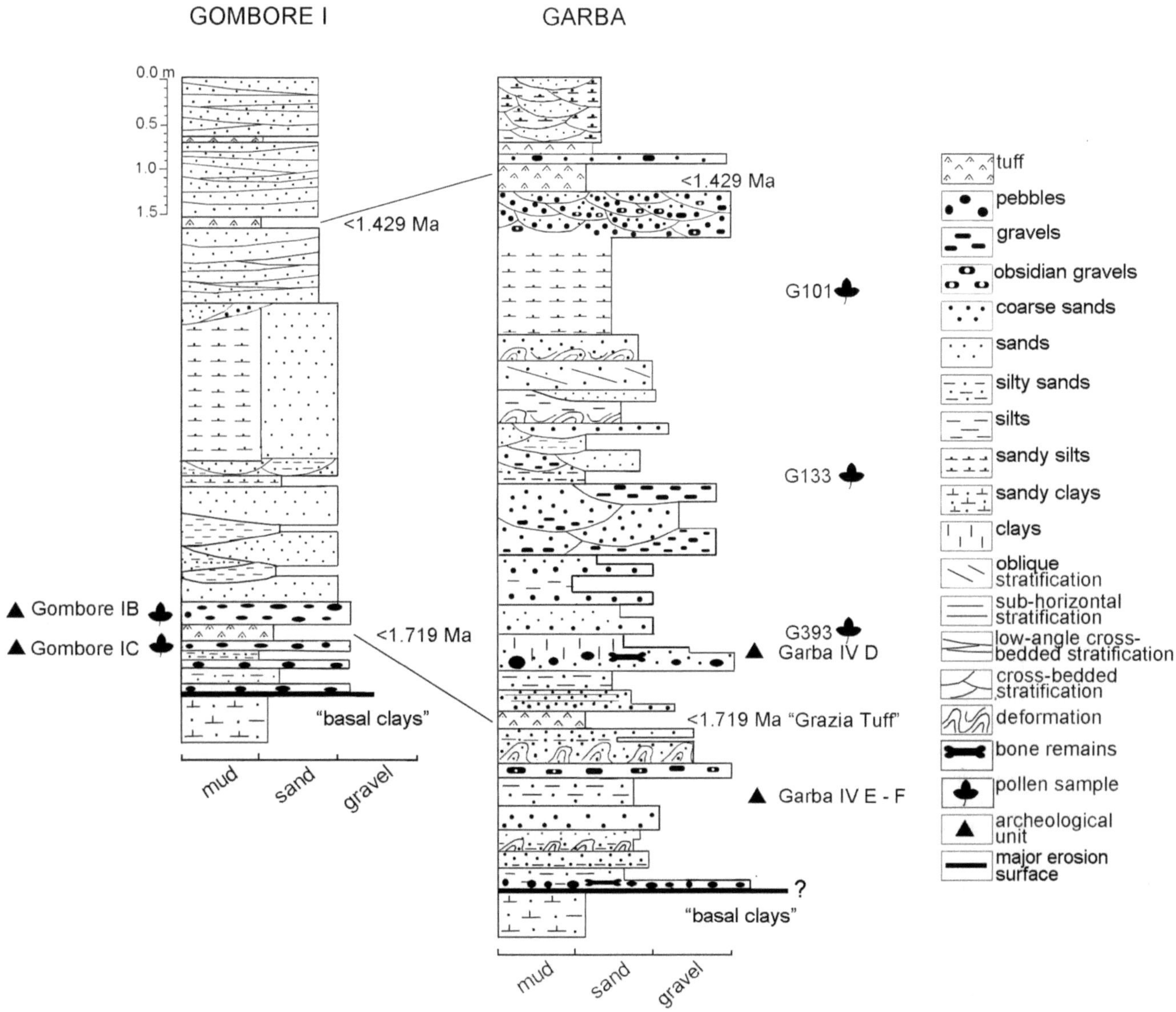

Fig. 5.5 Stratigraphic position and correlation between the Early Pleistocene (1.8–1.6 Ma) pollen samples collected at Gombore and Garba, with respect to dated volcanic deposits

Table 5.1 Chronostratigraphical constraints of the Early Pleistocene pollen samples according to the position versus dated tuffs (Morgan et al. 2012), the magnetostratigraphy (Tamrat et al. 2014) and the results of new fieldwork

Kella	Gombore	Garba
		G133
		Tuff <1.6; >1.4 Ma
	Gombore IB sample *Gombore IB—early Acheulean level*	**G393** *Garba IVD—early Acheulean level*
Tuff 1.666 ± 0.009 Ma	***Undated tuff* cfr *Grazia Tuff***	***Grazia Tuff* <1.719 ± 0.199** Ma
	Gombore IC sample *Gombore IC—Oldowan level*	
K242		
K269		
K265		
K263		
Tuff 1.874 ± 0.012 Ma		
eroded basal clays, predating the upper limit of the Olduvai event at 1.778 Ma		

Table 5.2 Detailed pollen counts of Melka Kunture samples dating ~1.8–1.6 Ma. AP: arboreal pollen from trees; S: shrubs; C: climbers. NAP: non-arboreal pollen (herbs); Ind.: indetermined. * = pollen from similar but different species or genera have the same morphology, producing uncertainty in the attribution to a specific genus or species

Plant form	Pollen Taxa	K 263	K 265	K 269	K 242	Gomb IC	Gomb IB	G 393	G 133
AP	Acacia (2 sp.)	103	7	3	–	–	–	–	1
NAP	Achyranthes aspera	–	–	20	1	–	–	–	–
S	Amaranthaceae	–	–	–	1	–	–	–	–
NAP	Anthospermum	–	–	–	–	–	3	–	–
NAP	Apiaceae	–	1	1	2	1	2	–	–
NAP	Arabis	–	–	–	–	–	1	–	–
AP	Arecaceae Palmae	–	–	–	16	–	–	–	–
NAP	Asteraceae tubuliflorae	63	179	7	7	1	2	10	6
NAP	Asteraceae cichoriae	22	13	–	–	2	–	–	–
NAP	Brassicae	–	–	–	3	–	–	–	–
AP	Cadia	–	–	–	1	–	–	–	–
NAP	Caylusea	58	26	–	–	–	–	–	–
AP	Calpurnia	–	–	–	1	–	–	–	–
NAP	Carduus	–	–	–	–	–	1	–	–
AP	Carissa edulis	–	–	–	–	–	1	–	–
NAP	Catananche	–	9	–	–	–	–	–	–
AP	Celtis	–	–	–	–	–	–	–	5
NAP	Chenopodiaceae	–	1	–	7	–	4	–	1
C	Cissus* quadrangularis	–	–	–	1	–	–	–	–
C	Clematis	–	–	–	1	18	1	–	–
AP	Combretaceae	–	–	1	1	–	–	–	–
AP	Cordia	–	1	–	–	–	–	–	–
NAP	Cyperaceae	–	–	–	5	–	11	–	2
S	Dodonaea angustifolia	–	–	8	4	1	1	–	3
NAP	Dyschoriste	–	–	–	3	–	–	–	–
AP	Encephalartos	–	10	–	32	–	–	–	–
AP	Euclea	–	–	1	2	–	–	–	–
NAP	Euphorbiaceae	–	–	–	1	–	–	–	–
AP	Fabaceae	5	69	–	3	–	–	–	–
AP	Hagenia abyssinica	–	–	–	1	–	–	–	–
AP	Heteromorpha	–	–	–	–	–	2	–	–
AP	Hypericum	–	75	–	–	–	2	–	–
AP	Juniperus procera	–	–	–	–	3	100	–	–
AP	Loganiaceae	–	–	–	1	–	–	–	–
AP	Myrica salicifolia	–	–	–	2	–	1	–	2
AP	Myrsine africana	–	–	1	1	–	–	1	–
AP	Olea sp. (2 sp.)	–	1	2	6	–	2	–	–
AP	Piliostigma thonningii	–	1	–	1	–	–	–	–
AP	Pistacia	–	–	–	3	–	–	–	–
NAP	Plantago africana*	–	–	1	17	5	6	2	2
NAP	Plectranthus	–	–	–	–	–	1	–	–
NAP	Poaceae	–	11	38	169	86	290	96	87
AP	Podocarpus	–	–	4	11	19	11	1	–
AP	Polyscias ferruginea	–	–	–	–	–	1	–	–
AP	Psychotria	–	–	–	5	–	–	–	–
NAP	Pteridophyta	–	–	–	2	1	9	1	–
AP	Rhamnaceae	–	–	123	–	–	–	–	–
C	Rhynchosia	4	–	–	–	–	1	–	–
S	Rumex	–	–	–	2	–	1	–	–
AP	Syzygium guineense	–	–	–	8	–	–	–	–
AP	Tamarix	–	–	–	2	–	–	–	–

(continued)

Table 5.2 (continued)

Plant form	Pollen Taxa	K 263	K 265	K 269	K 242	Gomb IC	Gomb IB	G 393	G 133
AP	Trema	–	–	–	–	–	–	–	3
NAP	Typha	–	–	–	–	1	1	–	–
	Not Determined	–	2	2	8	2	1	–	1
Sum	**Total**	**255**	**406**	**212**	**331**	**140**	**456**	**111**	**113**
	Total AP	**108**	**164**	**135**	**103**	**22**	**120**	**2**	**11**
	% AP	**42.4**	**40.4**	**63.7**	**31.1**	**15.7**	**26.3**	**1.8**	**9.7**

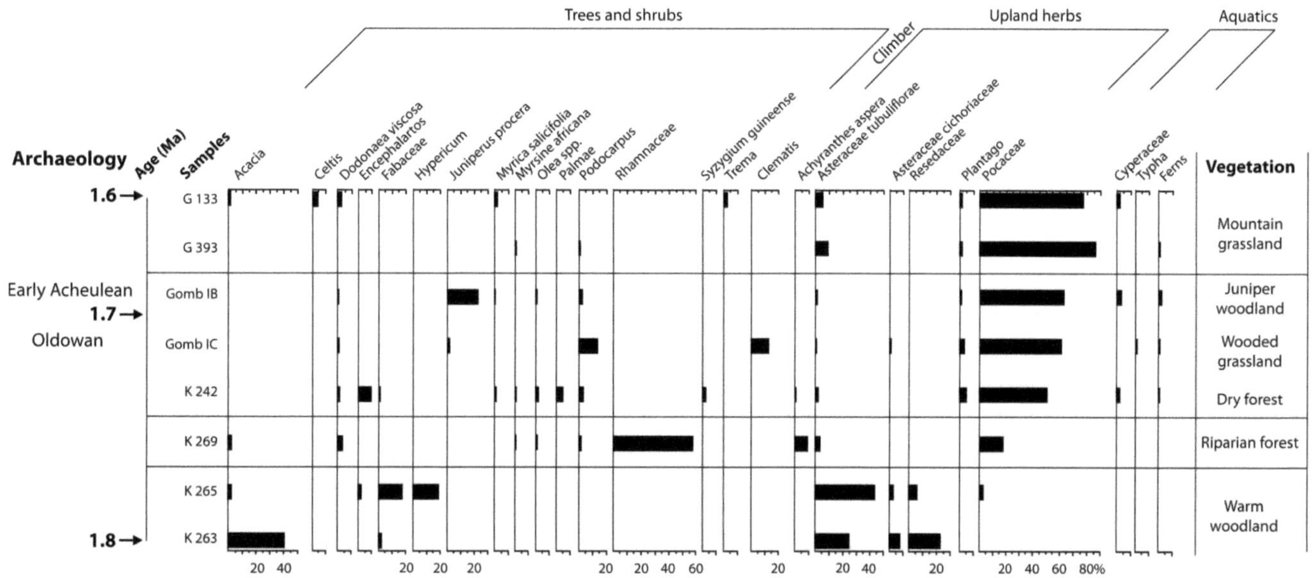

Fig. 5.6 Percentage distribution for selected pollen taxa identified in samples of the Early Pleistocene (1.8–1.6 Ma), with reference to the corresponding archaeological levels and the proposed interpretation of past vegetation. Relative pollen % are calculated versus the total pollen sum. Pollen not represented in this graph are found in Table 5.2. Abbreviations: G for Garba, Gomb for Gombore, K for Kella. New name for *Dodonaea viscosa* is *Dodonaea angustifolia*

proportion of grass pollen (Poaceae) points to a local abundance of herbaceous plants (Fig. 5.6).

The oldest K263 pollen spectrum (n = 255) shows an unusual distribution of pollen among Asteraceae, Resedaceae and Fabaceae (including *Rhynchosia*) and *Acacia*, mostly insect pollinated plant taxa. No proportion of *Acacia* as high as 40% is known from any of the many soil samples analyzed from different vegetation types in tropical Africa (Bonnefille 2007). However, the association of *Acacia tortilis* and Fabaceae among trees, together with *Caylusea*, an herb of the Resedaceae family, characterizes the vegetation along wadies on the Tibesti slopes in the Sahara (White 1983). Particular events such as instantaneous ash flooding of the Kella River may have concentrated pollen from plants growing along streambed before the ashfall.

The next K265 sample (n = 406) contains a few more pollen taxa than those found in the previous sample. The abundance of Asteraceae pollen together with the occurrence of Resedaceae and the increase in Fabaceae are noteworthy. The abundant Asteraceae pollen includes *Tarchonanthus*, a tall shrub widespread in Africa (White 1983) and abundant in dry *Juniperus-Olea* woodlands or forests, which are common in southern Ethiopian highlands close to Somalia (Friis 1992). This interpretation would be consistent with the finding of a single grain of *Olea* and of *Piliostigma*, whereas there is a much smaller amount of *Acacia* pollen, and a much lower proportion of Poaceae.

Pollen assemblage K269 (n = 212) shows abundant and diversified tree pollen taxa, while Asteraceae pollen is less abundant. Among the tree and shrub taxa, the occurrence of Combretaceae, *Euclea, Dodonaea angustifolia, Podocarpus, Olea africana,* and *Myrsine africana* attests to a more diversified type of evergreen bushland with some forest trees. The abundance of a pollen type attributed to *Ziziphus* (Rhamnaceae) indicates a woody riparian vegetation, confirmed by the abundance of *Achyranthes aspera*, a shade-loving herb, frequent on river banks. The percentages of grasses increase. A few additional pollen grains extracted from another uppermost sample confirm the occurrence of *Celtis, Syzygium, Trilepisium* (Table 5.2) associated with *Ricinus communis* and indicate enriched wooded riverine vegetation.

Pollen sample K242 (n = 331) includes a large number of pollen taxa. Among them there are various tree taxa, the majority of which can be attributed to the mountain dry forest or evergreen bushland, whereas a few of them belong to a riparian forest (Bonnefille 1968). However, the high percentage of grass pollen and the occurrence of herbs attributed to *Plantago*, Brassicaceae (former Cruciferae) or Apiaceae (former Umbelliferae) attest to the opening-up of the tree cover, interpreted as mountain grassland. At the headwaters of the Atebella, a left tributary of the Awash River, the Wochacha volcano, already formed 4 Ma ago (Chernet et al. 1998), rises to 3400 m asl (Fig. 5.1). At about 20 km from Melka Kunture, its slopes show a succession of vegetation types at different altitudinal zones. The various vegetation provided pollen input in sediments deposited at Melka Kunture. There were also plants such as Arecaceae (former Palmae) (15%) and *Encephalartos* (10%) that belongs to the botanical family Zamiaceae of the Cycadae order, among the Gymnosperms. *Encephalartos* pollen is distinguished by its large size and elongated shape (30 to 40 μm), smooth exine, and invaginated *sulcus* (Fig. 5.7). The pollen does not disperse very well, like in the case of palms. This fossil pollen was compared to modern pollen of several known African species and tentatively attributed to *Encephalartos hildebrandtii* (Bonnefille 1975). The genus *Encephalartos* is native to Africa and several of its species are found in South Africa, Congo, Tanzania, Kenya, Sudan, etc. Commonly known as "bread palm" because a bread-like food can be prepared from its pith, *E. hildebrandtii* grows in the Usambara mountains in Tanzania. Associated with *Brachylaena*, it was an important component of dry woodlands near the coast, north of Lamu, in the Lunghi forest of Kenya (Pichi-Sermolli 1957). A few isolated *Encephalartos* tree have been described on remote hills, for instance at Ololokwe, in central Kenya. Therefore, the fossil *Encephalartos* pollen recorded at Melka Kunture may have come from a relic of Pliocene vegetation whose geographic distribution once extended to Ethiopia, north of its present-day range. In sample K242, a larger proportion of *Encephalartos* is accompanied by pollen from palm trees, associated with *Olea, Hagenia abyssinica, Pistacia, Psychotria, Polyscias*, Loganiaceae, *Myrsine africana,* and *Dodonaea angustifolia*; all of them are components of drier types of mountain forest and bushland. We conclude that 1.7 Ma ago a diversified type of dry forest, including *Syzygium guineense*, *Podocarpus* and *Myrica salicifolia*, Combretaceae and the dry Sinopteridaceae ferns, grew not far from Melka Kunture. However, the percentage of grass pollen (>50%) remains high, indicating that extensive areas were still occupied by mountain woodland or dry forest, with various herbaceous plants such as *Plantago*, *Dyschoriste*, Apiaceae, Brassicae (Cruciferae) in large clearings near Melka Kunture.

The number of pollen grains (n = 140) extracted from archaeological layer Gombore IC is smaller than in previous samples, hence a smaller number of plant taxa have been identified. Among trees, abundant *Podocarpus* is accompanied by small amounts of pollen from *Juniperus* and from some shrubs, including *Dodonaea angustifolia* and the climber *Clematis*. The herbaceous component is dominated by grasses (61%), with *Plantago* and Apiaceae of open grassland and the aquatic *Typha*. Most of these pollen are known to disperse widely, a phenomenon that leads to pollen deposition on sub-aerial surfaces rather than in fluvial deposits. The vegetation of the area was an open grassland. Few forest trees grew not far away from the site.

Pollen sample Gombore IB (n = 456) yielded more substantial results. It contains a greater amount of *Juniperus* (>20%) associated with *Podocarpus, Olea, Polyscias, Hypericum,* and *Myrica*; all of these plants are mostly encountered in highland forest at altitudes far above 2800 m. The percentage of arboreal pollen is as high as 27%, indicating that the dry conifer forest was close to Melka Kunture at that time, because *Juniperus* does not disperse its pollen very far. The proximity of bushland or forest was later confirmed when a wood fragment identified as *Caesalpinioxylon* was recovered from the excavated surface of Gombore IB (Chavaillon and Koeninger 1970). A juniper-dominated forest whose composition was similar to that of the modern Managasha forest expanded from the northern slopes of the Wochacha volcano toward Melka Kunture. The large proportion of grass pollen (Poaceae 63%) indicates that this forest was broken up by large open spaces in the vegetation, probably near the river. A great variety of grass species characterizes mountain grasslands. We are unable to distinguish among them because the pollen morphology is uniform throughout the various genera of the Poaceae family. However pollen from different botanical tribes is grouped in size classes. The fossil pollen is more abundantly distributed in two of the size classes. Pollen from Eragrostideae and Chlorideae grasses is in the 25–35 μm class, whereas Andropogoneae pollen is larger (35–40 μm; Bonnefille and Riollet 1980a). The size distribution of grass pollen measured in the Gombore IB sample ranges from 15 to 55 μm. The pattern of grass pollen size distribution in this fossil sample is similar to the pattern in a modern soil sample collected in the evergreen juniper open woodland at 2600 m asl (Fig. 5.8). This similarity confirms that the early Acheulean site was indeed surrounded by this type of vegetation.

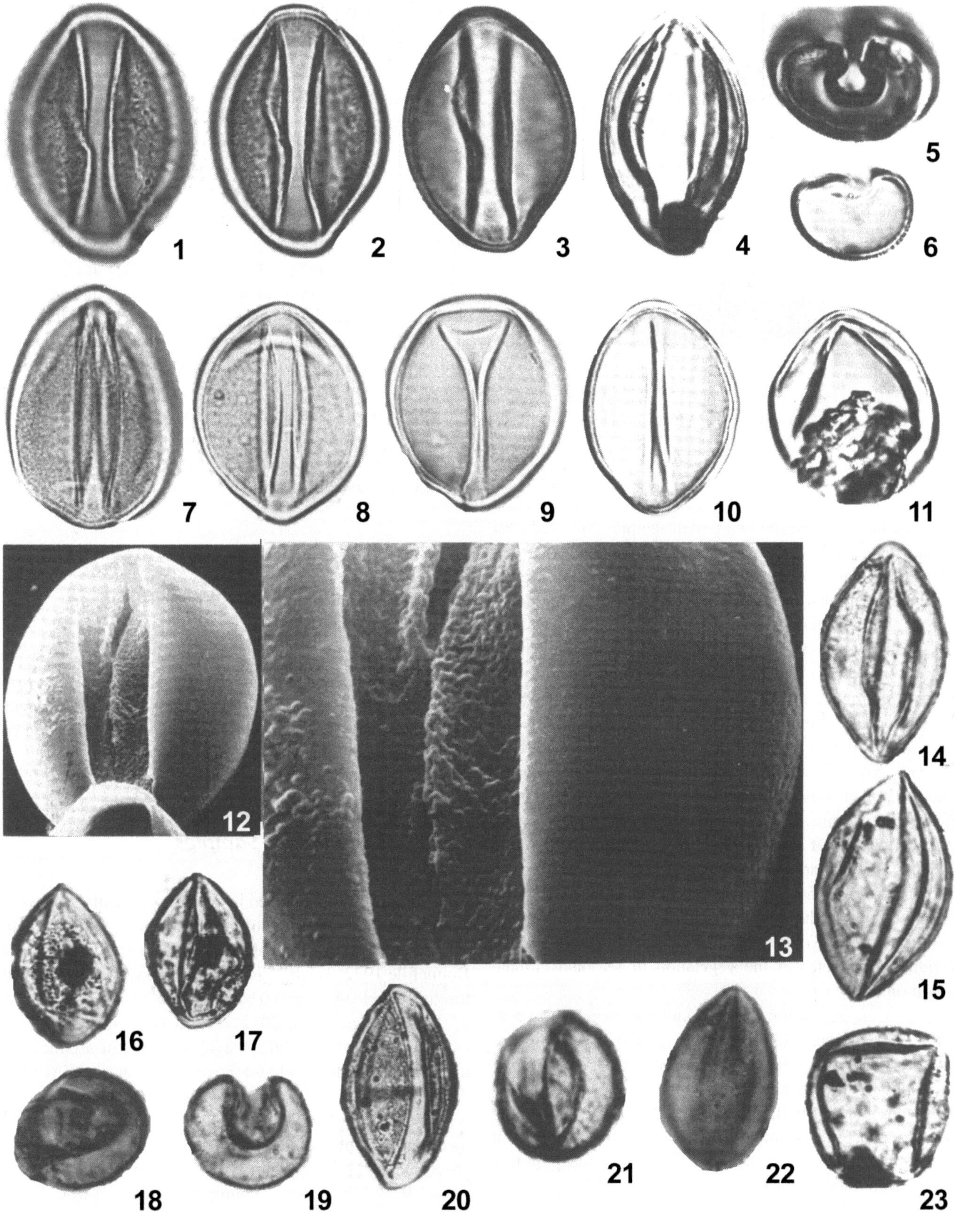

Fig. 5.7 Modern and fossil pollen. 1–5: *Encephalartos bubalinus*; 6: optical section of *Phoenix reclinata* (Arecaceae) for comparison showing columella in the exine; 7–13: *Encephalartos hildebrandtii*, showing an invaginated *sulcus*; 12 and 13 SEM photographs: 12 (×2000) and 13 (×5000) showing granulated *sulcus* membrane: 14–23: fossil pollen grains extracted from Melka Kunture Pleistocene sample K242, and related to *Encephalartos hildebrandtii*. *Encephalartos* pollen is distinguished from that of *Cycas* by a thicker exine membrane and granulations visible on the proximal side. The asiatic genus *Gingko* shows an undulated border of the *sulcus*

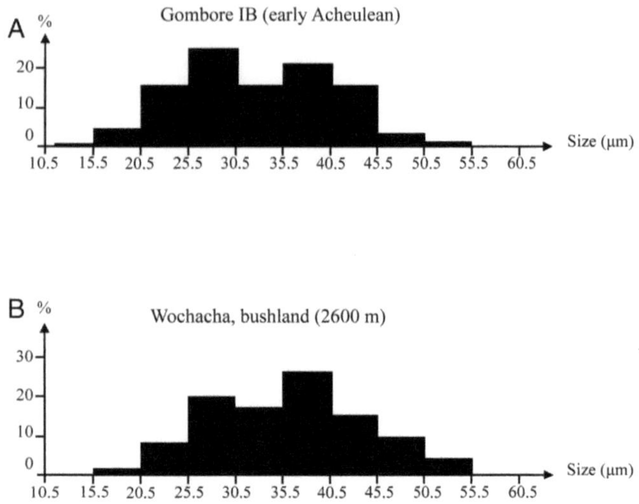

Fig. 5.8 Measurements of size of grass pollen grains: **A** in a high-elevation modern sample; **B** in the fossil sample from the early Acheulean horizon Gombore IB

In the Garba gully, the two pollen samples collected near archaeological layer Garba IVD and near the ~1.7 Ma "Grazia tuff" are not rich in pollen. Counts on the order of one hundred grains provide a rough estimate of tree cover density, but not enough details on the diversity of tree species. Pollen assemblage G393 (n = 111) is dominated by grass pollen (86%) indicating a high-elevation grassland in which *Plantago* and various species of Asteraceae were present together. Remnants of mountain forest are attested to by a single pollen grain of *Podocarpus* accompanied by *Myrsine africana*, a shrub that most likely grew nearby, since its pollen does not disperse over long distances, unlike that of *Podocarpus*. Pollen sample G133 (n = 113) shows a high percentage of grass pollen (76%), and only four tree species. The total arboreal percentage is 12%, indicating the proximity of wooded riparian vegetation characterized by the presence of the highland riparian *Myrica salicifolia* associated with *Celtis* and *Trema*, and attesting to the deciduous character of the vegetation in seasonal climate conditions.

In conclusion, during this 200 ka-long period, the vegetation surrounding Melka Kunture underwent significant changes. They are reflected in the pollen composition by variations in the total abundance of tree taxa and changes in the predominance of different species. At ~1.7 Ma, we note the appearance of the modern Afromontane character of all the trees that belong to present-day mountain vegetation that grows from 2000 to 2800 m asl.

5.3.3 *Pollen Data of the Early Pleistocene, ~1.4 Ma*

One pollen sample, G101 (n = 111), comes from a sandy silt layer overlying a consolidated Mn Fer crust in the Garba gully, ~3 m above Garba IVD level. Its stratigraphic position is below volcanic ash 27-19 dated <1.429 ± 0.029 Ma, validated as a minimum age (Morgan et al. 2012) (Fig. 5.5). Another volcanic ash (ash 27-16), higher up in the stratigraphy, is dated to 1.037 ± 0.088 Ma. The pollen spectrum is dominated by grasses (77%) and includes *Plantago*, an herbaceous plant that produces abundant pollen in mountain grasslands. The presence of *Achyranthes*, a shade-loving erect herb abundant on river banks, together with more spores from Pteridophytae (3 distinct types), points to local humidity conditions that encouraged the growth of ferns. The surrounding vegetation can be interpreted as mountain woodland and wooded grassland. A few *Podocarpus, Euclea, Dodonaea angustifolia, Syzygium,* and *Acacia* trees were also present

5.4 Pollen Data of the Early/Mid-Pleistocene, ~0.9–0.6 Ma

Six pollen samples document past vegetation between ~0.9 and ~0.6 Ma. They were collected from exposed sediments along the Garba (G) and Gombore (Gomb) gullies, where several archaeological levels have been excavated.

5.4.1 *Chronological Placement of the Pollen Samples*

Four samples were collected in good stratigraphic succession from the westernmost part of the Garba gully, in a two-meter section atop a gray blue volcanic ash layer ("cinerite 106" in Bonnefille 1972; Taieb 1974). We correlate this cinerite with the tuff dated 0.869 ± 0.02 Ma in the Garba gully (sample 27-13 in Morgan et al. 2012) (Fig. 5.9). Sample G372 was extracted from the medullar cavity contents of a bone fragment found in the sand-and-gravel bed that contains some Acheulean stone tools, and collected at the lower contact with the dated tuff. Pollen sample G136, extracted from the medullar infill of another long bone fragment, came from the cross-bedded yellow sand lying 100 cm above the tuff, hence it is significantly younger than 0.869 Ma. Pollen

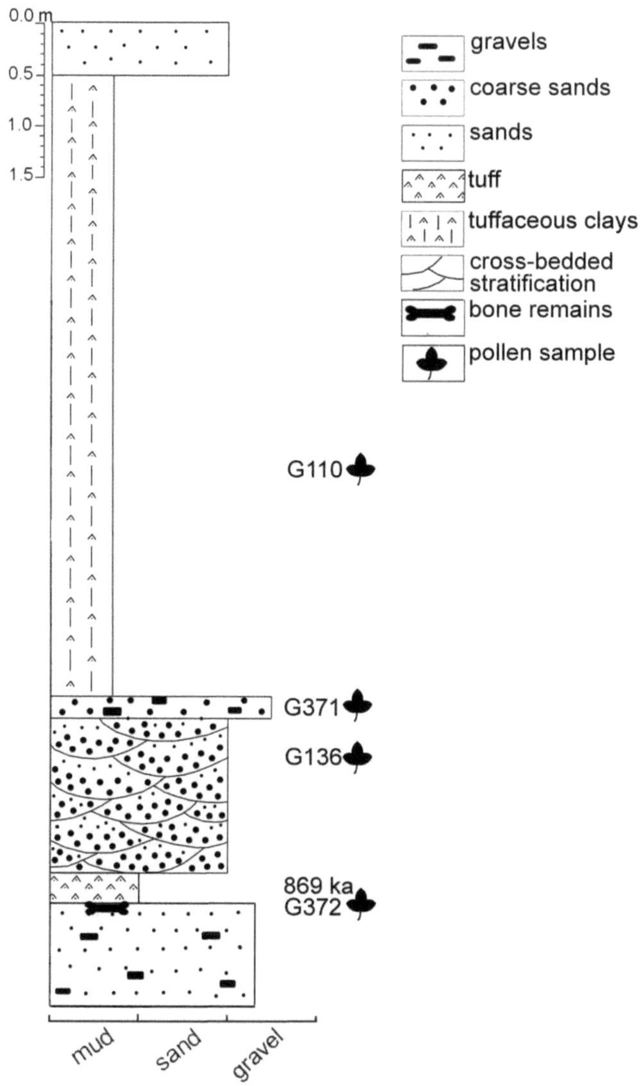

Fig. 5.9 Stratigraphic position of the Early/Mid-Pleistocene (0.9–0.6 Ma) pollen samples collected at Garba, with respect to a dated volcanic deposit (redrawn after Bonnefille 1972, Fig. 37)

sample G371 was taken from the sand-and-gravel layer outcropping 150 cm above the tuff, hence it too is younger than G136. The last pollen sample of this group, G110, comes from the 4-m-thick tuffaceous clay that forms an abrupt cliff 200 cm above the dated tuff.

Two other samples come from the Gombore gully, situated a hundred meters west of the Garba gully (Fig. 5.3). This section has two main archaeological horizons, bracketing the Early/Middle Pleistocene boundary (Fig. 5.10). Pollen sample Gomb173 was collected at the base of a silty–sandy deposit, 2.5 m above Gombore II-1, a middle Acheulean archaeological site (Gallotti et al. 2010) overlying a tuff dated at 0.875 ± 0.010 Ma (sample 27-08, Morgan et al. 2012). We correlate this tuff with the one dated at 0.869 ± 0.02 Ma in the Garba gully, a result coincident within 1σ. The second main archaeological horizon, nearly 1 m above sample Gomb173, is Gombore II-2, the so-called "Hippo Butchery Site" (Mussi et al. 2016), which is capped by a volcanic ash dated 0.709 ± 0.013 Ma (sample 27-09, Morgan et al. 2012). Pollen sample Gomb173, which lies between Gombore II-1 and Gombore II-2, and below the Matuyama-Brunhes boundary, as identified by Tamrat et al. (2014), is slightly younger than 0.8 Ma.

Pollen sample Gomb349 was extracted from the infill sediment of a long bone fragment collected in a clayish tuffaceous deposit above Gombore II-2 (Mussi et al. 2016). Its estimated age is close to ~0.6 Ma.

The six pollen samples span more than 200 ka (Figs. 5.9 and 5.10).

5.4.2 Pollen Results, ~0.9–0.6 Ma

The results of pollen counts obtained from the six pollen samples included in the 0.9–0.6 Ma period are shown in Table 5.3. Among the 71 recognized pollen taxa, 39 (55%) belong to tall trees and 9 (12%) to shrubs or climbers, a proportion indicating highly diversified arboreal and forb components in the past vegetation. Identification of the pollen indicates that all the taxa belong to the Dry Afromontane forest, i.e. they are currently encountered in the vegetation of the Ethiopian highlands (Fig. 5.11).

The rich pollen spectrum G372 (n = 1084) is dominated by grasses (93%). *Plantago* is associated with *Artemisia*, a small shrub abundant today in the Afroalpine grassland above the forest limit at ~3000 m asl. A few forest tree taxa are nevertheless present: *Myrica salicifolia* and *Salix* from riparian vegetation encountered today at higher altitudes, above 2500 m asl. At ~0.9 Ma, the vegetation surrounding Melka Kunture can be reconstructed as an Afroalpine grassland with scattered forest trees.

G136 is an abundant pollen spectrum with a total pollen count close to 1000 pollen grains and a great number of identified pollen taxa. There is still a high percentage of grasses (84%), although it is smaller than in sample G372. The tree taxa are more numerous and more diversified. The shrubs, including *Dodonaea angustifolia*, *Rumex*, *Jasminum*, *Rhus*, and *Myrsine africana* indicate a dense evergreen bushland. The increasing proportion of *Podocarpus*, and the occurrence of *Juniperus*, *Olea*, and Rutaceae, indicate that the dry mountain forest was more diverse and closer to Melka Kunture. *Acacia* associated to *Faidherbia albida* (syn. *Acacia albida*), a tall Legume tree, may have been present locally (Bonnefille 1975). The landscape was more wooded and richer in species including *Euphorbia* and Rubiaceae. *Typha* (cattail) and *Phragmites* (reed) occupied locally humid flatlands surrounding quiet freshwater ponds, possibly pointing to wetter climate conditions.

The pollen results from sample G371 (n = 421) show an increase in the total arboreal pollen (22%) in which

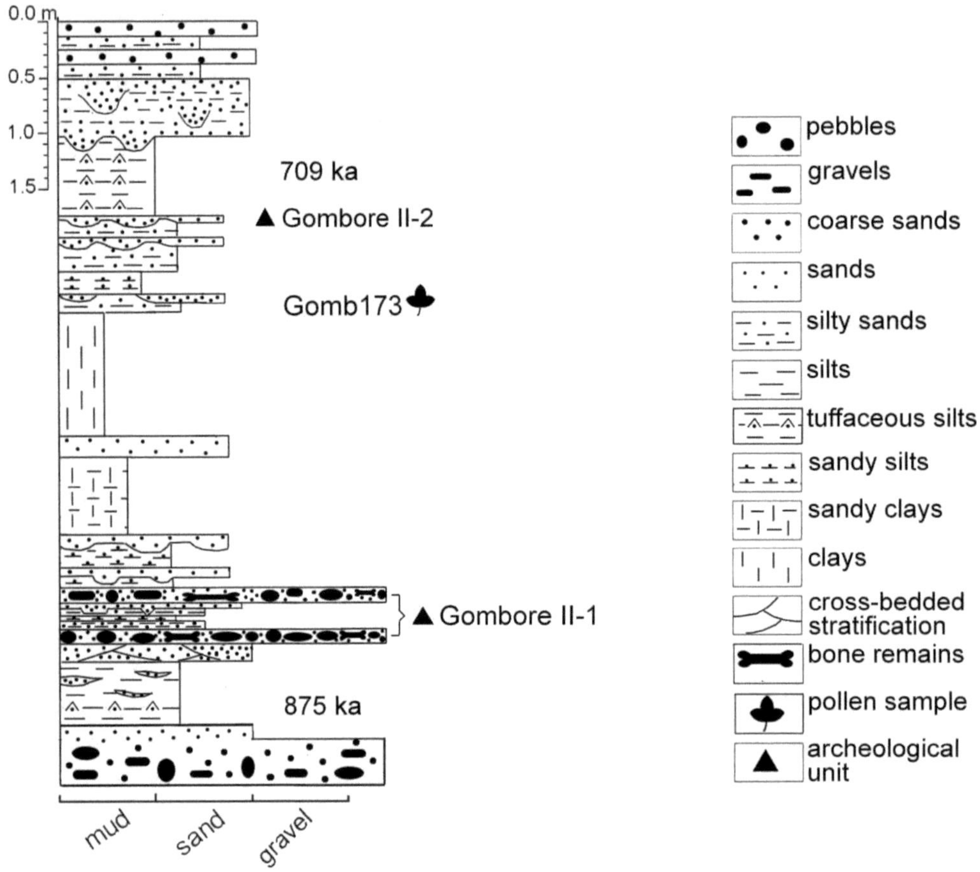

Fig. 5.10 Stratigraphic position of the Early/Mid-Pleistocene (0.9–0.6 Ma) pollen samples collected at Gombore, with respect to dated volcanic deposits. The log is the outcome of the excavations carried out at Gombore since 2012, which do not include Gombore III. Gombore III has been relocated above the uppermost part of the sequence illustrated in the figure

Juniperus becomes more abundant (6.5%). The number of tree pollen taxa is considerable, including Ulmaceae associated with Apocynaceae, *Pygeum* and Celastraceae whose pollen does not disperse over long distances. There is less pollen attributed to *Acacia*. Pollen taxa from herbaceous plants include Labiatae and Cruciferae together with *Ricinus communis*. Sedges decrease. Taken as a whole the pollen results indicate that a different type of diversified forest/woodland with dominant *Juniperus* had developed near Melka Kunture.

Pollen spectrum G110 (n = 718) contains a great number of pollen taxa although the total of arboreal pollen has decreased (15%). *Juniperus* is not present. The beautiful tree *Hagenia abyssinica*, well known in Ethiopia as *Kosso*, is found together with *Anthospermum*, a common herb of high-elevation grasslands. The notable proportion of Asteraceae (4.6%) is consistent with an Afroalpine vegetation that now occupies areas above the forest line. Pollen from Legume trees such as Caesalpiniaceae (2%) and *Acacia* are associated with *Combretum*, trees or climbers normally abundant in deciduous vegetation growing in climates characerized by seasonal rainfall. The pollen found in this deposit attests to types of vegetation that grow in drier and colder climates. Highland forest and local deciduous woodland existed somewhere whereas freshwater swamp vegetation including *Typha* and sedges was locally present.

Gomb173 provided a rich pollen spectrum (n = 573). Pollen from trees and shrubs accounts for 10% of the total. Almost 20 shrubs or trees taxa have been identified, thus attesting to high diversity among forest trees. *Juniperus* pollen is present but not abundant. We note the occurrence of *Maesa*, a common tree in the Shewa forests and of Ericaceae, the mountain heath. More abundant are Ebenaceae, *Croton, Pterolobium*, together with *Acacia* and *Carissa edulis*, associated with the shrubby *Heteromorpha* of the Apiaceae (Umbelliferae) family. These finding point to a dense and diversified woody bushland or shrubland in the area. Asteraceae pollen representing several species accounts for 67% of the total count and dominates the herbaceous pollen (75%), whereas grass pollen amounts to only 15%. This finding is consistent with an attribution to the evergreen bushland that is normally found above the forest belt and points to a climate that has become significantly drier and colder than it was before.

Table 5.3 Detailed pollen counts of Melka Kunture samples dating ~0.9–0.6 Ma. AP: arboreal pollen from trees; S: shrubs; C: climbers. NAP: non-arboreal pollen (herbs); Ind.: indetermined. * = pollen from similar but different species or genera have the same morphology, producing uncertainty in the attribution to a specific genus or species

Plant form	Pollen Taxa	G 372	G 136	G 371	G 110	Gomb 173	Gomb 349
AP	Acacia (2 sp.)	1	8	4	8	2	5
NAP	Achyranthes aspera	–	1	1	–	1	–
Ind.	Amaranthaceae	–	2	–	–	–	–
NAP	Anthospermum	–	–	–	1	–	–
NAP	Apiaceae	–	–	–	4	–	–
NAP	Artemisia	1	–	–	–	–	–
Ind.	Asteraceae tubuliflorae	4	18	3	32	383	5
NAP	Asteraceae cichoriae	–	1	1	1	3	–
NAP	Brassicae	–	–	3	1	–	2
AP	Brucea antidysenterica	–	–	–	–	–	1
AP	Caesalpiniacae	–	–	–	4	–	1
AP	Capparaceae	–	3	–	–	–	–
AP	Carissa edulis	–	–	–	–	1	–
NAP	Caryophyllaceae	–	–	–	–	–	1
AP	Celtis	–	9	4	–	1	1
NAP	Chenopodiaceae	1	5	1	6	–	2
C	Clematis	1	–	13	1	18	–
AP	Combretaceae	–	–	–	2	–	–
AP	Croton	–	–	–	–	4	–
NAP	Cyperaceae	2	5	1	4	2	2
AP	Diospyros	–	–	–	–	–	2
S	Dodonaea angustifolia	13	63	16	28	6	44
AP	Ebenaceae	–	–	–	–	4	1
AP	Erica arborea	–	–	–	–	1	–
AP	Euclea	2	–	2	4	1	2
S	Euphorbia	–	1	1	–	–	1
AP	Fabaceae	–	–	–	–	1	1
AP	Fagaropsis angolensis	–	–	–	–	–	1
AP	Faidherbia albida	–	3	–	–	–	–
NAP	Geranium	–	–	–	–	1	–
AP	Hagenia abyssinica	–	–	–	2	–	4
AP	Heteromorpha	–	–	–	–	2	–
AP	Hymenocardia acida	–	1	–	–	–	–
AP	Hypericum	–	–	–	4	1	–
C	Jasminum	–	3	–	–	2	–
AP	Juniperus procera	–	1	27	–	1	2
NAP	Liliaceae	–	–	–	–	1	–
NAP	Linum	–	–	–	1	–	–
AP	Macaranga	–	–	–	1	–	–
AP	Maesa lancelota	–	–	–	–	1	–
AP	Myrica salicifolia	1	3	1	–	–	–
S	Myrsine africana	–	3	–	–	–	–
AP	Olea sp. (2 sp.)	1	5	3	6		7
NAP	Osyris	–	–	–	–	–	1
NAP	Paronychia*	1	–	–	1	–	–
NAP	Pentas	–	1	–	–	–	–
NAP	Phragmites	–	3	–	–	–	–
AP	Piliostigma thonningii	–	–	–	–	–	3
NAP	Plantago africana*	18	12	18	9	–	6
NAP	Plectranthus	–	–	1	–	–	–
NAP	Poaceae	1015	770	293	542	116	442

(continued)

Table 5.3 (continued)

Plant form	Pollen Taxa	G 372	G 136	G 371	G 110	Gomb 173	Gomb 349
AP	Podocarpus	8	37	10	33	11	10
NAP	Pteridophyta	7	–	1	–	–	1
AP	Pterolobium stellatum	–	–	1	3	3	–
AP	Pygeum africanum	–	–	1	–	–	1
AP	Rapanea	–	–	–	1	–	–
AP	Rosa abyssinica	–	1	–	2	–	–
AP	Rhamnaceae	–	–	–	–	–	1
AP	Rhus*	–	3	–	1	–	1
S	Rumex	1	8	2	–	1	2
AP	Rutaceae	–	1	–	–	–	–
AP	Salix* subserrata	1	–	–	–	–	–
AP	Sapium* ellipticum	–	–	–	–	–	3
S	Solanum	1	–	–	1	–	–
AP	Syzygium guineense	2	1	3	6	–	2
AP	Trema	–	1	–	–	–	–
NAP	Tribulus	–	–	–	1	–	–
AP	Trilepisium			2		2	5
NAP	Typha	–	4	–	5	–	–
S	Vernonia	–	–	6	–	–	–
	Not determined	3	5	2	3	3	2
	TOTAL	**1084**	**982**	**421**	**718**	**573**	**565**
	AP +S	**16**	**77**	**58**	**77**	**36**	**54**
	% AP	**1.5**	**7.8**	**13.8**	**10.7**	**6.3**	**9.6**

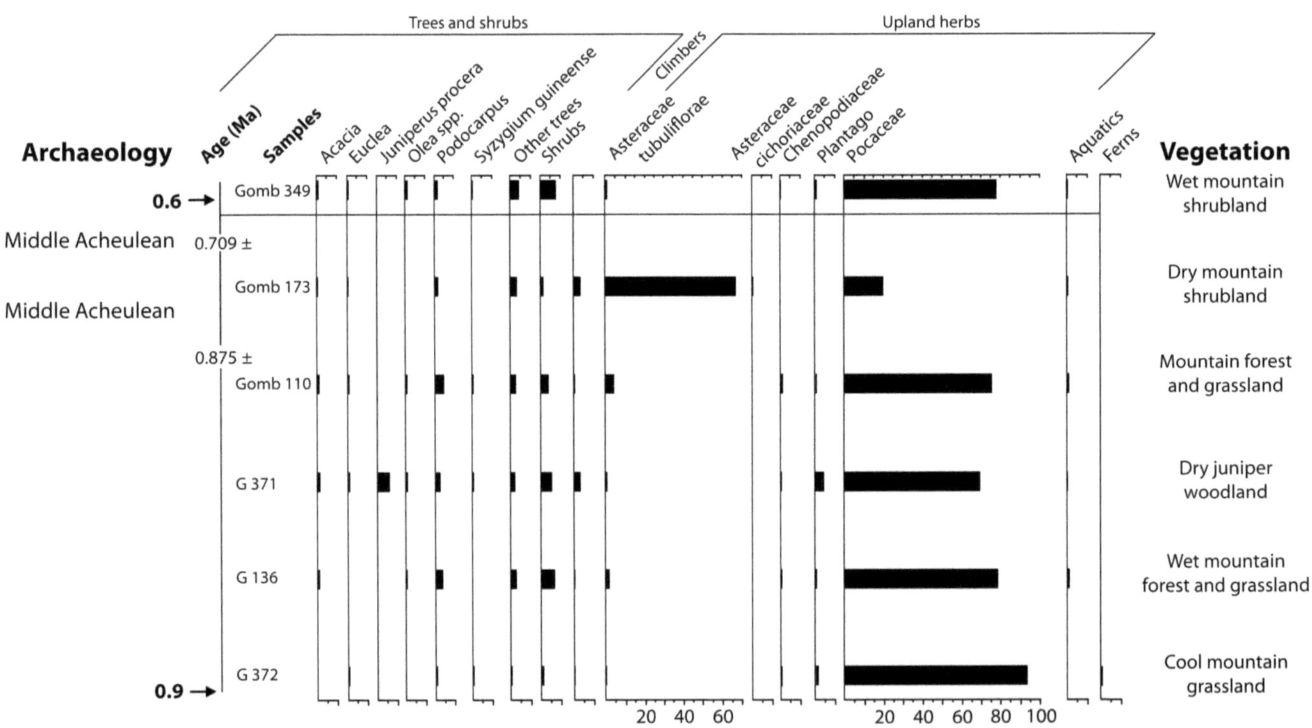

Fig. 5.11 Distribution of percentages for selected pollen taxa identified in samples dated ~0.9–0.6 Ma corresponding to the MPT period, related to archaeological evidence and to the proposed interpretation of past vegetation. Relative pollen % calculated as in Fig. 5.5. Pollen not represented in this graph are indicated in Table 5.3

Gomb349 pollen spectrum (n = 565) at ~0.6 Ma, belongs to a late phase of the Mid-Pleistocene Transition (MPT). It contains a rich and diversified proportion of 28 tree and shrub taxa. Most of the trees are the same as in sample Gomb173; the presence of *Olea* points to warmer climatic conditions. With *Sapium*, among the Sapotaceae, *Fagaropsis* among the Rutaceae, *Diospyros* among the Ebenaceae and *Osyris* (Santalaceae), a tall herb, the pollen composition indicates a diversified forest with a dense canopy of trees and shrubs (17.5%).

During the Mid-Pleistocene Transition, at Melka Kunture past vegetation remains permanently in the Dry evergreen Afromontane Domain already established during the Early Pleistocene. All the plant taxa are of mountain vegetation. However, the six pollen assemblages documenting two-three hundred thousand years from ~0.9 to ~0.6 Ma show important changes in the floristic composition of past vegetation, notably in the relative abundance of various forest trees pointing to the influence of warmer or wetter conditions, while some others indicate slightly cooler conditions.

5.5 Discussion

With the exception of *Encephalartos* in samples K265 and K242 (Fig. 5.7), all the fossil pollen taxa refer to plants that are present today in the flora of Ethiopia (Friis 1992). A number of modern pollen taxa such as *Astragalus*, *Bridelia*, *Cassia*, Commelinaceae, *Crotalaria*, *Diospyros*, *Drognetia*, *Galiniera coffeoïdes*, *Gardenia*, *Kohautia aspera*, Malvaceae, *Pterocarpus*, *Rubia cordifolia* and *Sterculia*, all of which were identified in a mud sample collected in 1968 from the Awash River (Bonnefille 1969), are not found in the fossil pollen assemblages (Tables 5.2 and 5.3). However, since most of these plant taxa are herbaceous plants or are represented in the assemblages by only one grain, their absence from the fossil record is not statistically significant and cannot be regarded as indicating that during the Pleistocene plant diversity was lower than it is today.

5.5.1 *Afromontane Character of the Past Vegetation at Melka Kunture since 1.8 Ma*

In the overall fossil pollen counts (n = 2024 for 1.8–1.6 Ma; n = 4343 for 0.9–0.6 Ma) 114 different fossil pollen taxa were identified, including 70 that were not found in a modern sample from the Awash River (n = 2865; Bonnefille 1969). The fact that modern and fossil pollen assemblages do not overlap has two implications. First it confirms that the fossil assemblages were not contaminated by any modern pollen input, as can happen when rivers overflood. On the other hand, it indicates that vegetation was more diverse in the past that it is today. Indeed, the impoverishment of present-day vegetation is expected at Melka Kunture as the result of long-lasting human impact. Forests were already rare during the sixteenth century, when travelers wrote the first detailed descriptions of the landscape (Zerihun 1999). Nevertheless, except for *Encephalartos*, all fossil taxa are related to genera or species encountered somewhere in the modern vegetation of the northern Ethiopian highlands.

Of the 114 fossil pollen taxa identified in this study, 69 (60%) are attributed to trees, shrubs or climbers. This high number attests to a very diverse arboreal (woody) component of the past vegetation. Twenty-two fossil pollen taxa (18%) belong to herbaceous plants, whereas 23 others (22%) cannot be attributed to a well-defined plant form. Trees and shrubs taxa that belong to the montane floristic domain identified as the Dry evergreen Afromontane forest and grassland complex (DAF) dominate the pollen assemblages (~80%) (Tables 5.2 and 5.3). We conclude that highland vegetation as it is known today was already established in the Awash basin at ~1.8 Ma. Accordingly, we suggest that all hominins locally associated with the Oldowan and Acheulean had already adapted to mountain climate conditions.

5.5.2 *Environmental Changes*

In all samples, the percentages of grass pollen dominate because grasses produce large amounts of pollen. However, the relative abundance of each of the pollen taxa listed in Figs. 5.6 and 5.11 is subject to significant fluctuations. At Melka Kunture, Middle Pleistocene vegetation differs sharply from that of the Early Pleistocene. The Early Pleistocene has a variable but lower proportion of grass pollen, a higher proportion of arboreal pollen and greater diversity of trees and shrubs. Relict palms and cycadales are associated with « warm » taxa such as *Acacia, Tamarix*, Resedaceae. During the Mid-Pleistocene Transition, the pattern is different. More extensive grasslands and less tree cover are associated with a few « colder » Afroalpine grassland taxa such as *Artemisia*, *Anthospermum*, *Erica arborea* and *Plantago*.

In general, species diversity increases with total pollen counts and with the number of analyzed samples. The total pollen counts (n = 2024) for the Early Pleistocene (EP) is smaller than that of the six samples from the MPT (n = 4343).

Therefore the greater number of identified pollen taxa recorded for the MPT (71 versus 58 for the EP) might not reflect greater plant diversity. However, in each period large variations are observed among the samples. Stratigraphic constraints suggest that such changes occurred at intervals on the order of a few tens of thousands of years.

5.5.3 Early Pleistocene Wet–Dry Variability (1.8–1.6 Ma)

A well-marked vegetation change is observed between the ~1.8 Ma Kella results on the one hand, and the ~1.6 Ma samples related to Garba IVD and Gombore IB archaeological sites on the other. Bushy woodland, possibly existing under warmer and drier conditions, differs sharply from the extended grasslands that dominated the landscape in the more recent samples from Garba IVD, at the time of the emergence of the Acheulean. Mountain forest trees and shrubs became more abundant at Gombore I; a real *Juniper* woodland/forest grew near the Gombore IB archaeological site, close in age to Garba IVD (Gallotti and Mussi 2018), although locally open vegetation left room for an herbaceous cover and some periaquatic plants. *Juniperus* percentages did not reach the same values that they did under a forest cover such as prevails today in the Managesha forest, on the Wochacha volcano (Fig. 5.2). Our interpretation is that a real forest existed in the vicinity, but that locally at the site open space was covered by herbaceous vegetation. Conversely, during and after the early Acheulean occupation at Garba IVD (Gallotti 2013), mountain grassland expanded greatly. The presence of *Plantago*, *Anthospermum*, and Apiaceae indicate affinities with high-elevation grasslands, now found above the forest zone. This context explains the presence of wind-transported pollen of *Podocarpus*, frequently observed in the Afroalpine zone (Umer et al. 2007). In this cool grassland herbivores were present as shown by the fossil fauna, which also includes the highland rodent *Tachyoryctes* (Geraads et al. in press). The expansion of mountain grassland implies a significant lowering of vegetation belts explained by a colder and drier climate.

At Melka Kunture, environmental conditions differed sharply from those of the savanna existing elsewhere in the Rift Valley, in the warmer climate inferred at archaeological sites located at lower altitude. Nevertheless, vegetation changes documented in the Ethiopian northern highlands are contemporary to vegetation changes evidenced by previous pollen studies at Olduvai from ~1.8 Ma to ~1.6 Ma, during Bed I/lower Bed II times (Bonnefille and Riollet 1980a; Bonnefille et al. 1982). Olduvai is located at lower elevation in a warmer wooded grassland and woodland near a permanent salt lake (Barboni 2014). At this site, past vegetation changes also preceded and followed the emergence of the Acheulean, dated at 1.664 ± 0.019 Ma at the FLKW site (Diez-Martin et al. 2015). At both the Ethiopian and the Tanzanian sites, the availability of freshwater, the density of trees in mountain wooded grassland and the proximity of mountain forests provided various plant resources to hominins producing at first the Oldowan, and later the early Acheulean. However, at Melka Kunture, sedges, *Typha* or ferns were rare whereas at Olduvai the grassland surrounding a permanent salt lake and the periodically flooded adjacent marshes are thought to have produced abundant freshwater food (Magill et al. 2013) and a C_4 diet exclusive to *Paranthropus boisei* (Ungar and Sponheimer 2011). The past vegetation of Melka Kunture did not provide abundant aquatic food, but probably did offer more resources from nearby forests, such as fruit trees.

5.5.4 Mid-Pleistocene Transition Variability

Pollen assemblages between ~0.9 and 0.6 Ma are more homogenous, with a proportion of grass pollen exceeding 70%, except in one sample characterized by overrepresentation of Asteraceae. At that time, mountain shrub grasslands were permanently established with forests and Afroalpine vegetation in the background. The expansion of cold mountain grassland at 0.9 Ma is followed by the downward expansion of the mountain juniper forest related to more humid conditions. The return of cooler conditions and the re-expansion of grasslands occur at a time close to the Bruhnes/Matuyama boundary. At ~0.6 Ma, the establishment of a rich and diversified humid forest closer to the site attests to a warmer and probably more humid climate.

5.5.5 Past Vegetation Changes linked to Global Orbital Forcing

In the tropics, variability in past vegetation is most likely explained by changes in precipitation, rainfall or available moisture. Between ~1.8–1.6 Ma, wet–dry climate variability most probably triggered the environmental changes observed at Melka Kunture. Rainfall fluctuated in response to orbital changes combined to astronomical forcing linked to precession cycles. Precessional global climate forcing dominates the tropical climate system during the Early Pleistocene (deMenocal 2004). Precession monsoon cycles have been evidenced at Olduvai, on the basis of sedimentological and isotopic studies that show several wet and dry cycles (Hay and Kyser 2001; Sikes and Ashley 2007). At Melka Kunture, the necessary high-resolution sampling at regular intervals cannot be done with the same dating accuracy. However, changes documented seven times within a ~200 ka time-span would attest to cyclicity on the order of 20 ka. This is consistent with precession change cycles evidenced by isotopic studies at Turkana, during the 2–1.85 Ma interval (Joordens et al. 2011) and at Olduvai, from 1.877 to 1.803 Ma (Deino 2012).

Following the onset and development of the continental arctic icesheet, the ocean circulation changed suddenly at 1.6 Ma (Lisiecki and Raymo 2005), and precession cyclicity was replaced by 41 ka obliquity cyclicity.

At 0.9 Ma, when the icesheet had reached a greater height and volume, the tropical climate changed to a system characterized by less influence of precession cycles but greater influence of high-amplitude glacial/interglacial successions. During the Middle Pleistocene Transition (MPT), pollen data from Melka Kunture show the response of past mountain vegetation to longer and more marked climate cycles. At the onset of the MPT, drier and cooler climate conditions indicated at Melka Kunture by pollen studies appear at the same time as greater icesheet thickness in the northern hemisphere, and as temperature decrease in an Antarctica ice core at ~750 ka (Jouzel et al. 2007). During the MPT interval, three glacial stages, MIS 22 (0.9 Ma), MIS 20 (0.8 Ma) and MIS 18 (before 0.7 Ma) have been identified in the oceanic record (Lisiecki and Raymo 2005). These glacial stages are linked to lower solar insolation at 65°N, lower amplitude in precession cycles and lower pCO_2 in the atmosphere (Almogi-Labin 2011). The global climate was greatly modified, as recorded in long terrestrial sedimentary sequences such as those in the eastern Mediterranean region (Almogi-Labin 2011). In palynology, past vegetation changes are traditionally based on the fluctuations of total arboreal pollen (AP) versus non-arboreal pollen (NAP). In the Levant and Greece at the time of MIS 22, the terrestrial pollen record of lacustrine sequences shows a minimum of AP, and the extinction of a few relict Tertiary species; whereas during MIS 16 there was a steppe environment, after which a forest of modern composition with *Quercus* and *Carpinus* dominates (Tzedakis et al. 2006). The extreme glacial conditions of both MIS 22 and 16 appear to have played a major role in the disappearance of a large number of Tertiary relict species at that time. During the MPT, the absence of sapropel in Mediterranean cores reflects the weakening of the tropical monsoon in East Africa, resulting in less rainfall; the increased dryness encouraged the development of grasslands at the expense of forest. The marine pollen record offshore from the mouth of the Congo River indicates that the equatorial region experienced cyclically predominantly warm and dry conditions during interglacials, but cool and humid conditions during glacial periods (Dupont et al. 2001). In South America, during glacial periods the long terrestrial pollen record spanning MIS 22 to 19 highlights the lowering and oscillations of the upper forest line (Hooghiemstra et al. 1993).

Pollen data from long cores is not available in Africa. The Melka Kunture pollen data are the first terrestrial data for the MPT period. Colder and drier conditions documented in the Ethiopian highlands correspond to glacial stages 22, 20 and 18; whereas warmer more humid conditions, more favorable to forest development, correspond to interglacial stage 21, 19 and 17. Past vegetation changes are consistent with global glacial/interglacial cycles. We discuss elsewhere weather in the highlands the hominins were able, or not, to withstand both interglacial and glacial climates (Mussi et al. 2016).

5.6 Conclusion

The study presented here, which is based on a new assessment of the chronostratigraphy at Melka Kunture, and a new interpretation of the pollen data, emphasizes ecological variability and ecosystems changes that followed global climatic changes during the Early and Middle Pleistocene in the Ethiopian highlands. During a critical period, close in time to the Acheulean emergence (~1.7 Ma), the pattern of fluctuations related to the precessional monsoon cycle had already appeared in the Turkana basin and at Olduvai. Although the stratigraphy of Melka Kunture does not allow much precision in the high resolution, the pollen record presented here is consistent with such precession climate variability. *Homo erectus s.l.* adapted to mountain climate conditions characterized by large diurnal temperature ranges. Later, during the MPT transition, hominins of a species close to that of *Homo heidelbergensis* – the knappers of the middle Acheulean lithic industries – experienced the impact of changing environmental conditions during glacial/interglacial cycles.

5.7 Note About Palynological Processing

Sediment samples used for pollen analysis were collected on cleaned outcrops or deep trenches dug out from exposures. The initial stratigraphic position of each pollen sample (Bonnefille 1972) can be found on the same lithostratigraphic sections as geological studies (Taieb 1974). A few additional samples collected in the Simbiro gully, and at Melka Gila (Djilla) did not yield pollen. One pollen assemblage from the Tuka gully (Bonnefille 1972) is not included here because the sample cannot be securely placed in the updated chronostratigraphy.

Pollen extraction from the sediment (up to 100 g) is performed according to the classical chemical treatment, attacking the sediment with HCl and HF acids to remove the mineral fraction, followed by hot-water rinses and by treatment with diluted KOH to remove humic acids. The residue is then stained with diluted safranine and kept in glycerin. It is mounted on slides in glycerin allowing to turn pollen grains and facilitating the observation under the microscope of the pollen wall structure. The identification of pollen was performed by comparing them with a 7,000-specimen reference collection of modern pollen stored at Cerege, Aix-en-Provence. Descriptions of pollen from most common trees and plants are available (Bonnefille 1969, 1971a, b; Riollet and Bonnefille 1976; Bonnefille and Riollet 1980b), with photo in the African Pollen Database. 15% of the processed samples yielded fossil pollen grains (Bonnefille 1972). The pollen assemblages from bones infills show species distribution similar to the one in a modern sample from the Awash River. Taphonomic preservation of bone remains (Fiore and Tagliacozzo 2004) on the banks of a meandering Awash River may explain this exceptional pollen preservation. We consider pollen assemblages extracted from bone cavities to be reliable.

Acknowledgements The recent research at Melka Kunture was supported by grants from Sapienza Università di Roma ("Grandi scavi archeologici") and from the Italian Foreign Ministry awarded to MM. We thank the Authority for Research & Conservation of Cultural Heritage of the Ministry of Culture & Tourism, the National Museum of Addis Ababa, and the Oromia Culture and Tourism Bureau for permits for fieldwork and access to the lithic collections. RB did the paleobotanical analysis and reexamined the environmental interpretation, RTM established the stratigraphic sequence and MM designed, encouraged and coordinated the project. The late Nicole Chavaillon and her daughters generously entrusted Jean Chavaillon's archives to MM. We thank Guillaume Jouve (CEREGE) et Morteza Djamali (IMBE) for their help in drawing the pollen diagrams. Rosalia Gallotti participated in the discussion with the three authors. Flavio Altamura provided important information related to the stratigraphic position of samples Gomb173 and Gomb349.

References

Almogi-Labin, A. (2011). The paleoclimate of the Eastern Mediterranean during the transition from early to mid Pleistocene (900 to 700 ka) based on marine and non-marine records: An integrated overview. *Journal of Human Evolution, 60,* 428–436.

Bailloud, G. (1965). *Les gisements paléolithiques de Melka-Kontouré.* Cahier n°1, Institut Ethiopien d'Archéologie.

Barboni, D. (2014). Vegetation of Northern Tanzania during the Plio-Pleistocene: A synthesis of the paleobotanical evidences from Laetoli, Olduvai, and Peninj hominin sites. *Quaternary International, 322–323,* 264–276.

Beyene, Y., Katoh, S., WoldeGabriel, G., Hart, W. K., Uto, K., Sudo, M., et al. (2013). The characteristics and chronology of the earliest Acheulean at Konso, Ethiopia. *Proceedings of the National Academy of Sciences, 110,* 1584–1591.

Bonnefille, R. (1968). Contribution de l'étude de la flore d'un niveau pléistocène de la haute vallée de l'Awash (Ethiopie). *Comptes Rendus de l'Académie des Sciences Paris, D, 266,* 1229–1232.

Bonnefille, R. (1969). Analyse d'un sediment recent : vases actuelles de la rivière Awash (Ethiopie). *Pollen et Spores, 11,* 7–16.

Bonnefille, R. (1971a). Atlas des pollens d'Ethiopie. Principales espèces des forêts de montagne. *Pollen et Spores, 13,* 15–72.

Bonnefille, R. (1971b). Atlas des pollens d'Ethiopie. Pollens actuels de la basse vallée de l'Omo, récoltes botaniques 1968. *Adansonia, 2* (11), 463–518.

Bonnefille, R. (1972). Association polliniques actuelles et quaternaires en Ethiopie (vallées de l'Awash et de l'Omo). Ph.D. Dissertation, Université de Paris.

Bonnefille, R. (1975). Présence de pollens fossiles d'*Acacia albida* Del. et de Cycadales dans les échantillons pléistocènes de Melka Kunturé (Ethiopie). *Boissiera, 24,* 33–42.

Bonnefille, R. (2007). Rain forest responses to past climatic changes in tropical Africa. In J. Flenley & M. Bush (Eds.), *Tropical Rain Forest Responses to Climatic Change* (pp. 118–170). New York: Springer.

Bonnefille, R., & Riollet, G. (1980a). Palynologie, végétation et climats de Bed I et Bed II à Olduvai, Tanzanie. In R. E. Leakey & B. A. Ogot (Eds.), *Proceedings of the 8th Panafrican Congress of Prehistory and Quaternary Studies, Nairobi, September 1977* (pp. 123–127). Nairobi: TILLMIAP.

Bonnefille, R., & Riollet, G. (1980b). *Pollens des savanes d'Afrique orientale.* Paris: CNRS.

Bonnefille, R., Lobreau, D., & Riollet, G. (1982). Pollen fossile de *Ximenia* (Olacaceae) dans le Pléistocène inférieur d'Olduvai en Tanzanie: implications paléoécologiques. *Journal of Biogeography, 9,* 469–486.

Cerling, T. E., Wynn, J. G., Andanje, S. A., Bird, M. I., Korir, D. K., Levin, N. E., et al. (2011). Woody cover and hominin environments in the past 6 million years. *Nature, 476,* 51–56.

Chavaillon, J. (1971). Les habitats préhistoriques en Ethiopie : Melka Kunture. *Sciences (Paris), 73,* 44–52.

Chavaillon, J. (1973). Chronologie des niveaux paléolithiques de Melka-Kunturé. *Comptes Rendus de l'Académie des Sciences Paris, D, 276,* 1533–1536.

Chavaillon, J. (1979). Stratigraphie du site archéologique de Melka-Kunturé (Ethiopie). *Bulletin de la Société Géologique Française, 7° série, XXI, 3,* 227–232.

Chavaillon, J. (1980). Chronologie archéologique de Melka-Kunturé (Ethiopie). In R. E. Leakey & B. A. Ogot (Eds.), *Proceedings of the 8th Panafrican Congress of Prehistory and Quaternary Studies, Nairobi, September 1977* (pp. 200–201). Nairobi: TILLMIAP.

Chavaillon, J., & Berthelet, A. (2004). The archaeological sites of Melka Kunture. In J. Chavaillon & M. Piperno (Eds.), *Studies on the Early Paleolithic Site of Melka Kunture, Ethiopia* (pp. 25–80). Florence: Origines.

Chavaillon, J., & Coppens, Y. (1975). Découverte d'Hominidé dans un site acheuléen de Melka-Kunturé. *Bulletin et Mémoires de la Société d'Anthropologie, Paris, 2,* 125–128.

Chavaillon, J., & Coppens, Y. (1986). Nouvelle découverte d'*Homo erectus* à Melka-Kunturé. *Comptes Rendus de l'Académie des Sciences Paris, II, 1*(303), 99–104.

Chavaillon, J., & Koeniguer, J.-C. (1970). Découverte d'un fragment de liane dans le sol oldowayen de Melka Konturé (Ethiopie). *Comptes Rendus de l'Académie des Sciences Paris, D, 270,* 1969–1972.

Chavaillon, J., & Piperno, M. (Eds.). (2004). *Studies on the Early Paleolithic site of Melka Kunture, Ethiopia.* Florence: Origines.

Chavaillon, J., Brahimi, C., & Coppens, Y. (1974). Première découverte d'Hominidé dans l'un des sites acheuléens de Melka-Kunturé (Ethiopie). *Comptes Rendus de l'Académie des Sciences Paris, D, 278,* 3299–3302.

Chavaillon, J., Chavaillon, N., Coppens, Y., & Senut, B. (1977). Présence d'Hominidé dans le site oldowayen de Gomboré I à Melka-Kunturé. *Comptes Rendus de l'Académie des Sciences Paris, D, 285,* 961–963.

Chernet, T., Hart, W. K., James, L., Aronson, J. L., Robert, C., & Walter, R. C. (1998). New age constraints on the timing of volcanism and tectonism in the northern Main Ethiopian Rift—southern Afar transition zone (Ethiopia). *Journal of Volcanology and Geothermal Research, 80,* 267–280.

Clark, J. D., & Kurashina, H. (1979). An Analysis of Earlier Stone Age Bifaces from Gadeb (Locality 8E), Northern Bale Highlands, Ethiopia. *The South African Archaeological Bulletin, 34,* 93–109.

Condemi, S. (2004). The Garba IV E mandible. In J. Chavaillon & M. Piperno (Eds.), *Studies on the Early Paleolithic Site of Melka Kunture, Ethiopia* (pp. 687–701). Florence: Origines.

Cressier, P. (1980). Magnétostratigraphie du gisement pléistocène de Melka-Kunturé (Ethiopie). Datation des niveaux oldowayens et acheuléens. Ph.D. Dissertation, Université Louis Pasteur, Strasbourg.

de Heinzelin, J., Clark, J. D., Schick, K., & Gilbert, W. H. E. (2002). *The Acheulean and the Plio-Pleistocene Deposits of the Middle Awash Valley, Ethiopia.* Tervuren: Musée Royal de l'Afrique Centrale.

de la Torre, I. (2004). Omo revisited. *Current Anthropology, 45,* 439–465.

de la Torre, I. (2011). The Early Stone Age lithic assemblages of Gadeb (Ethiopia) and the developed Oldowan/early Acheulean in East Africa. *Journal of Human Evolution, 60,* 768–812.

de la Torre, I., Mora, R., & Martínez-Moreno, J. (2008). The early Acheulean in Peninj (Lake Natron, Tanzania). *Journal of Anthropological Archaeology, 27,* 244–264.

Deino, A. L. (2012). $^{40}Ar/^{39}Ar$ dating of Bed I, Olduvai Gorge, Tanzania, and the chronology of early Pleistocene climate change. *Journal of Human Evolution, 63,* 251–273.

deMenocal, P. (2004). African climate change and faunal evolution during the Pliocene–Pleistocene. *Earth and Planetary Science Letters, 220,* 3–24.

Di Vincenzo, F., Rodriguez, L., Carretero, J. M., Collina, C., Geraads, D., Piperno, M., et al. (2015). The massive fossil humerus from the Oldowan horizon of Gombore I, Melka Kunture (Ethiopia, >1.39 Ma). *Quaternary Science Reviews, 122,* 207–221.

Diez-Martín, F., Sánchez Yustos, P., Uribelarrea, D., Baquedano, E., Mark, D. F., Mabulla, A., et al. (2015). The Origin of the Acheulean: The 1.7 Million-Year-Old Site of FLK West, Olduvai Gorge (Tanzania). *Nature Scientific Reports, 5,* 17839.

Dupont, L. M., Donner, B., Schneider, R., & Wefer, G. (2001). Mid-Pleistocene environmental change in tropical Africa began as early as 1.05 Ma. *Geology, 29,* 195–198.

Fiore, I., & Tagliacozzo, A. (2004). Taphonomic analysis of the bone remains from the Oldowan site of Garba IV. In J. Chavaillon & M. Piperno (Eds.), *Studies on the Early Paleolithic site of Melka Kunture, Ethiopia* (pp. 639–682). Florence: Origines.

Friis, I. (1992). Forest and forest trees of Northeast Tropical Africa: Their natural habitats and distribution pattern in Ethiopia, Djibouti and Somalia. *Kew Bulletin Additional Series, 15,* 1–396.

Friis, I., & Sebsebe, D. (2001). Vegetation maps of Ethiopia and Eritrea. A review of the existing maps and the need for a new map for the Flora of Ethiopia and Eritrea. *Biologiske Skrifter, 54,* 399–439.

Friis, I., Demissew, S., & Van Breugel, P. (2010). Atlas of the Potential Vegetation of Ethiopia. *Biologiske Skrifter, 58,* 1–307.

Gallotti, R. (2013). An older origin for the Acheulean at Melka Kunture (Upper Awash, Ethiopia). Techno-economic behaviours at Garba IVD. *Journal of Human Evolution, 65,* 594–620.

Gallotti, R., & Mussi, M. (2015). The Unknown Oldowan: ~1.7-Million-Year-Old Standardized Obsidian Small Tools from Garba IV, Melka Kunture, Ethiopia. *PLoS ONE, 10,* e0145101.

Gallotti, R., & Mussi, M. (2018). Before, During, and After the Early Acheulean at Melka Kunture (Upper Awash, Ethiopia): A Techno-economic Comparative Analysis. In R. Gallotti & M. Mussi (Eds.), *The Emergence of the Acheulean in East Africa and Beyond. Contributions in Honor of Jean Chavaillon.* Cham: Springer.

Gallotti, R., Collina, C., Raynal, J.-P., Kieffer, G., Geraads, D., & Piperno, M. (2010). The Early Middle Pleistocene site of Gombore II (Melka Kunture, Upper Awash, Ethiopia) and the issue of Acheulean Bifacial Shaping Strategies. *African Archaeological Review, 27,* 291–322.

Gallotti, R., Raynal, J.-P., Geraads, D., & Mussi, M. (2014). Garba XIII (Melka Kunture, Upper Awash, Ethiopia): A new Acheulean site of the late Lower Pleistocene. *Quaternary International, 343,* 17–27.

Geraads, D., Gallotti, R., Raynal, J. P., Bonnefille, R., & Mussi, M. (in press). Melka Kunture, Ethiopia: Lower Pleistocene faunas of the Ethiopian highlands. In S. C. Reynolds & R. Bobe (Eds.), *African Paleoecology and Human Evolution.* Cambridge: Cambridge University Press.

Gibbard, P. L., Head, M. J., Walker, M. J. C., & the Subcommission on Quaternary Stratigraphy (2010). Formal ratification of the Quaternary System/Period and the Pleistocene Series/Epoch with a base at 2.58 Ma. *Journal of Quaternary Sciences, 25,* 96–102.

Goren-Inbar, N., Melamed, Y., Zohar, I., Akhilesh, K., & Pappu, S. (2014). Beneath Still Waters—Multistage Aquatic Exploitation of *Euryale ferox* (Salisb.) during the Acheulian. In R. Fernandes & J. Meadows (Eds.), *Human Exploitation of Aquatic Landscapes, special issue.* Internet Archaeology. https://doi.org/10.11141/ia.37.1.

Goren-Inbar, N., Sharon, G., Melamed, Y., & Kislev, M. (2007). Nuts, nut cracking, and pitted stones at Gesher Benot Ya'aqov, Israel. *Proceedings of the National Academy of Sciences, 99,* 2455–2460.

Hay, R. L., & Kyser, T. K. (2001). Chemical sedimentology and paleoenvironmental history of Lake Olduvai, a Pleistocene lake in northern Tanzania. *Geological Society of America Bulletin, 113,* 1505–1521.

Hooghiemstra, H., Melice, J. L., Berger, A., & Shackleton, N. J. (1993). Frequency spectra and paleoclimatic variability of the high-resolution 30–145 ka Funza I pollen record (eastern Cordillera Colombia). *Quaternary Science Reviews, 12,* 141–156.

Isaac, G. Ll., & Isaac, B. (Eds.). (1997). *Koobi Fora Research Project. Plio-Pleistocene Archaeology* (Vol. 5). Oxford: Clarendon Press.

Joordens, J. C. A., Vonhof, H. B., Feibel, C. S., Lourens, L. J., Dupont-Nivet, G., van der Lubbe, J. H. J. L., et al. (2011). An astronomically-tuned climate framework for hominins in the Turkana Basin. *Earth and Planetary Science Letters, 307,* 1–8.

Jouzel, J., Masson-Delmotte, V., Cattani, O., Dreyfus, G., Falourd, S., Hoffmann, G., et al. (2007). Orbital and millennial Antarctic climate variability over the past 800,000 years. *Science, 317,* 793–797.

Kieffer, G., Raynal, J.-P., & Bardin, G., 2002. Cadre structural et volcanologiques des sites du Paléolithique ancien de Melka Kunture (Awash, Ethiopie): Premiers résultats. In J.-P. Raynal, C. Albore-Livadie & M. Piperno (Eds.), *Hommes et Volcans. De l'éruption à l'objet, Les Dossiers de l'Archéologie n° 2* (pp. 77–92). Liège: Archéo-Logis/CDERAD.

Leakey, M. D. (1971). *Olduvai Gorge. Excavations in Bed I and II, 1960–1963* (Vol. 3). Cambridge: Cambridge University Press.

Lepre, C. J., Roche, H., Kent, D. V., Harmand, S., Quinn, R. L., Brugal, J.-P., et al. (2011). An earlier origin for the Acheulian. *Nature, 477,* 82–85.

Lisiecki, L. E., & Raymo, M. E. (2005). A Pliocene-Pleistocene stack of 57 globally distributed benthic $\delta^{18}O$ records. *Paleoceanography, 20,* PA1003.

Magill, C. R., Ashley, G. M., & Freeman, K. H. (2013). Ecosystem variability and early human habitats in eastern Africa. *Proceedings of the National Academy of Sciences, 110,* 1167–1174.

Mohr, P. A. (1971). Ethiopian rift and plateaus: Some volcanic petrochemical differences. *Journal of Geophysical Research, 76,* 1967–1984.

Morgan, L. E., Renne, P. R., Kieffer, G., Piperno, M., Gallotti, R., & Raynal, J.-P. (2012). A chronological framework for a long and persistent archaeological record: Melka Kunture, Ethiopia. *Journal of Human Evolution, 62,* 104–115.

Mussi, M., Altamura, F., Bonnefille, R., De Rita, D., & Melis, R. T. (2016). The environment of the Ethiopian highlands at the Mid

Pleistocene transition: Fauna, flora and hominins in the 850–700 ka sequence of Gombore II (Melka Kunture). *Quaternary Science Reviews, 149,* 259–268.

Mussi, M., Altamura, F., Macchiarelli, R., Melis, R. T., & Spinapolice, E. (2014). Garba III (Melka Kunture, Ethiopia): A MSA site with archaic *Homo sapiens* remains revisited. *Quaternary International, 343,* 28–39.

Pichi-Sermolli, R. E. G. (1957). Une carta geobotanica dell'Africa orientale (Eritrea Etiopia, Somalia). *Webbia, 13,* 15–132.

Piperno, M., Collina, C., Gallotti, R., Raynal, J.-P., Kieffer, G., Le Bourdonnec, F.-X., et al. (2009). Obsidian exploitation and utilization during the Oldowan at Melka Kunture (Ethiopia). In E. Hovers & D. R. Braun (Eds.), *Interdisciplinary Approaches to the Oldowan* (pp. 111–128). Dordrecht: Springer.

Profico, A., Di Vincenzo, F., Gagliardi, L., Piperno, M., & Manzi, G. (2016). Filling the gap. Human cranial remains from Gombore II (Melka Kunture, Ethiopia; ca. 850 ka) and the origin of *Homo heidelbergensis. Journal of Anthropological Sciences, 94,* 1–24.

Raynal, J.-P., & Kieffer, G. (2004). Lithology, dynamism and volcanic successions at Melka Kunture (Upper Awash, Ethiopia). In J. Chavaillon & M. Piperno (Eds.), *Studies on the Early Paleolithic Site of Melka Kunture, Ethiopia* (pp. 111–135). Florence: Origines.

Roche, H., Brugal, J.-P., Delagnes, A., Feibel, C., Harmand, S., Kibunjia, M., et al. (2003). Les sites archéologiques plio-pléistocènes de la formation de Nachukui, Ouest-Turkana, Kenya : bilan synthétique 1997–2001. *Comptes Rendus Palevol, 2,* 663–673.

Riollet, G., & Bonnefille, R. (1976). Pollen des Amaranthacées du Lac Rodolphe (Afrique orientale), déterminations générique et spécifique. Pollen et spores, 18/1, 67–92, 6 pl.

Sikes, N. E., & Ashley, G. M. (2007). Stable isotopic signatures of pedogenic carbonates as indicators of paleoecology in the Plio-Pleistocene (upper Bed I) western margin of Olduvai Basin, Tanzania. *Journal of Human Evolution, 53,* 574–594.

Taieb, M. (1974). Évolution quaternaire du bassin de l'Awash. Ph.D. Dissertation, Université de Paris.

Tamrat, E., Thouveny, N., Taieb, M., & Brugal, J. P. (2014). Magnetostratigraphic study of the Melka Kunture archaeological site (Ethiopia) and its chronological implications. *Quaternary International, 343,* 5–16.

Teaford, M. F., & Ungar, P. S. (2000). Diet and the evolution of the earliest human ancestors. *Proceedings of the National Academy of Sciences, 97,* 13506–13511.

Terwilliger, V., Eshetu, Z., Huang, Y., Alexandre, M., Umer, M., & Gebru, T. (2011). Local variation in climate and land use during the time of the major kingdoms of the Tigray Plateau in Ethiopia and Eritrea. *Catena, 85,* 130–143.

Tzedakis, P. C., Hooghiemstra, H., & Pälike, H. (2006). The last 1.35 million years at Tenaghi Philippon: Revised chronostratigraphy and long-term vegetation trends. *Quaternary Science Reviews, 25,* 3416–3430.

Umer, M., Lamb, H. F., Bonnefille, R., Lézine, A.-M., Tiercelin, J.-J., Gibert, E., et al. (2007). Late Pleistocene and Holocene vegetation history of the Bale Mountains, Ethiopia. *Quaternary Science Reviews, 26,* 2229–2246.

Ungar, P. S., & Sponheimer, M. (2011). The Diets of Early Hominins. *Science, 334,* 190–193.

Wesphal, M., Chavaillon, J., & Jaeger, J.-J. (1979). Magnétostratigraphie des dépôts pléistocènes de Melka-Kunturé (Ethiopie), premières données. *Bulletin de la Société Géologique Française, 7ème série, XXI*(3), 237–241.

White, F. (1983). *The Vegetation of Africa.* Paris: Unesco.

Zerihun, W. (1999). *Vegetation Map of Ethiopia.* Addis Ababa: Addis Ababa University.

Zerihun, W., & Backéus, I. (1991). The shrubland vegetation in western Shewa, Ethiopia, and its possible recovery. *Journal of Vegetation Science, 2,* 173–180.

Zilberman, U., Smith, P., Piperno, N., & Condemi, S. (2004). Evidence of amelogenesis imperfecta in an early African *Homo erectus. Journal of Human Evolution, 46,* 647–653.

Chapter 6
The Early Acheulean ~1.6–1.2 Ma from Gona, Ethiopia: Issues related to the Emergence of the Acheulean in Africa

Sileshi Semaw, Michael J. Rogers, Isabel Cáceres, Dietrich Stout, and Amanda C. Leiss

Abstract Konso in Ethiopia and Kokiselei in Kenya, both dated to ~1.7 million years ago (Ma), and FLK West, a recently reported site from Olduvai dated to 1.7 Ma, are the earliest Acheulean sites known in East Africa. Ongoing archaeological investigations at Gona, in the Afar Depression of Ethiopia, have also produced early Acheulean stone assemblages at several sites, estimated to ~1.6–1.2 Ma. A number of sites, including BSN-12 and OGS-12, have yielded archaeological materials comparable to the earliest Konso artifacts. The stone assemblages from the Gona sites consist of crudely made handaxes, cleavers, and picks, as well as Mode I (Oldowan) cores, and débitage. A variety of raw materials were exploited at Gona, with trachyte, rhyolite, and basalt being the most common.

Our understanding of the behavioral and ecological background for the emergence of the Acheulean is still limited. Preliminary comparisons of BSN-12 and OGS-12 with other early Acheulean sites demonstrate variability in paleoecological settings as well as raw material use. Current archaeological evidence indicates that early *Homo erectus/ergaster* use of this new technology was already in place in East Africa ~1.75 Ma. At Gona and elsewhere in Africa, continued survey and excavations are needed to document sites with potential for yielding archaeological traces that will help our understanding of the Oldowan–Acheulean transition, the identity of the toolmakers, and the function of the early Acheulean Large Cutting Tools (LCTs).

Keywords Ethiopia • Gona • Technical behaviors • Oldowan–Acheulean transition

6.1 Introduction

Based on current evidence, the Acheulean stone technology emerged in East Africa ~1.75/1.7 Ma (Lepre et al. 2011; Beyene et al. 2013; Diez-Martín et al. 2015). The preceding Oldowan, the earliest established stone technology, characterized by simple core/flake traditions (2.6–1.7 Ma), is well-documented in the archaeological record in East, North, and South Africa (Leakey 1971; Harris and Isaac 1997; Semaw et al. 1997, 2003, 2009a; de Heinzelin et al. 1999; Plummer et al. 1999; Semaw 2000; Sahnouni et al. 2011; Diez-Martín et al. 2014, 2015; Domínguez-Rodrigo et al. 2014; Páres et al. 2014; Granger et al. 2015, and references therein). Based on a recent discovery made in Ethiopia (Di Maggio et al. 2015; Villmoare et al. 2015), early *Homo* appears to have been present by 2.8 Ma and may be the most

S. Semaw (✉)
Centro Nacional de Investigación sobre la Evolución Humana (CENIEH), Sierra de Atapuerca 3, 09002 Burgos, Spain
e-mail: sileshi.semaw@cenieh.es

S. Semaw
CRAFT, Stone Age Institute, Indiana University, 1392 W. Dittemore Rd., Gosport, IN 47433, USA

M.J. Rogers
Department of Anthropology, Southern Connecticut State University, 501 Crescent Street, New Haven, CT 06515-1355, USA
e-mail: rogersm1@southernct.edu

I. Cáceres
Universitat Rovira i Virgili (URV), Avinguda de Catalunya 35, 43002 Tarragona, Spain
e-mail: icaceres@iphes.cat

I. Cáceres
IPHES, Institut Català de Paleoecologia Humana i Evolució Social, Zona Educacional 4 – Campus Sescelades URV (Edifici W3), 43007 Tarragona, Spain

D. Stout
Department of Anthropology, Emory University, 1557 Dickey Drive, Atlanta, GA 30322, USA
e-mail: dwstout@emory.edu

A.C. Leiss
Department of Anthropology, Yale University, 10 Sachem Street, New Haven, CT 06511-3707, USA
e-mail: amanda.leiss@yale.edu

R. Gallotti and M. Mussi (eds.), *The Emergence of the Acheulean in East Africa and Beyond: Contributions in Honor of Jean Chavaillon*, Vertebrate Paleobiology and Paleoanthropology,
https://doi.org/10.1007/978-3-319-75985-2_6

likely candidate for beginning the systematic manipulation of stones characteristic of the earliest Oldowan industry. According to another recent report from Lomekwi in Kenya (Harmand et al. 2015), the beginnings of ancestral hominin stone manipulation have been pushed back to 3.3 Ma with the discovery of stones with scars and pitting marks interpreted as evidence of "battering activities" by *Kenyanthropus platyops*. According to Harmand et al. (2015), these stones are technologically significantly different from the earliest Oldowan. The Oldowan (2.6–1.7 Ma) is a simple technology, but often made with systematic and patterned stone working techniques employed for creating sharp-edged cutting stones used primarily for processing animal carcasses for meat and bone marrow extractions, and probably other functions (Roche et al. 1999; Semaw et al. 2003; Domínguez-Rodrigo et al. 2005; Cáceres et al. 2017).

Lithic technologies are often described as relatively "simple" or "complex", but it is not always clear what these terms mean. Following Deacon (2012), Stout (2013) identified two dimensions of complexity: diversity and structure. Thus, one lithic technology might be considered more complex than another if it involved a greater variety of technical operations and/or if these operations were related to one another in a more structured way. Perreault and colleagues (2013) recently employed a diversity-based approach counting the number of procedural units (mutually exclusive manufacturing steps) to quantify the increasing complexity of Oldowan, Acheulean, and Middle Stone Age technologies. An example of structural complexity would be the nesting of individual operations in increasingly deep hierarchies of goals and sub-goals. Such structural analysis also indicates the greater complexity of Acheulean over Oldowan technology (Stout 2011).

Whereas Oldowan knapping is clearly a demanding perceptual-motor skill (Bril et al. 2015), there is some debate regarding the structural complexity of action sequences involved (e.g., Delagnes and Roche 2005; de la Torre and Mora 2005; Wynn et al. 2011). We have previously presented evidence (Stout et al. 2010) that Oldowan knappers at Gona displayed group-level biases toward particular reduction strategies (unifacial vs. bifacial), indicating the presence of structure beyond direct reaction to immediate core affordances. However, leaving aside reduction methods such as polyhedral (Roche 2005) and bifacial hierarchical centripetal flaking (de la Torre et al. 2003), which may not actually be characteristic of the Oldowan (de la Torre 2009; Stout et al. 2010), the minimal required structure that may be inferred remains limited to a simple chain in which the location of the next removal is determined from the previous one according to a local rule (e.g., vertically adjacent, horizontally adjacent, alternate face).

Early Acheulean technology includes all aspects of Oldowan knapping, for example preserving simple débitage on small cores, while adding more structured core reduction and shaping methods (de la Torre 2011; Stout 2011). Thus, early Acheulean technology may be considered more complex in terms of both procedural diversity (e.g., addition of large flake blank production) and structural organization (e.g., superordinate goal of core shaping). Cognitively, this increased structural complexity implies increased abstraction, maintenance, and manipulation of goal representations (Stout 2011), while behaviorally it implies greater contingency between actions (and thus greater skill). Whereas the limited dependency between sequential actions in Oldowan flaking means that unexpected outcomes and sub-optimal choices are easily accommodated if basic requirements for forceful, accurate percussion are met, errors during Acheulean shaping are more liable to compromise the intended outcome. This has been supported in experimental comparisons of Oldowan and later Acheulean-style knapping (Stout et al. 2015; Stout and Khreisheh 2015) which have found that the latter takes longer to learn and is more dependent on abilities to accurately predict action outcomes and make appropriate strategic choices. These abilities are in turn associated with increased neural activity and functional connectivity in the prefrontal cortex (Stout et al. 2015). Further experimental work is needed to test the degree to which these findings apply to early Acheulean technology.

The earliest Oldowan at Gona has been characterized based on thousands of stone artifacts excavated within fine-grained sediments, which have been recovered from ten archaeological sites distributed over a wide area (>4 km apart). All of these archaeological sites at Gona have been securely dated to 2.6–2.5 Ma by a combination of $^{40}Ar/^{39}Ar$ and paleomagnetic dating techniques (e.g., Semaw et al. 1997, 2003, 2009a). Perhaps surprisingly, our understanding of the more recent Oldowan–Acheulean transition remains more limited despite decades of systematic surveys and excavations of many sites dated to 1.8–1.4 Ma across East Africa (see Semaw et al. 2009b for discussions).

At Gona, several early Acheulean sites are documented preserving stone assemblages and associated fossil fauna estimated to $\sim$1.7–1.2 Ma (Quade et al. 2004). Unfortunately, these sites still lack precise absolute radiometric dates, and geological/geochronological studies are underway for securing more precise minimum ages. The early Acheulean sites at Gona are distributed over a wide area within the Dana Aoule North, Ounda Gona South, and Busidima North drainages (Fig. 6.1). A summary of the preliminary geological contexts and general characteristics of the stone assemblages from Busidima North 12 (BSN-12) and Ounda Gona South 12 (OGS-12) (among the most important sites) will be presented here, in the hopes that this continuing work can help shed light on this important watershed in human evolution.

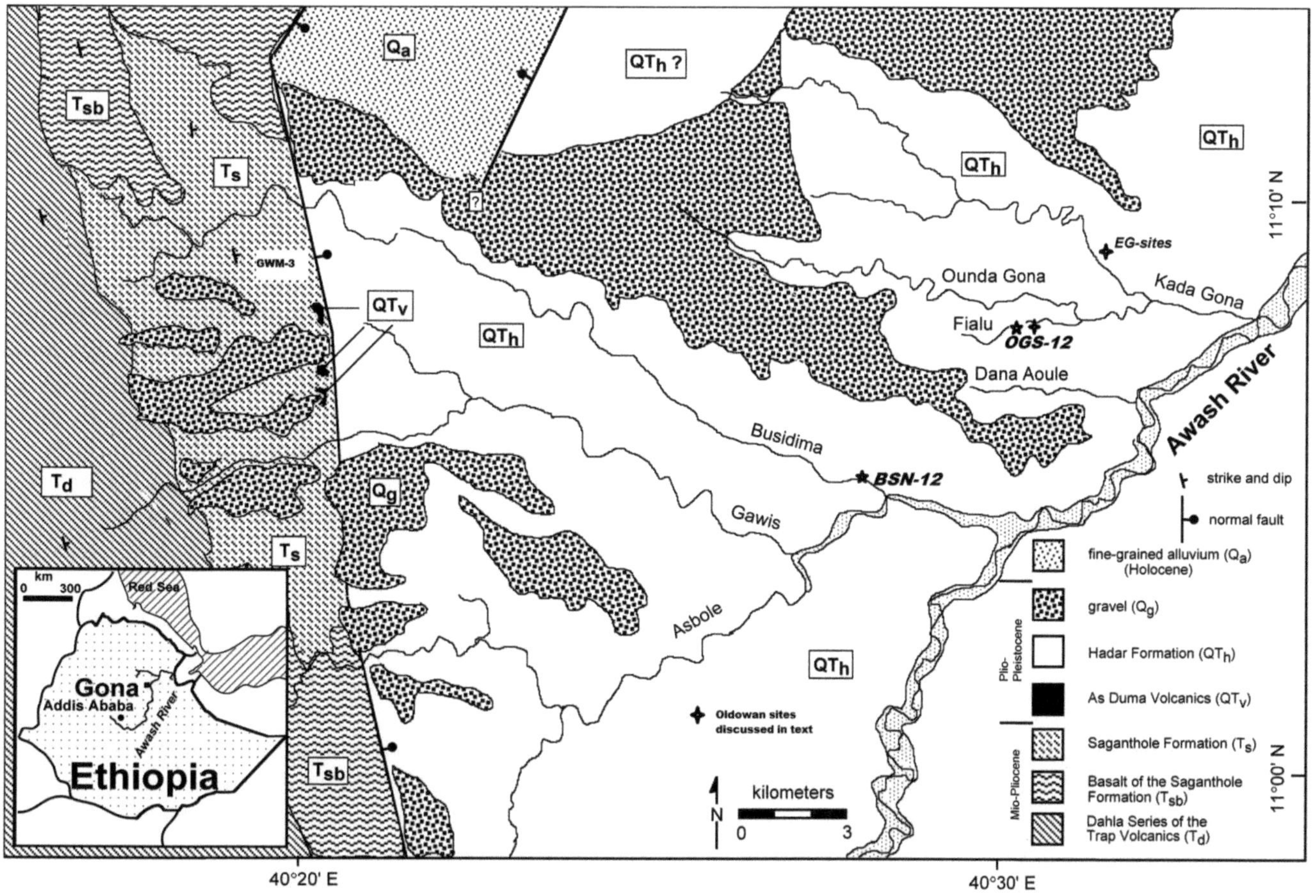

Fig. 6.1 Map of the Gona Project area showing the geology and the location of the Acheulean and some of the Oldowan sites discussed in the text (Figure modified after Quade et al. 2004)

6.2 The Oldowan–Acheulean Transition

Over the past fifty or so years, a large number of early Paleolithic sites, within the time interval between ~2.6 and 1.5 Ma, have been documented in much of East, North, and South Africa (e.g., Leakey 1971; Howell et al. 1987; Kimbel et al. 1996; Isaac and Harris 1997; Semaw et al. 1997, 2003, 2009a; de Heinzelin et al. 1999; Plummer et al. 1999; Roche et al. 1999; Chavaillon and Piperno 2004; de la Torre and Mora 2005, 2014; Boisserie et al. 2008; de la Torre et al. 2008, 2012; Delagnes et al. 2011; Lepre et al. 2011; Blumenschine et al. 2012; Beyene et al. 2013; Sahnouni et al. 2013a, b; Gallotti 2013; Diez-Martín et al. 2014, 2015; Domínguez-Rodrigo et al. 2014; Granger et al. 2015, and references therein). However, the nature and characteristics of the archaeological transition from the Oldowan to the Acheulean, as well as the identity of the makers of the earliest Acheulean and contemporaneous Oldowan in East Africa, are still poorly understood. The remarkable work by Leakey (1971) could be singled out as the most seminal in attempting to show the behavioral evolution of Oldowan–Acheulean ancestors. Based on the materials she excavated at Olduvai, Leakey saw the "Developed Oldowan" as an intermediary between the two stone industries. However, based on our current state of knowledge (with the ~1.75 Ma Konso and Kokiselei, and the 1.7 Ma FLK West discoveries), it appears that the earliest Acheulean was contemporaneous with, or actually preceded, Leakey's Developed Oldowan (with the earliest such assemblages from Olduvai dated ~1.7 Ma), making this "artifact tradition" unlikely to have been the transitional phase toward the Acheulean (Stiles 1979; de la Torre and Mora 2005; see Semaw et al. 2009b for details).

Beyond questions related to changes in the artifact forms themselves, the question of why ancestral hominins began making purposefully shaped large cutting tools (LCTs) beginning ~1.75 Ma, is also poorly understood (Lepre et al. 2011; Beyene et al. 2013). Although the early Acheulean LCTs are labeled as handaxes, picks, and cleavers, archaeologists are still grappling with basic questions regarding their respective functions. Much of our knowledge of the function of the early Acheulean was primarily derived from experimental butchery studies (e.g., Schick and Toth 1993; Jones 1994). Earlier investigations on microwear studies have shown meat and plant processing on Karari implements dated to 1.5 Ma (Keeley and Toth 1981). Further, woodworking has been proposed based on phytoliths traced on LCTs excavated at Peninj, in Tanzania (Domínguez-Rodrigo et al. 2001). However, exactly the sorts of woodworking/plant-processing activities accomplished either with the Karari or using these LCTs during the early Acheulean, and how these products were utilized, have yet to be unequivocally demonstrated based on the archaeological evidence. Advances in fat-residue and use-wear studies on late Acheulean bifaces and scrapers from Revadim in Israel have shown evidence of elephant butchery for meat (Solodenko et al. 2015).

A recent report from FLK West has provided spatially associated Acheulean stone assemblages and fauna suggesting exploitation of meat resources (Diez-Martín et al. 2015). At Gona, studies on bone surface modifications on faunas associated with early Acheulean LCTs at OGS-12 have shown exploitation of meat from small- and medium-sized animals. Nonetheless, the LCTs we refer to as picks have an indeterminate function, though some have suggested digging. Beyene et al. (2013) have presented diachronic morphological evidence to support the idea that they really are a distinct artifact class, showing little technological change through time (1.75–1.0 Ma).

For a long time the appearance of the Acheulean has been considered to have coincided with a larger-brained hominin, otherwise generally referred to as *Homo erectus/ergaster* (Klein 2009), or "*Homo erectus*-like" hominin, according to Beyene et al. (2013). With recent possible revisions in the age of the earliest *H. erectus/ergaster* fossils (McDougall et al. 2012), the new findings at Olduvai (Domínguez-Rodrigo et al. 2015), and with unresolved issues relating to dating uncertainties and taxonomic status of some fossils (see Antón 2012), the association between the onset of the Acheulean and the emergence of *H. erectus* remains unclear.

6.3 The Early Acheulean at Gona

The Acheulean is clearly distinguished from Oldowan stone technology with the creation of purposefully shaped large cutting tools (LCTs) made with great emphasis on large size and heavy weight. The beginning of the Acheulean marks the onset of human imposition of form on stone artifacts created with intended designs such as handaxes, cleavers, and picks. Acheulean artifacts during the initial phase were crudely worked, but attained symmetry and standardized shapes through time (e.g., Beyene et al. 2013; Sahnouni et al. 2013a, b). Our glimpse into the earliest Acheulean comes mainly from the two sites of Konso and Kokiselei. Now with ongoing studies at Gona, a number of additional excavated sites are yielding important data from ~1.7 to 1.2 Ma. At Gona, the main early Acheulean sites are found within the Busidima drainages, at Dana Aoule North and in the Ounda Gona South area (Fig. 6.1).

At Busidima, one of the most important sites is BSN-12, located near the present bank of the Busidima River, about 700 m upstream from its confluence with the Asbole River. In most cases at Gona, archaeological sites (both Oldowan and early Acheulean) are located stratigraphically just above major conglomeratic layers (characterized by a fining-upward sequence), interpreted as axial paleo-Awash gravels (Type I/II gravel, see Quade et al. 2004). At BSN-12, Acheulean artifacts have been found associated with the Boolihinan Tuff, which may date to ~1.2 Ma (Quade et al. 2004). Typical Acheulean artifacts, including picks and handaxes, were collected from the surface at BSN-12. Although Oldowan-type cores/flakes were found buried within the tuff itself, none of the Acheulean handaxes have been recovered *in situ*, except for one pick pulled out of the BHT tuff itself. Several other specimens (of Oldowan character) were collected with the tuff actually adhering on the artifacts as matrix, which we refer to here as "in context" (see Table 6.1 for the composition of the BSN-12 and OGS-12 stone assemblages). The association of the BSN-12 Acheulean site with cobble conglomerates (Type I/II channel) indicates that the makers ranged close to the paleo-Awash river, where stone raw materials and possible food resources would have been plentiful. As shown in Table 6.1, a low density of materials, especially LCTs, was recovered at the site.

In the Ounda Gona South area, Acheulean sites, including OGS-12, were found associated with a small channel (Type II) feeding into the paleo-Awash. At OGS-12 the materials were deposited within and just above such a channel, with a tuff dated to 1.64 Ma situated a few meters stratigraphically below the site (Quade et al. 2004). The site is estimated to 1.6–1.5 Ma, and laboratory work is underway to resolve the age of OGS-12 and other early Acheulean sites at Gona. The OGS-12 stone tool assemblage (Table 6.1), which consists of large flakes and handaxes made mainly of trachyte, rhyolite, and basalt, was found above a small pebble channel fill.

Examples of some of the LCTs from OGS-12 are shown in Figs. 6.2 and 6.3. The reworked pedogenic carbonate nodules and the small size of the channel suggest a small drainage with Type II gravel, a habitat near a stream tributary to the Awash. The recovery of fossil fauna including crocodiles, freshwater clams, and marsh cane rat (*Thryonomys swinderianus*) suggests that the channel was perennial. The location of OGS-12 on a Type II channel indicates that early *Homo erectus* spent considerable time in habitats some

Table 6.1 Composition of the OGS-12 and BSN-12 early Acheulean stone artifact assemblages

	BSN-12		OGS-12	
	Surface	In context	Surface	In context
Handaxes	3	0	12	1
Picks	0	1	11	2
Cleavers	0	0	1	0
Cores/choppers	25	7	10	5
Discoids	5	0	0	0
Whole flakes	95	4	77	31
Broken flakes and Angular fragments	49	0	56	25
Modified flakes	0	0	4	4
Modified cobbles	0	3	3	7
Unmodified cobbles	0	3	0	1
Split cobbles	0	0	1	0
Hammerstones	0	0	2	1
Total	177	18	177	77

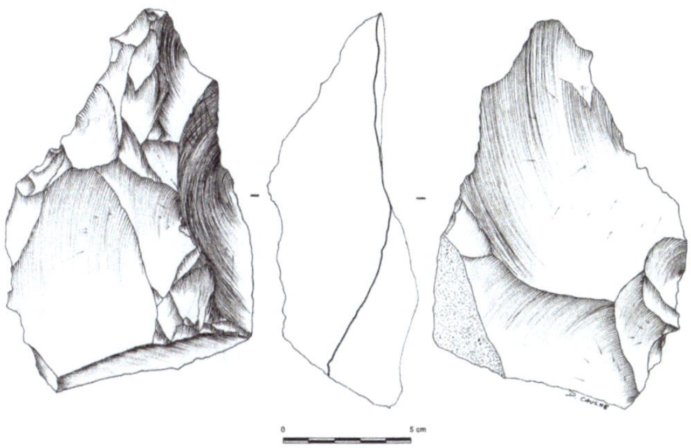

Fig. 6.2 An early Acheulean biface excavated from OGS-12

distance from the main paleo-Awash axial system. This contrasts with the settings for all of the earliest Oldowan sites at Gona dated to 2.6 Ma, which are associated with a Type I gravel interpreted as the paleo-Awash channel (Quade et al. 2004). Such different site distributions, i.e., proximity of the Oldowan sites next to or near the main paleo-Awash channel, and Acheulean occupations ranging close to but also away from the main axial river, was also a pattern documented at Koobi Fora in Kenya (Rogers et al. 1994). Interestingly, the OGS sites represent a geological context, with presence of permanent water in a tributary floodplain setting that is rare in the upper Busidima Formation. This may help explain the scarcity of archaeological materials in these younger deposits. Such isolated and

Fig. 6.3 Dorsal and ventral faces of a biface/handaxe (OGS-12-13) and a unifacial pick (OGS-12-63) collected from OGS-12, both made from side-struck flakes. The LCTs were recovered freshly eroding out of the excavation wall

localized perennial tributaries to the paleo-Awash with possible "wet and mixed open/closed" environments may have been the favored habitats for early *Homo erectus* at Gona. Several early Acheulean assemblages in the BSN and OGS drainages have been excavated, and research is in progress to refine the age of these sites and to precisely date these localities with both radiometric and non-radiometric dating techniques. Further, additional research is needed to understand better the habitat preferences of Acheulean toolmakers.

6.3.1 The BSN-12 and OGS-12 Stone Tool Assemblages

A total of 195 artifacts (177 surface and 18 "in context") from BSN-12, and a total of 254 (177 surface and 77 *in situ*) stone artifacts were recovered at OGS-12 (Table 6.1). Those referred to as "in context" from BSN-12 are artifacts recovered within the BHT or with the tuff adhering to the specimens as matrix; at OGS-12 the *in situ* artifacts were recovered from a small excavation (4 m × 3 m).

The metric and spatial data from both sites are still being analyzed, and only the assemblage compositions of the early Acheulean stone artifacts and raw material types used are discussed here. The main early Acheulean artifact types at both sites consist of crudely made handaxes, picks, and one cleaver (Tables 6.1, 6.2, 6.3, Figs. 6.2, 6.3). These are the emergent stone artifacts, the earliest of which are documented at 1.75 Ma at Konso and Kokiselei (Lepre et al. 2011; Beyene et al. 2013). Oldowan-type cores, whole and broken flakes, and fragments were recovered along with the Acheulean stone assemblages (Table 6.4). It is important here to note that the Acheulean did not replace the Oldowan; rather, Oldowan-type artifacts co-occurred with the early Acheulean, actually remaining ubiquitous throughout the Paleolithic (e.g., Clark et al. 1994). At the Busidima and Ounda Gona sites and elsewhere in East Africa, handaxes were made on large flake blanks (>12 cm) as well as on large cobbles. Picks (sometimes trihedral) were also made on both large flakes and cobbles.

A majority of the artifacts from BSN-12 and OGS-12 are worked on trachyte and rhyolite, with a substantial representation of basalt at OGS-12 (Tables 6.2, 6.3, 6.4, 6.5, 6.6). Whereas trachyte and rhyolite cobbles were easily accessible from the nearby cobble conglomerate associated with BSN-12, research is ongoing to trace the sources of the stone raw materials accessed by the OGS-12 toolmakers, some distance away. Our identification of the raw material types used for making the stone artifacts was conservative, but most of those included in the indeterminate category, particularly for the OGS-12 handaxes and picks (Tables 6.2, 6.3) were most likely basalt, also clearly reflected on the raw material composition of the cores/choppers, whole flakes,

Table 6.2 Stone Raw Materials, Handaxes, BSN-12, and OGS-12 (surface and "in context" combined)

	BSN-12	%	OGS-12	%
Trachyte	1	33.33	2	15.38
Rhyolite	1	33.33	2	15.38
Basalt	1	33.33	1	7.69
Vitreous volcanic	0	0.00	0	0.00
Other	0	0.00	2	15.38
Indeterminate	0	0.00	6	46.15
Total	3	100.00	13	100.00

Table 6.3 Stone Raw Materials, Picks, BSN-12, and OGS-12 (*Note* there were no picks recovered at BSN-12 from the surface)

	BSN-12				OGS-12	
	In context	%	Surface	%	In context	%
Trachyte	0	0.00	4	36.36	0	0.00
Rhyolite	0	0.00	2	18.18	0	0.00
Latite	0	0.00	0	0.00	0	0.00
Quartz latite	0	0.00	0	0.00	0	0.00
Aphanitic	0	0.00	1	9.09	0	0.00
Basalt	1	100.00	3	27.27	2	100.00
Vitreous volcanics	0	0.00	0	0.00	0	0.00
Other	0	0.00	0	0.00	0	0.00
Indeterminate	0	0.00	1	9.09	0	0.00
Total	1	100.00	11	100.00	2	100.00

Table 6.4 Stone Raw Materials, Cores/Choppers, BSN-12, and OGS-12

	BSN-12				OGS-12			
	Surface	%	In context	%	Surface	%	In context	%
Trachyte	11	44.00	0	0	2	20.00	2	40.00
Rhyolite	3	12.00	0	0	3	30.00	2	40.00
Latite	3	12.00	1	33.33	1	10.00	0	0.00
Quartz latite	0	0.00	0	0.00	1	10.00	1	20.00
Aphanitic	0	0.00	0	0.00	0	0.00	0	0.00
Basalt	5	20.00	0	0.00	3	30.00	0	0.00
Vitreous volcanics	3	12.00	1	33.33	0	0.00	0	0.00
Other	0	0.00	1	33.33	0	0.00	0	0.00
Indeterminate	0	0.00	0	0.00	0	0.00	0	0.00
Total	25	100.00	3	100.00	10	100.00	5	100.00

Table 6.5 Stone Raw Materials, Whole Flakes, BSN-12, and OGS-12

	BSN-12				OGS-12			
	Surface	%	In context	%	Surface	%	In context	%
Trachyte	38	40.00	0	0.00	9	11.69	10	32.26
Rhyolite	19	20.00	2	50.00	3	3.90	5	16.13
Latite	4	4.21	1	25.00	4	5.19	2	6.45
Quartz latite	0	0.00	0	0.00	0	0.00	1	3.23
Aphanitic	3	3.16	0	0.00	1	1.30	2	6.45
Basalt	18	18.95	1	25.00	53	68.83	10	32.26
Vitreous volcanics	3	3.16	0	0.00	0	0.00	0	0.00
Other	8	8.42	0	0.00	0	0.00	0	0.00
Indeterminate	2	2.11	0	0.00	7	9.09	1	3.23
Total	95	100.00	4	100.00	77	100.00	31	100.00

Table 6.6 Stone Raw Materials, Broken Flakes and Angular Fragments (*Note* there were no "in context" broken flakes and angular fragments recovered from BSN-12)

	BSN-12				OGS-12	
	Surface	%	In context	%	Surface	%
Trachyte	16	32.65	9	16.07	7	28.00
Rhyolite	4	8.16	2	3.57	3	12.00
Latite	3	6.12	2	3.57	0	0.00
Quartz latite	0	0.00	0	0.00	0	0.00
Aphanitic	0	0.00	3	5.36	4	16.00
Basalt	18	36.73	30	53.57	6	24.00
Vitreous volcanics	2	4.08	1	1.79	0	0.00
Other	5	10.20	0	0.00	0	0.00
Indeterminate	1	2.04	9	16.07	5	20.00
Total	49	100.00	56	100.00	25	100.00

and angular fragments recovered at the site (Tables 6.3, 6.4, 6.5, 6.6). Also interesting is the fact that the single pick pulled out of the BHT at BSN-12 was also worked on basalt.

Trachyte, rhyolite, and basalt were equally represented among the OGS-12 handaxes (Table 6.1). Remarkably, a large number of picks (~50% of the LCTs) were recovered at OGS-12, with only one cleaver collected at the site from the surface (Table 6.1). At BSN-12, a substantial number of cores/choppers (Table 6.4) as well as a piece identified as a polyhedron (Flaked Pieces of Isaac et al. 1981) were found "in context", i.e., recovered within the BHT or with the tuff adhering on the pieces as matrix. Trachyte, rhyolite, and basalt are well represented in both the BSN-12 and OGS-12 stone assemblages (Tables 6.3, 6.4, 6.5, 6.6), with more artifacts made of basalt identified at OGS-12. The Gona handaxes and cleavers appear to be equally worked on large cobbles as well as large flake blanks, although identifications of the original blanks for several of the specimens were difficult to determine.

6.3.2 *Taphonomic Approach to OGS-12 Bone Assemblage*

The fauna at BSN-12 includes numerous articulated fossils, suggesting that the Boolihinan ashfall may have led to their death. Further examination and analyses are in progress for this site. Here we present preliminary results from bone surface modification studies of the archaeofauna recovered in association with OGS-12. About 315 faunal remains (bone, teeth, and horn) have been recovered at OGS-12. The taphonomic and zooarchaeological analysis took into account the anatomical and taxonomic identification, grouping fossils by size and weight categories using the six classes provided by Bunn (1986). To assess the integrity of the assemblages, several quantification methods have been used: number of identified specimens (NISP), minimum number of elements (MNE), minimum number of individuals (MNI), and skeletal survival rate (%SSR) (Brain 1981). Fossils were analyzed microscopically to identify taphonomic bone modifications. The main bone alterations include anthropic activity (with evidence of cutmarks and bone breakage), carnivore tooth marks, and presence/absence of post-depositional modifications (manganese oxide, root etching, cracks, and cementation). Casts with silicone (Provil Novo Heraeus Light) and polyurethane resin (Feropur PR-55/E-55) have been made to analyze some modifications with scanning electron microscope (SEM Jeol-6400) and Hirox KH-8700 (Table 6.7).

The cutmarks identified were slicing marks. We took into account the location on bone surfaces, distributions, and orientations (Blumenschine et al. 1996; Lyman 2008; Domínguez-Rodrigo et al. 2009) to identify the activity of butchery processes carried out by hominins. Hominin-induced bone breakage has been identified by the presence and location of percussion pits (Blumenschine and Selvaggio 1988), percussion notches (Pickering and Egeland 2006), conchoidal scars, and bone flakes (Díez et al. 1999; Fernández-Jalvo et al. 1999). Carnivore tooth marks have been analyzed following the methodology established by several authors that consider the type of marks (pits, scores) and types of tissues (cancellous or cortical tissue) and their dimensions in maximum and minimum axis (Selvaggio and Wilder 2001; Domínguez-Rodrigo and Piqueras 2003; Delaney-Rivera et al. 2009; Andrés et al. 2012; Saladié et al. 2014).

The faunal assemblage is taxonomically dominated by Bovidae (40.3%) and unidentified fossils (50.8%). The remaining 8.9% comprise remains of Rodentia (3.5%), Crocodylidae (1.6%), Elephantidae (1%), Carnivora (1%),

Table 6.7 OGS-12 Faunal Assemblage NISP and MNE grouped by size, and weight categories following class size of animals established by Bunn (1986). [a]Crocodile, fish, and turtle remains are not included

NISP/(MNE)	Class 1 (MNI = 4)	Class 2 (MNI = 2)	Class 3 (MNI = 3)	Class 4 (MNI = 1)	Class 5 (MNI = 2)	Class 6 (MNI = 1)	Indeterminate	Total
Horn		1 (1)	2 (2)					3 (3)
Skull		1 (1)	2 (1)		2 (–)		3 (–)	8 (2)
Maxilla		2 (1)	1 (1)					3 (2)
Mandible	3 (3)	5 (3)	5 (5)		1 (1)		4 (–)	18 (12)
Hioides			1 (1)					1 (1)
Isolated teeth	3 (–)	3 (–)	8 (–)		4 (–)	1 (–)	18 (–)	37 (–)
Vertebra	2 (2)	12 (10)	5 (2)				15 (–)	34 (14)
Rib	2 (1)	10 (3)	3 (1)	1 (1)	1 (1)		12 (1)	29 (7)
Scapula		2 (2)						2 (2)
Humerus		1 (1)	1 (1)				1 (–)	3 (2)
Radius	1 (1)	3 (3)	3 (2)	1 (1)			1 (–)	9 (7)
Ulna		1 (1)	2 (2)				1 (–)	4 (3)
Carpal		1 (1)						1 (1)
Coxal	1 (1)							1 (1)
Femur		4 (3)	2 (2)					6 (5)
Tibia		6 (4)	1 (1)					7 (5)
Tarsal	7 (7)	5 (5)	2 (2)	1 (1)				15 (15)
Metapodial	3 (3)	5 (4)	4 (2)				1 (–)	13 (9)
Phalanx	4 (4)	11 (10)	3 (3)				2 (–)	20 (17)
Long bone	3 (–)	3 (–)		1 (–)			31 (–)	38 (–)
Indeterminate		1 (–)	2 (–)				52 (–)	55 (–)
Total	29 (22)	76 (53)	47 (28)	4 (3)	8 (2)	1 (–)	141 (–)	307[a] (108)

Rhinocerotidae (0.3%), Hippopotamidae (0.3%), Cercopithecidae (0.3%), Testudines (0.3%), and two fish remains (0.6%).

According to NISP (and MNE) (Table 6.7), most of the remains belong to small- and medium-size animals, with size Class 2 (24.4%), Class 3 (14.9%), and Class 1 (9.2%) comprising the majority of identifiable remains, in descending order of abundance and skeletal completeness. The large classes (4, 5 and 6) were rare, with the most abundant element being isolated teeth. The %SSR shows the incompleteness of the individuals recovered. In a general view, the small and medium classes (1, 2, and 3) have a high skeletal bias with low percentages. Thus, Class 1 has a % SSR of 4.4% similar to Class 3 (5.6%), while Class 2 has provided a higher value (10.4%).

Most of the skeletal elements belong to appendicular segments (37.5%) followed by cranial (24.1%) and axial (20.3%) elements. A majority of the appendicular remains are assigned to the lower limbs (metapodial, phalanx, carpal/tarsal) with a 15.6% of representation, while upper limbs (femur, humerus) and intermediate limbs (tibia, radius/ulna) show lower values, with 3.5% and 6.3%, respectively. The isolated teeth are the elements better represented on cranial segments.

In general, the bone assemblages were well-preserved and the main post-depositional modifications were manganese oxide pigmentations (74.6%), cracks (18.4%), cementations (12.3%), and chemical corrosion related to plant activity (4.1%). These modifications have allowed us to distinguish natural taphonomic damage from bone damage related to carnivore and hominin activities.

Carnivore tooth scores on cortical bone have been identified in two ribs and one long bone, and pits in cancellous bone in one indeterminate fragment. Six scores show dimensions for major axis (max = 3.52 mm; min = 1.55 mm; mean = 2.26 mm; s.d. = 0.73 mm) and minor axis (max = 0.82 mm; min = 0.17 mm; mean = 0.42 mm; s.d. = 0.24 mm) that suggest the intervention of a small carnivore. Only one pit was identified, which is insufficient to identify the carnivore responsible for the damage, but the dimensions (2.6 × 2.55 mm) suggest a small size.

Hominin activities have been identified in nine fossils (Fig. 6.4), six with cutmarks, and three bones showing evidence of intentional bone breakage. The cutmarks are slicing marks and are characterized by linear striae with V-shaped cross-section and internal microstriations. The fossils with cutmarks belong to small bovids (Class 2) and indeterminate fragments: one mandible with an isolated and transversal slicing mark on the ascending ramus; one vertebra with two cutmarks with transversal and oblique orientation concentrated close to the caudal articular process; one rib with two oblique marks concentrated on the ventral face; one mid-shaft tibia fragment with an isolated transversal mark on the diaphysis; and finally two indeterminate long bones with oblique marks concentrated on the mid-shaft.

The evidence of hominin marrow exploitation was identified by the presence of percussion pits on long bone

Fig. 6.4 Bones with cutmarks from OGS-12

fragments of Class 4-sized animals, one bone flake and a radial element of a Class 2 animal with medullary extraction. The evidence for hominin exploitation of meat may be limited, but this study has identified numerous activities related to butchery. Carcass-processing activities identified at OGS-12 include: evisceration, disarticulation, defleshing, and marrow exploitation. Evisceration has been identified through cutmarks identified on small bovid ribs. The location of marks on the ventral side of bones relates unequivocally to the consumption of viscera (Nilssen 2000). The marks on vertebrae, mandible, and limb bones, mainly on small bovids (Class 2), clearly indicate disarticulation and defleshing activities.

The evidence from OGS-12 suggests that hominins exploited animal resources from all skeletal segments (skull, trunk, and limbs). Furthermore, the cutmarks identified suggest that hominins had early access to prey at least on small bovids. Considering that viscera are among the first portions consumed by carnivores, they must have been consumed at an early stage (Domínguez-Rodrigo et al. 2005). Moreover, the location of marks in mid-shafts of limb bones is probably related to the butchery of fully fleshed bones, and as a result, also implies early access to carcasses (Bunn 1986, 2001; Domínguez-Rodrigo and Pickering 2003; Pickering and Domínguez-Rodrigo 2006; Pickering and Egeland 2009; Sahnouni et al. 2013a, b).

The intentional bone breakages for marrow exploitations have been identified only in animals belonging to size Classes 3 and 4. In these medium and large ungulates no cutmarks have been identified. Breakage marks on the bones were scarce, and thus did not provide adequate understating of the role of marrow consumption for these hominins. Since no carnivore tooth marks have been observed in these carcasses, we can suggest that carnivores were not involved in large animal acquisition and we can point out that hominins occasionally could have had access to large animal carcasses.

6.4 Issues Pertaining to the Earliest Acheulean

The nature of the archaeological transition from the Oldowan to the Acheulean—e.g., either gradual or abrupt—is still among the least understood issues in early Paleolithic studies. Some suggest the inception of the Acheulean could be traced back to the beginning of discoidal/spherical shaping documented ~1.9–1.8 Ma (Roche et al. 2009). Although few in number, some of the exhaustively worked cores made of fine-grained raw materials (e.g., vitreous volcanics) at Gona (2.6 Ma) could be identified as discoids, throwing some doubt on such a possibility. For example, a majority of the specimens identified as discoids and spheroids at Olduvai and at Aïn Hanech were made of quartz/quartzite, and experimental work has shown that with continuous flaking and use as a hammerstone, quartz/quartzite angular pieces tend to attain a rounded shape (Schick and Toth 1994; Sahnouni et al. 1997). Therefore, further experimental work is needed to determine the impact of raw materials on Mode I artifact forms for understanding whether any such relationship could have been possible between discoidal/spherical shaping and the emergence of the Acheulean. Isaac (1969) proposed that the emergence of the Acheulean was abrupt, and this remains a probable interpretation of current evidence, and we believe that such a rapid transition was likely. Such a conclusion, however, should await further field and laboratory investigations. Among the questions currently being investigated at Gona are issues related to understanding the nature and characteristics of the technological transition from the Oldowan to the Acheulean industry in Africa.

The remarkable number of fossil hominins dated to 1.8 Ma discovered in the Caucasus in Georgia were initially assigned to *H. ergaster/erectus* and recently to *Homo e. e. georgicus* (Lordkipanidze et al. 2013, and references therein). The artifacts associated with the Dmanisi hominins are simple core/flake Mode I (Ferring et al. 2011; Mgeladze et al. 2011), and the evidence for the arrival of early *H. erectus* in the Caucasus is broadly contemporaneous with the earliest Acheulean in Africa. Thus, the Acheulean provided no adaptive role in the expansion of early *Homo* out of Africa, contrary to earlier views on the initial hominin expansion outside Africa.

6.5 Discussion

Regarding the co-occurrence of Oldowan-type cores/flakes with Acheulean tool types, Lepre et al. (2011) hypothesize that different hominin groups may have been engaged in different tool activities. According to Beyene et al. (2013) the same early *Homo* species responsible for the early Acheulean could have also made and used Oldowan-type artifacts, which seems a plausible scenario. Oldowan-type cores/flakes co-occurred with the Acheulean, and actually persisted throughout the Paleolithic in the form of "expedient tools". The Acheulean LCTs are the new emergent tools, while Oldowan-type artifacts remain ubiquitous throughout the Paleolithic (e.g., Clark et al. 1994). As suggested by Beyene et al. (2013), the Acheulean represents an advanced stone industry, with LCTs utilized for new activities or created for providing a more efficient exploitation of the same activities in which early hominins engaged. The role that Mode 1 cores and flakes had within the Acheulean toolkit, and therefore the technological distinction between Oldowan and Acheulean, will be much better understood once the functions of the various stone tools are determined.

The rare small carnivore activity and the absence of damage related to large carnivores within the OGS-12 faunal assemblage suggests that carnivores had minor impact on the OGS-12 assemblages. We can suggest that hominin-carnivore competition for animal resources was rare or null, at least at this site. As some authors have suggested (e.g., Domínguez-Rodrigo et al. 2007; Pickering and Egeland 2009), hominins from 1.8 Ma were successful with regard to acquisition and processing of animal carcasses, indicating early and regular access to animal resources. OGS-12 shows that hominins around 1.6–1.5 Ma may have had greater cognitive and/or social capacities for accessing meat resources. The Acheulean technology also likely provided hominins the capabilities needed to gain an advantage over other predators for obtaining animal resources.

Compared to the Oldowan, the Acheulean is technologically advanced showing greater hierarchical planning depth and skilled execution of stone crafting (e.g., Stout 2011). The Oldowan is a core/flake tradition created by using the hand-held, direct percussion/bipolar stone working techniques for creating sharp-edged cutting flakes, primarily used for processing animal carcasses. In contrast, the making of the Acheulean demands increased motor skills and cognitive control for executing complex operational sequences involving knocking off large blanks (>12 cm, from our observations) from giant cores or large cobbles and then purposefully shaping these into handaxes with preconceived form (Sharon 2008; Stout et al. 2008, 2015; Stout 2011; Sahnouni et al. 2013a, b). This added complexity would have been associated with increased learning challenges, potentially implicating enhanced self-control for deliberate practice (Stout 2011) and more robust social support for learners (Stout 2002), possibly including intentional teaching (Morgan et al. 2015). This in turn could have important implications for the coevolution of language and technology (Morgan et al. 2015), as well as more direct influences on brain evolution through phenotypic accommodation (the "Baldwin effect"; Hecht et al. 2014). Our ongoing studies at Gona have great potential for answering some of the pressing questions on the mode and tempo of the transition from the Oldowan to the Acheulean, the paleoecological background for the emergence of the Acheulean, and its adaptive role in hominin lifeways.

Acknowledgements We would like to thank the Authority for Research and Conservation and Cultural Heritage (ARCCH) of the Ministry of Culture and Tourism of Ethiopia for research permits. The Culture and Tourism Bureau at Semera, the capital of the Afar Regional State, provided local permits and assistance for the Gona fieldwork. The L.S.B. Leakey Foundation has provided continuous funding for our field and laboratory research, and we are grateful. Additional funding was provided by grants from the Ministry of Economy and Competitiveness (MINECO) of the Spanish Government, (Project No HAR2013-41351-P, and Project No CGL2012-38434-C03-03), and the Catalonian Government, project 2014 SGR 899, Marie Curie (EU), the National Science Foundation, the Wenner-Gren Foundation, and the National Geographic Society. MJR thanks the Connecticut State University System for supporting this research through the CSU Research Grant program. The hard work in the field by our Afar colleagues is very much appreciated. The drawings were made by Dr. Dominique Cauche, and thank you.

Finally, we would like to thank Rosalia Gallotti and Margherita Mussi for inviting one of us (SS) to participate in the international workshop on "The Emergence of the Acheulean in East Africa", held at the University di Roma, Sapienza (September 12–13, 2013). It was a great pleasure to participate in the workshop organized for celebrating the 50th anniversary of the discovery of Melka Kunture (1963–2013), also honoring the lifetime achievement of the late Professor Jean Chavaillon and his contributions working at this important Paleolithic site in Ethiopia.

References

Andrés, M., Gidna, A. O., Yravedra, J., & Domínguez-Rodrigo, M. (2012). A study of dimensional differences of tooth marks (pits and scores) on bones modified by small and large carnivores. *Archaeological and Anthropological Science, 4,* 209–219.

Antón, S. C. (2012). Early *Homo*: Who, When, and Where. *Current Anthropology, 53,* S278–S298.

Beyene, Y., Katoh, S., WoldeGabriel, G., Hart, W. K., Uto, K., Sudo, M., et al. (2013). The characteristics and chronology of the earliest Acheulean at Konso, Ethiopia. *Proceedings of the National Academy of Sciences, 110,* 1584–1591.

Blumenschine, R. J., & Selvaggio, M. (1988). Percussion marks on bone surfaces as a new diagnostic of hominid behaviour. *Nature, 333,* 763–765.

Blumenschine, R. J., Marean, C. W., & Capaldo, S. D. (1996). Blind tests of inter-analyst correspondence and accuracy in the identification of cut marks, percussion marks, and carnivore toothmarks on bone surfaces. *Journal of Archaeological Science, 23,* 493–507.

Blumenschine, R. J., Stanistreet, I. G., Njau, J. K., Bamford, M. K., Masao, F. T., Albert, M. L., et al. (2012). Environments and hominin activities across the FLK Peninsula during *Zinjanthropus* times (1.84 Ma), Olduvai Gorge, Tanzania. *Journal of Human Evolution, 63,* 364–383.

Boisserie, J.-R., Guy, F., Delagnes, A., Hlusko, L. J., Bibi, F., Beyene, Y., et al. (2008). New palaeoanthropological research in the Plio-Pleistocene Omo Group, Lower Omo Valley, SNNPR (Southern Nations and Nationalities and People Regions), Ethiopia. *Comptes Rendus Palevol, 7,* 429–439.

Brain, C. K. (1981). The Hunters or the Hunted? An Introduction to African Cave Taphonomy. Chicago: Chicago University Press.

Bril, B., Parry, R., & Dietrich, G. (2015). How similar are nut-cracking and stone-flaking? A functional approach to percussive technology. *Philosophical Transactions of the Royal Society B, 370,* 20140355.

Bunn, H. T. (1986). Patterns of skeletal representation and hominid subsistence activities at Olduvai Gorge, Tanzania, and Koobi Fora, Kenya. *Journal of Human Evolution, 15,* 673–690.

Bunn, H. T. (2001). Hunting, power scavenging, and butchering by Hadza foragers and by Plio-Pleistocene *Homo*. In C. B. Stanford & H. T. Bunn (Eds.), *Meat-Eating and Human Evolution* (pp. 199–218). New York: Oxford University Press.

Cáceres, I., Rogers, M., Semaw, S. & Leiss, A. (2017). Hominin exploitation of animal resources in the Gona Plio-Pleistocene sites

(Afar, Ethiopia). In: M. Sahnouni, S. Semaw & J. Rios (Eds.), Proceedings of the II Meeting of African Prehistory (pp, 199–217). Consorcio CENIEH.

Chavaillon, J., & Piperno, M. (Eds.). (2004). *Studies on the Early Paleolithic site of Melka Kunture, Ethiopia*. Florence: Origines.

Clark, J. D., de Heinzelin, J., Schick, K. D., Hart, W. K., White, T. D., WoldeGabriel, G., et al. (1994). African *Homo erectus*: Old radiometric ages and young Oldowan assemblages in the Middle Awash Valley, Ethiopia. *Science, 264,* 1907–1909.

de Heinzelin, J., Clark, J. D., White, T. D., Hart, W., Renne, P., WoldeGabriel, G., et al. (1999). Environment and Behavior of 2.5 Million Year Old Bouri Hominids. *Science, 284,* 625–629.

de la Torre, I. (2009). Technological strategies in the Lower Pleistocene at Peninj (west of Lake Natron, Tanzania). In K. Schick & N. Toth (Eds.), *The Cutting Edge: New Approaches to the Archaeology of Human Origins*. Bloomington, IN: Stone Age Institute Press.

de la Torre, I. (2011). The Early Stone Age lithic assemblages of Gadeb (Ethiopia) and the Developed Oldowan/early Acheulean in East Africa. *Journal of Human Evolution, 60,* 768–812.

de la Torre, I., & Mora, R. (2005). *Technological Strategies in the Lower Pleistocene at Olduvai Beds I & II*. Liege: ERAUL.

de la Torre, I., Mora, R., Domínguez-Rodrigo, M., de Luque, L., & Alcala, L. (2003). The Oldowan industry of Peninj and its bearing on the reconstruction of the technological skills of lower Pleistocene hominids. *Journal of Human Evolution, 44,* 203–224.

de la Torre, I., Mora, R., & Martinez-Moreno, J. (2008). The Early Acheulian in Peninj (Lake Natron, Tanzania). *Journal of Anthropological Archaeology, 27,* 244–264.

de la Torre, I., McHenry, L., Njau, J., & Pante, M. (2011–2012). The origins of the Acheulean at Olduvai Gorge (Tanzania): A new paleoanthropological Project in East Africa. *Archaeology International, 15,* 89–98.

de la Torre, I., & Mora, R. (2014). The transition to the Acheulean in East Africa: An assessment of paradigms and evidence from Olduvai Gorge (Tanzania). *Journal of Archaeological Method and Theory, 21,* 781–823.

Deacon, T. W. (2012). *Incomplete nature: How mind emerged from matter*. New York: W.W. Norton & Co.

Delagnes, A., & Roche, H. (2005). Late Pliocene hominid knapping skills: The case of Lokalalei 2C, West Turkana, Kenya. *Journal of Human Evolution, 48,* 435–472.

Delagnes, A., Boisserie, J.-R., Beyene, Y., Chuniaud, K., Guillemot, C., & Schuster, M. (2011). Archaeological investigations in the Lower Omo Valley (Shungura Formation, Ethiopia): New data and perspectives. *Journal of Human Evolution, 61,* 215–222.

Delaney-Rivera, C., Plummer, T. W., Hodgson, J. A., Forrest, F., Hertel, F., & Oliver, J. S. (2009). Pits and pitfalls: Taxonomic variability and patterning in tooth mark dimensions. *Journal of Archaeological Science, 36,* 2597–2608.

Díez, J. C., Fernández Jalvo, Y., Rosell, J., & Cáceres, I. (1999). Zooarchaeology and Taphonomy of Aurora Stratum (Gran Dolina, Sierra de Atapuerca, Spain). *Journal of Human Evolution, 37,* 623–657.

Diez-Martín, F., Sanchez-Yustos, P., Gómez de la Rüa, D., Gómez Gonzalez, J. A., Luque, L., & Barba, R. (2014). The early Acheulean technology at ES2-Lepolosi (ancient MHS-Bayasi) in Peninj (Lake Natron, Tanzania). *Quaternary International, 322–323,* 209–236.

Diez-Martín, F., Sánchez-Yustos, P., Uribelarrea, D., Baquedano, E., Mark, D. F., Mabulla, A., et al. (2015). The Origin of the Acheulean: The 1.7 Million-Year-Old site of FLK West, Olduvai Gorge (Tanzania). *Scientific Reports, 5,* 17839.

Di Maggio, E. N., Campisano, C. J., Rowan, J., Dupont-Nivet, G., Deino, A. L., Bibi, F., et al. (2015). Late Pliocene fossiliferous sedimentary record and the environmental context of early *Homo* from Afar, Ethiopia. *Science, 347,* 1355–1359.

Domínguez-Rodrigo, M., & Pickering, T. R. (2003). Early hominid hunting and scavenging: A zooarchaeological review. *Evolutionary Anthropology, 12,* 275–282.

Dominguez-Rodrigo, M., & Piqueras, A. (2003). The use of tooth pits to identify carnivore taxa in tooth-marked archaeofaunas and their relevance to reconstruct hominid carcass processing behaviours. *Journal of Archaeological Science, 30,* 1385–1391.

Domínguez-Rodrigo, M., Egeland, C. P., & Pickering, T. R. (2007). Models of passive scavenging by early hominids: Problems arising from equifinality in carnivore tooth mark frequencies and the extended concept of archaeological palimpsests. In T. R. Pickering, K. Schick, & N. Toth (Eds.), *Breathing Life into Fossils: Taphonomic Papers in Honor of C.K. (Bob) Brain* (pp. 255–267). Gosport, IN: Stone Age Institute Press.

Domínguez-Rodrigo, M., de Juana, S., Galán, A. B., & Rodríguez, M. (2009). A new protocol to differentiate trampling marks from butchery cut marks. *Journal of Archaeological Science, 36,* 2643–2654.

Domínguez-Rodrigo, M., Díez-Martin, F., Mabulla, A., Baquedano, E., Bunn, H. T., & Musiba, C. (2014). The evolution of hominin behavior during the Oldowan-Acheulian transition: Recent evidence from Olduvai Gorge and Peninj. *Quaternary International, 322–323,* 1–6.

Domínguez-Rodrigo, M., Pickering, T. R., Almécija, S., Heaton, J. L., Baquedano, E., Mabulla, A., et al. (2015). Earliest modern human-like hand bone from a new >1.84-million-year-old site at Olduvai in Tanzania. *Nature Communications, 6,* 7987.

Domínguez-Rodrigo, M., Pickering, T. R., Semaw, S., & Rogers, M. J. (2005). Cutmarked bones from Pliocene archaeological sites at Gona, Afar, Ethiopia: Implications for the function of the world's oldest stone tools. *Journal of Human Evolution, 48,* 109–121.

Domínguez-Rodrigo, M., Serrallonga, J., Juan-Tresserras, J., Alcala, L., & Luque, L. (2001). Woodworking activities by early humans: A plant residue analysis on Acheulian stone tools from Peninj (Tanzania). *Journal of Human Evolution, 40,* 289–299.

Fernández-Jalvo, Y., Díez, J. D., Cáceres, I., & Rosell, J. (1999). Human Cannibalism in the Early Pleistocene of Europe (Gran Dolina, Sierra de Atapuerca, Burgos, Spain). *Journal of Human Evolution, 37,* 591–622.

Ferring, R., Oms, O., Agusti, J., Berna, F., Nioradze, M., Shelia, T., et al. (2011). Earliest human occupations at Dmanisi (Georgia Caucasus) dated to 1.85–1.78 Ma. *Proceedings of the National Academy of Sciences, 108,* 10432–10436.

Gallotti, R. (2013). An older origin for the Acheulean at Melka Kunture (Upper Awash, Ethiopia): Techno-economic behaviours at Garba IVD. *Journal of Human Evolution, 65,* 594–620.

Granger, D. E., Gibbon, R. J., Kuman, K., Clarke, R. J., Bruxelles, L., & Caffee, M. W. (2015). New cosmogenic burial ages for Sterkfontein Member 2 *Australopithecus* and Member 5 Oldowan. *Nature, 522,* 85–88.

Harmand, S., Lewis, J. E., Feibel, C. S., Lepre, C. J., Prat, S., Lenoble, A., et al. (2015). 3.3-million-year-old stone tools from Lomekwi 3, West Turkana. Kenya. *Nature, 521,* 310–315.

Harris, J. W. K., & Isaac, G. L. (1997). Sites in the Upper KBS, Okote, and Chari Members: Reports. In G. L. Isaac & B. Isaac (Eds.), *Koobi Fora Research Project, Volume 5: Plio-Pleistocene Archaeology* (pp. 115–236). Cambridge: Cambridge University Press.

Hecht, E. E., Gutman, D. A., Khreisheh, N., Taylor, S. V., Kilner, J., Faisal, A. A., et al. (2014). Acquisition of Paleolithic toolmaking abilities involves structural remodeling to inferior frontoparietal regions. *Brain Structure and Function, 220,* 2315–2331.

Howell, F. C., Haeserts, P. & de Heinzelin, J. (1987). Depositional environments, archeological occurrences and hominids from Members E and F of the Shungura Formation (Omo basin, Ethiopia). *Journal of Human Evolution, 16,* 665–700.

Isaac, G. L. (1969). Studies of early culture in East Africa. *World Archaeology, 1,* 1–28.

Isaac, G. Ll., Harris J. W. K., & Marshall, F. (1981). Small is informative: The application of the study of mini-sites and least effort criteria in the interpretation of the Early Pleistocene archaeological record at Koobi Fora, Kenya. In *Proceedings, Union Internacional de Ciencias Prehistoricas Y Protohistoricas, X Congress* (pp. 101–119). Mexico City.

Jones, P. R. (1994). Results of Experimental Work in Relation to the Stone Industries of Olduvai Gorge. In M. D. Leakey & D. Roe (Eds.), *Olduvai Gorge - Excavations in Beds III, IV and the Masek Beds (1968–71)* (Vol. 5, pp. 254–298). Cambridge: Cambridge University Press.

Keeley, L. H. & Toth, N. (1981). Microwear polishes on Early Stone Tools from Koobi Fora, Kenya. *Nature, 293,* 464–465.

Kimbel, W. H., Walter, R. C., Johanson, D. C., Reed, K. E., Aronson, J. L., Assefa, Z., et al. (1996). Late Pliocene *Homo* and Oldowan tools from the Hadar Formation (Kada Hadar Member), Ethiopia. *Journal of Human Evolution, 31,* 549–561.

Klein, R. G. (2009). *The Human Career* (3rd ed.). Chicago: University of Chicago Press.

Leakey, M. D. (1971). *Olduvai Gorge, Vol. III, Excavations in Beds I and II, 1960–1963.* London: Cambridge University Press.

Lepre, C., Roche, H., Kent, D. V., Harmand, S., Quinn, R. L., Brugal, J.-P., et al. (2011). An earlier origin for the Acheulian. *Nature, 477,* 82–85.

Lordkipanidze, D., Ponce de León, M. S., Margvelashvili, A., Rak, Y., Rightmire, G. P., Vekua, A., et al. (2013). A complete skull from Dmanisi, Georgia, and the evolutionary biology of early *Homo. Science, 342,* 326–331.

Lyman, R. L. (2008). *Quantitative Paleozoology.* Cambridge: Cambridge University Press.

McDougall, I., Brown, F. H., Vasconcelos, P. M., Cohen, B. E., Thiede, D. S., & Buchanan, M. J. (2012). New Single ^{40}Ar/^{39}Ar ages improve time scale for deposition of the Omo Group, Omo-Turkana Basin, East Africa. *Journal of Geological Society, London, 169,* 213–236.

Mgeladze, A., Lordkipanidze, D., Moncel, M.-H., Despriee, J., Chagelishvili, R., Inoradse, M., et al. (2011). Hominin occupations at the Dmanisi site, Georgia, Southern Caucasus: Raw materials and technical behaviors of Europe's first hominins. *Journal of Human Evolution, 60,* 571–596.

Morgan, T. J. H., Uomini, N. T., Rendell, L. E., Chouinard-Thuly, L., Street, S. E., et al. (2015). Experimental evidence for the co-evolution of hominin tool-making teaching and language. *Nature Communications, 6,* 6029.

Nilssen, P. J. (2000). An Actualistic Butchery Study in South Africa and Its Implications for Reconstructing Hominid Strategies of Carcass Acquisition and Butchery in the Upper Pleistocene and Plio-Pleistocene. Ph.D. Dissertation, University of Cape Town.

Páres, J. M., Sahnouni, M., der Made, J. V., Pérez-González, A., Harichane, Z., Derradji, A., et al. (2014). Early human settlements in Northern Africa: Paleomagnetic evidence from the Ain Hanech Formation (northeastern Algeria). *Quaternary Science Reviews, 99,* 203–209.

Perreault, C., Brantingham, P. J., Kuhn, S. L., Wurz, S., & Gao, X. (2013). Measuring the complexity of lithic technology. *Current Anthropology, 54,* S397–S406.

Pickering, T. R., & Egeland, C. P. (2006). Experimental patterns of hammerstone percussion damage on bones: Implications for inferences of carcass processing by humans. *Journal of Archaeological Science, 33,* 459–469.

Pickering, T. R., & Domínguez-Rodrigo, M. (2006). The acquisition and use of large mammal carcasses by Oldowan hominins in eastern and southern Africa: A selected review and assessment. In N. Toth & K. Schick (Eds.), *The Oldowan: Studies into the Origins of Human Technology* (pp. 113–128). Gosport, IN: Stone Age Institute Press.

Pickering, T. R., & Egeland, C. P. (2009). Experimental Zooarchaeology and its role in defining the investigative parameters of the behavior of Early Stone Age hominids. In K. Shick & N. Toth (Eds.), *The Cutting Edge: New Approaches to the Archaeology of Human Origins* (pp. 171–177). Gosport, IN: Stone Age Institute Press.

Plummer, T., Bishop, L. C., Ditchfield, P., & Hicks, J. (1999). Research on late Pliocene Oldowan sites at Kanjera South, Kenya. *Journal of Human Evolution, 36,* 151–170.

Quade, J., Levin, N., Semaw, S., Renne, P., Rogers, M. J., Simpson, S., et al. (2004). Paleoenvironments of the earliest stone toolmakers, Gona, Ethiopia. *Geological Society of America Bulletin, 16,* 1529–1544.

Roche, H. (2005). From simple flaking to shaping: Stone knapping evolution among early hominins. In V. Roux & B. Bril (Eds.), *Stone Knapping: The necessary conditions for a uniquely hominin behaviour* (pp. 35–48). Cambridge: McDonald Institute for Archaeological Research.

Roche, H., Delagnes, A., Feibel, C., Kibunjia, M., Mourre, V., & Texier, P.-J. (1999). Early hominid stone tool production and technical skill 2.34 Myr ago in West Turkana, Kenya. *Nature, 399,* 57–60.

Roche, H., Blumenschine, J., & Shea, J. J. (2009). Origins and adaptation of early *Homo*: What archeology tells us. In F. E. Grine, J. G. Fleagle, & E. Richard (Eds.), *The first Humans: Origins and Early Evolution of the Genus* Homo (pp. 135–147). New York: Springer.

Rogers, M. J., Feibel, C. S., & Harris, J. W. K. (1994). Changing Patterns of Land Use by Plio-Pleistocene Hominids in the Lake Turkana Basin. *Journal of Human Evolution, 27,* 139–158.

Sahnouni, M., Van der Made, J., & Everett, M. (2011). Ecological background to Plio-Pleistocene hominin occupation in North Africa: The vertebrate faunas from Ain Boucherit, Ain Hanech and El-Kherba, and paleosol stable-carbon-isotope studies from El-Kherba, Algeria. *Quaternary Science Reviews, 30,* 1303–1317.

Sahnouni, M., Rosell, J., Made, J. Van der, Vergès, J.M., Ollé, A., Kandi, N., et al. (2013). The first evidence of cut marks and usewear traces from the Plio-Pleistocene locality of El-Kherba (Ain Hanech), Algeria: Implications for early hominin subsistence activities circa 1.8 Ma. *Journal of Human Evolution, 64,* 137–150.

Sahnouni, M., Schick, K., & Toth, N. (1997). An experimental investigation into the nature of facetted limestone "spheroids" in the Early Palaeolithic. *Journal of Archaeological Science, 24,* 701–713.

Sahnouni, M., Semaw, S., & Rogers, M. (2013b). The African Acheulean: An archaeological summary. In P. Mitchell & P. Lane (Eds.), *The Oxford Handbook of African Archaeology* (pp. 307–323). Oxford: Oxford University Press.

Saladié, P., Rodríguez-Hidalgo, A., Huguet, R., Cáceres, I., Díez, C., Vallverdú, J., et al. (2014). The role of carnivores and their relationship to hominin settlements in the TD6-2 level from Gran Dolina (Sierra de Atapuerca, Spain). *Quaternary Science Reviews, 93,* 47–66.

Schick, K., & Toth, N. (1993). *Making silent stones speak: Human evolution and the dawn of technology.* New York: Simon and Schuster.

Schick, K., & Toth, N. (1994). Early Stone Tools and Case Study into the Nature and Function of Spheroids and Subspheroids. In R. S. Corruccini & R. L. Ciochon (Eds.), *Integrative Paths to the Past: Palaeoanthropological Advances in Honor of F. Clark Howell* (pp. 429–449). New Jersey, Prentice Hall: Englewood Cliffs.

Selvaggio, M. M., & Wilder, J. (2001). Identifying the Involvement of Multiple Carnivore Taxa with Archaeological Bone Assemblages. *Journal of Archaeological Science, 28,* 465–470.
Semaw, S. (2000). The World's oldest stone artifacts from Gona, Ethiopia: Their implications for understanding stone technology and patterns of human evolution between 2.6–2.5 million years ago. *Journal of Archaeological Science, 27,* 1197–1214.
Semaw, S., Renne, P., Harris, J. W. K., Feibel, C. S., Bernor, R. L., Fesseha, N., et al. (1997). 2.5-million-year-old stone tools from Gona, Ethiopia. *Nature, 385,* 333–336.
Semaw, S., Rogers, M. J., Quade, J., Renne, P. R., Butler, R. F., Domínguez-Rodrigo, M., et al. (2003). 2.6-Million-year-old stone tools and associated bones from OGS-6 and OGS-7, Gona, Afar, Ethiopia. *Journal of Human Evolution, 45,* 169–177.
Semaw, S., Rogers, M., & Stout, D. (2009a). Insights into Late Pliocene lithic assemblage variability: The East Gona and Ounda Gona South Oldowan Archaeology (2.6 Million years ago), Afar, Ethiopia. In K. Schick, & N. Toth (Eds.), *The Cutting Edge: New Approaches to the Archaeology of Human Origins* (pp. 211–246). Bloomington, IN: Stone Age Institute Press.
Semaw, S., Rogers, M., & Stout, D. (2009b). The Oldowan-Acheulian transition: Is there a "Developed Oldowan" artifact tradition? In M. Camps & P. Chauhan (Eds.), *Sourcebook of Paleolithic Transitions, Methods, Theories, and Interpretations* (pp. 173–193). New York: Springer.
Sharon, G. (2008). The impact of raw material on Acheulian large flake production. *Journal of Archaeological Science, 35,* 1329–1344.
Solodenko, N., Zupancich, A., Cesaro, S. N., Marder, O., Lemorini, C., & Barkai, R. (2015). Fat residue and use-wear found on Acheulian biface and scraper associated with butchered elephant remains at the site of Revadim, Israel. *PLoS ONE, 10,* e0118572.
Stiles, D. (1979). Early Acheulian and Developed Oldowan. *Current Anthropology, 20,* 126–129.
Stout, D. (2002). Skill and cognition in stone tool production: An ethnographic case study from Irian Jaya. *Current Anthropology, 45,* 693–722.
Stout, D. (2011). Stone toolmaking and the evolution of human culture and cognition. *Philosophical Transactions of the Royal Society B: Biological Sciences, 366,* 1050–1059.
Stout, D. (2013). Neuroscience of technology. In P. J. Richerson & M. Christiansen (Eds.), *Cultural Evolution: Society, Technology, Language, and Religion* (pp. 157–173). Cambridge, MA: MIT Press.
Stout, D., & Khreisheh, N. (2015). Skill Learning and Human Brain Evolution: An Experimental Approach. *Cambridge Archaeological Journal, 25,* 867–875.
Stout, D., Hecht, E., Khreisheh, N., Bradley, B., & Chaminade, T. (2015). Cognitive Demands of Lower Paleolithic Toolmaking. *PLoS ONE, 10,* e0121804.
Stout, D., Semaw, S., Rogers, M. J., & Cauche, D. (2010). Technological variation in the earliest Oldowan from Gona, Afar, Ethiopia. *Journal of Human Evolution, 58,* 474–491.
Stout, D., Toth, N., Schick, K., & Chaminade, T. (2008). Neural correlates of Early Stone Age toolmaking: Technology, language and cognition in human evolution. *Philosophical Transactions of the Royal Society B: Biological Sciences, 363,* 1939–1949.
Villmoare, B., Kimbel, W. H., Seyoum, C., Campisano, C. J., DiMaggio, E. N., Rowan, J., et al. (2015). Early *Homo* at 2.8 Ma from Ledi-Geraru, Afar, Ethiopia. *Science, 347,* 1352–1355.
Wynn, T., Hernandez-Aguilar, R. A., Marchant, L. F., & McGrew, W. C. (2011). "An ape's view of the Oldowan" revisited. *Evolutionary Anthropology: Issues, News, and Reviews, 20,* 181–197.

Chapter 7
The East African Early Acheulean of Peninj (Lake Natron, Tanzania)

Fernando Diez-Martín, Policarpo Sánchez-Yustos, and Luis de Luque

Abstract The Pleistocene record of Peninj, dated to 1.5–1.4 Ma and located on the Western shore of Lake Natron (Tanzania), is one of the classic archaeo-paleontological sources for the study of the early Acheulean in Africa. Beginning with the seminal project led by Glynn Isaac in the decades of 1960s and 1980s, other research programs have been carried out in Peninj since then, particularly the landscape archaeology approach undertaken by M. Domínguez-Rodrigo between 1995 and 2005. In 2007, fieldwork was resumed in the area and a new project is currently in progress. As a result of this long-lasting scientific effort, the variety of geological, contextual, technological, and spatial information gathered so far can shed light on a number of aspects related to the early Acheulean record identified in the three different archaeological areas of Peninj (the Type Section, the North and the South Escarpments). This paper presents a synthesis of the history of research in Lake Natron and the geology of the Peninj Group. It also reviews some of the main discussions related to the Type Section technology, the bifacial hierarchical centripetal method hypothesis, and the Oldowan–Acheulean dichotomy for the attribution of the lithic samples in the framework of the archaeological record of Peninj. The paper includes a synthesis of the new data gathered in the Acheulean sites of the Escarpments in the course of the present research project and, finally, a regional interpretation of the early Acheulean of the Lake Natron.

F. Diez-Martín (✉) · P. Sánchez-Yustos
Department of Prehistory and Archaeology, University of Valladolid, Plaza del Campus s/n, 47011 Valladolid, Spain
e-mail: fernando.diez.martin@uva.es

P. Sánchez-Yustos
e-mail: policarpo.sanchez@uva.es

L. de Luque
Fundación Conjunto Paleontológico de Teruel-Dinópolis, Avenida de Sagunto s/n, 44002 Teruel, Spain
e-mail: luisluque.geologo@gmail.com

Keywords Lake Natron • Technology • LCTs • Landscape Type Section • Escarpments

7.1 History of Scientific Research in Lake Natron

Lake Natron is in northern Tanzania near the border with Kenya (Fig. 7.1). It is but one of many shallow, salty, and alkaline water bodies located in the Eastern branch of the Great Rift Valley. Along with Lake Magadi (Kenya), Lake Natron is part of a Basin covering some 1000 km^2 (Luque 1996). The intense tectonic activity that characterizes the Great Rift Valley has led to Lake Natron being surrounded by a number of volcanoes (Luque et al. 2009a). Those of Sambu and Shompole lie to the north, Gelai to the east, and Lengai and Kerimasi to the south. Lake Natron is at the bottom of a pronounced Basin at an altitude of 610 m, surrounded by Escarpments that reach up 2942 m in the case of the Gelai volcano, or 1800 m for the Serengeti Plateau. This topography gives rise to an exceptionally dry climate. In this extreme ecological context the predominant vegetation is savannah-mosaic with intercalations of shrubby prairies, and with a gradation toward more or less open woodland with acacias.

Lake Natron was not explored by Westerners until well into the 19th century. It was mainly German explorers who studied the geography, cartography, and geology of the Natron Basin. Between 1882 and 1883, the German explorer Gustav Adolf Fischer undertook a voyage for the Hamburg Geographic Society into Maasai territory. The Ol Doinyo Lengai volcano attracted the attention of the biologist O. Neumann in 1894, and of the M. Schöller expedition in

R. Gallotti and M. Mussi (eds.), *The Emergence of the Acheulean in East Africa and Beyond: Contributions in Honor of Jean Chavaillon*, Vertebrate Paleobiology and Paleoanthropology,
https://doi.org/10.1007/978-3-319-75985-2_7

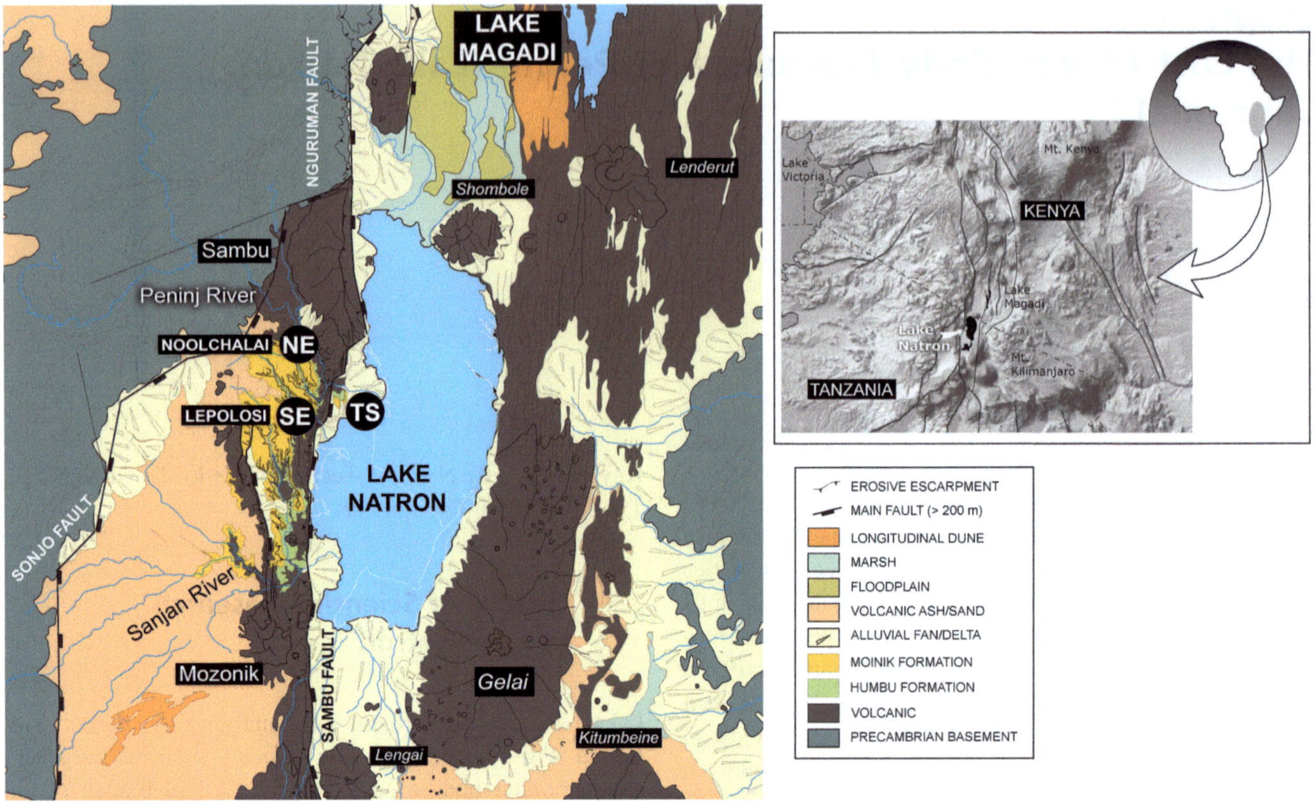

Fig. 7.1 Geological map of the Lake Natron and location of the three fossiliferous areas: Type Section (TS), North Escarpment (NE), South Escarpment (SE)

1896–1897 to equatorial East Africa and Uganda.[1] Of particular importance were the expeditions of the German geologists F. Jaeger and C. Uhlig. Between 1904 and 1910, under the auspices of the Otto Winter Foundation, they undertook a series of geological studies and made a map of the region (Uhlig and Jaeger 1942).

After World War I, a period of systematic paleoanthropological exploration in African emerged (Gowlett 1990: 18). However, during these years the archaeological potential of Lake Natron remained unknown, even though geological, petrological, and hydrochemical work continued in the Natron Basin (Guest 1953). Lake Natron would, however, burst onto the world paleoanthropological stage via the discovery of *Zinjanthropus boisei* at Olduvai Gorge (Leakey 1959). At the beginning of August 1959, while journeying to Nairobi from Olduvai for the scientific announcement of her historical discovery, Mary Leakey became aware of the potential archaeological interest of some of the sediments outcropping along the West bank of Lake Natron. The busy years that followed, devoted to the excavation of FLK and media exposure to the fossils discovered in Olduvai, kept Mary and Louis Leakey away from any exploration around Lake Natron (Morell 1996). In 1963, however, Richard Leakey "rediscovered" the same deposits while flying in a small plane from Nairobi to Olduvai. This led to a first brief exploration by Richard and Glynn Isaac. That first fieldwork visit led to the recognition of the outcrops they sought at the mouth of the River Peninj (the area later known as Maritanane and Type Section), and the collection of a handful of fossils.

In early January 1964, Leakey and Isaac set out for Lake Natron on the first archaeological survey of the area. On 11 January the exceptional find of a *Paranthropus* mandible took place (Leakey and Leakey 1964). A second expedition by Isaac and Leakey to Lake Natron occurred between the months of July and September 1964 (Isaac 1964). This second expedition undertook a preliminary geological study of the Lake Natron Basin and provided a description of a regional sedimentary sequence that Isaac named the Peninj Group (Isaac 1965, 1967). Two important Acheulean sites were also discovered and excavated at the top of the Sambu Escarpment, which were then given the names MHS and RHS (Isaac 1969).

Though Peninj appeared to be a propitious place to begin examining the Acheulean period (Isaac and Curtis 1974), events would divert attention elsewhere. Richard Leakey

[1]1901: *Mitteilungen über meine Reise nach Äquatorial-Ostafrika und Uganda 1896–1897.* [A report of my journey in Eastern Equatorial Africa and Uganda between 1896 and 1897].

joined a new French-American project in the valley of the River Omo in Ethiopia (Morel 1996). Isaac also had obligations that would take him away from Lake Natron for nearly 17 years (Isaac 1977; Isaac and Isaac 1997). He took renewed interest in Peninj in 1981, when he led a brief reconnoitering expedition (Isaac 1981–1982). A new project started up in 1982, with an international team co-led by Isaac, Amini Mturi, and Maurice Taieb. The geological and paleontological studies undertaken by Taieb between 1981 and 1984 were the most productive of all those in this period (Taieb and Fritz 1987). Unfortunately, only superficial archaeological studies could be undertaken between 1981 and 1982 (Isaac 1982). The sudden death of Isaac in 1985 was a significant setback to their endeavors. Thanks to the field notes and other unpublished documentation, including a detailed proposal for an archaeological intervention sent to the National Science Foundation (Isaac 1982), we know that Isaac's research was conceived as an integrative landscape archaeology project.

In 1994, a new round of paleoanthropological research began in Peninj, led by M. Domínguez-Rodrigo. The project conducted extensive research until 2005, following the same precepts of landscape archaeology that Isaac designed for this enclave (that included both site and regional or off-site fieldwork). The geological and stratigraphic data for the Peninj Group were meticulously reviewed and updated, along with paleoecological analyses. Fieldwork was also resumed in the three archaeological areas documented in Peninj: Maritanane or Type Section, MHS-Bayasi (South Escarpment), and RHS-Mugulud (North Escarpment). Abundant sites in all these three areas were extensively excavated, and spatial, technological, taphonomic, and functional studies were performed (Luque 1995, 1996; Domínguez-Rodrigo 1996; Domínguez-Rodrigo et al. 2001a, b, 2002, 2005, 2009a; Downey and Domínguez-Rodrigo 2002–2003; de la Torre et al. 2003, 2008; de la Torre 2009). Fieldwork has continued in the area since 2007, with a new team working on extending and complementing this earlier work (Diez-Martín 2008; Diez-Martín et al. 2009a, 2010, 2011, 2012, 2014a, b).

7.2 The Peninj Group: Geology, Dating, and Archaeological Areas

The Pleistocene sediments located in the southwestern area of the Natron Basin were described systematically by Glynn Isaac in a sequence known as the Peninj Group (1965, 1967). The geology and geomorphology of the Peninj Group were subsequently refined by Luque et al. (2009a, b). Although sediments of the Peninj Group crop out in three different areas (North and South Escarpments, and Type Section) (Fig. 7.1), the reference for the Peninj Group was provided by the badland outcrops located around the mouth of the modern Peninj River delta, which cover a surface of some 1 km^2 (the area known as Maritanane and Type Section; Fig. 7.2; Luque et al. 2009b). This area, intensely modified by tectonic and erosive activity (Foster et al. 1997), constitutes a complex web of gullies exposing most of the geological sequence:

Fig. 7.2 Panoramic view of the Type Section area of Peninj

- The series defined by Isaac lies on older Plio-Pleistocene volcanic materials. At the base of the Peninj Group are the Sambu lavas, 400 m of volcanic materials deposited when the Sambu volcano entered periods of great eruptive intensity, bracketed between 3.5 and 2 Ma by means of K/Ar datings (Isaac 1967; Isaac and Curtis 1974).
- Above this basalt basement lie the Hajaro Beds, a set of sandy, clayey and basaltic materials deposited before the existence of the Peninj River and its drainage network, in about 2 Ma (Thouveny and Taieb 1987).
- Above the Sambu and Hajaro sediments lie the materials of the Peninj Group. These reach 80 m in thickness and contain the fertile layers with paleontological and archaeological remains. The Peninj Group is divided into two ~40 m thick units (Isaac 1965, 1967):
- At the base, the Humbu Formation consists largely of sandstone and alluvial deposits, intercalated with a complex sequence of tufaceous materials. In the middle part of the Humbu sequence is the Main Tuff, a thick volcanic deposit that includes the Wa Mbugu basalt interbedded (Isaac and Curtis 1974).
- The Main Tuff constitutes the reference for dividing the Humbu Formation into three members, from bottom to top: the Basal Sands and Clays (BSC), the initial detritic infill of the Basin and the unit where the *Paranthropus* jaw was discovered (Leakey and Leakey 1964); the Main Tuff itself (MT).
- The Upper Sands and Clays (USC) where most of the archaeological sites discovered so far are located (Domínguez-Rodrigo et al. 2009a; Fig. 7.3).

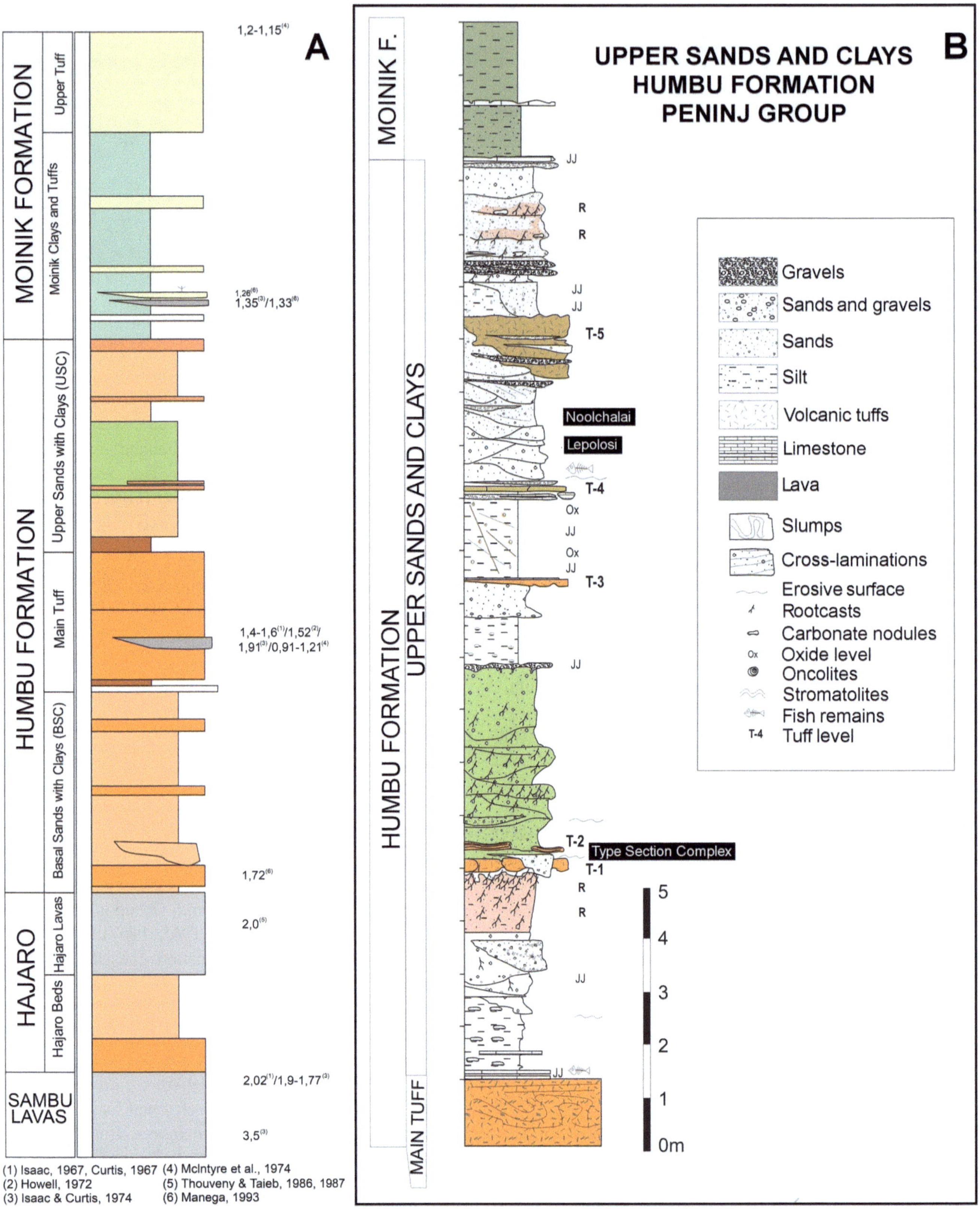

Fig. 7.3 **A** Type stratigraphic column of the Peninj Group; **B** type stratigraphic column of the Upper Sands with Clays member (USC) in the Humbu Formation of the Peninj Group, and situation of the main archaeological horizons

- On top of the Humbu Formation, the Peninj Group ends with the Moinik Formation, a 40-m-thick deposit of lacustrine sediments, where some Acheulean sites have also been discovered (Diez-Martín et al. 2009b).

Precise dates for the archaeological evidence of Peninj remain controversial. The diagenetic alteration undergone by the tuffs exposed in the Peninj Group prevent reliable datings (Luque et al. 2009a, b; McHenry et al. 2011). The poor quality of the tephra layers is responsible for the contradictory results obtained by the different dating efforts carried out in Peninj. Based on the various radiometric (K/Ar and Ar/Ar) analyses undertaken in the Wa Mbugu basalt in the Main Tuff (Howell 1972; Isaac and Curtis 1974; McIntyre et al. 1974) and in the base of the Moinik Formation (Isaac and Curtis 1974; Manega 1993), and supported by paleomagnetic correlations (Thouveny and Taieb 1987), the archaeological evidence deposited within the USC unit has been preferentially bracketed between 1.7 and 1.4 Ma (Isaac and Curtis 1974; Domínguez-Rodrigo et al. 2009a, 2002). However, recent $^{40}Ar/^{39}Ar$ dating of the Peninj Group carried out by Deino et al. (2006) suggests a slightly different chronological scenario. According to this new data, the Wa Mbugu basalt would be dated to 1.19 Ma, while the Upper Moinik Tuff would close the Peninj Group at about 1.01 Ma. Thus, the archaeological sites preserved within the USC would have been deposited within a short period of time, between 1.2 and 1.1 Ma, significantly younger on average than previously assumed. de la Torre et al. (2008) have already pointed out severe contradictions between the new chronological framework proposed by Deino and colleagues, the techno-typological characteristics of the Peninj Acheulean, and its close correlation with the Olduvai Acheulean, dated to 1.5 Ma (Leakey 1971). While bearing in mind the stratigraphic inconsistence produced by the chronological discrepancies between old and new datings and the need of further geochronological research in the area (McHenry et al. 2011), we agree with de la Torre et al. (2008) when they claim for a closer link of the Peninj Acheulean to the chronological framework originally proposed by Isaac and Curtis (1974).

The USC alluvial sediments in which all the archaeological and paleontological remains have been preserved crop out in three different areas: Lepolosi (South Escarpment) and Noolchalai (North Escarpment) on the Sambu Escarpment, and Maritanane or Type Section, in the Peninj River delta (Fig. 7.1). All three were discovered and preliminarily studied between 1963 and 1964 by the Leakey–Isaac expedition. It is important to remark that the archaeological evidence contained within the USC mostly appears to result from two discrete depositional episodes (Fig. 7.3).

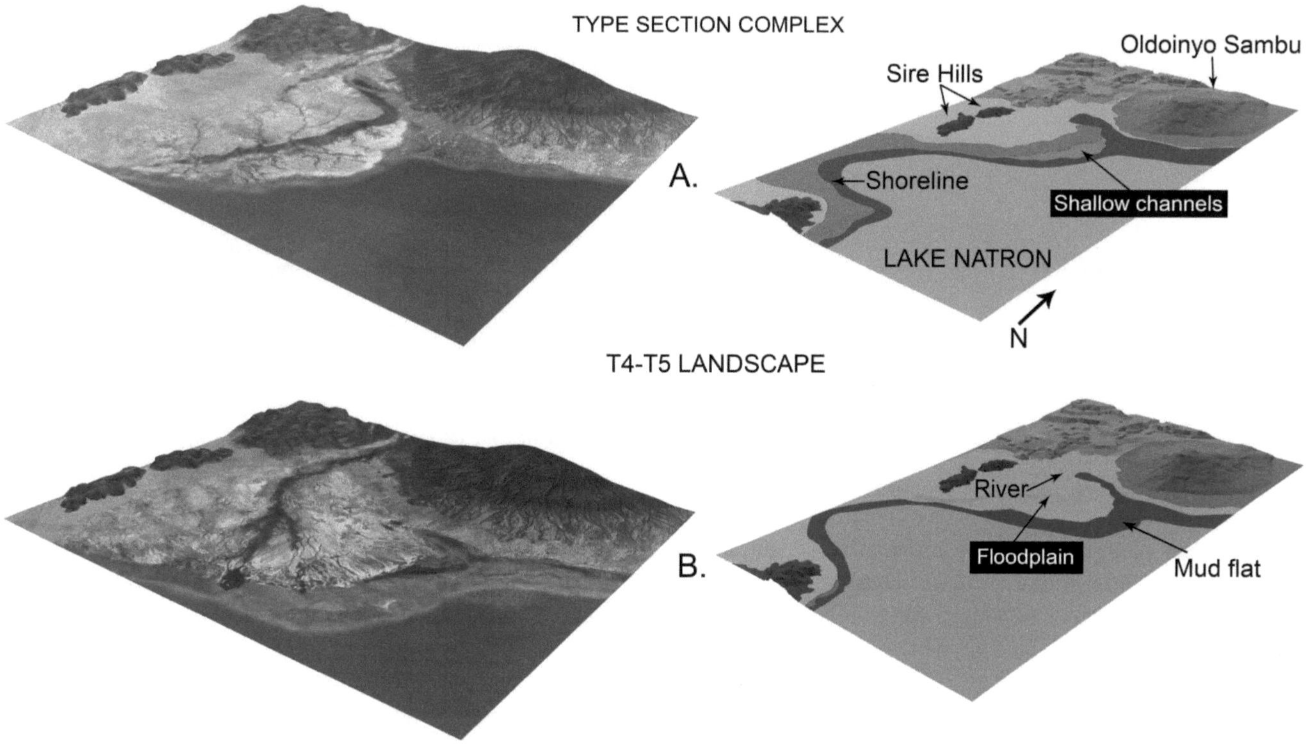

Fig. 7.4 Paleogeographical reconstruction of the Natron Basin during: **A** The Type Section Complex and **B** the T4–T5 stratigraphic interval, within the USC of the Humbu Formation

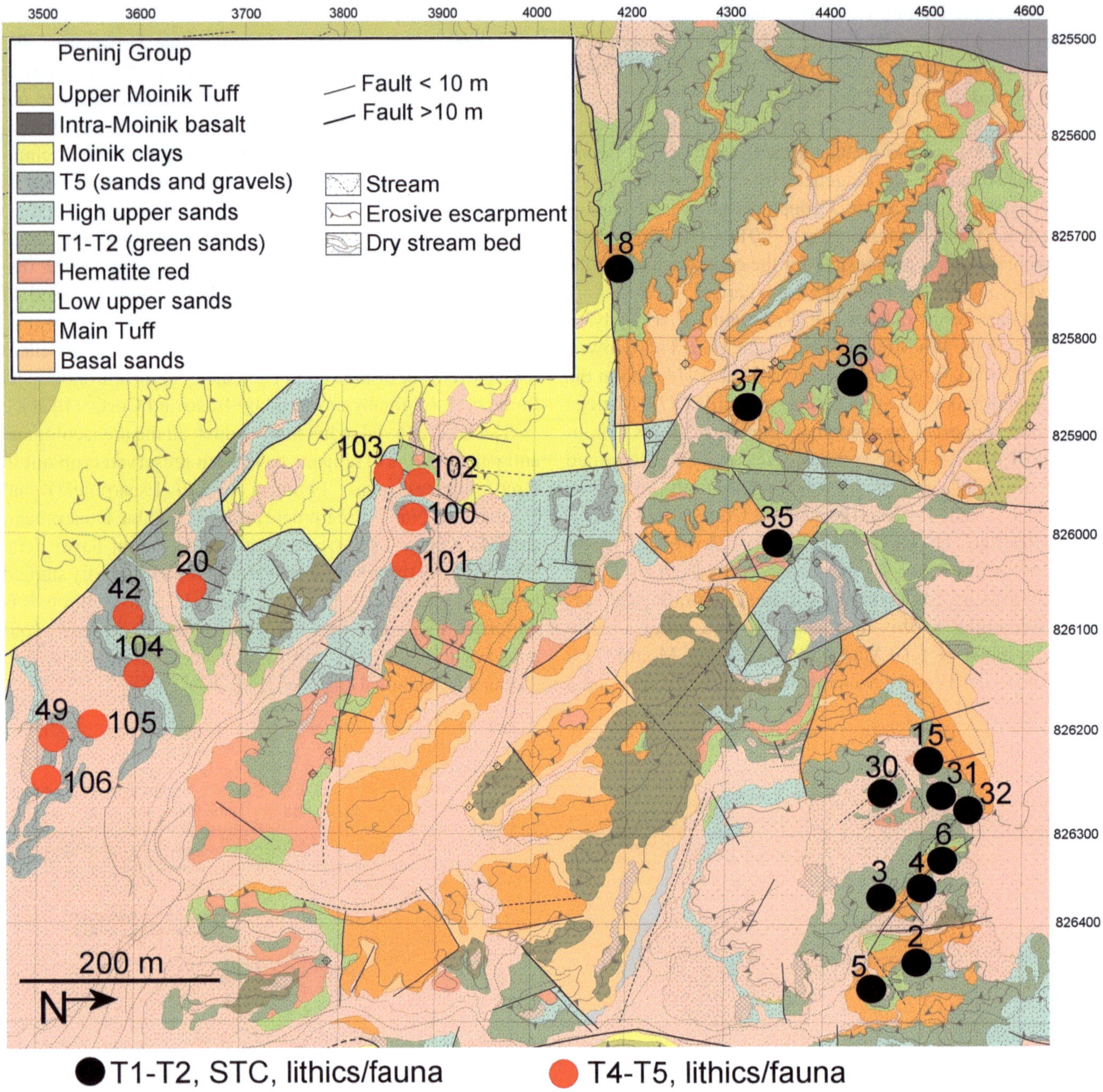

Fig. 7.5 Geological map of the Type Section area of Peninj and location of the main sites with lithics and/or fauna in the T1–T2 and T4–T5 depositional intervals

The oldest is represented by a paleosol of sands and clays between Tuffs 1 and 2[2]. This layer is very rich in lithic and faunal remains which were deposited in the context of an alluvial fan in which shallow channels drained toward the lake (Fig. 7.4: A) (Domínguez-Rodrigo et al. 2009b). This layer is particularly well preserved in the Type Section and, conversely, not represented in the South and North Escarpments. Thus, in the Type Section area, this paleosurface has been the subject of intense study and spatial cataloguing within the 1995–2005 landscape archaeology project. The many localities recognized here fall within what is known as the Type Section Complex (Domínguez-Rodrigo et al. 2002, 2009b; Fig. 7.5). The second of these depositional episodes roughly appears

[2]The USC member has a thickness that varies between 4 and 20 m. It consists of alluvial facies constituted by clays, silts, sands, and dolomitic carbonates. These alluvial materials are interbedded with five volcanic tuffs (from bottom to top, T1 to T5) (Luque et al. 2009b) (Fig. 7.3)

between Tuffs 4 and 5, toward the top of the Humbu Formation. It is associated with a different landscape context, related to relatively large, high-energy fluvial channels over the floodplain (Fig. 7.4: B). This depositional episode crops out particularly well in the North and South Escarpment areas (Domíngez-Rodrigo et al. 2009c, d). Conversely, it is poorly persevered in the Type Section area (Fig. 7.5). Thus, it is important to note that in Peninj there is a heterogeneous and biased preservation of archaeological resources in the three study areas: the TSC bracketed between T1 and T2 is preserved in the Type Section area, while the archaeological sites related to the interval between T4 and T5 are much better represented in sites located on the Sambu Escarpment area. This bias in terms of differential preservation has consequences in the regional interpretations of hominin landscape use in the Basin.

7.3 The Archaeology of the Type Section Complex: Current Issues

As stated above, the Type Section Complex (TSC) is a fertile horizon deposited between T1 and T2, in a penecontemporaneous context of a deltaic channel system near the lake margin. Although archaeological materials have also been discovered in other stratigraphic positions in the Type Section area (de la Torre 2009; Domínguez-Rodrigo et al. 2009a), most of the remains, including a wealth of stone tools and fossil bones, come from this exceptionally well-preserved paleolandscape window (Domínguez-Rodrigo et al. 2002, 2009b). Based on the lithic categories predominantly represented in the TSC lithic collections (hammerstones, cores, small and medium-sized flakes, scarce retouched flakes), this industry was originally labelled as Oldowan (de la Torre et al. 2003; de la Torre 2009). However, the identification of complex core exploitation models that included core hierarchization and flake predetermination (the so-called bifacial hierarchical centripetal model, BHCM, that accounted for 30% of the core collection retrieved from the TSC) introduced new elements to the debate on the cognitive capabilities of the purported "Oldowan" hominins (de la Torre et al. 2003) and had a significant impact in subsequent literature (Davidson and McGrew 2005; Delagnes and Roche 2005; Harmand 2007; Braun et al. 2008; Semaw et al. 2009; White et al. 2011). Recent reappraisals of the TSC industry, along with a comprehensive set of experimental studies aimed at testing the validity of the BHCM at Peninj, have forced a revision of various issues related to the TSC industry (Diez-Martín et al. 2012).

7.3.1 On the Bifacial Hierarchical Centripetal Method

Following de la Torre et al. (2003: 218), cores related to the BHCM, as identified in the TSC of Peninj, consisted of: (1) two surfaces separated by a plane of intersection, in which (2) the relation between surfaces was hierarchical, with the main surface acting as the exploitation area (aimed at obtaining flakes) and the subordinate surface serving as a preparation surface (aimed at preparing striking platforms). The hierarchical relation between both surfaces (preparation/exploitation) seemed non-interchangeable, as their role was maintained stable through all the flaking process. (3) The maintenance of this volumetric structure was aimed at obtaining predetermined flakes. (4) Flakes detached from the main flaking surface were parallel or sub-parallel with respect to the intersection plane. (5) The preparation surface produced secant flakes with respect to the intersection plane between both surfaces.

It was argued also (de la Torre et al. 2003; de la Torre and Mora 2009: 182) that many of the cores retrieved from the TSC represented the continuity of a single technological sequence and that this sequence could be reconstructed. Based on the fact that the dimensional relation between unifacial centripetal, hierarchical centripetal, and multifacial irregular cores (the three most abundant core types identified at the Type Section) seemed to be constant, in a decreasing trend from the former to the latter, these authors (de la Torre and Mora 2009: 180–183) created a theoretical model that explained in detail the way in which the sequence was performed by hominins. This model encompassed six consecutive phases (de la Torre and Mora 2009: Fig. 7.52), in which several archaeological specimens could actually be inserted (ibid: Fig. 7.53): (1) The core would initially be exploited centripetally on one surface. (2) As the unifacial centripetal exploitation continued, the core would lose its peripheral convexities. (3) The loss of the required convexities would make exploitation difficult, and in order to reactivate the convex volume of this striking surface, it would be necessary to start preparation in the sagittal plane. (4) This reactivation would produce a hierarchical core, as explained above. (5) Once the hierarchical pattern was established, the model would continue in successive series. (6) Core exhaustion would imply a final irregular multifacial form.

A recent analysis of a sample of 46 cores included in the TSC industry (most of them used as referential specimens for exemplifying the BHCM by de la Torre et al. 2003 and de la Torre and Mora 2009) has pointed out remarkable discrepancies with previous interpretations of the core sample (Diez-Martín et al. 2012; Figs. 7.6 and 7.7). The new set of diacritical diagrams showed that different reduction sequences were carried out in the Type Section and that this operational variety correlates with the different blank shapes available in the area (spherical and sub-spherical, hemispherical, and angular). According to these core blanks the collection could be divided into two different exploitation groups: (a) massive blanks (spherical, sub-spherical and polyhedral shape), for which reduction was based on different unipolar series intersected orthogonally, although in few cases an incipient centripetal organization of negative scars can be documented, or pairs of negative scars that intersect perpendicularly and tend to overlap as a result of core rotation (Fig. 7.6: A and 7.6: B); (b) medium-sized hemispherical and flake blanks. In these cases, exploitation starts with parallel short series (two or three negative scars) that converge in a bipolar manner. The exploitation continues first with alternate isolated detachments and then with new unipolar series that, in some cases, converge again in a bipolar way. Later phases of the reduction sequence show centripetal and orthogonal schemes (Fig. 7.7: B).

Raw material quality seems to have played an important role in the operational schemes carried out by hominins in the Type Section. The following associations have been identified (Diez-Martín et al. 2012; Fig. 7.8): (1) Group 1: Poor quality basalts (mostly related to massive blanks) show a partial alternation of negative scars and a remarkably low degree of exploitation. (2) Group 2: Other basalts and nephelinites of medium to good quality (hemispherical and flake blanks) show a medium degree of exploitation through unipolar and orthogonal series. (3) Group 3: Very good and optimal basalts were intensively exploited in centripetal, orthogonal, and orthogonal–polyhedral sequences. Group 1 seems to have been unconnected with the others: spherical and sub-spherical relatively heavy volumes show a limited exploitation (less than 10% of core mass) and could have been related, among other tasks, to percussion activities. Several medium-sized nephelinite cobbles were used in percussion activities after being successfully exploited as cores. It is worth bearing in mind that some of the cores included in the earliest stages of the BHCM by previous researchers actually belong to this Group 1. Groups 2 and 3 might be interconnected and might represent different parts of the same reduction scheme (initial reduction phase and a full exploitation phase).

In sum, taking into account the constrains imposed by core blank morphology and raw material quality, recent interpretations of the core sample retrieved from the TSC industry show that exploitation strategies were driven by the technical principle of discontinuous alternation of different surfaces (preferentially two) in the knapping process, rather than by a hierarchization of the volumes. Knappers were here systematically undertaking the unipolar exploitation of

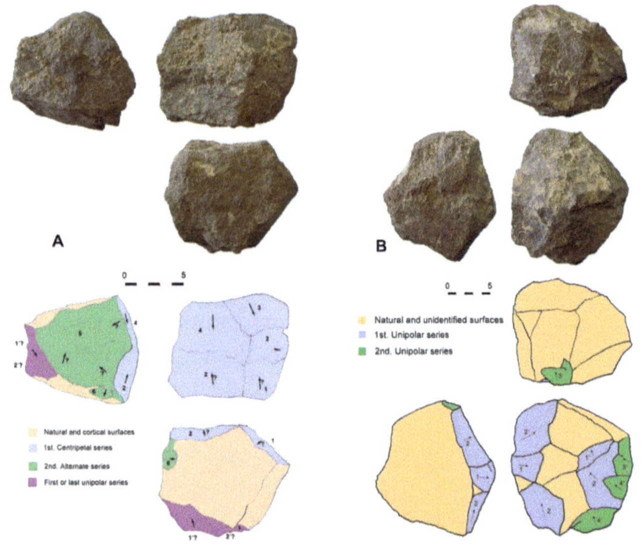

Fig. 7.6 Photographs and diacritical schemes of cores selected as representative examples of the various phases hypothesized for the BHCM (de la Torre and Mora 2009). **A** ST31 A-28 (Example of Stage 1). Large basalt cobble of very low quality showing orthogonal series of detachments on two different surfaces. Note the polyhedral-like morphology and the absence of a unifacial centripetal pattern. **B** ST4 U0-33 (Example of Stage 3). Very low quality and irregular basalt blank showing unifacial orthogonal series on one surface. Note the polyhedral shape and the lack of preparation on transverse and sagittal planes

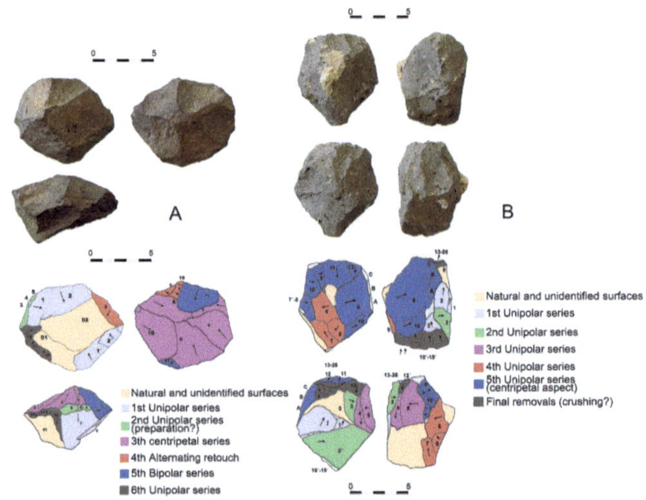

Fig. 7.7 Photographs and diacritical schemes of cores selected as representative examples of the various phases hypothesized for the BHCM (de la Torre and Mora 2009). **A** ST46-A17 (between Tuffs 4–5 in Type Section). Core on good quality basalt showing orthogonal/discoid-like exploitation by the combination of unipolar and bipolar/centripetal series. **B** ST32-S2 (Example of Stage 4) Core on good quality nephelinite showing orthogonal/discoid-like exploitation by the combination of unipolar series

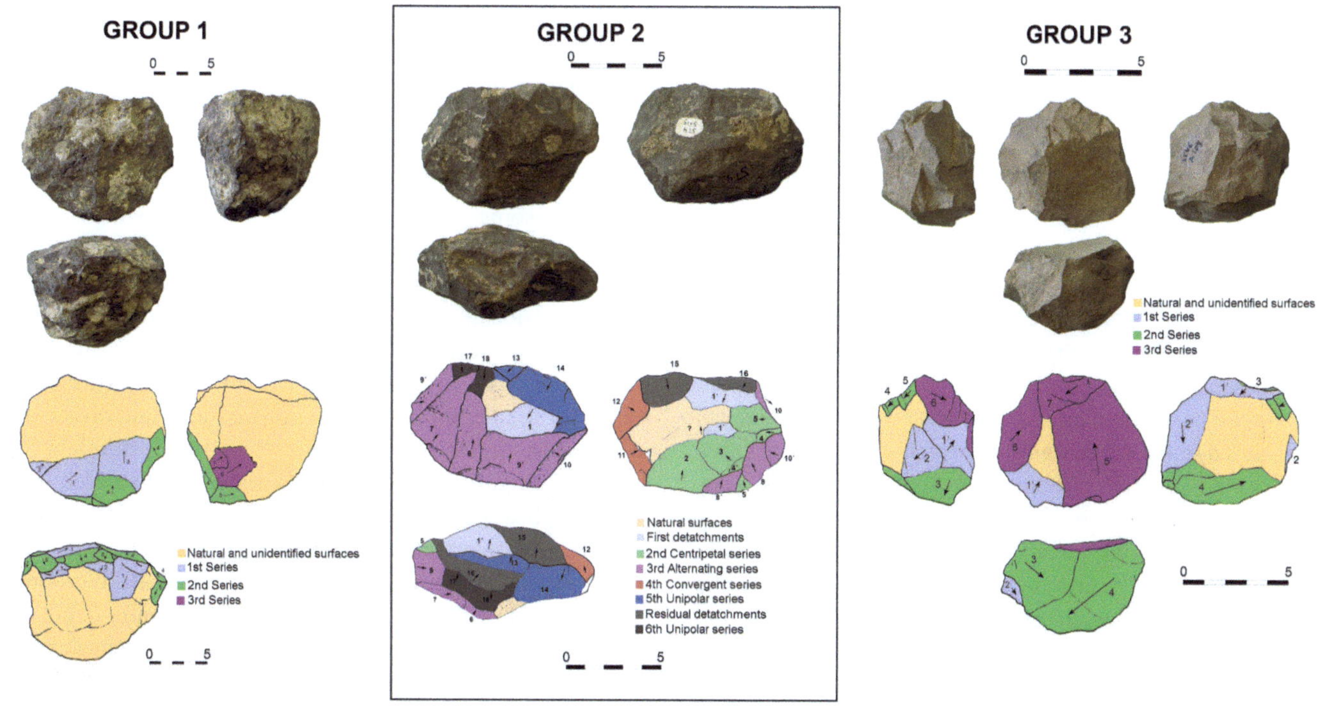

Fig. 7.8 Examples, photographs, and diacritical schemes of the different associations identified in the core sample of the Type Section: **A** Group 1: Poor quality basalts showing a partial alternation of negative scars and a remarkably low degree of exploitation (ST2-12). **B** Group 2: Basalts and nephelinites of medium/good quality hemispherical and flake blanks showing a medium degree of exploitation through unipolar and orthogonal series (ST4-S15). **C** Group 3: Very good/optimal basalts, intensively exploited in centripetal, orthogonal, and orthogonal–polyhedral sequences (ST46-A104, between Tuff 4 and 5, at the Type Section)

appropriate striking surfaces as intensively as possible. In the course of this reduction scheme, the generation of new appropriate striking platforms favored the exploitation of new, adjacent surfaces. This technical principle differs substantially from the discontinuous discoid technique (where each new blow strikes on the edge of previous negative scars) or the Levallois technique (where striking platform preparation shows a complex set of technical gestures).

In the Type Section, the use of this discontinuous alternation has resulted in orthogonal or opposed reduction strategies, although centripetal discoid in appearance (see Moore and Perston 2016 for a similar conclusion). Although some cores show a clear centripetal organization of negative scars, the only reduction pattern that seems to show consistency is constituted by long knapping series of discontinuous alternation.

7.3.2 *Oldowan or Acheulean? The Cultural Attribution of the TSC Industry*

The TSC industry shares a quite homogeneous stratigraphic position, on a paleosol directly located on the surface of Tuff 1 (Luque et al. 2009a, b), and was deposited in a relatively short period of time and within the same environmental context: an alluvial area in a deltaic environment at the intersection of river channels (Domínguez-Rodrigo et al. 2009b: 105). This environment was repeatedly visited by hominins to process carcasses obtained in the vicinity of the alluvial area. The absence of high-density lithic patches, the predominance of scatters over the landscape and the composition of the lithic aggregates would suggest sporadic hominin incursion in this area (Domínguez-Rodrigo et al. 2005). The lithic collections retrieved from this paleosurface, in agreement with a scenario of low anthropogenic impact and high raw material flow, are characterized by the production of small to medium-sized flakes retaining no cortex on their dorsal surfaces, with a very low percentage of retouched tools and few cores, hammerstones and unmodified cobbles. Due to their composition, the TSC assemblages were first defined as belonging to the Oldowan techno-complex (de la Torre et al. 2003), in a moment in which it overlapped chronologically with the first Acheulean in other parts of East Africa (Semaw et al. 2009).

However, a recent regional reinterpretation of the Lake Natron archaeological evidence has claimed that this industry fits much better within the Acheulean techno-complex and not with the Oldowan as previously stated (de la Torre 2009). From this new perspective, the core and flake component of the Type Section industry would represent a functional–economic–technical adaptation to an alluvial environment of the same humans that produced the more evident Acheulean sites located on top of the Sambu Escarpment, placed in a more distal position from the lake floodplain (Domínguez-Rodrigo et al. 2009c, d). There are strong reasons to support this new cultural attribution (Diez-Martín et al. 2012): (1) Some of the flakes recovered in the TSC assemblages have been identified as handaxes or Large Cutting Tool (LCT) resharpening/configuration flakes. Following the experimental work by Goren-Inbar and Sharon (2006), a number of small, relatively long, and thin flakes show remnants of what has been interpreted as a invasive edge on the margins or present plain dorsal faces. Handaxe flow (input and output) in the Type Section complex industry would be implicit through the presence of these objects in some assemblages; (2) large flakes (about 10 cm long) have been retrieved in sites such as ST2. Furthermore, recent analysis of the core sample suggests that the knappers of the Type Section were aware of the advantages of knapping large hemispherical flakes, as they provided a good natural interaction between two surfaces. Although large flakes found in the Acheulean sites of the Escarpment are significantly larger (Diez-Martín et al. 2014a, b), the production of this type of blank, either for large tool configuration or (as should be the case in the Type Section) for exploitation, has been considered to be a representative technological trait of the early Acheulean (Isaac 1984, 1986). (3) The Acheulean at Peninj, profusely documented in the North and South Escarpments, is stratigraphically related to a slightly younger depositional event in the USC member (post-Tuff 4) of the Humbu Formation (Domínguez-Rodrigo et al. 2009c, d). Clearly defined Acheulean sites (including large flakes and various types of LCTs) located in the same stratigraphic position have also been found in archaeological aggregates in the Type Section area (e.g. at ST 23, 28, 46, 48, 75, 76) (Figs. 7.9 and 7.10), as well as in other younger positions of the Moinik Formation, as at ST 69 (Diez-Martín et al. 2009a). Furthermore, the oldest Acheulean site documented in the Peninj region (PEES1, where, among the 21 specimens retrieved, 16 detached objects, 1 core, 1 chopper, 2 polyhedrons, and 1 handaxe were included) has been found in the South Escarpment directly on the Main Tuff and, thus, older than the sites studied here (Domínguez-Rodrigo et al. 2009c). It seems apparent then, that the TSC is bracketed between clearly defined Acheulean sites. This evidence advocates that, beyond typological variability, the different archaeological areas in the Lake Natron region should be considered sub-systems of a regional Acheulean system interconnected with and driven by different environmental, locational, economic and functional interests and/or constraints. New interpretations of the core assemblage retrieved in the Type Section favors a scenario in which hominins were maximizing raw material and intensively exploiting some pieces (specifically those showing a final morphology similar to the discoid method *sensu lato*). This fact, together with the high percentage of flakes without cortex on dorsal areas and striking platforms (Type 6), supports the idea that some good quality basalts and nephelinites were quarried at a certain distance (the mid-section of the Peninj River area, for instance) and discarded in the delta of the Peninj River (the Type Section) in an advanced stage of reduction. This reinforces the idea that rock supplies were intensively flowing along an interdependent and interconnected landscape. Formal variability observed in the Natron area (e.g., core and flake assemblages versus assemblages where large flake configuration is observed) would be related to some sort of synchronic (functional or environmental) and not diachronic (Oldowan–Acheulean) variability (Isaac 1977: 98), although the hypothetical presence of groups with different technological traditions cannot be ruled out.

de la Torre (2009) has suggested that the key trait that supports a link between the various industries retrieved from the Lake Natron area is precisely represented by the BHCM, found both in the TSC (where LCTs are formally lacking) and the Acheulean assemblages recovered from post-T4 sediments in both the Escarpments and the Type Section (where abundant LCTs have been found). Following this perspective, de la Torre (2009: 103) suggests that "the ability to exploit the entire volume of a piece through a structured bifacial method… which is what defines the ST Site Complex cores-shares the same technical scheme usually attributed to the Acheulean". A number of authors have already remarked on the technological and conceptual affinities between Acheulean handaxe production and complex hierarchical reduction strategies (Rolland 1995; Schick 1998; DeBono and Goren-Inbar 2001; Tryon et al. 2006; Lycett et al. 2010). Thus, from this perspective, the BHCM would be the expression of the same technical model and the humans that inhabited the region and produced assemblages both with and without handaxes shared the same technological concept and knapping structure (the technical skills and methods shared by a human community, as defined by Boëda (1991) and Pelegrin (1985).

Recent technological reassessment of the reduction strategies documented in Peninj (Diez-Martín et al. 2012,

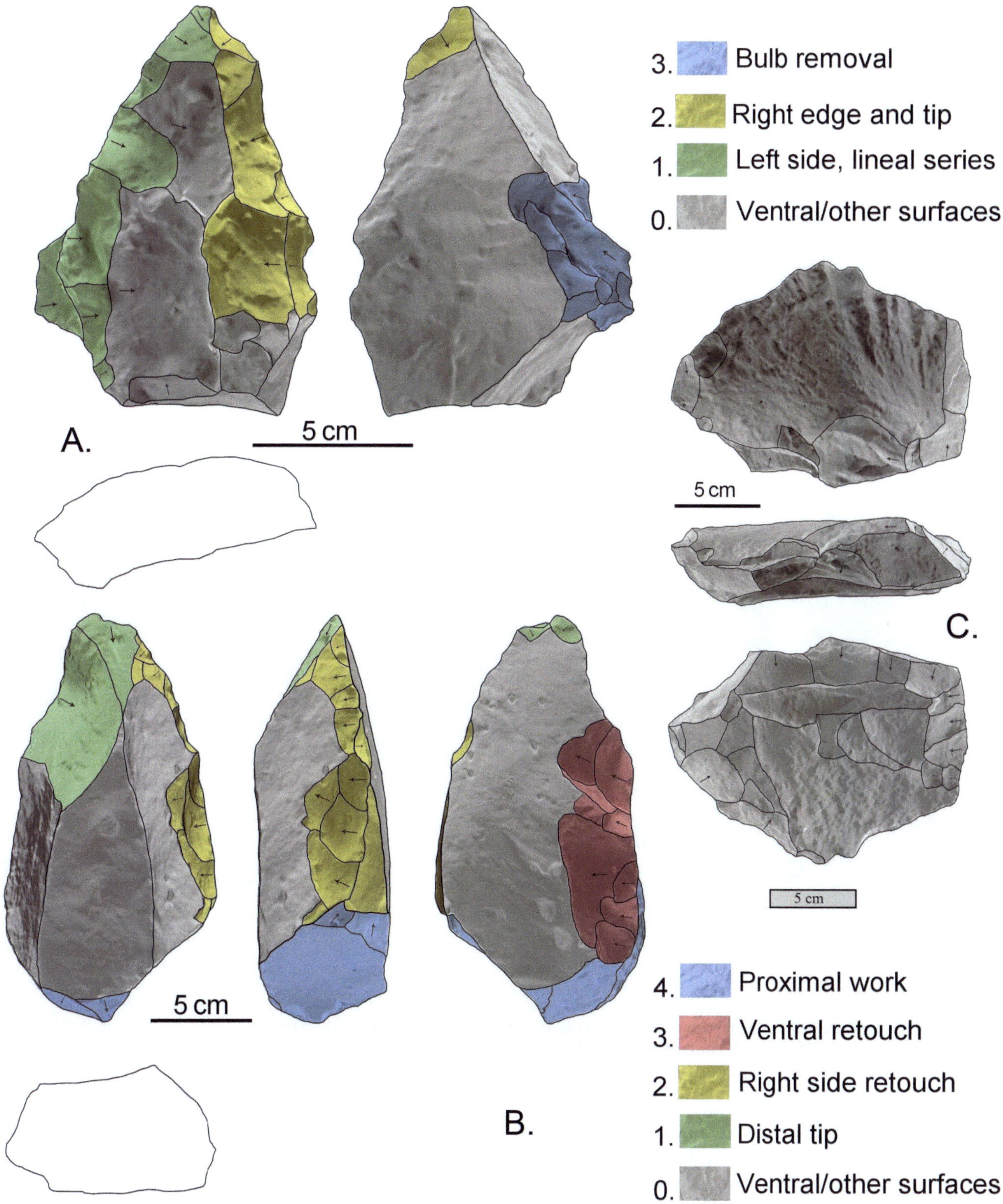

Fig. 7.9 Lithic implements recovered from various sites in the Type Section (T4–T5 stratigraphic interval): **A** LCT (ST23-2) on a basalt large flake blank showing two unifacial series converging in a distal tip, that is reinforced through a final ventral detachment. **B** LCT (ST46-A174) on a basalt large and thick flake blank, showing two alternate side series plus a third series aimed at distal tip configuration. **C** Basalt large flake (ST48-A15) with dorsal retouch and ventral removal of the butt area

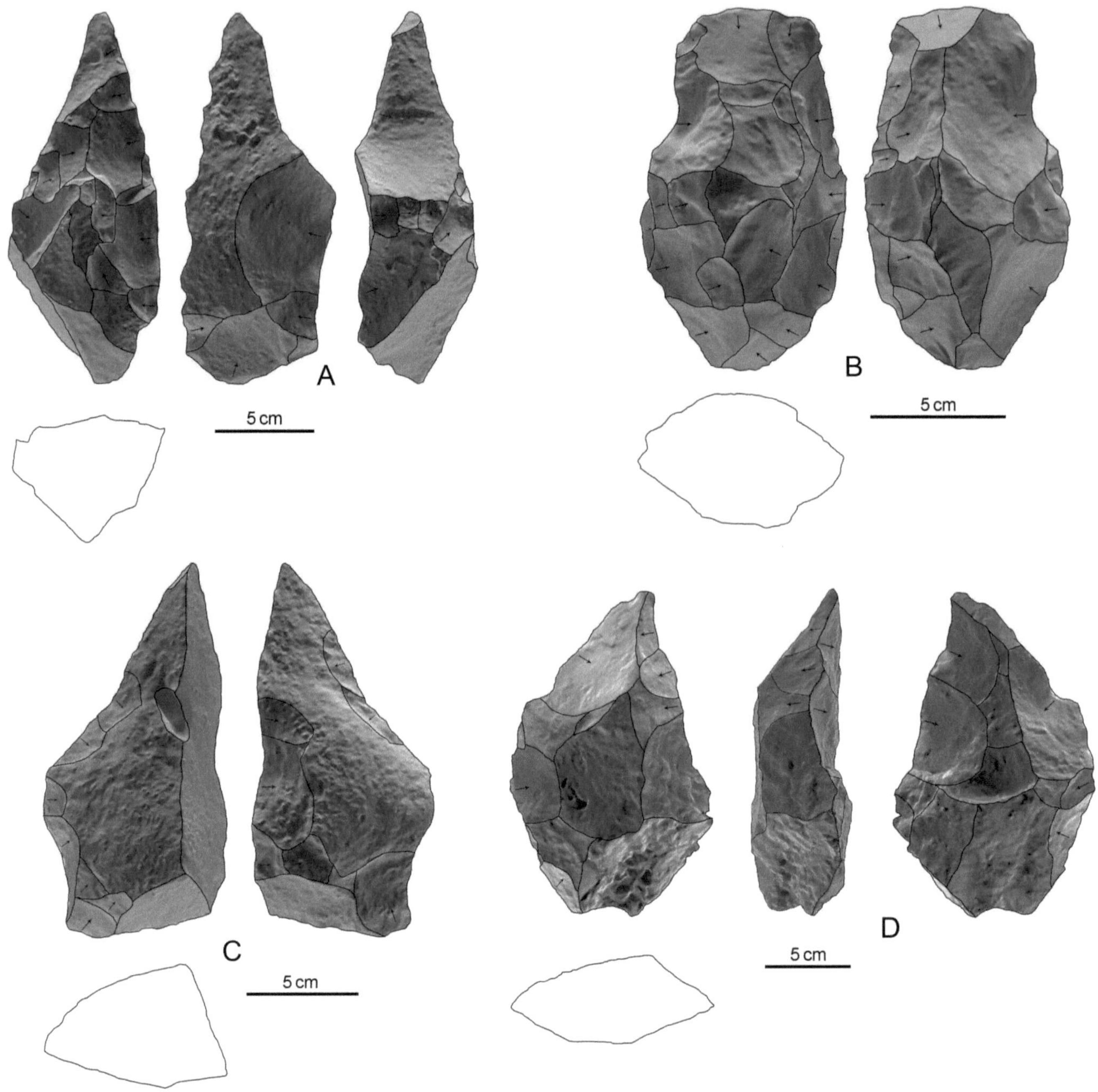

Fig. 7.10 Lithic implements recovered from various sites in the Type Section (T4–T5 stratigraphic interval): **A** and **C** Pointed/trihedral LCTs on large basalt flakes from ST28. **B** Basalt cleaver from ST23. **D** Pointed bifacial LCT on basalt from ST104

2014a, b) casts doubt on the role played by the reduction methods for the definition of such a shared knapping structure among the human populations that inhabited the Natron Basin during the Early Pleistocene. Current information does not fully support the existence of the so-called BCHM among the Type Section knappers (Diez-Martín et al. 2012); thus it cannot be the link between this industry and the Escarpment industries, where the BCHM shows similar interpretative shortcomings (Diez-Martín et al. 2014a, b). If the same species was responsible for the production of the varied technical behaviors displayed in the Natron area, the links connecting such a variety of technical solutions expressed within the regional structure of the Basin must be found elsewhere.

7.4 The Acheulean in the Escarpments

7.4.1 Lepolosi (Former MHS-Bayasi, South Escarpment)

Lepolosi, as it is known by the local Maasai people, is the most important archaeological site in the South Escarpment area. The site was discovered by Margaret Cropper in 1964, during the second paleoanthropological expedition undertaken by Glynn Isaac and Richard Leakey to Lake Natron. The site, which was given the arbitrary name of MHS, was the subject of both an unsystematic collection of surface items and a 13 m^2 step excavation. This step trench rendered 38 Acheulean specimens (Isaac 1965, 1967). After a new visit to the site, now renamed Bayasi, by Isaac in 1981 (Isaac 1982), archaeological fieldwork resumed in Natron at MHS-Bayasi in 1996 (Domínguez-Rodrigo et al. 2009c). Together with a new geological study of the area (ibid.: 205–210), in 1996, 2000, and 2002, a total surface of $\sim$20 m^2 was opened in the site, unearthing a large collection of artifacts, including significant clusters of Large Cutting Tools (ibid.: Figs. 9.19, 9.20). In the course of this new fieldwork, sampling carried out in both the site's paleosol and upon stone tools allowed the identification of different types of phytoliths preserved on artifact cutting edges (from the genus *Acacia*) versus the surrounding soil (mostly grasses) (Domínguez-Rodrigo et al. 2001b). This sharp difference was interpreted as a clear functional indicator, suggesting that the sampled large tools were used for activities involving woodworking. Between 2008 and 2011, in the framework of the current research project, another $\sim$43 m^2 were excavated in this site (Diez-Martín et al. 2014a).

Lepolosi is located on the Sambu Escarpment, 500 m above Lake Natron. The Peninj Group outcrops in this area, overlying the Sambu lavas, and is mostly represented by the Main Tuff member and the USC (8–10 m thick) in the Humbu Formation overlain by dozens of meters of silty and volcanic sediments pertaining to the Moinik Formation. From a stratigraphic point of view, it is placed around 3 m on top of the T4 tephra layer and below a fluvial level of sand and quartz gravels around 2 m thick. The T5 tephra layer, stratigraphically deposited on top of T4, has not been preserved in the vicinity of the site. The archaeological horizon is formed of approximately 1-m-thick greenish muddy sandstone interbedded laterally with rootmarked sandstone lenses (Domínguez-Rodrigo et al. 2001b, 2009c). Current paleoenvironmental interpretation suggests that the site was deposited in the context of low-energy distributary channels on a floodplain related to a swampy fluvial–alluvial environment (Domínguez-Rodrigo et al. 2009c; Luque et al. 2009a).

7.4.2 The Site of Noolchalai (Former RHS-Mugulud, North Escarpment)

Noolchalai, as it is called by local Maasai people, is located in the North Escarpment area, in a densely vegetated area on the southern hillsides of the Ol Doinyio Sambu volcano. Originally named RHS, it was discovered by Richard Leakey in 1964. After a survey, in the course of which 161 basalt artifacts were recovered, a $\sim$48 m^2 excavation unearthed 215 tools on a bank adjacent to a small watercourse that seemed to be comparable to EF-HR in Bed II at Olduvai Gorge (Isaac 1965: 118, 1982). In 1981 Isaac undertook further archaeological work in the site, now renamed Mugulud. In 1995, 2001, and 2002, fieldwork was resumed in the North Escarpment (Domínguez-Rodrigo et al. 2009d). This work included an excavation that opened an additional 38 m^2 adjacent to Isaac's grid. In the course of their archaeological work, Domínguez-Rodrigo and colleagues retrieved 352 lithic artifacts, 197 from the surface, and 155 from their excavation. A total of 126 small (<3 cm) and heavily weathered bone fragments were also unearthed from the excavation that seemed to have no connection with the lithic sample (ibid.: 241). Between 2009 and 2013, new fieldwork was undertaken in Noolchalai and the surrounding area, which included an archaeological excavation in the site and in other high-density patches discovered in the vicinity plus a complete set of geological, sedimentological, and geoarchaeological studies (Diez-Martín et al. 2014b).

The Pleistocene outcrops in the North Escarpment area include materials of Humbu and Moinik Formations of the Peninj Group. Humbu sediments are relatively scarce, while Moinik materials are abundant and thick. The sediments outcropping around the archaeological site of Noolchalai include the uppermost part of the USC of the Humbu Formation, a discordant contact with the Moinik Formation and the first few meters of the Moinik alluvial sequence. Initially, Isaac (1965) located the stratigraphic position of Noolchalai in the uppermost part of the USC. Later observations (Diez-Martín et al. 2014b) led to the conclusion that the site is stratigraphically located at the base of the Moinik Formation that, discordant after an intense erosive period, had incorporated archaeological materials originally sedimented on top of the Humbu Formation. Due to a more energetic context and a more irregular paleolandscape toward the base of the site, the first steps of this reworking of Humbu materials include the heaviest and largest lithic specimens of Acheulean assemblages originally deposited in the Humbu Formation. Therefore, the lithic industry recovered from the base of Noolchalai (Moinik Formation) is the by-product of an intense erosive process that transported sediments previously deposited in the USM of the Humbu Formation to their current location (Diez-Martín et al. 2014b).

7.4.3 The Early Acheulean Technology from the Escarpments

The abundant lithic collections retrieved in the course of the different research projects undertaken in the North and South Escarpment areas (Table 7.1), particularly significant in the case of LCTs, addresses some of the currently open issues related to the technological characterization of the East African early Acheulean (Diez-Martin and Eren 2012). At both sites most of the technological processes are related to the manipulation or transformation of basalt raw material (Diez-Martín et al. 2014a, b). Fine-grained and good quality boulders originated in the vicinity of the Sambu volcano were also available in the middle section of the Peninj River watercourse. Acheulean toolmakers took advantage of the different natural morphologies of cobbles and boulders to undertake different tasks. Medium-sized basalt cobbles were used for both percussion activities and flake production purposes. Basalt and quartz core exploitation processes represented at Lepolosi and Noolchalai show that toolmakers were selecting specimens of both raw materials for a systematic production of medium-sized flakes using a variety of unifacial, bifacial, and multifacial exploitation patterns. Multifacial/polyhedral cores, particularly in the relatively high numbers of quartz polyhedrals that have been retrieved from Lepolosi (Diez-Martín et al. 2014a), were also intensively used in percussion activities, as revealed by the presence of battered and intensively blunted ridges. Although examples of bifacial centripetal cores have been found in both sites (i.e., discoids in which a bifacial continuous alternation model was undertaken for the detachment of flakes), they constitute a residual pattern. Confirming again a previous diagnosis of the Type Section reduction patterns (Diez-Martín et al. 2012), the current interpretation does not support the presence of a reduction model in which a recurrent successive exploitation/preparation of surfaces with asymmetrical and non-interchangeable areas occur (i.e., the BHCM; Fig. 7.11: A). The main goal of these reduction

Table 7.1 Distribution of the lithic samples retrieved from Lepolosi and Noolchalai sorted by lithic category

Lithic category	Collection	
	Lepolosi (n)	Noolchalai (n)
Cobbles	32	8
Hammerstones	26	7
Cores	60	59
Detached products	376	260
Fragments	32	47
LCTs	139	202
Undetermined/chunks	45	–
Total	710	583

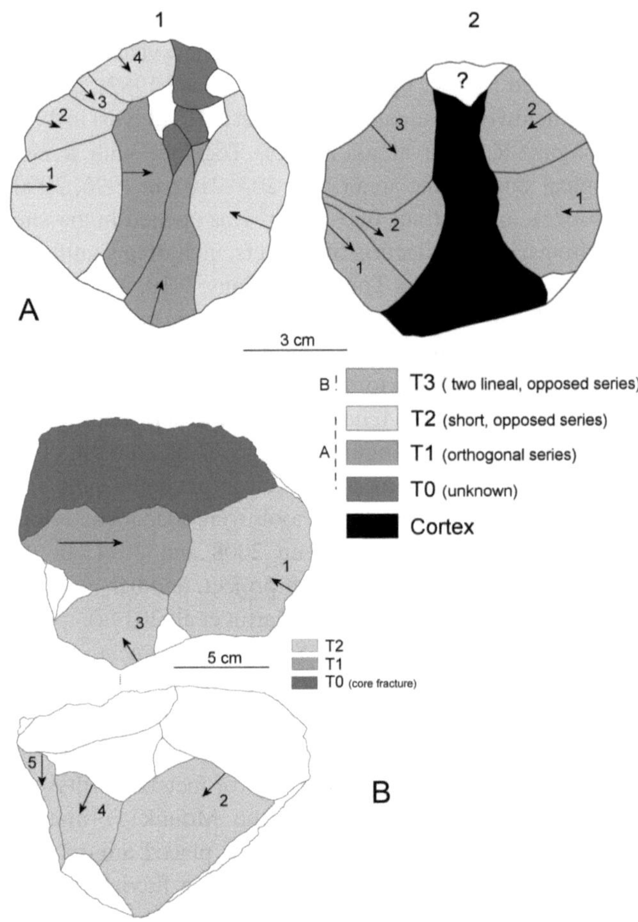

Fig. 7.11 **A** Diacritical scheme of a core retrieved from Noolchalai and previously identified as an example of BHCM (de la Torre 2009: Figure 11). The chronological reconstruction of the knapping process is as follows: after previous undetermined reduction sequence (Time 0), face **A** shows two subsequent series of (T1) orthogonal and (T2) lineal, short, and opposed detachments. Reduction then continues on face **B** (T3) with two opposed series of lineal detachments. **B** Diacritical reconstruction of a core found on surface in Noolchalai and previously identified as an example of large flake core (de la Torre et al. 2008: Figure 9A). The specimen shows a clear fracture area affecting part of the volume and masking part of the knapping structure. The detachment included in T1 is the remnant of a previous series. T2 represents a bifacial alternation of detachments using the same striking platform

models mainly related to lineal and orthogonal knapping gestures (as seen on the flake dorsal patterns identified) was the production of usable medium-sized flakes that, on few occasions, were subject to rather unsystematic retouch.

An undetermined fraction of the flake sample must be the desired by-products of the aforementioned reduction strategies, although another undetermined fraction of flakes must be related to core rejuvenation and LCT configuration processes. To a certain extent different fractions of different operational sequences (i.e., those related to exploitation and shaping activities) must have been produced on-site. Apart from the medium-sized cores themselves, this is confirmed by a wide range of detached products, including edge core flakes and LCT shaping flakes recovered from both Lepolosi and Noolchalai.

Along with those operational processes aimed at the production and marginal retouch of medium-sized flakes, the production and subsequent transformation into LCTs of large flake blanks (>10 cm) constitute a relevant technological trait of these Acheulean sites. Very large and thick flakes were being detached in the vicinity and transported on-site fully or partially shaped in their final desired forms. While empirical data confirm, at least at Lepolosi, a partial resharpening of a number of LCTs, arguing for an on-site large blank production is more contentious. We have identified a number of cores showing negative scars larger than the average flakes detached from medium-sized cores. However, these specimens show no clear exploitation pattern and it is impossible to rely on them for the interpretation of the operational sequences used by the Acheulean toolmakers to produce the huge flakes they were using as LCT blanks. As in many other Acheulean cases (Sharon 2007), no clear large cores have been retrieved so far in excavation from the Acheulean sites in Lake Natron. This is true for Lepolosi, where our excavated large flake cores are invalid for an assessment of the operational processes undertaken for the production of these blanks. This is also true for Noolchalai, where no unambiguous large flake cores have been unearthed from an archaeological context. At Noolchalai, it has been claimed

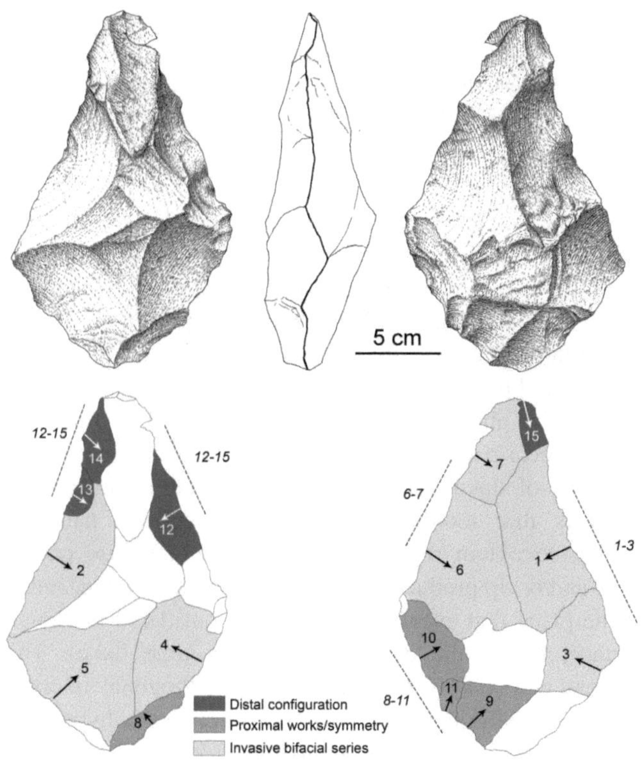

Fig. 7.12 Bifacial LCT from Lepolosi and diacritical scheme showing the chronological arrangement of three different sequences: 1. Invasive bifacial series; 2. Proximal reduction aimed at symmetrical enhancement; 3. Distal configuration aimed at tip shaping

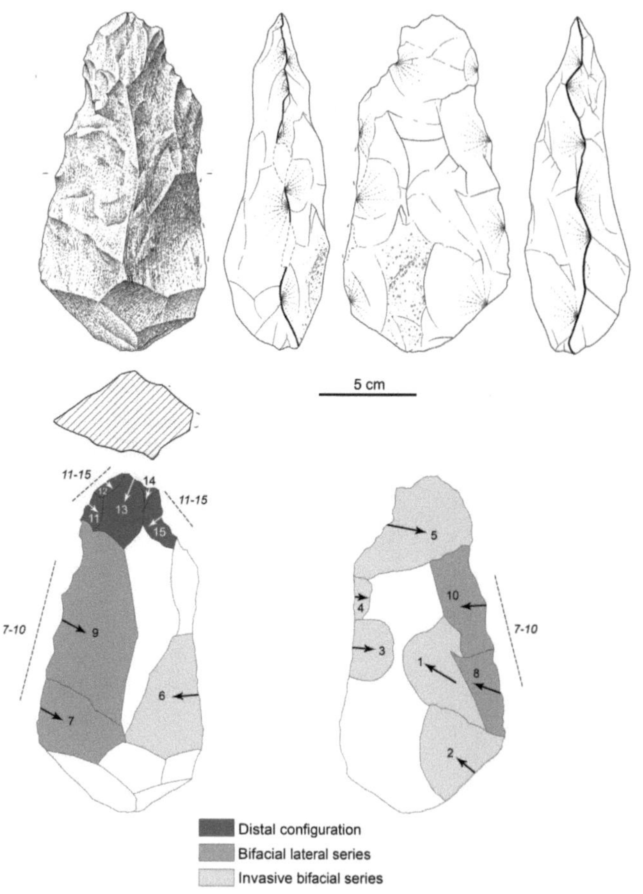

Fig. 7.13 Bifacial LCT from Lepolosi and diacritical scheme showing the chronological arrangement of three different sequences: 1. Invasive bifacial series; 2. Lateral bifacial knapping; 3. Distal configuration aimed at tip shaping

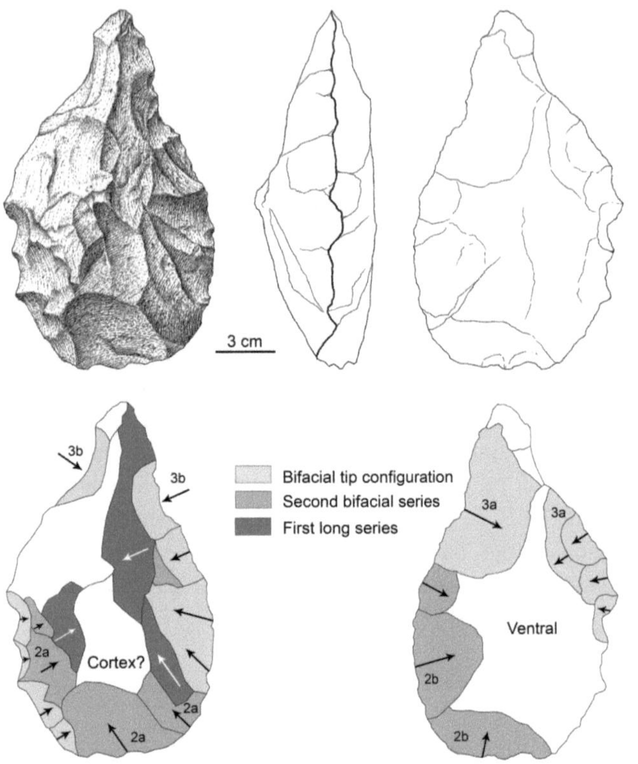

Fig. 7.14 Bifacial LCT from Noolchalai, and diacritical scheme showing a bifacial treatment of the volume (through a long and invasive series) followed by an intense bifacial configuration of the distal tip

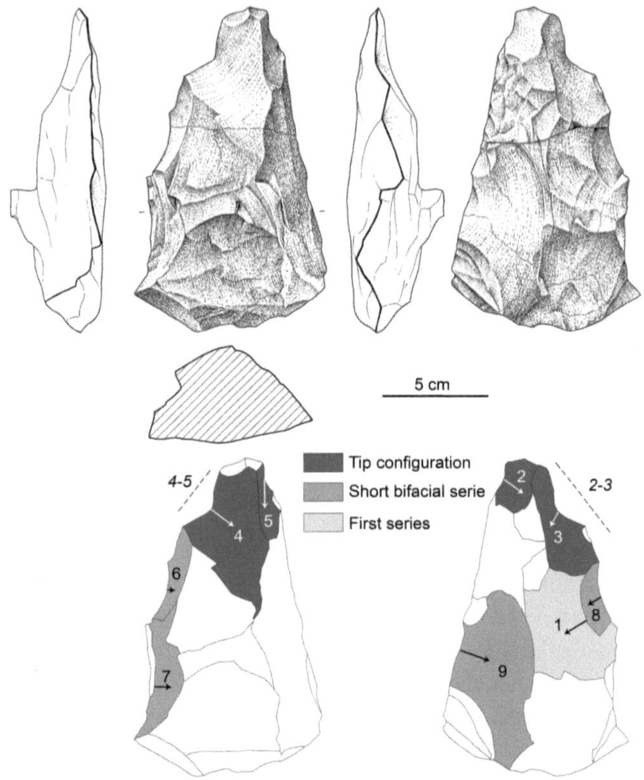

Fig. 7.16 Pointed LCT from Lepolosi and diacritical scheme showing the chronological arrangement of three different sequences: 1. Invasive series; 2. Bifacial short and sinuous (denticulate) lateral series (with no apparent functional meaning); 3. Tip configuration

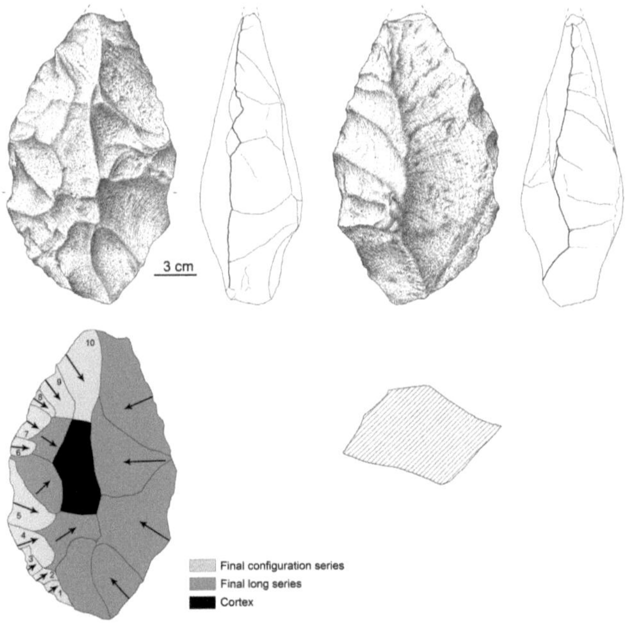

Fig. 7.15 Bifacial LCT from Noolchalai, and diacritical scheme showing the following sequence: invasive series on the dorsal surface, aimed at the volumetric treatment of the artifact; second alternation of detachments on both surfaces aimed at the final configuration

that some surface specimens are good proxies for the definition of large flake production in the Peninj area (de la Torre et al. 2008: Fig. 9.19). However, these items show ambiguous technical patterns (Dominguez-Rodrigo et al. 2009d; Diez-Martin et al. 2014b; Fig. 7.11: B).

In the Peninj Acheulean, the only available data to hypothesize about large core reduction systems come from the technical patterns seen in LCTs or in plain large flakes. Analysis of flake dorsal patterns and section morphology suggests that toolmakers were using bifacial patterned models to obtain regular blanks and to maximize cutting edge and/or tip production (via the production of concavities in the plane that eventually would be considered the dorsal surface, before the detachment of the large flake). This process was mainly obtained through orthogonal intersections (Diez-Martín et al. 2014a, b). Although this hypothesis remains conjectural, the technological patterns described here (Texier and Roche 1995) would imply a considerable amount of standardization in their large core production strategies. The places where boulders were originally processed, the stage of reduction in which the cores were transported and the spatial implications of large core management are issues that remain completely elusive.

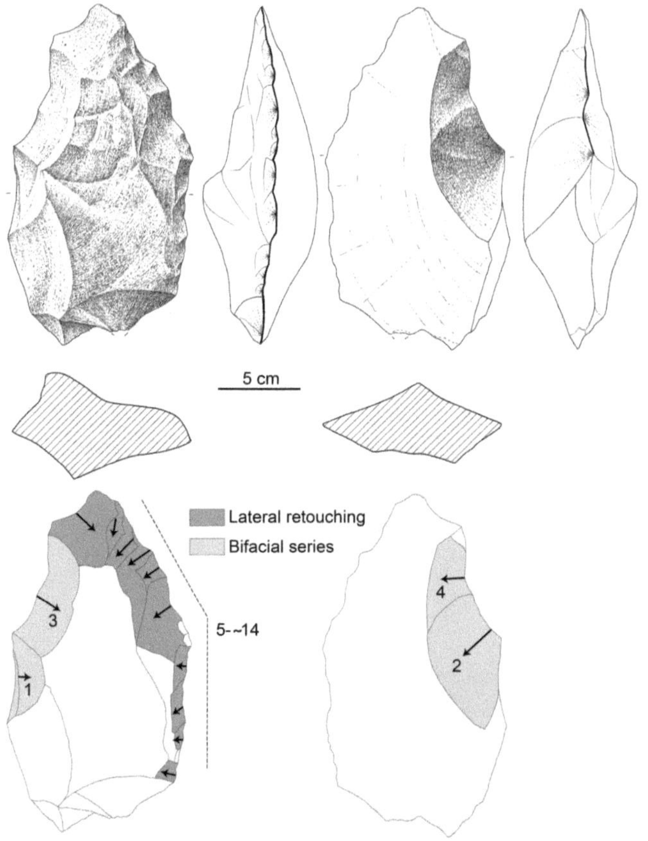

Fig. 7.17 LCT from Lepolosi and diacritical scheme showing the chronological arrangement of two different sequences: 1. Bifacial series aimed at butt thinning and probably grasp; 2. Unifacial marginal and sinuous (denticulate) series

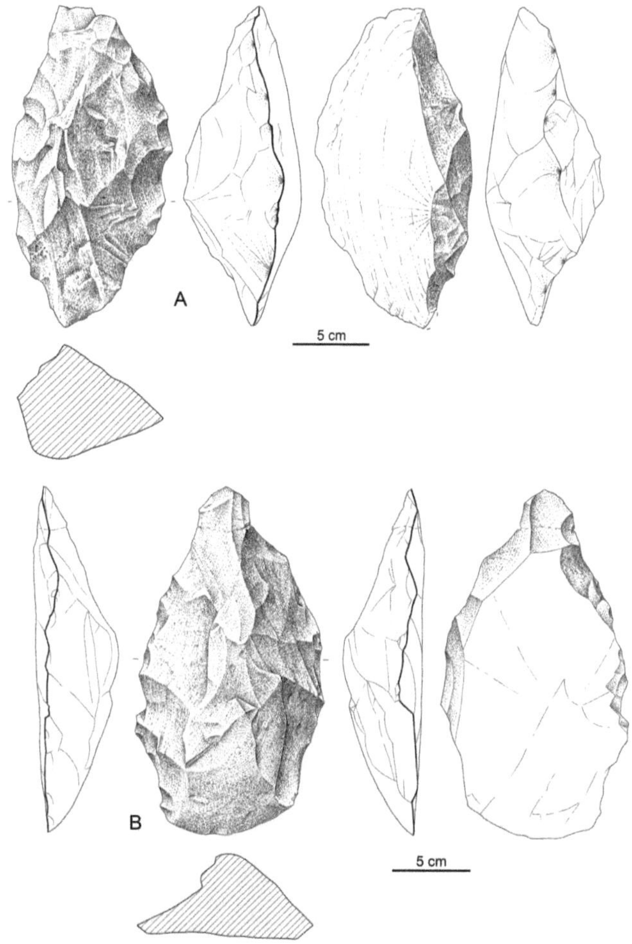

Fig. 7.18 Pointed LCTs from Lepolosi: **A** Knife. **B** LCT with marginal bilateral retouch associated to a distal tip and enhancement through marginal and bilateral notch retouch of proximal natural cutting edge

According to this volumetric awareness, in the Escarpments Acheulean toolmakers had the precise knowledge to produce normative handaxes (Diez-Martín et al. 2014a, b). Through long, invasive, and bifacial knapping series they were able to symmetrically transform the original blanks and to create rather biconvex symmetrical volumes. A number of examples retrieved from excavation confirm this pattern (Figs. 7.12, 7.13, 7.14, 7.15). However, at Peninj, LCT shaping interests seemed to have been driven by more functional and less time-consuming factors. Along the collection of normative bifaces and cleavers, the bulk of the LCT sample is defined by more ad hoc and casual shaping processes. Toolmakers were interested in more superficial transformation of large flake blanks in order to produce points and/or active cutting tools. Rather than driven by the volumetric principles of symmetry and bifacial shaping, configuration processes are full of examples of practical solutions for achieving their functional goals: creating distal or lateral abrupt areas for ergonomic purposes, retouching precise areas to create or simply enhance tips, delimiting areas to enhance natural acute cutting edges, or briefly retouching long edges. This set of solutions, apparently casual on occasions, seem effective configuration strategies to add an extra functional meaning to the massive and heavy flake blanks. Certainly these tools were created to combine strength and shape in order to serve for precise tasks. This array of solutions has created a quite diverse group of final forms, among which knives, picks or heavy-duty scraper-like forms can be identified (Figs. 7.16, 7.17, 7.18). A quite diversified pattern for the production of LCTs emerges, in which, on one hand, the ability or interest in configuring symmetric bifacial volumes and shaping normative cleavers and, on the other hand, maximization of functional solutions are combined (Fig. 7.19).

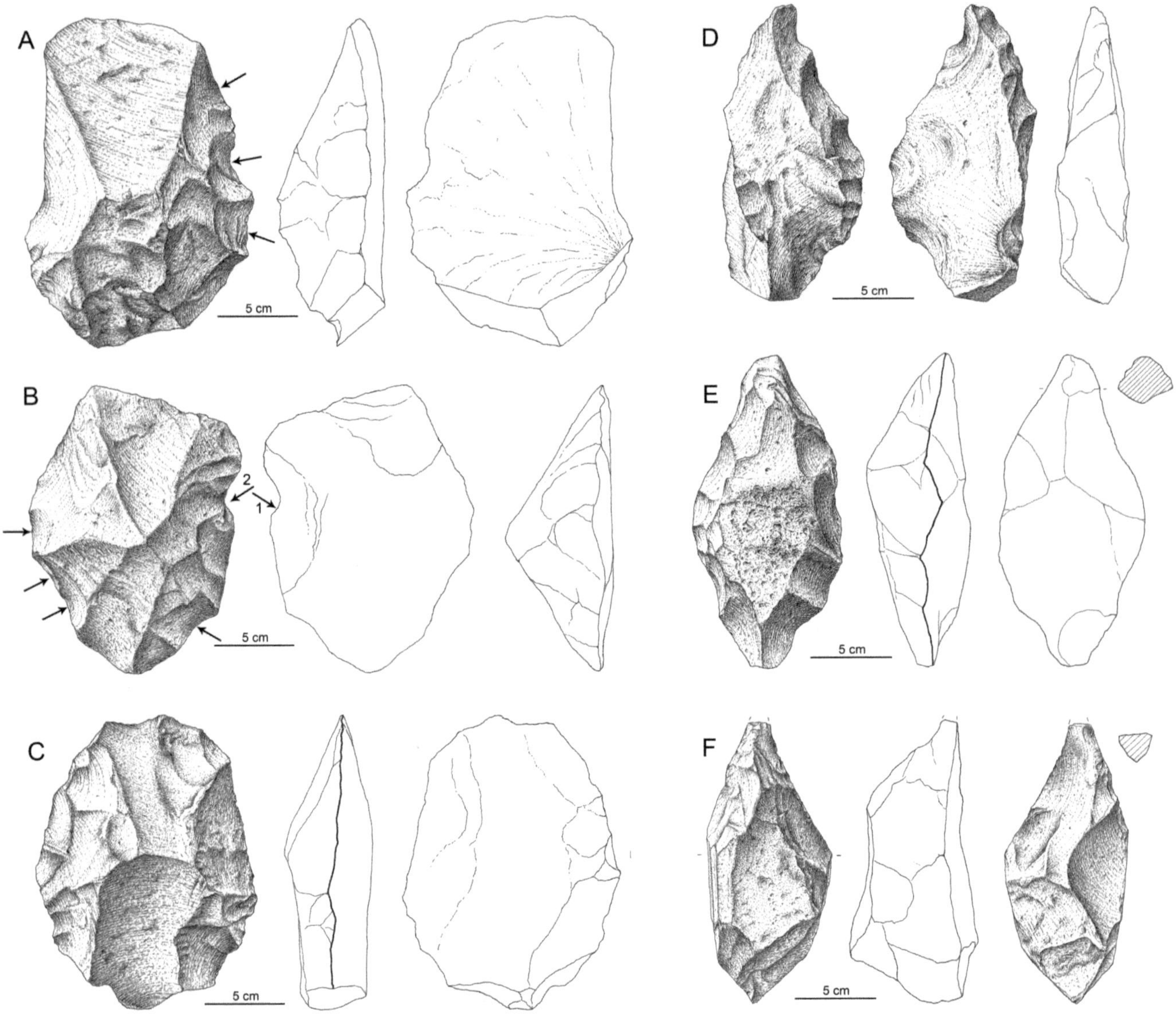

Fig. 7.19 LCTs from Noolchalai: **A–C** cleavers, **D–F** pointed and pick-like specimens

7.5 A Regional Interpretation for the Acheulean of the Natron Basin

Taking into consideration the disparate integrity of Noolchalai (reworked in a new Moinik location from an unknown Humbu depositional context) and Lepolosi (a low-energy floodplain context with limited, mostly vertical, post-sedimentary disturbance), both sites reproduce a similar technological pattern, particularly evident in the case of the LCT shaping processes. It seems plausible that both locations in the uppermost section of the USC member of the Humbu Formation would represent similar responses to similar landscape constraints and/or functional frameworks. Glynn Isaac (1982) was aware of the importance of the regional component in the Peninj record. The regional interconnection of the Peninj record was enlarged (Domínguez-Rodrigo et al. 2009a), particularly through the "ecological hypothesis" for the origin of the Acheulean (Domínguez-Rodrigo et al. 2005). This hypothesis suggested that the archaeological record located in the different fertile areas at Peninj (TSC in the Type Section, North Escarpment, and South Escarpment) showed traces of spatial behaviors that were related to each other. The assemblages deposited in the TSC corresponded to an alluvial area in a deltaic environment close to the lake shore in which hominins processed carcasses and discarded few and relatively small stone tools, probably in the framework of sporadic incursions in the area. Conversely, the accumulation of LCTs in the Escarpments, away from the lacustrine environment, would suggest alternative and complementary tasks for these aggregates. These alternative tasks would not involve carcass consumption or manipulation, as these Acheulean sites are devoid of fossil bones in connection with the lithic materials. It has been argued that the absence of fauna in these sites must bear a behavioral meaning because it cannot be explained through taphonomic constrains (Domínguez-Rodrigo et al. 2009c, d). Both the fact that the Acheulean in the TSC (T1–T2) and the Escarpments (T4–T5) are not penecontemporaneous and the evidence in the Type Section area of a meager, but significant, number of localities with LCTs in sediments with the same stratigraphic position as those recovered in the Escarpments (T4–T5) could challenge some aspects of the ecological hypothesis, showing that: (a) a chronological, landscape, and environmental gap existed between the TSC industry and the Escarpment industry, and b) that geographic location cannot be considered a determinant factor for the regional archaeological structure (the locational antagonism proximal/distal to the lake margin serving as motor of different techno-economic behaviors).

Nonetheless, recent interpretations of the Peninj record (de la Torre 2009; Diez-Martín et al. 2012) stress the fact that the diverse technological behaviors observed in Peninj are interconnected fractions of the Acheulean techno-complex and that formal differences among them can be interpreted from a regional perspective. Thus, beyond variability, the different archaeological areas of the Lake Natron should be considered as sub-systems of a regional Acheulean system interconnected with and driven by different environmental, locational, economic, and functional interests. From this perspective, the set of technological patterns observed in the main sites of the Escarpments can add new data to this regional interpretation. The diagram represented in Fig. 7.20 aims to constitute an update of the ecological hypothesis for the Acheulean in Peninj, showing a hypothetical interaction of the different types of localities recorded or envisioned in Peninj and their functional meaning.

Place A is represented as an area devoted to the exploitation of large cores for large flake procurement, in which fractions of configuration processes cannot be certainly excluded. These extractive quarries have been recognized by a number of authors in the archaeological record (see contributions in Goren-Inbar and Sharon 2006), while a number of experimental works have provided insights into the way large flakes might have been produced in these quarry areas (Jones 1994; Toth 2001; Madsen and Goren-Inbar 2004). In the Acheulean of Lake Natron such extractive places have not been recognized yet. Despite the presence of cores from which relatively large flakes have been detached (≤ 10 cm) in both Lepolosi and Noolchalai, to date we have no empiric information about the actual cores from which very large flakes were being detached. Mean maximum length of large flakes in the Peninj Acheulean is 155 mm in Noolchalai and 163 mm in Lepolosi. Such heavy implements would require the exploitation of massive cores, currently absent in the Peninj archaeological record.

Place B is a site in which large implement configuration and use took place. Large flakes would reach these places in order to be transformed and used in specific tasks, although the production of usable medium-size flakes would also occur. We find reasons to suggest that Noolchalai, Lepolosi, and other high-density patches recently discovered on top of the Sambu Escarpment (Fig. 7.21) would fit within this type of places. Due to the very bad preservation of T4–T5 sediments in the Type Section area, and the meager lithic collections retrieved from this interval, it would be unwise to include these sites within this theoretical model. The

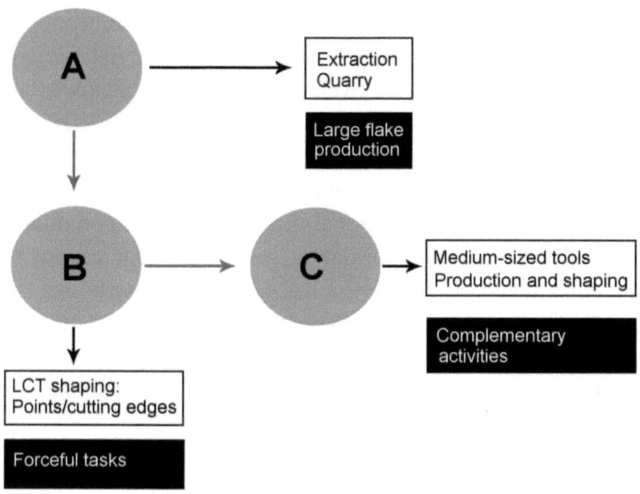

Fig. 7.20 Hypothetical model for the regional archaeological system of the Acheulean in the Natron Basin

Fig. 7.21 High-density patch of Acheulean tools in the Noolchalai area (Noolchalai-2 site)

functional meaning envisioned for Place B sites is corroborated by the following facts: (a) large flake transfer is reinforced by the lack of correlation between LCT maximum length (specimens that, due to their scarce transformation, tend to preserve indications of their original size) and the core collection retrieved from these sites; (b) evidence of LCT shaping has been found in both sites, particularly through the recognition of configuration flakes within the lithic collections. This fact would support the idea that, at least to some extent, resharpening processes would have taken place in-site; (c) although formal Acheulean implements have been found in both sites (handaxes, cleavers, and trihedral picks), the bulk of the LCT collection at the two sites is formed by specimens casually transformed in order to produce very precise morpho-functional patterns: pointed areas and/or long segments. On a number of occasions the functional interest of these active areas is reinforced by the configuration of prehensile areas via abrupt reconditioning of opposed areas. Despite being aware of the principles of volumetric bifacial transformation and geometric symmetry, proved by the existence of normative Acheulean implements, the toolmakers were recurrently stressing their interest in shaping specific active forms in massive and heavy artifacts. This persistent goal would suggest a particular functional meaning to be undertaken in the place: tasks in which specific interactive actions (related to the action of tips and robust cutting edges) would need the concurrence of heavy and massive volumes. The combination of strength and very precise morpho-functional areas envision a number of economic activities, among which woodworking (suggested by the phytoliths retrieved on LCT implements from ES2-Lepolosi; Domínguez-Rodrigo et al. 2001a) should still be taken into consideration; (d) combined with the LCT shaping and use, the exploitation of cores for the production of usable medium-sized flakes also took place in the Escarpments Acheulean sites. Here we find the presence of a variety of medium-size core reduction models, including unipolar, orthogonal, and multifacial. Core rejuvenation flakes and a diverse variety of medium-sized flakes indicate that complementary knapping activities would have taken place here.

In this model, Place C represents a site in which alternative activities to those undertaken in Place B took place. An indicative sign of alternative functional meaning would be the absence of the conspicuous LCT accumulation seen in Noolchalai, Lepolosi, and other sites in the Escarpments. Although the first example is more contentious (as it is located in a derived secondary context), the significant accumulation of large implements in a specific location of the low-energy floodplain in ES2-Lepolosi constitutes an exceptional patch in the paleolandscape. This accumulation of both raw material and large implements in the Escarpments markedly contrasts with what we see in the Type Section area. We agree with other authors (Downey and Domínguez-Rodrigo 2002–2003; Domínguez-Rodrigo et al. 2005; de la Torre et al. 2008), when they claim for a certain behavioral pattern to explain this differences in the archaeological record. In this framework, we consider that the archaeological sites located in the TSC could fit well within Type C sites, in which hammerstones, medium-sized cores, medium-sized flakes and a number of retouched flakes could indicate processing tasks complementary to those undertaken in Type B sites. In the Type Section sites, associated with the processing of herbivore carcasses, the lithic assemblages are characterized by the production of small to medium-sized flakes, with a very low percentage of retouched tools, cores, hammerstones, and unmodified cobbles (de la Torre and Mora 2009; Diez-Martín et al. 2012). Furthermore, LCT resharpening flakes and relatively large flakes (≤ 10 cm) indicate some sort of regional raw material and tool flow in which Acheulean toolmakers were maximizing raw material and intensively exploiting a number of core specimens.

This theoretical model is an updated version of the ecological hypothesis for the Acheulean (Domínguez-Rodrigo et al. 2005), in an attempt to overcome an excessively rigid conception of the locational constraints, after the evidence that the proximal versus distal to the lake margin parameter was not determinant to explain the technological variability in the Peninj archaeological record (i.e., LCTs recovered from the Type Section). At this point, it seems more plausible that the variability seen in the Acheulean of the Natron Basin (sites with abundant LCTs versus sites dominated by small and medium-sized flakes and cores), rather than being driven by locational parameters is related to environmental/functional/ecological/economic constraints: those based on the dichotomy alluvial fan and shallow channels (T1–T2) versus high-energy channels and floodplains (T4–T5). Nevertheless, in the Lake Natron Acheulean most regional information available to us has a technological character. Although more contextual information is needed at a regional scale, the spatial component of the technological information gathered so far cannot be underestimated. The coherent technological patterns seen in the Escarpments and the Type Section are indicating the dramatic influence of functional and economic parameters in the development of the Acheulean techno-complex in East Africa.

References

Boëda, E. (1991). Approche de la variabilité des systèmes de production lithique des industries du Paléolithique inférieur et moyen : chronique d'une variabilité attendue. *Techniques et Culture, 17–18,* 37–79.

Braun, D. R., Rogers, M. J., Harris, J. W. K., & Walker, S. J. (2008). Landscape-scale variation in hominin tool use: Evidence from the Developed Oldowan. *Journal of Human Evolution, 55,* 1053–1063.

Davidson, I., & McGrew, W. C. (2005). Stone tools and the uniqueness of human culture. *Journal of the Royal Anthropological Institute, 11,* 793–817.

de la Torre, I. (2009). Technological strategies in the Lower Pleistocene at Peninj (West of Lake Natron, Tanzania). In K. Schick & N. Toth (Eds.), *The Cutting Edge: New Approaches to the Archaeology of Human Origins* (pp. 93–113). Goshomore: The Stone Age Press.

de la Torre, I., & Mora, R. (2009). The technology of the ST site complex. In M. Domínguez-Rodrigo, L. Alcalá, & L. Luque (Eds.), *Peninj. A Research Project on the Archaeology of Human Origins (1995–2005)* (pp. 145–189). Oxford: Oxbow.

de la Torre, I., Mora, R., Domínguez-Rodrigo, M., Luque, L., & Alcalá, L. (2003). The Oldowan industry of Peninj and its bearing on the reconstruction of the technological skills of Lower Pleistocene hominids. *Journal of Human Evolution, 44,* 203–224.

de la Torre, I., Mora, R., & Martínez-Moreno, J. (2008). The early Acheulean in Peninj (Lake Natron, Tanzania). *Journal of Anthropological Archaeology, 27,* 244–268.

DeBono, H., & Goren-Inbar, N. (2001). Note of a link between Acheulian handaxes and the Levallois method. *Journal of the Israel Prehistoric Society, 31,* 9–23.

Delagnes, A., & Roche, H. (2005). Late Pliocene hominid knapping skills: The case of Lokalalei 2C, West Turkana, Kenya. *Journal of Human Evolution, 48,* 435–472.

Deino, A., Domínguez-Rodrigo, M., & Luque, L. (2006). ^{40}Ar/^{39}Ar dating of the Pleistocene Peninj Group, Lake Natron, Tanzania. Eos Trans. AGU, *87*(52), Fall Meeting. Supplement, Abstracts, C1771.

Diez-Martín, F. (2008). El origen del achelense en Peninj (Lago Natrón, Tanzania). Adaptación territorial e innovación tecnológica durante el Pleistoceno inferior. *Informes y Trabajos, 1,* 169–175.

Diez-Martín, F., & Eren, M. I. (2012). The Early Acheulean in Africa: Past paradigms, current ideas, and future directions. In M. Domínguez-Rodrigo (Ed.), *Stone tools and fossil bones. Debates in the archaeology of human origins* (pp. 310–357). Cambridge: Cambridge University Press.

Diez-Martín, F., Barba, R., Gómez, J. A., & Luque, L. (2009a). El achelense de Peninj (Lago Natron, Tanzania): excavación arqueológica en ES2-Lepolosi en la campaña de 2008. *Informes y Trabajos, 3,* 237–244.

Diez-Martín, F., Luque, L., & Domínguez-Rodrigo, M. (2009b). ST-69: An Acheulean assemblage in the Moinik Formation of Type Section of Peninj (Lake Natron, Tanzania). In M. Domínguez-Rodrigo, L. Alcalá, & L. Luque (Eds.), *Peninj. A research project on the archaeology of human origins (1995–2005)* (pp. 192–204). Oxford: Oxbow.

Diez-Martín, F., Cuartero, F., Sánchez Yustos, P., Baena, J., Domínguez-Rodrigo, M., & Rubio, D. (2012). Testing cognitive skills in Early Pleistocene hominins: An analysis of the concepts of hierarchization and predetermination in the lithic assemblages of Type Section (Peninj, Tanzania). In M. Domínguez-Rodrigo (Ed.), *Stone tools and fossil bones. Debates in the archaeology of human origins* (pp. 245–309). Cambridge: Cambridge University Press.

Diez-Martín, F., Luque, L., Gómez de la Rúa, D., & Sánchez, P. (2010). Campaña de investigación arqueológica de 2009 en el Lago Natron (Tanzania). Estudio geoarqueológico de los yacimientos documentados en el Escarpe Norte, EN1-Noolchalai. *Informes y Trabajos, 5,* 320–335.

Diez-Martín, F., Sánchez, P., & Cuartero, F. (2011). Estudio tecnológico y experimental del comportamiento técnico en la Sección Tipo de Peninj (Lago Natron, Tanzania). *Informes y Trabajos, 7,* 337–347.

Diez-Martín, F., Sánchez Yustos, P., Gómez de la Rúa, D., Gómez González, J. A., Luque, L., & Barba, R. (2014a). The early Acheulean technology at ES2-Lepolosi (ancient MHS-Bayasi) in Peninj (Lake Natron, Tanzania). *Quaternary International, 322–323,* 209–236.

Diez-Martín, F., Sánchez Yustos, P., Gómez González, J. A., Luque, L., Gómez de la Rúa, D., & Domínguez-Rodrigo, M. (2014b). Reassessment of the Early Acheulean at EN1-Noolchalai (ancient RHS-Mugulud) in Peninj (Lake Natron, Tanzania). *Quaternary International, 322–323,* 237–263.

Domínguez-Rodrigo, M. (1996). La cronología del Grupo Peninj, al oeste del Lago Natrón (Tanzania): revisión de la discordancia bioestratigráfica. *Complutum, 7,* 7–16.

Domínguez-Rodrigo, M., Alcalá, L., Luque, L., & Serrallonga, J. (2005). Quelques aperçus sur les significations paléoécologique et comportementale des sites oldowayens anciens et acheuléens du Peninj (Upper Humbu Formation, Ouest du Lac Natron, Tanzanie). In M. Sahnouni (Ed.), *Le Paléolithique en Afrique. L'Histoire la plus longue* (pp. 129–156). Paris: Editions Artcom.

Domínguez-Rodrigo, M., Alcalá, L., & Luque, L. (Eds.). (2009a). *Peninj. A Research Project on the Archaeology of Human Origins (1995–2005).* Oxford: Oxbow.

Domínguez-Rodrigo, M., Alcalá, L., & Luque, L. (2009b). A taphonomic study of the T1 paleosurface in Type Section (Maritanane): The ST Site Complex. In M. Domínguez-Rodrigo, L. Alcalá, & L. Luque (Eds.), *Peninj. A Research Project on the Archaeology of Human Origins (1995–2005)* (pp. 73–108). Oxford: Oxbow.

Domínguez-Rodrigo, M., Serrallonga, J., Luque, L., Diez-Martín, F., & Bushozi, P. (2009c). The archaeology of the Acheulean sites from South Escarpment. In M. Domínguez-Rodrigo, L. Alcalá, & L. Luque (Eds.), *Peninj. A Research Project on the Archaeology of Human Origins (1995–2005)* (pp. 205–226). Oxford: Oxbow.

Domínguez-Rodrigo, M., Diez-Martín, F., Luque, L., Alcalá, L., & Bushozi P. (2009d). The archaeology of the North Escarpment. In M. Domínguez-Rodrigo, L. Alcalá, & L. Luque (Eds.), *Peninj. A Research Project on the Archaeology of Human Origins (1995–2005)* (pp. 227–256). Oxford: Oxbow.

Domínguez-Rodrigo, M., López-Sáez, J. A., Vicens, A., Alcalá, L., Luque, L., & Serrallonga, J. (2001a). Fossil pollen from the Upper Humbu Formation of Peninj (Tanzania): Hominid adaptation to a dry open Plio-Pleistocene savanna environment. *Journal of Human Evolution, 40,* 151–157.

Domínguez-Rodrigo, M., Serrallonga, J., Juan-Treserras, J., Alcalá, L., & Luque, L. (2001b). Woodworking activities by early humans: A plant residue analysis on Acheulian stone tools from Peninj, Tanzania. *Journal of Human Evolution, 39,* 421–436.

Domínguez-Rodrigo, M., de la Torre, I., Luque, L., Alcalá, L., Mora, R., Serrallonga, J., et al. (2002). The ST site complex at Peninj, West Lake Natron, Tanzania: Implications for early hominid behavioral models. *Journal of Archaeological Science, 29,* 639–665.

Downey, C., & Domínguez-Rodrigo, M. (2002–2003). Palaeocological reconstruction and hominid land use of the Lake Natron basin during the Early Pleistocene. *Before Farming, 4,* 1–29.

Foster, A., Ebinger, C., Mbede, E., & Rex, D. (1997). Tectonic development of the northern Tanzanian sector of the East African Rift System. *Journal of the Geological Society of London, 154,* 689–700.

Guest, N. J. (1953). The geology and petrology of the Engaruka Oldoinyio Lengai-Lake Natron area of Northern Tanganyka territory. Ph.D. Dissertation, University of Sheffield.

Goren-Inbar, N., & Sharon, G. (2006). Invisible handaxes and visible Acheulean biface technology at Gesher Benot Ya'aqov, Israel. In N. Goren-Inbar & G. Sharon (Eds.), *Axe Age. Acheulean toolmaking from quarry to discard* (pp. 111–135). London: Equinox.

Gowlett, J. A. (1990). Archaeological studies of human origins and Early Prehistory in Africa. In P. Robertshaw (Ed.), *A history of African archaeology* (pp. 13–38). London: James Currey.

Harmand, S. (2007). Economic behaviors and cognitive capacities of early hominins between 2.34 Ma and 0.70 Ma in West Turkana, Kenya. *Mitteilungen der Gesellschaft für Urgeschichte, 16,* 11–23.

Howell, F. C. (1972). Pliocene/Pleistocene Hominidae in East Africa: Absolute and relative ages. In W. W. Bishop & J. A. Miller (Eds.), *Calibration of Hominoid Evolution* (pp. 331–339). Edinburgh: Scottish Academic Press.

Isaac, G. L. (1964). Natron field notebook. Unpublished manuscript, National Anthropological Archives, Smithsonian Institution.

Isaac, G. L. (1965). The stratigraphy of the Peninj Beds and the Provenance of the Natron Australopithecine Mandible. *Quaternaria, 7,* 101–130.

Isaac, G. L. (1967). The stratigraphy of the Peninj Group. Early middle Pleistocene formations West of Lake Natron, Tanzania. In W. W. Bishop & J. D. Clark (Eds.), *Background to evolution in Africa* (pp. 229–257). Chicago: University of Chicago Press.

Isaac, G. L. (1969). Two Lower Acheulian sites in the Humbu Formation near Lake Natron, Tanzania. Unpublished manuscript, National Anthropological Archives, Smithsonian Institution.

Isaac, G. L. (1977). *Olorgesailie. Archeological Studies of a Middle Pleistocene Lake Basin in Kenya.* Chicago: The University of Chicago Press.

Isaac, G. L. (1981–1982). Natron Site Notes. Unpublished manuscript, National Anthropological Archives, Smithsonian Institution.

Isaac, G. L. (1982). Investigation of Early Acheulian adaptation and land use patterns in the Lake Natron Basin, Tanzania. Research Grant Proposal, NSF, 1982–1984. Unpublished manuscript, National Anthropological Archives, Smithsonian Institution.

Isaac, G. L. (1984). The archaeology of human origins: Studies of the Lower Pleistocene in East Africa 1971–1981. In F. Wendorf & A. E. Close (Eds.), *Advances in World Archaeology* (pp. 1–87). Orlando: Academic Press.

Isaac, G. L. (1986). Foundation stones: Early artefacts as indicators of activities and abilities. In G. N. Bailey & P. Callow (Eds.), *Stone Age Prehistory: Studies in Memory of Charles McBurney* (pp. 221–241). Cambridge: Cambridge University Press.

Isaac, G. L., & Curtis, G. H. (1974). Age of early Acheulian industries from the Peninj Group, Tanzania. *Nature, 249,* 624–6627.

Isaac, G. L., & Isaac, B. (Eds.). (1997). *Koobi Fora Research Project 5.* Oxford: Clarendon Press.

Jones, P. R. (1994). Results of experimental work in relation to the stone industries of Olduvai Gorge. In M. D. Leakey & D. A. Roe (Eds.), *Olduvai Gorge 5, Excavations in Beds III, IV, and the Masek Beds, 1968–1979* (pp. 254–298). Cambridge: Cambridge University Press.

Leakey, L. S. B. (1959). A new fossil skull from Olduvai. *Nature, 184,* 491–493.

Leakey, L. S. B., & Leakey, M. D. (1964). Recent discoveries of Fossil Hominids in Tanganyika: At Olduvai and near Lake Natron. *Nature, 202,* 5–7.

Leakey, M. D. (1971). *Olduvai Gorge, vol. 3. Excavations in Bed I and II, 1960–63.* Cambridge: Cambridge University Press.

Luque, L. (1995). Los yacimientos Plio-Pleistocenos del Grupo Peninj (Oeste del Lago Natrón, Tanzania). *Geogaceta, 18,* 84–87.

Luque, L. (1996). Estratigrafía del Grupo Peninj y geomorfología del margen occidental del lago Natrón (Tanzania). M.Sc. Dissertation, Universidad Complutense de Madrid.

Luque, L., Alcalá, L., & Domínguez-Rodrigo, M. (2009a). The Peninj Group. Volcanism, and sedimentary paleoenvironments during the Lower Pleistocene in the Lake Natron Basin (Tanzania). In M. Domínguez-Rodrigo, L. Alcalá, & L. Luque (Eds.), *Peninj. A Research Project on the Archaeology of Human origins (1995–2005)* (pp. 15–48). Oxford: Oxbow.

Luque, L., Alcalá, L., & Domínguez-Rodrigo, M. (2009b). The Peninj Group in type section (Maritanane): An analysis of landscape evolution. In M. Domínguez-Rodrigo, L. Alcalá, & L. Luque (Eds.), *Peninj. A Research Project on the Archaeology of Human Origins (1995–2005)* (pp. 49–72). Oxford: Oxbow.

Lycett, S. J., von Cramon-Taubadel, N., & Gowlett, J. A. J. (2010). A comparative 3D geometric morphometric analysis of Victoria West cores: Implications for the origins of Levallois technology. *Journal of Human Evolution, 37,* 1110–1117.

MacIntyre, R. M., Mitchell, J. G., & Dawson, J. B. (1974). Age of Fault Movements in Tanzanian Sector of East African Rift System. *Nature, 247,* 354–356.

Madsen, B., & Goren-Inbar, N. (2004). Acheulean giant core technology and beyond: An archaeological and experimental case study. *Eurasian Prehistory, 2,* 3–52.

Manega, P. C. (1993). Geochronology, Geochemistry and Isotopic Study of the Plio-Pleistocene Hominid Sites and the Ngorongoro Volcanic Highland in Northern Tanzania. Ph.D. Dissertation, University of Chicago at Boulder.

McHenry, L., Luque, L., Gómez, J. A., & Diez-Martín, F. (2011). Promise and pitfalls for characterizing and correlating zeolitically altered tephra of the Pleistocene Peninj Group. *Quaternary Research, 74,* 708–720.

Moore, M., & Perston, Y. (2016). Experimental insights into the congitive significance of early stone tools. PlosOne 11, e0158803.

Morell, V. (1996). *Ancestral passions. The Leakey family and the quest for humankind's beginnings.* New York: Touchstone.

Pelegrin, J. (1985). Réflexions sur le comportement technique. In M. Otte (Ed.), *La signification culturelle des industries lithiques* (pp. 72–91). Oxford: BAR, International Series 239.

Rolland, N. (1995). Levallois technique emergence: Single or multiple? A review of the Euro-African record. In H. D. Dibble & O. Bar-Yosef (Eds.), *The Definition and Interpretation of Levallois technology* (pp. 333–359). Madison: Prehistory Press.

Schick, K. (1998). A comparative perspective on Paleolithic cultural patterns. In T. T. Akazawa, K. Aoki, & O. Bar-Yosef (Eds.), *Neandertals and Modern Humans in Western Asia* (pp. 449–460). New York: Plenum Press.

Semaw, S., Rogers, M., & Stout, D. (2009). The Oldowan-Acheulian transition: Is there a "Developed Oldowan" artifact tradition? In M. Camps & P. Chauhan (Eds.), *Sourcebook of Paleolithic Transitions. Methods, Theories and Interpretations* (pp. 173–193). New York: Springer.

Sharon, G. (2007). *Acheulean large flake industries: Technology, chronology, and significance.* Oxford: BAR, International Series 1701.

Taieb, M., & Fritz, B. (1987). *Lac Natron. Géologie, géochimie et paléontologie d'un bassin évaporitique du Rift est-africain.* Sciences Géologiques Bulletin 40. Estrasbourg: CNRS.

Texier, P.-J., & Roche, H. (1995). The impact of predetermination on the development of some Acheulean chaînes opératoires. In J. M. Bermúndez, J. L. Arsuaga, & E. Cabonell (Eds.), *Human evolution in Europe and the Atapuerca evidence II* (pp. 403–420). Valladolid: Junta de Castilla y León.

Thouveny, N., & Taieb, M. (1987). Étude paleomagnétique des formations du Plio-Pléistocène de la région de la Peninj (Ouest du lac Natron, Tanzanie). Limites de la interprétation magnétostratigraphique. *Science Géologique Bulletin, 40*(1–2), 57–70.

Toth, N. (2001). Experiments in quarrying large flakes at Kalambo Falls. In J. D. Clark (Ed.), *Kalambo Falls Prehistoric Site* (Vol. III, pp. 600–660). Cambridge: Cambridge University Press.

Tryon, C. A., McBrearty, S., & Texier, P. J. (2006). Levallois lithic technology from the Kapthurin Formation, Kenya: Acheulean origin and Middle Stone diversity. *African Archaeological Review, 22,* 199–229.

Uhlig, C., & Jaeger, F. (1942). *Die Ostafrikanische Bruchstufe.* Lepizig: Deutsches Institut für Länderkunde.

White, M., Ashton, N., & Scott, B. (2011). The emergence, diversity and significance of Mode 3 (prepared core) technologies. *Developments in Quaternary Science, 14,* 53–65.

Chapter 8
Bifacial Shaping at the TK Acheulean Site (Bed II, Olduvai Gorge, Tanzania): New Excavations 50 Years After Mary Leakey

Manuel Santonja, Susana Rubio-Jara, Joaquín Panera, Alfredo Pérez-González, Raquel Rojas-Mendoza, Manuel Domínguez-Rodrigo, Audax Z. P. Mabulla, and Enrique Baquedano

Abstract This paper presents a detailed analysis of the bifacial shaping and the spatial distribution of 85 bifaces recorded in an area of 51.9 m^2 on the Lower Floor of the TK site, located alongside the Trench I excavated by M. Leakey in 1963.

The repeated use of shaping schemes and patterns demonstrates that the knappers who produced these tools had a good command of the concept of bifacial reduction. These processes were adapted differently to fit the characteristics of the exploited raw material. Formal similarities observed among the handaxes seem to reflect preconceived formal schemes, i.e., mental templates. The presence of handaxe fragments and preforms shows that they were knapped at the site with the aim of being used right there. This tool assemblage was later abandoned without the site having undergone any major alterations after its formation.

This command of the bifacial shaping concept observed at the TK site, dated to ca. 1.353 ± 0.035 Ma, undermines the validity of M. Leakey's distinction between an early and a middle phase of the Acheulean techno-complex.

Keywords Early African Acheulean • Economy of raw materials • Lithic technology • Bifaces • Handaxes

M. Santonja (✉) · S. Rubio-Jara · J. Panera · A. Pérez-González
Centro Nacional de Investigación sobre la Evolución Humana (CENIEH), Sierra de Atapuerca 3, 09002 Burgos, Spain
e-mail: manuel.santonja@cenieh.es

S. Rubio-Jara
e-mail: susana.rubio@cenieh.es

J. Panera
e-mail: joaquin.panera@cenieh.es

A. Pérez-González
e-mail: alfredo.perez@cenieh.es

R. Rojas-Mendoza
Asociación "El Hombre y el Medio", Santo Tomás de Aquino, 21, 28982 Madrid, Spain
e-mail: raquelrojasm@gmail.com

M. Domínguez-Rodrigo
Department of Prehistory, Complutense University, Prof. Aranguren s/n, 28040 Madrid, Spain
e-mail: m.dominguez.rodrigo@gmail.com

M. Domínguez-Rodrigo · E. Baquedano
Institute of Evolution in Africa (I.D.E.A.), Museo de San Isidro, Plaza de San Andrés 2, 28005 Madrid, Spain
e-mail: enrique.baquedano@madrid.org

A.Z.P. Mabulla
Archaeology Unit, University of Dar es Salaam, P.O. Box 35050, Dar es Salaam, Tanzania
e-mail: aumab@udsm.ac.tz

E. Baquedano
Museo Arqueológico Regional, Plaza de las Bernardas s/n, 28801 Alcalá de Henares, Madrid, Spain

8.1 Introduction

The site known as TK (Thiongo Korongo; Fig. 8.1) is located in the upper part of Olduvai Bed II, close to where it is adjacent to Bed III. Its estimated dating is close to that of Tuff II^D, which recently provided a new $^{40}Ar/^{39}Ar$ age of 1.353 ± 0.035 Ma (Domínguez-Rodrigo et al. 2013). M. Leakey's 1963 excavations made it possible to identify two main levels, the TK Lower Floor (TKLF) and the TK Upper Floor (TKUF), both characterized by an important industrial component and a lesser presence of fauna (Leakey 1971: 172–197; Yravedra et al. 2016). M. Leakey interpreted the upper of these two levels as Developed Oldowan (DO) and the lower as Acheulean (Leakey 1975: 484, 1976: 31, 1978). Thereinafter, TK became a key reference in the discussion on the relationships between Oldowan and Acheulean, to which several other scholars contributed (Bower 1977; Stiles 1977, 1979; Willoughby 1987; Sahnouni 1991; Ludwig and Harris 1998; Kimura 2002). In time, they arrived at the conclusion that the industrial assemblages found in the TKLF and the TKUF were technologically similar and should both be

R. Gallotti and M. Mussi (eds.), *The Emergence of the Acheulean in East Africa and Beyond: Contributions in Honor of Jean Chavaillon*, Vertebrate Paleobiology and Paleoanthropology,
https://doi.org/10.1007/978-3-319-75985-2_8

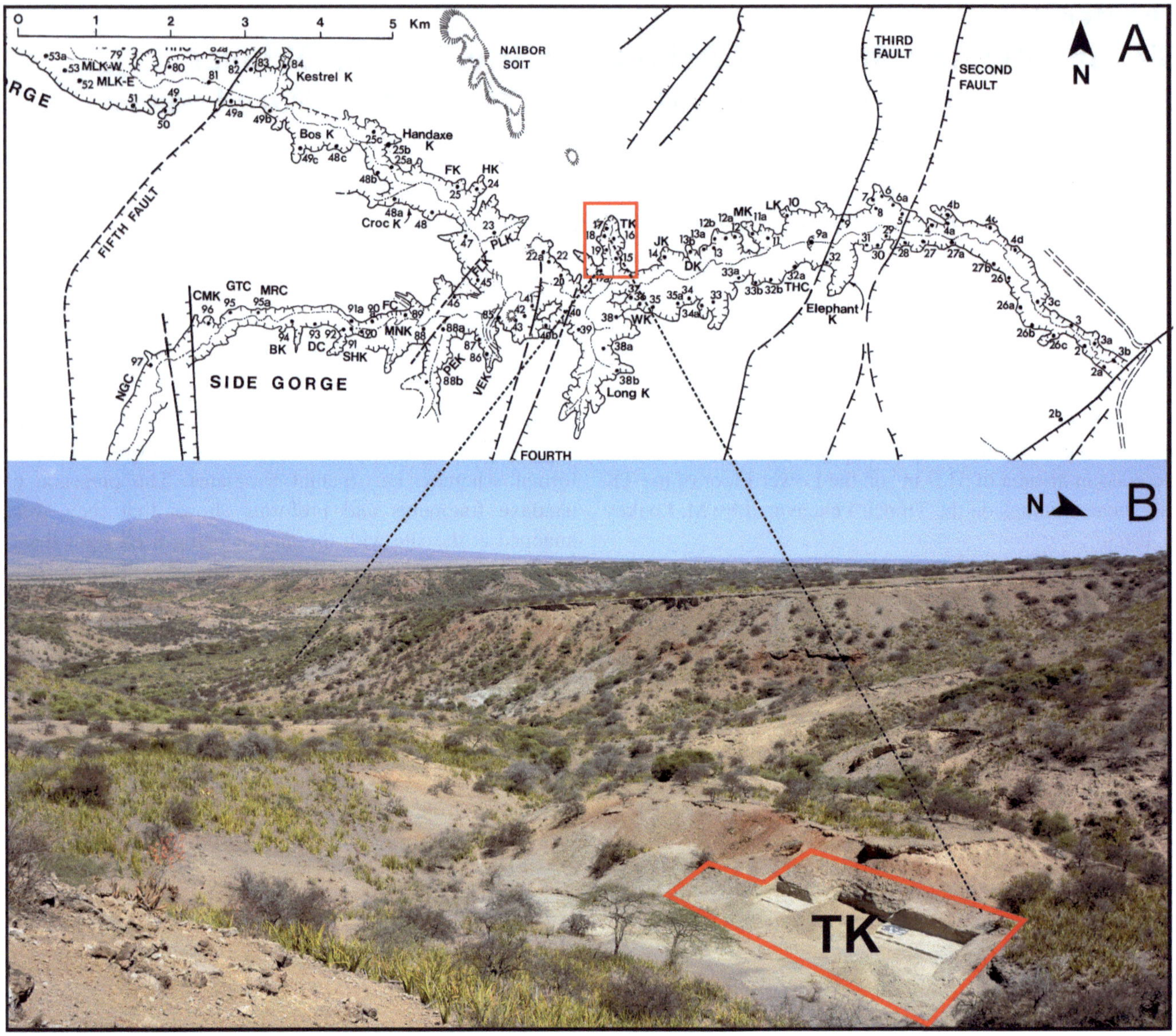

Fig. 8.1 **A** Location of the Thiongo and Long Korongos, and the TK site at Olduvai Gorge (modified after Hay, 1976). **B** View of the Thiongo and Long Korongos, showing the excavated areas at TK

attributed to the Acheulean techno-complex (de la Torre 2004; de la Torre and Mora 2005).

M. Leakey reported the presence of the Acheulean techno-complex (hereinafter AT) in the Olduvai Gorge, in Middle Bed II (Leakey 1971: 3, cf. Table 8.1). She noted its earliest occurrences at the CK, Elephant K and especially the EF-HR sites, in stratigraphic relationship with Tuff II^B, which is about 1.5 Ma, although we have no direct dating (Manega 1993; Stanistreet 2012). Progressing upward from Tuff II^A (Leakey 1971: 3, cf. Table 8.1), which is $^{40}Ar/^{39}Ar$ dated to around 1.72 Ma (Stanistreet 2012), M. Leakey situated industries that she attributed to the DO, in which she identified a phase A between Tuff II^A and II^B, a phase B starting from II^B continuing through the rest of Bed II, and also a phase C, in Beds III and IV. In her opinion, the DO represented a continuation of the Oldowan tradition observed upward from the base of Bed I. She thus concluded that the DO and the AT were two distinct technological entities. The DOA assemblages were characterized by the presence of choppers, spheroids, subspheroids, and protobifaces. The DOB added small bifaces to this toolkit, a smaller number of crude bifaces and also a small group of light-duty tools (Leakey 1971: 4–8).

The coexistence in Middle Bed II of assemblages placed in the two technological categories identified by Leakey gave rise to decades of debate. Some scholars have considered the DO to be a transitional phase between the Oldowan and the Acheulean (i.e., Clark 1970; Chavaillon et al. 1979; Clark and Kurashina 1979; Bar-Yosef 1994; Klein 1999). Others have linked the differences between the two categories to different uses of the knapped raw materials (Stiles 1979), to different states of tool reduction or mobility patterns (Jones 1994), to environmental factors (Hay 1976, 1990; Isaac 1984), or to site functionality (Gowlett 1986). More recent studies that focus on Olduvai (de la Torre and Mora 2005) or that also take into account other parts of the African continent (Semaw et al. 2009; Diez-Martín and Eren 2012; Sahnouni et al. 2013) agree in considering the DOB and the AT to be equivalent, arguing that their obvious contemporaneous existence at Olduvai is reason enough to drop the idea that the DOB was a transitional phase.

We do not intend to enter this debate here, because we think it has already been settled and that, as Diez-Martín and Eren report (2012: 312–315), it was inherited from an earlier time when the biface was considered the type fossil par excellence of the Acheulean: an assemblage had to contain a certain percentage of bifaces to be attributed to the Acheulean (Kleindienst 1962; Leakey 1971). Current consensus endorses techno-economic assessment of the chaînes opératoires (Soressi and Geneste 2011), and that is the point of view we take in this paper. In general, the Acheulean technology has been quite unlike the Oldowan (Isaac 1984; Wynn 1989, 2002; Schick and Toth 1993; Wenban-Smith 1998; Hodgson 2009), with striking differences in raw material procurement, different knapping techniques, and different technological aims (Sahnouni et al. 2013). The Acheulean must have emerged suddenly, in reaction to rapid (not gradual) change (Sahnouni et al. 2013) or perhaps also to invention automatisms, rather than to evolutionary trends, as has been suggested elsewhere (Semaw et al. 2009).

The Acheulean techno-complex, considered in its overall time period and its African geography, is often divided into three periods: early, late (or middle), and terminal (i.e., Clark 1994; Schick and Toth 2001; Bar-Yosef 2006; Diez-Martín and Eren 2012; Sahnouni et al. 2013). These three phases are identified primarily on the basis of the definition of shaped tools and especially the refinement of the bifaces—handaxes, cleavers, picks, and knives (cf. Diez-Martín and Eren 2012)—produced during each phase. This does indeed seem to be an after-effect of the "fossil approach". The initial appearance of the African Acheulean (Semaw et al. 2009: 185; Diez-Martín and Eren 2012: 325) has been recorded at specific sites—Konso (Asfaw et al. 1992; Beyene 2003; Suwa et al. 2007), Kokiselei 4 (Roche and Kibunja 1994; Lepre et al. 2011), Gona-Busidima (Quade et al. 2004; Semaw et al. 2009), Wonderwerk Cave and locations in the Vaal River basin (Chazan et al. 2008; Gibbon et al. 2009)—with chronologies clustered at about 1.7 Ma. The industry at these locations seems to be characterized by the presence of scarce and crude bifaces and large cutting tools, which demonstrates their makers' ability to obtain large flakes and shape tools from such blanks. This is the technological factor that makes it possible to identify a qualitative difference between the Acheulan and the Oldowan techno-complex (Isaac 1972; Roche et al. 2003). Nevertheless, we lack studies of complete series, and according to some authors, the current view of this industrial phase may turn out to be overly simplified (Semaw et al. 2009; Diez-Martín and Eren 2012). On the other hand, early Acheulean now seems to have lasted longer than previously thought, extending from 1.7 to 1.0 Ma, and to have spread throughout the African continent (cf. Diez-Martín and Eren 2012: 325 ff.; Sahnouni et al. 2013: 309 ff.). What distinguishes early from the late AT is the lesser definition of its bifacial tools (Sahnouni et al. 2013).

In the following pages, we will attempt to verify these approaches based on the results obtained in the excavations we have been carrying out yearly in TK since 2010 as part of The Olduvai Paleoanthropological and Paleoecological Project (TOPPP). These digs, which have explored areas adjacent to those excavated in 1963, have enabled us to reinterpret the site's stratigraphy. Our results invalidate the published comparisons between the industry in the TKLF and that in the TKUF, which Leakey and other authors (Leakey 1971, 1975, 1976, 1978; Bower 1977; Stiles 1977, 1979; Kimura 2002; de la Torre 2004; de la Torre and Mora 2005) took to be representative respectively of Acheulean and of Developed Oldowan B. The new lithic series we found at TK (Santonja et al. 2014) fill in some of the gaps noted in the collections formed at this location by M. Leakey (de la Torre and Mora 2005), in particular the underrepresentation of minor elements such as non-retouched flakes and shatter in general (Diez-Martín and Eren 2012: 329).

In the context of the chaînes opératoires identified in the TKLF, in the following sections we will present a detailed technological analysis of the bifaces (handaxes, cleavers, and trihedral picks) found at this level, in order to evaluate the shaping processes observed, in the context of current knowledge about the initial phase of the AT, as it has been recently considered by various scholars (cf. Diez-Martín and Eren 2012: 325 ff.; Sahnouni et al. 2013: 309 ff. and references therein).

8.2 Results

8.2.1 *Stratigraphic Interpretation of the TK Site and Problems Related to M. Leakey's TKUF and TKLF Levels*

Between 2010 and 2012 we excavated areas in the immediate vicinity of M. Leakey's Trench I (TI) and Trench II (TII), specifically the areas called Sector A and Sector B (Fig. 8.2). We started digging in SB, between M. Leakey's two trenches; our aim was to verify the stratigraphic and archaeological results that had been reported previously (Leakey 1971: 172 ff.). In both Sectors A and B we identified five levels containing archaeological remains. From the bottom up: the TK Lower Floor (TLKF); a loamy sand channel facies deposit; two tuff levels (corresponding to M. Leakey's Intermediate level); and the TK Upper Floor (TKUF) (Fig. 8.3).

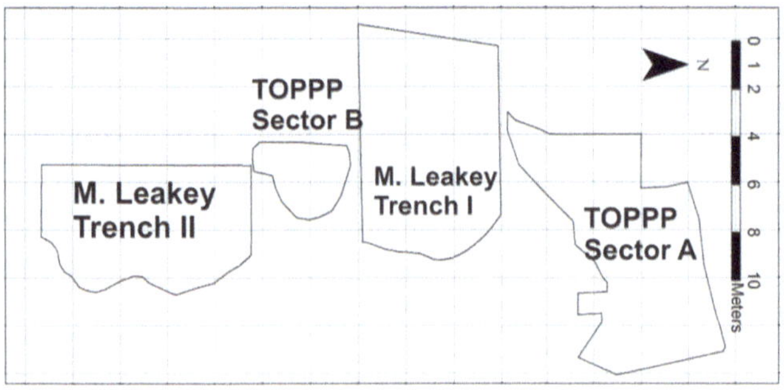

Fig. 8.2 Positions of the sectors excavated by TOPPP (2010–2012) and the trenches excavated by M. Leakey (1963)

We combined the results of our fieldwork with the west and north faces of the sections excavated by M. Leakey (1971: 172–174) to create a stratigraphic section (Santonja et al. 2014: 185, Fig. 8.4) using Hay's stratigraphic nomenclature (Hay 1976). The section's maximum thickness is about 8.90 m, of which 7.25 m correspond to Bed II, about 1.25 m to Bed III and 0.40 m to Bed IV.

Bed II starts with clays and a sandy clay loam channel facies. Above this, the TKLF is marked by a stratigraphic discontinuity and a flat topography whose western part consists of light brownish-gray clay, over which is a layer of calcrete that becomes thicker in the eastern part. The TKLF lies over both levels, stretching from west to east; in the central zone of Sector A it is partly covered by a loamy sand channel facies with a lamination structure created by low-energy water flows. The TKUF lies in stratigraphic discontinuity about one meter above the TKLF, between two tuffs. It is marked by a discontinuous calcrete layer up to 10–15 cm thick. The upper sequence of Bed II is made up of several clay sublevels containing carbonate nodules and root casts. The stratigraphic sequence of TK confirms the absence of major erosive processes, which, had they occurred, would have resulted in the presence of centimeter-size stones in the TKLF unit. Nevertheless, overland flow and/or erosive contact between the channel and the TKLF would have caused the rolling marks that we documented in the lithic industry, as well as the presence of lightweight faunal remains (Petraglia and Potts 1994; de la Torre 2004: 257–258; Yravedra et al. 2016).

Fig. 8.3 View of Trenches I and II (excavated by M. Leakey), and of Sectors B and C (excavated by TOPPP) as they appear today, showing the location of the TKUF and TKLF levels. The slope projection clearly shows that M. Leakey could have excavated the TKUF in only a small part of Trench II. Hence, most of the material found in Trench II actually corresponds to the TKLF

The area excavated by Leakey in Trench II (Leakey 1971: 186, Fig. 86), which is still perfectly recognizable today (Fig. 8.3), corresponds stratigraphically and topographically to the Lower Floor (LF), not the Upper Floor (UF), as the scholars who have studied the TK industry accepted (de la Torre 2004: 258 ff.). In the stratigraphic profile of the only remaining wall of the outcrop, the TKUF can be clearly identified on the western wall, ~30–40 cm above the TKLF. In this position, the TKUF would constitute only a small portion of the western area of Trench II, so the UF items found there were probably attributed to the upper tuff, which contained scattered artifacts.

The consequences of this are very significant. In all studies made to date, in the assemblages assigned to the Upper Occupation Floor, UF materials from TI are mixed up with materials attributed to the UF of TII but that actually correspond to the LF. Only the industry that M. Leakey attributed to the TKLF—all from Trench I—is homogeneous as regards its stratigraphic origin and is valid in comparative terms. However, since there are doubts about whether absolutely all lithic materials at the site were in fact collected during those excavations (de la Torre and Mora 2005; Diez-Martín and Eren 2012: 329), we based our study solely on the finds obtained in the excavations we carried out between 2010 and 2012.

8.2.2 General Characteristics of the Lithic Industry Present in the TKLF

Our study focuses on the 51.9 m^2 area of the TKLF. We excavated in this area, called Sector A, from 2010 to 2012, and we registrered 5,805 lithic items, which amounts to a density of 111.8 items per m^2 (38.4 per m^2 if we exclude shatter). Our analysis of the TKLF's lithic industry was based on the chaîne opératoire concept (Böeda et al. 1990; Inizan et al. 1992, 1995), from the procurement of raw material to the final phases of tool maintenance and abandonment. We differentiated these items according to the type of raw material employed (Table 8.1) and found that 92.4% are made of quartzite (81.1%, shatter excluded). It came from the Naibor Soit Ndogo inselberg whose base, at the time this site was established, was only 750 m away (Santonja et al. 2014).

Table 8.1 TKLF chaîne opératoire. 0. Procurement phase includes cobbles, vesicular lava cobbles, positives produced by percussion (basalt fragments with no percussion platform and bulb produced by a hammerstone), cobbles with pits and slabs with percussion marks. 1.1. First reduction phase produces highly cortical flakes (>90% cortex), cortical flakes (50–90% cortex), and cortical flake fragments when the flakes have small fractures which do not prevent determination of the cortex percentage. 1.2. Full reduction phase produces ordinary products: flakes with cortex removed (<10% cortex), partly cortical flakes (10–50% cortex), flake fragments with cortex removed, and Janus flakes. 1.3. Full reduction phase produces backed items: flakes and flake fragments with débitage back or cortical back. 1.4. Cores (on slab, cobble, flake, or undetermined blank). 2. Façonnage phase. 3. Shatter: undifferentiated products. They are by-products of the chaîne opératoire but are not specific to any particular phase, and can be originated by both bipolar and freehand knapping. Shatter includes debris, chunks, and angular fragments, mostly of NQ slabs. Debris is produced during knapping and does not present a defined butt or bulb. Angular fragments are polyhedral, of any size, and were produced during the knapping of NQ slabs because it generated a large number of uncontrolled fractures. Chunks are core waste that display removal traces but nevertheless cannot be assigned to any category of cores

TKLF chaîne opératoire									
Phases identified in Naibor quartzite (**a**) volcanic rocks (**b**) and other rocks (**c**)									
Chaîne opératoire phases	Unworked items			Used, retouched or shaped specimens			Total		
	a	b	c	a	b	c	a	b	c
0. Procurement phase	22	112	4	7	69	5	29	181	9
1.1. First reduction phase	228	14	2	5	0	0	233	14	2
1.2. Full reduction phase	956	89	5	37	2	1	993	91	6
1.3. Full reduction phase/back products	71	3	0	8	0	0	79	3	0
1.4. Cores	162	22	3	71	4	6	233	26	9
Subtotal of production phase	1417	128	10	121	6	7	1538	134	17
2.1. Façonnage phase: handaxes, preforms and fragments				67	10		67	10	
2.2. Façonnage phase: trihedral picks				3	0		3	0	
2.3. Façonnage phase: cleavers on flakes				3	0		3	0	
2.4. Façonnage phase: large scrapers				1	1		1	1	
Subtotal of façonnage phase				74	11		74	11	
3. Shatter	3745	66	1				3745	66	1
Total	5184	306	15	202	86	12	5386	392	27
Total: 5805									

Naibor quartzite (NQ) is a metamorphic rock made up almost entirely of quartz (98%) and has a linear structure, hence it responds to knapping very differently from the other fine-grained quartzites present at Olduvai (Jones 1994). A total of 6.8% (16.4%, shatter excluded) of the items are "volcanic rocks" (VR)—phonolites, nephelinite, trachyte, and basalt (Hay 1976: 182 ff.; Kyara 1999)—which could have been obtained from nearby stream beds. 0.9% (2.6%, excluding shatter) of the remaining items are made on vesicular lava, non-Naibor quartzite (nNQ), gneiss, and flint.

NQ and VR are the only raw materials in respect of which the chaînes opératoires are represented in all their phases, and practically the only ones on which bifaces are shaped. Knapping activities at the TKLF site focused on producing flakes from cores and occasionally reshaping them by retouch, as well as fashioning large tools from slabs, cobbles, and even large flakes. Direct freehand percussion (FHP) exploitation schemes have been identified on blanks obtained from these two rock types. Bipolar percussion (BP), on the other hand, was observed only on NQ, though it must be noted that NQ cores were exploited by both BP and FHP.

Almost half of all the VR items are cobbles (46.2%). They were mainly used as mobile hammerstones, which indicates both the importance of percussion activities and the versatility of these blanks, which were also exploited as cores.

The TKLF cores were exploited opportunistically: single surfaces with monopolar or bipolar removals, multipolar, and bifacial cores, and surfaces with centripetal removals. All the identified exploitation systems are recorded in both NQ and VR in very similar percentages, with the exception of the peripheral unipolar system (similar to the Quina concept documented in Europe, Turq 2000: 316), which is frequent in NQ and is characterized by secant or perpendicular exploitation of a main guiding surface, which may be cortical or flaked. The exploitation is carried out continuously around the initial surface. The latter is a unique characteristic of the TKLF and had not been previously recognized despite its high percentage (21.1%). Of the various production schemes observed, discoids are the most complex. We have not found any cores related to a Levallois concept. The FHP exploitation method clearly prevailed over BP (respectively 83.6% and 16.4%).

A total of 1,421 flakes (NQ: 91.8%; VR: 7.6%; nNQ: 0.4%; gneiss: 0.1%; flint: 0.1%) were identified. The ratio between the numbers of NQ and VR flakes and cores, the relationship between the sizes of flake scars on the cores and the sizes of the flakes, and the presence of cortical and full reduction flakes of all categories in similar proportions for both NQ and VR show beyond doubt that the VR and NQ cores were exploited *in situ* (Santonja et al. 2014).

Production of retouched tools was not a primary objective at the TKLF (only 3.8% of the NQ flakes and 1.8% of the VR flakes). Among the tools we found mainly scrapers, denticulates, becs, awls, and backed knives. The presence of débitage or naturally backed knives (5.7% of total flakes and fragments) points to intentional production of flakes that enabled immediate use of natural edges.

Façonnage was a widely used process at the TKLF, accounting for a total of 85 bifaces, of which handaxes are the majority; trihedral picks, cleavers, flakes, and large scrapers are also represented, despite the difficulty of detaching large flakes on NQ.

There is a large number (3,812 items) of undifferentiated products or shatter (debris, chunks and angular fragments of NQ slab), which are by-products that are included in the chaîne opératoire but are not specific to any particular phase. They constitute 65.7% of the TKLF assemblage, of which 98.2% are on NQ and only 1.7% on VR (Santonja et al. 2014). The high occurrence of shatter may be due in part to bipolar knapping, which is confirmed to be on NQ for the TKLF cores and bifaces, though we do not dismiss the possibility that they could result from percussion activities (Diez-Martín et al. 2011; Sánchez-Yustos et al. 2012; de la Torre and Mora 2013).

The TKLF assemblage in Sector A weighs a total of 494.3 kg, 115 kg (23.3%) of which are made up of VR. Its distribution in chaînes opératoires is uneven. Cobbles, commonly used as hammerstones, account for 16.8% and are generally produced on VR (94.7%). Cores and flakes constitute almost half of the total weight (48%) and bifaces 23.5%. In all these categories, VR constitutes 10% of the total weight, which indicates an interest in obtaining flakes and bifaces from these types of rock too. Finally, shatter (11.6%) is mostly made up of NQ (57 kg as opposed to 0.5 kg of VR), which suggests that VR was not always knapped entirely at the site.

8.2.3 Study of the Bifacial Shaping Processes in the TKLF

The 85 bifaces account for 4.3% of the TKLF lithic assemblage (excluding shatter). Most are produced from NQ; only eleven were produced from VR. There are no significant differences in the proportions of bifaces made of the two types of raw material (4.5% and 3.3%, respectively). We identified 77 handaxes, three trihedral picks, three cleavers on flakes, and two large scrapers.

8.2.3.1 Handaxes

A collection of 77 specimens has been classified in the handaxe category (3.9% of the total sample of NQ and VR, excluding shatter). Besides whole handaxes, it includes tips, basal fragments, and preforms corresponding to an intermediate stage of shaping (Table 8.2).

The distribution of raw materials among these artifacts is similar to their distribution in the total TKLF sample. The quantity of NQ is slightly higher (89.6%) than in the entire assemblage (82.3%), while the exact opposite occurs for VR, which amount to 10.4%, as opposed to 16.4% in the total assemblage.

As for blank types, all the NQ handaxes were produced from slabs (except two cases in which flakes were used). VR handaxes, on the other hand, were knapped on flakes or cobbles, though there is a similar percentage of undetermined blanks due to the greater intensity of bifacial reduction (Table 8.2). Cores from which large flakes could have been extracted have not been documented either in NQ or in VR.

On average, the whole handaxes found in the TKLF measure over 20 cm in length and over 10 cm in width (Table 8.3). The results of linear regression indicate that the ratio of length to width has very good linear predictability ($r^2 = 0.71$). Based on their sizes, almost all of the whole handaxes can be fitted into two large groups of 24 specimens each (Fig. 8.4A). Six specimens do not fit in either of the two groups: three are larger (between 241 and 288 mm) and three smaller (between 167 and 222 mm). The raw materials are the same, but the NQ items are the larger ones. 50% of the handaxes weigh between 1,082 and 2,002 g. Four specimens weigh over 2.8 kg and two less than 400 g (Fig. 8.4B).

We based our technological study on the bifacial shaping concept (Inizan et al. 1992: 41 ff., 1995: 43 ff.), whereby bifacial façonnage can be broken down into two phases: first, volumetric reduction; second, bilateral symmetry. The first step was to sort the TKLF handaxes according to their silhouette shape and the reduction method used. We thus obtained several basic sets.

Pointed handaxes, with amygdaloid silhouettes, make up the largest set in the TKLF (38 specimens). Their sides are more or less convex and converge into pointed tips (their rounded appearance is most likely due to wear), while their bases are often thick.

A second set, smaller and clearly differentiated, consists of 10 items whose silhouettes consist essentially of a continuous line that defines an oval or elliptical format.

The third set consists of specimens whose silhouettes tend to be rectangular or oval and are characterized by a transverse cutting edge; it also includes three other items.

Included in the TKLF handaxe assemblage are 14 unfinished or preform items whose knapping process was interrupted by accident or for some other reason. Finally, there are nine smaller fragments that were detached from finished pieces or may have been by-products of the manufacturing process. Since the former do not preserve enough characteristics to be subjected to a complete analysis of their manufacturing process, the two types cannot be distinguished from each other.

Of these three TLKF sets, pointed handaxes are larger and thinner, while the dimensions of the artifacts in the oval handaxe set are more uniform (Fig. 8.5). The transverse edge handaxes are the smallest and the thickest.

Pointed Handaxes (Figs. 8.6, 8.7, 8.8, 8.9, 8.10)

Table 8.2 TKLF handaxes classified by blank type

Blanks	NQ			VR			Total
	Slab	Flake	Undet.	Cobble	Flake	Undet.	
Whole handaxes	45	2	1	1	4	1	54
Tips and basal fragments	3		5			1	9
Preforms	11			1		2	14
Total	59	2	6	2	4	4	77

Table 8.3 Average size (mm) and weight (g) of whole handaxes (NQ n = 48; VR n = 6)

	Ranges		Average		Standard deviation	
	NQ	VR	NQ	VR	NQ	VR
Length	327–97	284–167	229.6	210.7	50.3	52.2
Width	166–53	129–87	111.7	104	21.8	16.4
Thickness	79–30	71–36	51.2	55	11.1	14.5
Weight	3115–219	2085–746	1667.1	1240.7	628.6	585.3

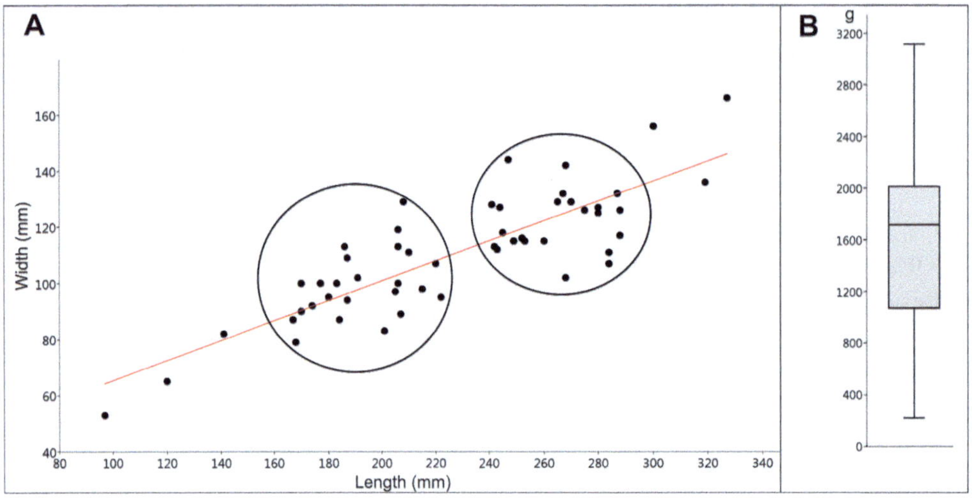

Fig. 8.4 **A** Correlations between handaxe lengths and widths (R2 Linear = 0.709). **B** Boxplot showing handaxe weights

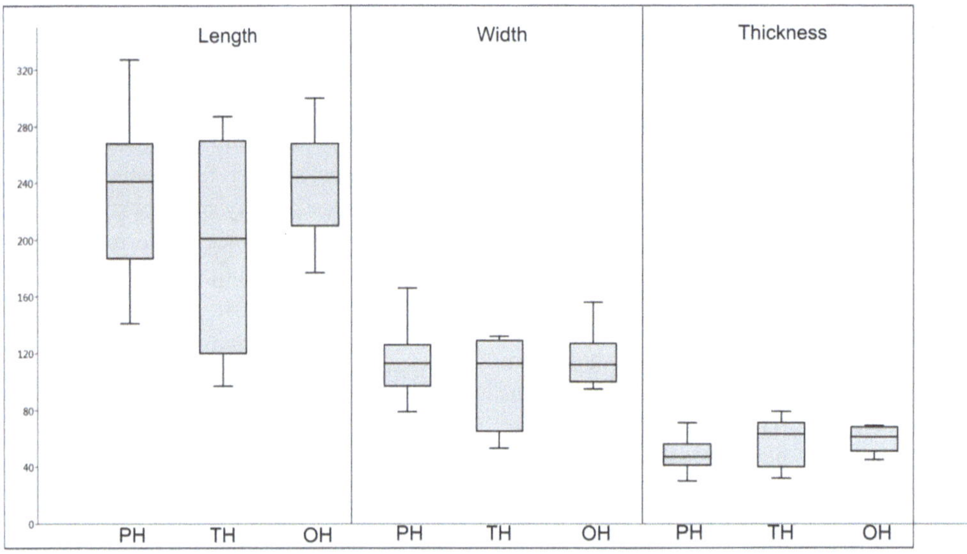

Fig. 8.5 Boxplot showing measurements (mm) of the 38 pointed handaxes (PH), the six transverse edge handaxes (TH) and the 10 oval handaxes (OH). TH are handaxes whose silhouette tends to be rectangular or oval with a transversal edge

First, we recorded an assemblage consisting of 15 items in NQ and 2 in VR, all defined by their bilateral shaping (almost always bifacial), which extends around practically the entire contour, including the basal area. In three specimens (Fig. 8.8: A-A1, F, H-H1), one of the sides had been worked by unifacial knapping. In another two (Fig. 8.7: C-C1; Fig. 8.8: C), the base and a small portion of the peripheral sector preserve cortical areas. These handaxes—over 25 cm long and weighing about 1,700 g—are the largest in the TKLF assemblage (Table 8.4).

The two VR specimens (Fig. 8.6: A; Fig. 8.8: B; Fig. 8.9: J)—one probably manufactured from a large flake, the other made on a cobble—have plainly undergone an initial phase of intense volumetric reduction. Likewise, an initial reduction phase intended to decrease the thickness of the apical third can also be observed in five of the specimens shaped on a NQ slab (Fig. 8.7: A-A1, D-D1, F-F1; Fig. 8.8: A-A1, I). In any case, this initial reduction is less intense in the NQ specimens than in the two VR handaxes, though in one of them (Fig. 8.8: I) the process has completely eliminated all cortex residues from one of the main planes.

Most (14) of the items in this set preserve traces of a finishing phase which defined the apical end either by lateral and peripheral removals (Fig. 8.6: A and Fig. 8.8: B; Fig. 8.7: A-A1, E-E1; Fig. 8.8: G-G1), or including (in 10 cases) zenithal sharpening blows struck in an axial direction (Fig. 8.6: B and Fig. 8.8: E; Fig. 8.7: B-B1, C-C1, D-D1, F-F1; Fig. 8.8: A-A1, H-H1, I, J; Fig. 8.9: J). The bases of

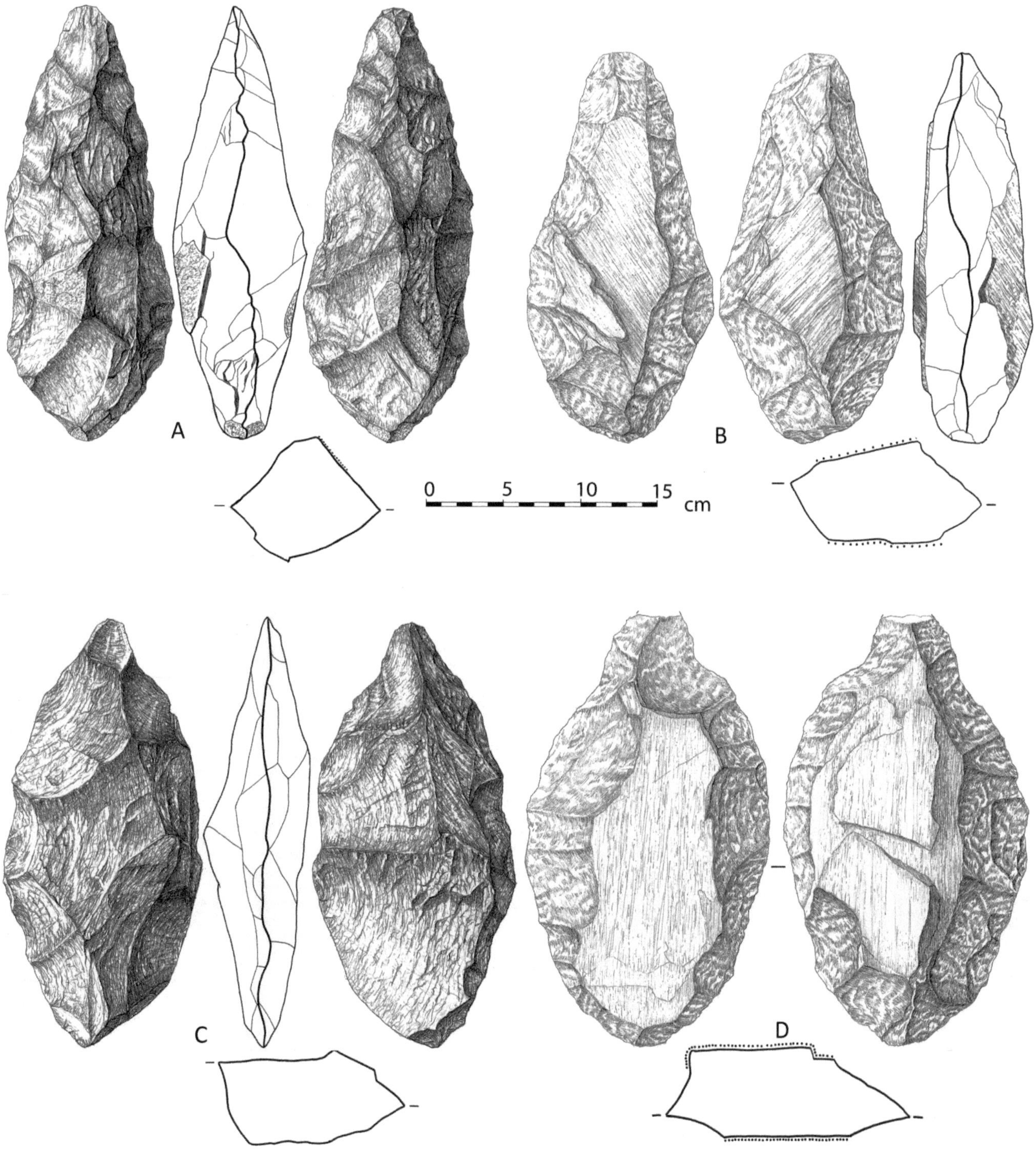

Fig. 8.6 Pointed handaxes on VR (A) and NQ (B). Oval handaxes on VR (C) and NQ (D). The diacritic diagrams of their flake removal sequences are shown in the following figures: A = Fig. 8.8-B; B = Fig. 8.8-E; C = Fig. 8.11-A; D = Fig. 8.11-B. Drawings by R. Rojas-Mendoza

five of these handaxes underwent special shaping (Fig. 8.6: B; Fig. 8.7: F-F1; Fig. 8.8: A-A1, J; Fig. 8.9: J), though this finishing gave a true cutting nature to only two of them (italicized in the above list). Since signs of a final finishing meant to give a regular shape to the artifact are observed in only a very few cases, bilateral shaping seems to have generally sufficed to achieve silhouette symmetry (Fig. 8.6: A and Fig. 8.8: B; Fig. 8.9: B).

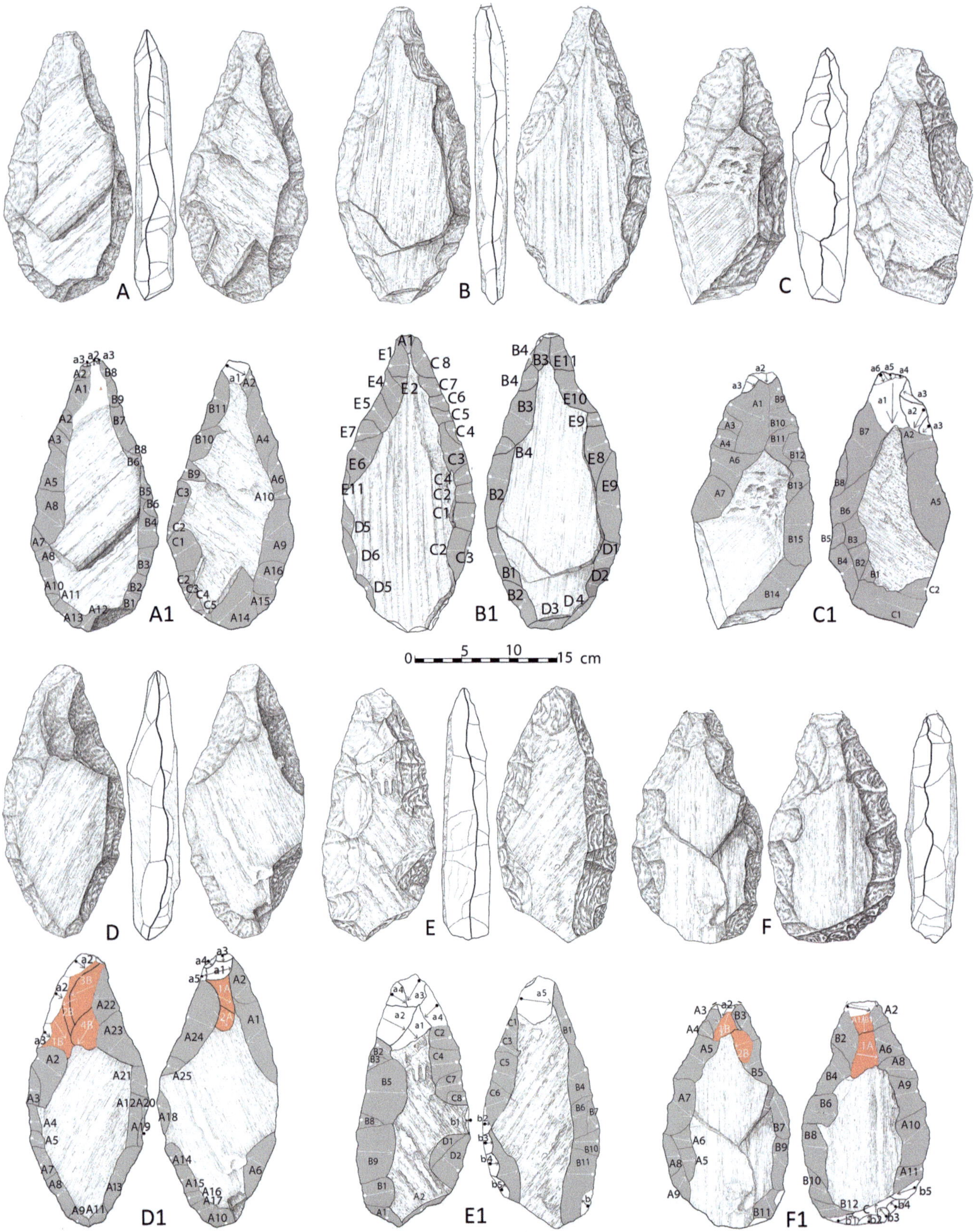

Fig. 8.7 Pointed handaxes on NQ (**A** to **F**) and diacritic diagrams showing the shaping sequence applied in each specimen and the chronological order of flake removals (**A1** to **F1**). For the key to symbols and codes, see Fig. 8.13. Drawings by R. Rojas-Mendoza

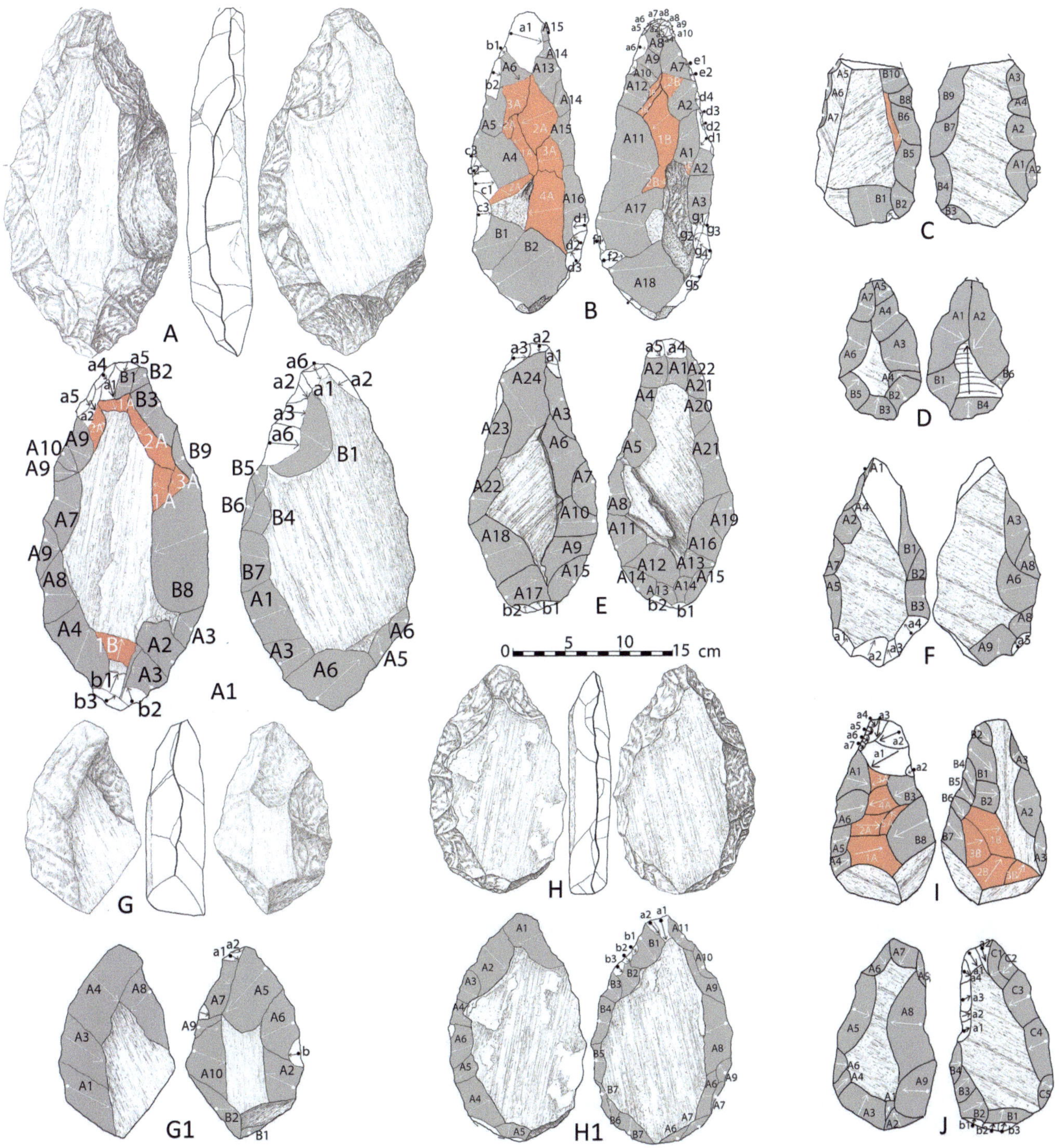

Fig. 8.8 Pointed handaxes on NQ (**A**, **G**, **H**) and diacritic diagrams showing their shaping and flake removal sequences (VR: **B**; NQ: **A1**, **C** to **F**; **G1**, **H1**, **I,** and **J**). For the key to symbols and codes, see Fig. 8.13. Drawings by R. Rojas-Mendoza

The next assemblage, though not as homogeneous as the previous one, comprises specimens likewise characterized by an amygdaloid silhouette which, however, is obtained through bilateral shaping that is not generalized or is applied exclusively to the apical third. This group too contains 17 handaxes, but their dimensions are smaller and more homogeneous than those of the ones described above (Table 8.5). Six have a straight base, which is cortical in some (Fig. 8.9: I; K; Fig. 8.10: G) and in others is defined by knapping (Fig. 8.9: A, C, E). The bilateral definition of one of these handaxes is solely unifacial, with intense final shaping of the tip (Fig. 8.9: E), while the bilateral shaping of the other five is more highly defined, partially bifacial and with lateral and apical finishing that in some cases is rather intense (Fig. 8.9:

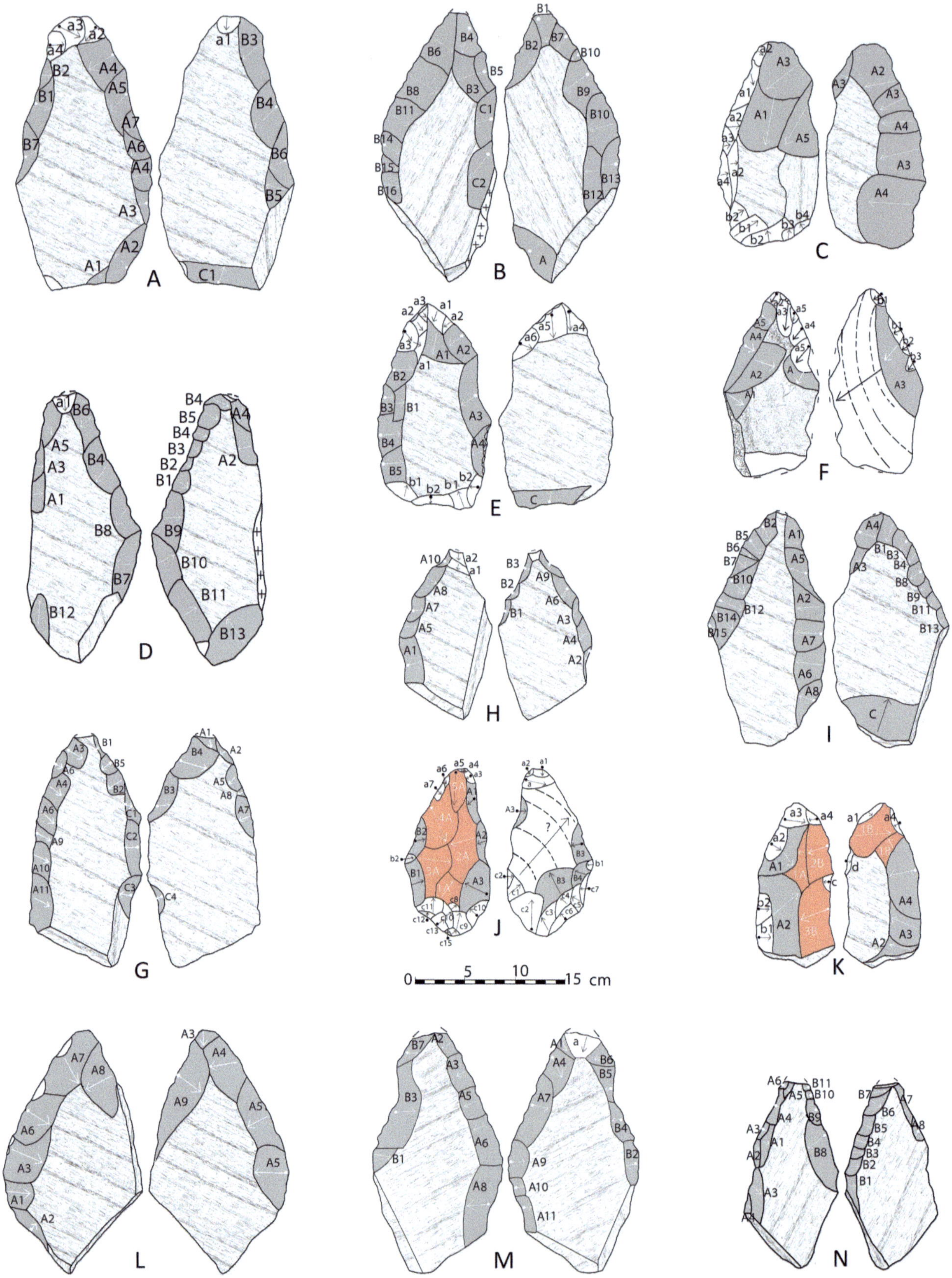

Fig. 8.9 Diacritic diagrams of other pointed handaxes on NQ (**A** to **E**, **G** to **I** and **K** to **N**) and on VR (**F** and **J**) showing their shaping and flake removal sequences. For the key to symbols and codes, see Fig. 8.13. Drawings by R. Rojas-Mendoza

Fig. 8.10 Other pointed handaxes (**A**, **B**, **F**) and their respective diacritic diagrams (**A1**, **B1,** and **F1**), and diacritic diagrams of similar items on NQ (**D**, **E**, **G,** and **H**) and on VR (**C** and **C1**). For the key to symbols and codes, see Fig. 8.13. Drawings by R. Rojas-Mendoza

Table 8.4 Average size (mm) and weight (g) of pointed handaxes with shaping around the entire contour (NQ n = 15; VR n = 2)

	Ranges	Average	Standard deviation
Length	327–141	235.6	57.4
Width	166–82	115	21.6
Thickness	71–30	51.6	11.8
Weight	3115–618	1598.4	610.9

Table 8.5 Average size (mm) and weight (g) of pointed handaxes with non-generalized bilateral shaping (NQ n = 16; VR n = 1)

	Ranges	Average	Standard deviation
Length	288–168	229.6	41.1
Width	144–79	110.0	18.7
Thickness	62–36	46.6	7.8
Weight	2289–746	1604.2	539.1

Table 8.6 Average size (mm) and weight (g) of oval handaxes (NQ n = 9; VR n = 1)

	Ranges	Average	Standard deviation
Length	300–177	239.3	36.2
Width	156–95	114.7	18.4
Thickness	69–45	59.8	8.2
Weight	2827–1426	1907.8	413.8

A, C, K). A handaxe made on a VR flake, which originated from a cleaver on flake whose cutting edge and basal area were reshaped to obtain the final symmetry of an amygdaloid handaxe, can be added to this group (Fig. 8.10: C-C1).

The other ten specimens underwent medium-intensity bilateral reduction; they have thick wedge-shaped bases resulting from the intersection of two cortical planes. In some cases bifacial knapping extends over more than half of the contour (Fig. 8.9: L, M; Fig. 8.10: F-F1), while in others it appears one side and part of the other (Fig. 8.9: A, D), or partially on both sides (Fig. 8.9: G, N), or on one side only, while the other has been worked with unifacial knapping (Fig. 8.9: H; Fig. 8.10: A-A1, B-B1). In four handaxes the point was particularly well shaped, either by axial removals made during the bilateral shaping phase (Fig. 8.9: B, L; Fig. 8.10: B-B1) or by a final finishing phase (Fig. 8.9: M). Only in one case do we observe an initial volumetric reduction that thinned apical end of the slab (Fig. 8.10: A-A1).

Lastly, we have four handaxes in which bilateral reduction is limited to the apical third. Three are in NQ, and one was produced on a thick VR slab (Fig. 8.9: F). All four are smaller than the handaxes in the other sets, measuring on average 206 × 109 × 50 mm and weighing about 1,319 g. Two have straight thick bases (Fig. 8.10: D, H), one has a wedge-shaped base (Fig. 8.10: E), and the base of the fourth is fragmented (Fig. 8.9: F). Two of these specimens show initial volumetric reduction (Fig. 8.10: D, E). Bilateral shaping is bifacial in all four, while two of them also present final apical and lateral finishing (Fig. 8.9: F; Fig. 8.10: D).

Oval Handaxes (Fig. 8.11)

A second set of handaxes consists of six specimens with an oval silhouette (Fig. 8.6: C and Fig. 8.11: A; Fig. 8.6: D and Fig. 8.11: B; Fig. 8.11: E-E1, F, I, J), plus four with elliptical shapes (Fig. 8.11: C-C1, D, G, H). All of them were produced on NQ except for one on VR (Fig. 8.6: C and Fig. 8.11: A). This is the group where specimen measurements are more clustered (Table 8.6) and average dimensions are larger, especially thickness (over 8 mm) and weight (over 309 g). Among the blanks is a very large NQ flake (size: 244 × 127 × 64 mm; weight: 2,123 g). Slabs were used for all the other specimens, but in two cases, including the VR item, the type of blank could not be determined.

Three specimens stand out in this set: two are pointed at both ends (Fig. 8.6: C; Fig. 8.11: G), while the point of the third juts out from the contour at one end (Fig. 8.6: D). Several of them, especially the VR specimen, had undergone intense initial volumetric reduction on both surfaces as well, as had a NQ specimen of with one face and a good part of the other were fully fashioned (Fig. 8.11: G).

Bilateral shaping is mostly bifacial. In four cases it affects the whole outline (Fig. 8.11: A, B, G, I, fragmented). In three specimens, some parts of it are preserved on the base (Fig. 8.11: D, F, wedge-shaped: J). In two specimens, the shaping—unifacial in one (Fig. 8.11: C-C1) and partly bifacial in the other (Fig. 8.11: H)—is combined with bipolar reduction in some portions. This knapping method is also observed on the base of one of the above-mentioned items (Fig. 8.11: D).

The handaxe fashioned on a NQ flake (Fig. 8.11: E-E1) was bilaterally defined by unifacial and bifacial knapping that, together with the final finishing retouch, affects the entire contour except for a natural flat facet pertaining to the slab blank itself, which also forms an edge and was adjusted by retouch. This final phase extends along the sides and reaches one of the two ends, where axial removal was used to shape a bevel.

Two items in this set (Fig. 8.6: C, D and Fig. 8.11: A, B) have pointed ends that jut out from the overall oval shape or simply produce a cutting finish (Fig. 8.11: G, I, J).

Transverse Edge Handaxes (Fig. 8.12: A-A1, C-C1, E-E1, F, G, and I)

This set comprises six handaxes whose silhouette tends to be rectangular or oval. These items are the smallest ones of the group and the thickest (Fig. 8.4; Table 8.7). The only one produced on VR is actually an intermediate form, something between a handaxe and a cleaver (Fig. 8.12: G), for it was first manufactured as a cleaver on a cortical flake (type 0, Tixier 1956); after the cutting edge broke, it was knapped again and the artifact became similar to a transverse edge handaxe.

Two pairs of items on NQ slabs are particularly interesting. The two specimens in each pair have similar shapes. One pair is very large (270 and 281 mm; Fig. 8.12: A-A1, C-C1), and the other is shorter (97 and 120 mm, Fig. 8.12: E-E1; I). Although only one of the two large handaxes (Fig. 8.12: A-A1) preserves traces of an initial volumetric

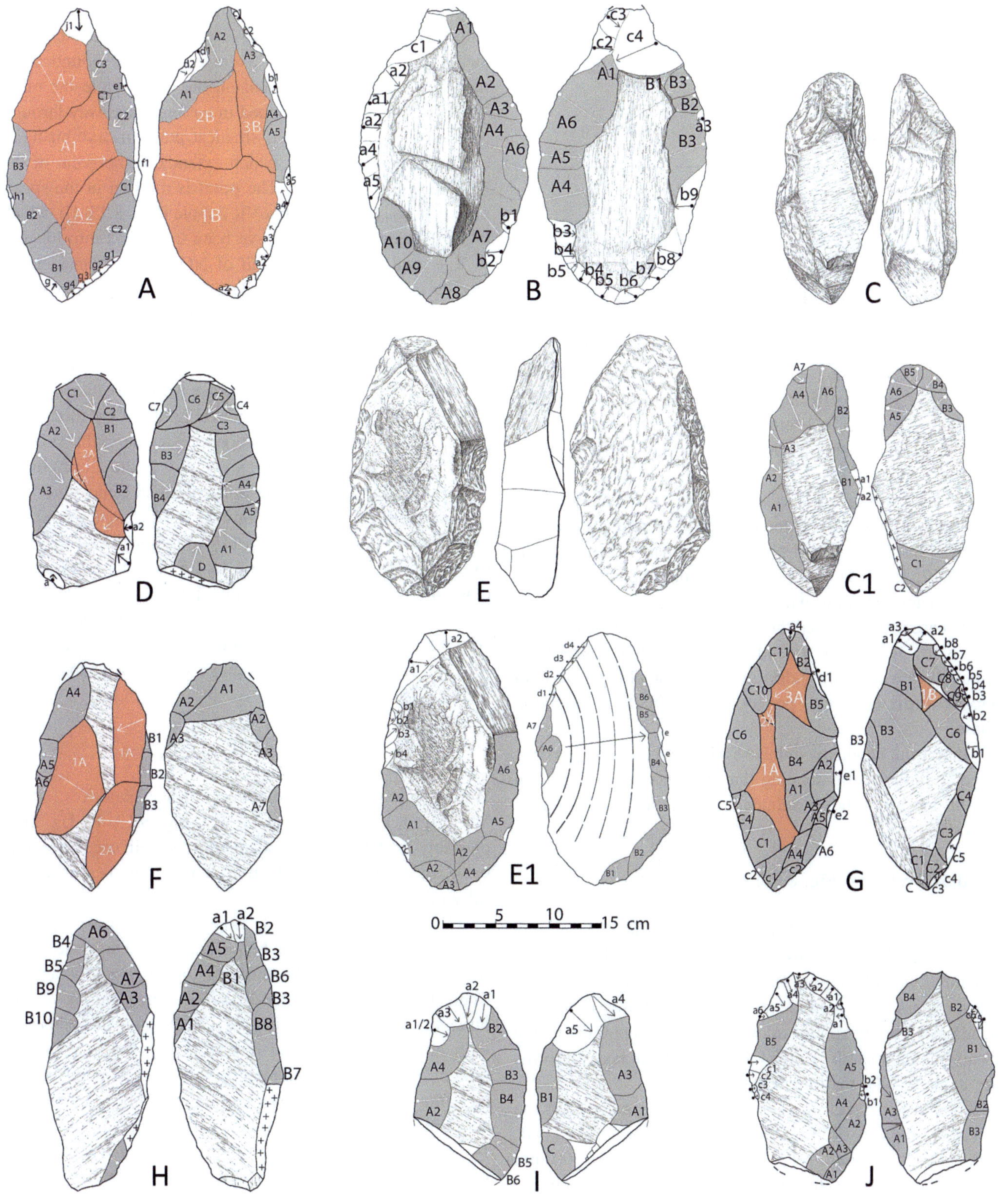

Fig. 8.11 Oval handaxes on NQ (**C**, **E**) and their respective diacritic diagrams (**C1**, **E1**), and diacritic diagrams of similar items (**B**, **D**, **F**, **H**, **I**, and **J**). Diacritic diagrams of similar items on VR (**A** and **G**). For the key to symbols and codes, see Fig. 8.13. Drawings by R. Rojas-Mendoza

Table 8.7 Average size (mm) and weight (g) of transverse edge handaxes (NQ n = 5; VR n = 1)

	Ranges	Average	Standard deviation
Length	287–97	193.5	76.7
Width	132–53	95.8	33.6
Thickness	79–32	55.0	18.7
Weight	2965–219	1473.7	1184.0

reduction, both were given a bifacial and bilateral shape that affects the entire contour; in one, chained bifacial flaking gestures were used (Fig. 8.12: A-A1), in the other (Fig. 8.12: C-C1), and each surface and the base were knapped successively. Though the two large handaxes are similar in shape, the cutting edge of the first one was delineated by means of unifacial tranchet blows, that of the second by combining lateral and axial removals, some of them executed during a final finishing phase (Fig. 8.12: C-C1).

The morphology of the second pair of handaxes, both produced on NQ slabs, is more or less rectangular (which is common in cleavers) and preserves extensive cortical areas (Fig. 8.12: E-E1, I). Both specimens present alternate unifacial bilateral shaping which extends to the base and edge. In one case it was defined through extensive oblique removal (Fig. 8.12: E-E1), in the other through two consecutive removals in a coinciding direction, likewise oblique. In both handaxes, the edge was prepared in the bilateral reduction phase.

One of the handaxes in this set (Fig. 8.12: F-F1) presents bifacial and bilateral shaping that does not reach the basal area. It has a thin bevel defined by lateral and axial bifacial removals.

Handaxe Preforms (Fig. 8.13)

Besides the items described above, the group we studied includes another 14 items: handaxes that were being manufactured when knapping fractures made them unfit for use. Six of them have silhouettes similar to those of pointed handaxes, five are similar to oval handaxes, and three lack a clearly defined shape. Half of the pointed handaxe preforms are made of NQ, the others of VR.

Though these preforms are considerably smaller than the finished handaxes (Table 8.8), three specimens (two of VR, Fig. 8.13: F, G; one of NQ, Fig. 8.13: H) are over 200 mm long. The two basalt pieces show signs of volumetric reduction and of an intense bilateral shaping phase during which knapping accidents occurred, whereupon the blanks were abandoned. The NQ piece is fractured in the basal area; its bilateral shaping, which was still in its initial stage when it was interrupted, tends to bifacial; the pointed tip had been prepared through bifacial removals in an axial direction before the bilateral reduction was started. One other NQ piece (Fig. 8.13: B) is of average size and bounded by fractures around the base and on one side, while the other side and the point were shaped in an integrated way through bifacial knapping. The two smallest preforms exhibit bilateral and bifacial shaping which extends to the point of the NQ piece (Fig. 8.13: A). The point is missing in the specimen that was produced on a VR cobble (Fig. 8.13: E), which in its finished state could have been up to 132 mm long.

The oval specimens are all of NQ and are considerably smaller than the finished handaxes, even though they could have been more than 200 mm long (Table 8.9). Four are in an advanced stage of bilateral reduction, either bifacial or unifacial (Fig. 8.13: C, D, I, M). Of these, three also present bifacial shaping of the point in an axial direction; this is particularly evident in one of them (Fig. 8.13: D), though the point of this piece was knapped after the right side. In the fifth piece (Fig. 8.13: K), shaping was done only in the apical third and includes the tip, with axial removals.

Finally, three items made on NQ slabs (lengths from 164 to 219 mm; widths from 77 to 115 mm; thicknesses from 41 to 66 mm; weights from 1,261 to 1,572 g) have no clearly defined shape, either because bifacial reduction was to follow (Fig. 8.13: L) or because they were only roughed out (Fig. 8.13: T). One has a double fracture at each end (Fig. 8.13: J), so we cannot make out what shape it was meant to be.

Handaxe Fragments (Fig. 8.13)

We identified the points of three finished handaxes, one in VR (Fig. 8.13: P) and two in NQ (Fig. 8.13: R, U). They probably broke off by accident during their use. They preserve traces of bifacial reduction, which in two cases did not fully eliminate the cortex.

We also recorded six NQ fragments that probably came from specimens that were being knapped. Three are apical fragments, similar to the three in the previous paragraph. Their length ranges from 46 to 97 mm. One is shaped bifacially and bilaterally (Fig. 8.13: Q), and the other two with removals in an axial direction (Fig. 8.13: V, W)—a type of blow that appears often in the series. The other three NQ fragments display marks of intense bifacial reduction and are larger than the others, measuring from 106 to 146 mm. It is impossible to establish clearly whether these fragments are apical or basal (Fig. 8.13: N, O, S).

8.2.3.2 Trihedral Picks

The assemblage of pointed-tip bifaces includes three picks exhibiting a basic trihedral façonnage method (Inizan et al.

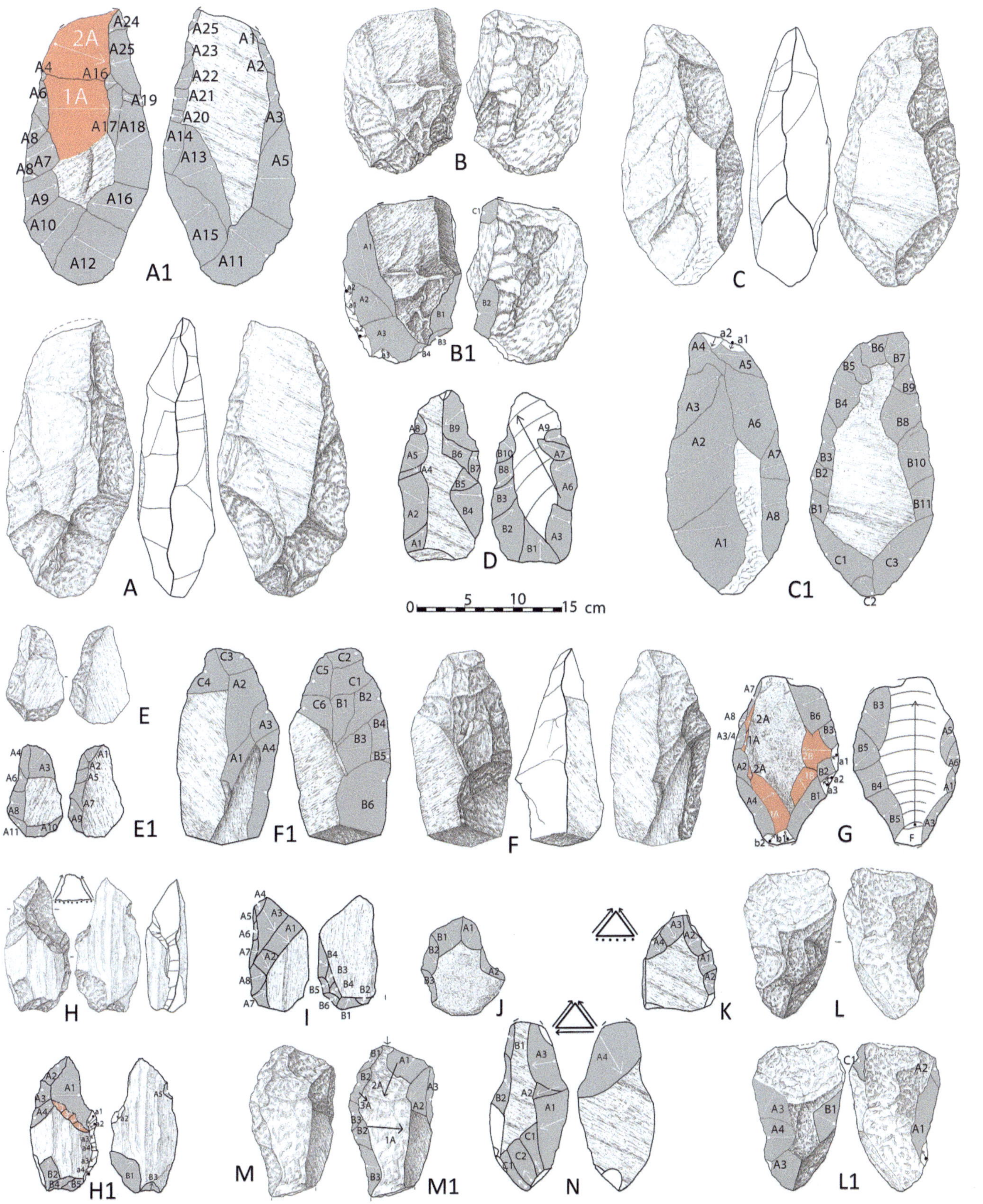

Fig. 8.12 Transverse edge handaxes on NQ (**A**, **C**, **E,** and **F**) and their respective diacritic diagrams (**A1**, **C1**, **E1,** and **F1**), and diacritic diagrams of similar items on NQ (9) and on VR flakes (7). Cleavers on NQ flakes: descriptive drawings and diacritic diagrams (**B-B1**; **D** and **L-L1**). Trihedral NQ picks: drawings and diacritic diagrams (**H-H1**; **K** and **N**). Large scrapers on NQ (**M-M1**) and on VR (**J**). For the key to symbols and codes, see Fig. 8.13. Drawings by R. Rojas-Mendoza

Table 8.8 Average size (mm) and weight (g) of pointed handaxe preforms (NQ n = 5; VR n = 1)

	Ranges	Average	Standard deviation
Length	269–119	190.2	52.6
Width	72–120	93.7	17.1
Thickness	96–47	71.0	18.8
Weight	1930–391	1332.0	614.2

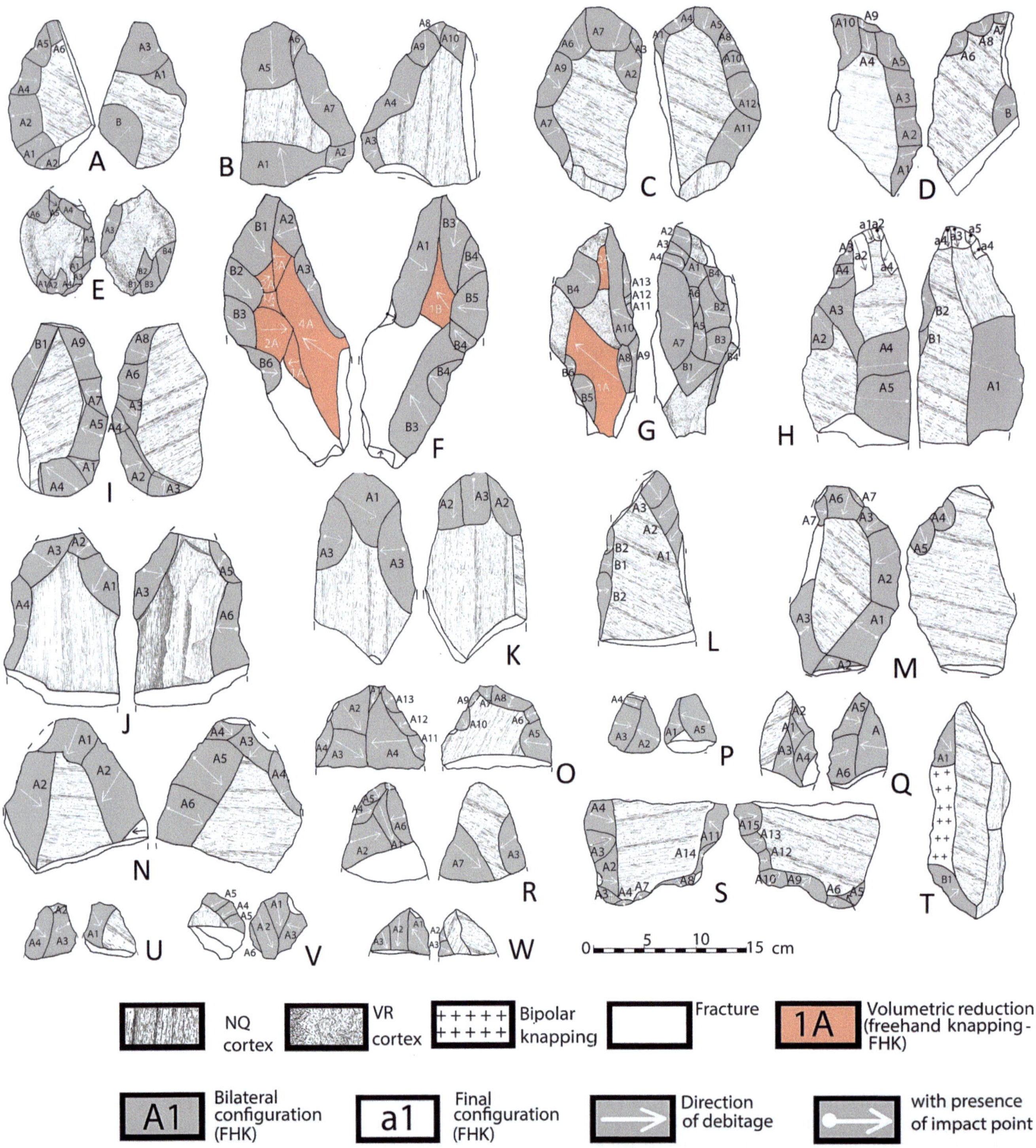

Fig. 8.13 Diacritic diagrams of handaxe preforms on NQ (**A** to **D**; **H** to **M** and **T**) and on VR (**E** to **G**). Apical handaxe fragments on NQ (**N**, **O**, **Q**, **R**, **U**, and **W**) and on VR (**P**). Basal handaxe fragment on NQ (**S**) and undefined handaxe fragment on NQ (**V**). Drawings by R. Rojas-Mendoza

Table 8.9 Average size (mm) and weight (g) of oval handaxe preforms (NQ n = 5)

	Ranges	Average	Standard deviation
Length	195–181	190.0	5.5
Width	106–90	98.2	7.5
Thickness	65–54	58.4	5.2
Weight	1563–1128	1416.3	197.4

1992: 44) (Fig. 8.12: H-H1, K, N). They are smaller than the handaxes, measuring from 174 to 103 mm in length and weighing between 1,375 and 372 g. They were made on NQ slabs, thick in two specimens and tending to flat in the third (Fig. 8.12: K). The knapping method used to produce these specimens is quite simple; it is applied on two converging surfaces whose intersection defines the apical ridge. Only one specimen presents traces of volumetric reduction and final lateral and basal finishing (Fig. 8.12: H-H1).

8.2.3.3 Cleavers on Flakes

The biface assemblage includes three cleavers on flake that closely fit Tixier's (1956) definition, among the most typical tools of the Acheulean techno-complex. They were produced by a combination of the débitage and façonnage methods (Mourre 2003).

One of these cleavers was made on a normal NQ flake and corresponds to Tixier's type II (Fig. 8.12: L-L1). One side was shaped bilaterally and bifacially, and the other exploits the planes provided by the butt of the blank flake and of a cortical area perpendicular to the plane of the flake. The contour was completed by means of isolated removals.

Each of the other two cleavers was manufactured on a cortical flake; they are among the smallest bifaces found in the TKLF (L = 150 and 188 mm; We = 835 and 1,549 g). One of them presents total bifacial lateral shaping that delimited a natural oblique cutting edge (Fig. 8.12: D) partly affected by a lateral tranchet blow. The bifacial knapping is invasive; therefore, it actually makes it impossible for us to determine whether the blank flake was fully cortical and could thus be classified as Tixier's type V. The second cleaver was made on a fully cortical flake and corresponds to type 0 (Fig. 8.12: B-B1). It exhibits bilateral, basically unifacial knapping and cortical segments, and was completed with final retouching on one side to give it a regular shape.

8.2.3.4 Large Scrapers

Two large flakes—an ordinary one in NQ (153 × 87 × 61 mm and 730 g, Fig. 8.12: M-M1) and a cortical one, in VR (102 × 90 × 33 mm and 311 g, Fig. 8.12: J)—show unifacial shaping that produced a cutting edge opposite the dorsal face (Fig. 8.12: M-M1) and a pointed end created by two converging retouched sides opposite a thick cortical base (Fig. 8.12: J).

8.2.4 *The Techno-functional Approach Applied to the Handaxes*

According to Boëda (2001), a techno-functional unit consists in the adaptation of the support to its desired or potential function, making the artifact operative of the wished task. As we saw in the previous section, the most evident techno-functional unit in the TKLF handaxes is their apical area. It was shaped as a tip in the 38 pointed handaxes and in at least two oval ones (Fig. 8.6: C, D and Fig. 8.11: A, B). There are also two other specimens with a fracture in their apical areas that seems to be related to a pointed end (Fig. 8.11: D, F). The presence of three-point fragments of finished handaxes would confirm that the point was very important in the use of these tools.

The shape of the apical area of six oval handaxes is almost the segment of a circle. Its morphology is clearly different from that of pointed handaxes, therefore making it possible to associate its function with a different task. Likewise, the six transverse edge handaxes were probably shaped for a different task than the ones just described.

The remaining portion of the contour of the handaxes can be divided into cutting-edge segments and backed segments. Traceological analysis is the ideal method for determining whether the TFUs in these tools were conceived considering only bilateral symmetry or if there was also a functional objective. This kind of study will be made soon. In the meantime, an analysis of the sections of the basal and apical areas of the handaxes makes it possible to give a tentative answer to this important question.

Based on the degree of corticality and on the type of shaping (unifacial or bifacial) and of percussion (bipolar or freehand), nine sections have been identified, some of which present variations. They have been ranked from least to most processed (Fig. 8.14: A). In sections A to G, shaping is limited to the bilateral finishing phase, while in sections H and I it also affects bifacial symmetry. The sides are formed by backs or cutting edges, or by a combination of both. The backs may be cortical (A, B, and C) or shaped by bipolar percussion (A1, B1, and C1), while the edges are shaped by unifacial (D, E) or bifacial freehand percussion (F, G, H, I). Sections with sides that were shaped in a different way are common (B1, C1, F, H).

We analyzed the sections of 54 whole handaxes, 48 of them in NQ and 6 in VR, including a VR specimen with a small

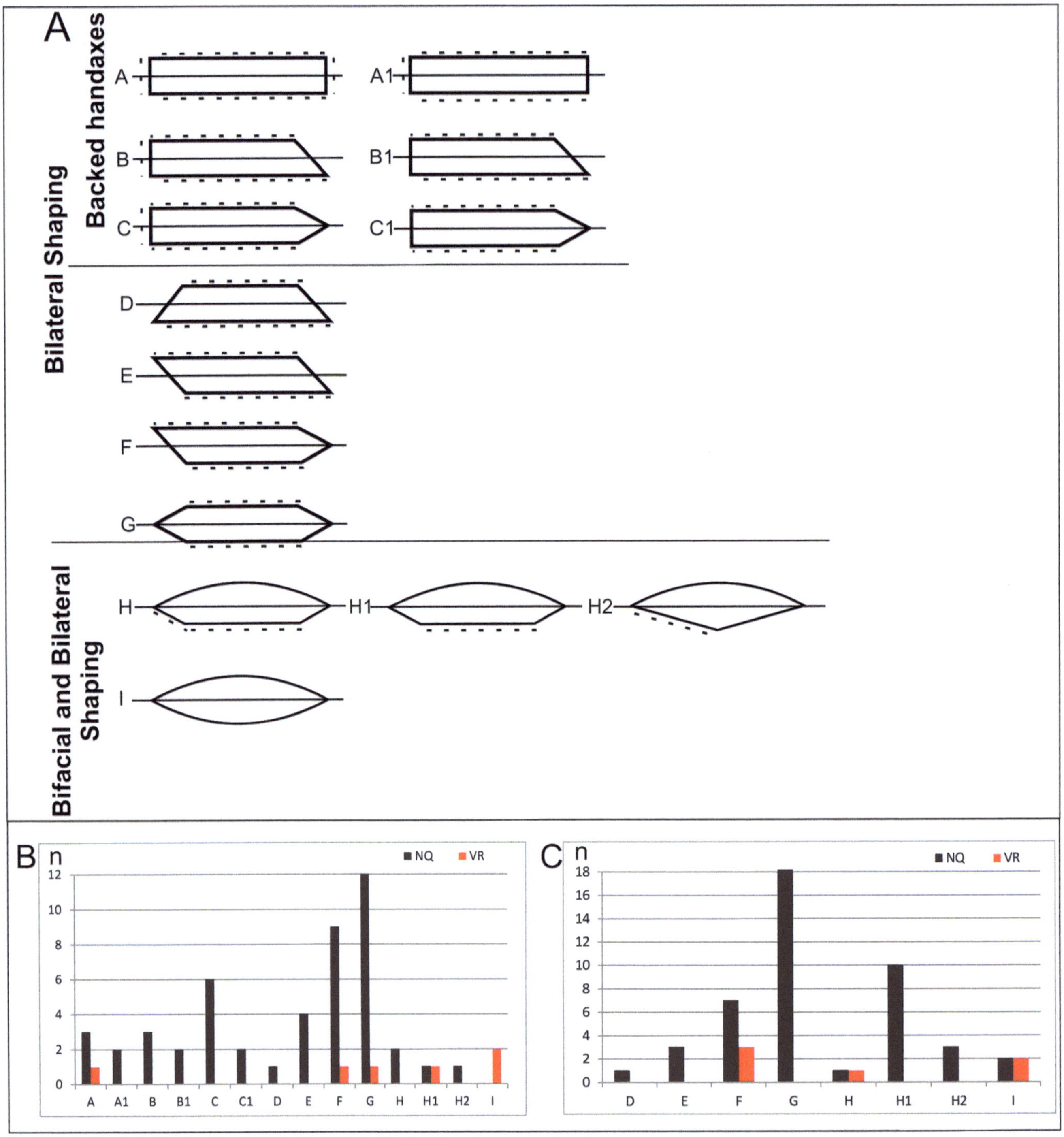

Fig. 8.14 **A** Cross-sections of the handaxes found in TKLF Sector A, showing their corticality, shaping (unifacial or bifacial) and type of percussion (bipolar or freehand). Dotted lines indicate cortical and/or bulbar areas. Bipolar percussion (BP) is represented by lines perpendicular to the bifacial symmetry plane (e.g., A1), freehand percussion (FHP) by lines at an angle with the bifacial symmetry plane (e.g., B). Bifacial shaping is represented by two slanting lines converging onto the bifacial symmetry plane (e.g., C). A Cortical faces and sides, or shaped by bipolar percussion. B Cortical faces and one side cortical or shaped by BP, the other side by unifacial FHP. C Cortical faces and one side cortical or shaped by BP, the other shaped by bifacial FHP. D Cortical faces and both sides shaped by unifacial FHP produced starting from the same face. E Cortical faces and sides shaped by unifacial FHP starting from opposite faces. F Cortical face, one side shaped by bifacial FHP and the other by unifacial FHP. G Cortical faces and sides shaped by bifacial FHP. H One face cortical and another shaped by FHP; sides shaped by unifacial or bifacial FHP. I Faces and sides shaped by bifacial FHP. **B** Cross-sections of the bases of the 54 whole handaxes. **C** Apical cross-sections of the 54 whole handaxes

fracture in its apical area. The significant number of backed handaxes stands out. The basal areas were shaped with a back in almost 40% of the specimens (19 out of 48; sections A, B and C). The presence of backs may be correlated with the ease with which the tools could be gripped. However, the fact that most of the backs are natural, except for six that were produced by bipolar knapping (sections A1, B1 and C1), does not allow us to rule out the possibility that they might have resulted from a preselection of suitably shaped blanks aimed at minimizing the shaping process. 21% of the backed NQ handaxes have backs on both sides, which makes it evident that the sole purpose of these pieces was to shape a tip. The reason for the unifacial knapping of the side opposite the natural back, observed in five items, may have been to give them bilateral symmetry, or to produce an active cutting edge (Fig. 8.9: A, G, I; Fig. 8.10: A-A1; Fig. 8.11: C-C1). However, in the case of the eight specimens whose cutting edge was shaped by bifacial knapping (sections C and C1), its purpose was most likely functional (Fig. 8.7: C-C1; Fig. 8.9: B, D, H, L, M; Fig. 8.10: B-B1).

The cutting edges of the 35 handaxes that have no backs take up at least a good portion of the contour and were shaped by unifacial or bifacial knapping or a combination of both (Fig. 8.14, sections D to I). In the unifacially fashioned segments (sections D and E, Fig. 8.14), it is impossible to tell whether the cutting edge was made to obtain bifacial symmetry or to create an active edge as well (Fig. 8.9: E; Fig. 8.11: D; Fig. 8.12: E-E1, F-F1, I). However, in the 30 specimens with bifacially shaped cutting-edge segments (F to I) the second reason is more likely. In this sense it seems significant that the three specimens longer than 300 mm have bifacial shaping on at least one side (Fig. 8.14, F in one case and G in two; Fig. 8.8: A-A1 and Fig. 8.6: D; Fig. 8.7: B-B1). This circumstance could be related to the fact that in these specimens, the cutting edges as well as the apical areas were functional.

There is only one VR backed handaxe (Fig. 8.8: C), and only one of the last five specimens (Fig. 8.10: C-C1) has one side shaped by unifacial knapping (section F). In these items, the cutting edge may therefore have been as important as the apical area (Fig. 8.14: B).

8.2.5 *Analysis of the Spatial Distribution of Bifaces in the TKLF*

The TK's stratigraphic sequence confirms that no major erosive processes occurred that could have shifted the bifaces from their original position; hence, the position in which these artifacts were found when they were unearthed may be considered to be the same or very close to their original one (Domínguez-Rodrigo et al. 2014; Santonja et al. 2014).

The biface spatial distribution is clearly heterogeneous (Fig. 8.15). In the 52 m^2 area excavated in Sector A, the average number of bifaces found was 1.6 per m^2 (including fragments and handaxe preforms), with a maximum density of 10 items per m^2 versus areas of up to 8 m^2 which did not contain any bifaces (gray-line polygons in Fig. 8.15). The distribution of these tools in two adjacent areas, both measuring 16 m^2 (blue-line polygons in Fig. 8.15), is significant. Sixty-three bifaces were found in the western area, but only seven in the eastern one; this implies a density of respectively 3.8 and 0.4 items per m^2. In other words, artifact density in the western area is 8.7 times higher than in the eastern one. If we were to consider only whole handaxes, artifact density in the two areas would be 2.7 versus 0.3 items per m^2 (respectively 45 and 4 items), that is, 11 times as many.

Spatial distribution becomes even more heterogeneous when we consider the different types of bifaces separately. The entire Sector A contains an average of 1 whole handaxe per m^2 (excluding fragments or preforms), while items in other categories amount to 0.2 items per m^2. However, the most interesting aspect of this differentiation is that practically all the bifaces that are not handaxes (with the exception of a large scraper) are concentrated in the more western part of Sector A, measuring about 10 m^2. Cleavers on flakes were documented in an area that is only 2 m^2, while two of the three trihedral picks were found next to each other, and the third only 3.5 m away. Lastly, another fact that draws attention is that 12 of the 14 handaxe preforms were found in an area of about 10 m^2, whereas five of the six handaxe fragments were located in an area of 9 m^2, but both clusters lay inside the area of maximum biface concentration. Therefore, the main area of biface activity would be situated in the central and western part of Sector A (X56-62/Y95-99, cf. Fig. 8.15).

8.3 Discussion

8.3.1 *Raw Materials and Bifacial Shaping*

As we have seen, the production of bifacial tools at TK depended largely on the specific morphology of the selected blank. The most frequently used raw material was quartzite from the Naibor Soit inselberg. When the TK site was established, the base of this mountain was probably around 750 m away (Santonja et al. 2014: 204, Fig. 27), but the deposits that later accumulated on Bed II might hide outcrops situated even closer to the site.

This quartzite, which is made up almost entirely of mineral quartz (Santonja et al. 2014: 188, Table 3), surfaces as outcrops. Its inner structure consists of slabs of variable width, which enabled knappers to select as blanks fragments

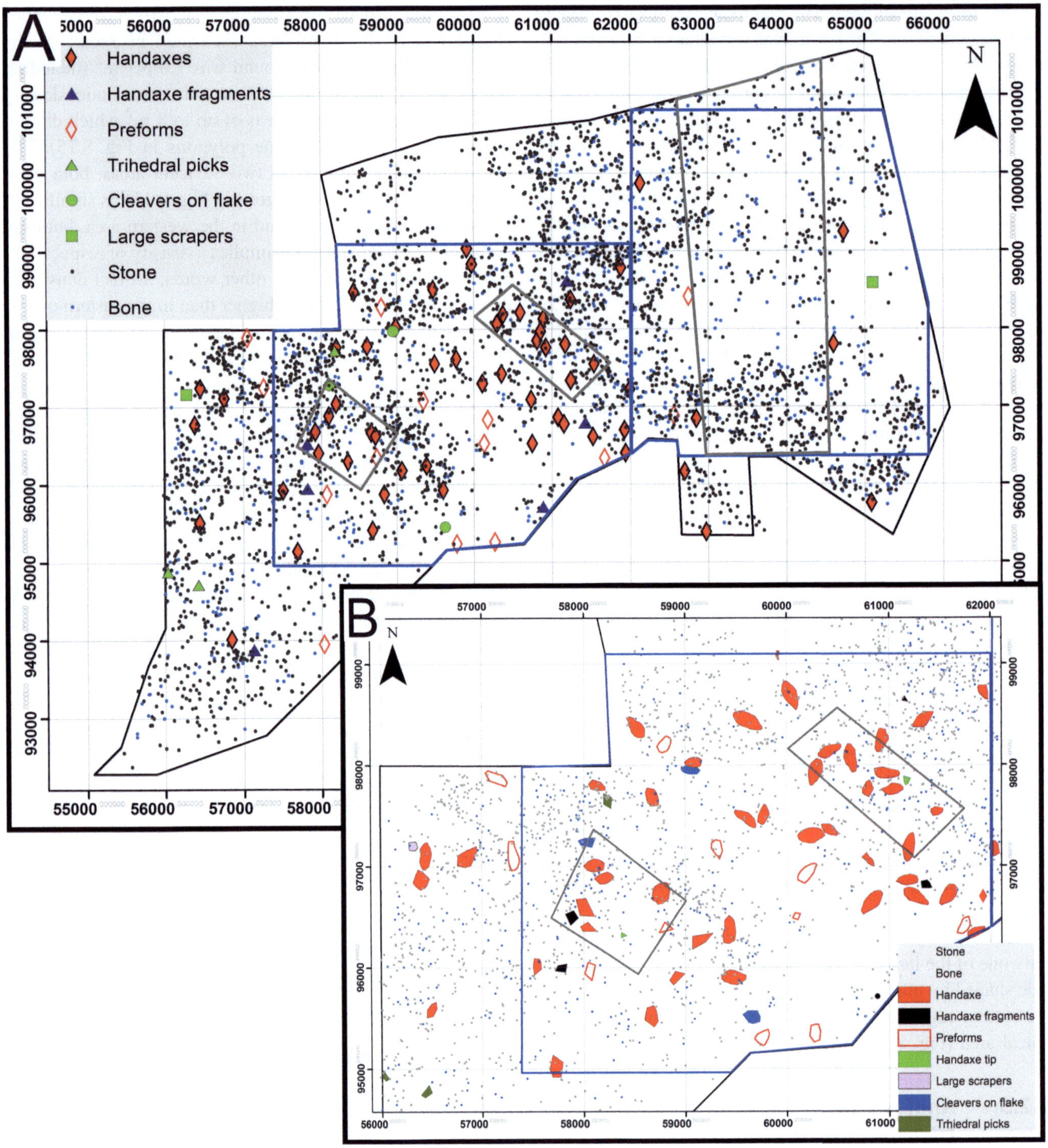

Fig. 8.15 **A** Spatial distribution of LCTs in TKLF Sector A. **B** Enlargement of central area with the largest concentration of LCTs

whose thicknesses would be most suitable for the tools that they were going to manufacture, thus diminishing or even eliminating the need of the first knapping phase—i.e., volumetric reduction. The selected blank already had two symmetrical surfaces parallel to the equatorial plane. The formal symmetry of this slab is comparable to that of the flakes, which made bifacial reduction easier (Gallotti et al. 2014). To obtain the desired shape of the tool and to fashion its cutting edges, knappers used the same balancing and bilateral reduction process.

An initial volumetric reduction phase has been documented in some specimens. It is particularly intense in four of the six VR handaxes (Fig. 8.6: A, C; Fig. 8.9: J; Fig. 8.12: G); it had not been necessary in the other two because they were made on flakes that were not very thick. This initial process is likewise observed in 10 handaxes made on NQ slabs, where it was employed to thin their apical third (Fig. 8.7: A-A1, D-D1, F-F1; Fig. 8.8: A-A1, I; Fig. 8.10: A-A1; Fig. 8.11: D, F, G; Fig. 8.12: A-A1). The most important façonnage phase in the bifaces we studied is bilateral shaping, especially in the handaxes, but also in cleavers, trihedral picks, and large scrapers. We have seen that it was responsible for the formal definition of practically the entire handaxe assemblage, though in four pointed handaxes it was concentrated only in the apical third. This shaping is mostly bifacial, completed with unifacially knapped segments and occasionally by bipolar knapping that provided backs that made it easier to grasp the tool (Fig. 8.9: B, D; Fig. 8.10: A-A1, D, E; Fig. 8.11: H; Fig. 8.13: T). Formal patterns are repeated in the studied series, even originating specimens that are practically identical to each other (Fig. 8.12: A-A1 and Fig. 8.12: C-C1, Fig. 8.12: E-E1 and Fig. 8.12: I, Fig. 8.7: A-A1 and Fig. 8.7: B-B1), thus proving that the bilateral shaping processes were aimed at adapting to predefined formats.

In more than half of the handaxes included in the series, the silhouette had been adjusted and corrected, particularly in the parts that seem to be more significant from the functional standpoint, such as the apical and basal areas or on the sides. This was the case for 12 of the 17 pointed handaxes with well-defined silhouettes (Fig. 8.6: A, B; Fig. 8.7: A-A1, C-C1, D-D1, F-F1; Fig. 8.8: A-A1, G-G1, H-H1, I, J; Fig. 8.9: J) and for 9 of the remaining 21 (Fig. 8.9: A, B, C, F, K, L, M; Fig. 8.10: B-B1, D); for 7 of the 10 oval-shaped handaxes (Fig. 8.11: A, B, E-E1, G, H, I, J); and for 1 of the 6 transverse edge handaxes (Fig. 8.12: C-C1). In a good number of cases, the apical end was shaped by removals from the whole periphery (cf., i.e., Figs. 8.7, 8.8, 8.9). Particularly striking are the cases in which this preparation was carried out in an axial direction, including with bifacial removals (Fig. 8.7: C-C1; Fig. 8.8: E, H-H1, J; Fig. 8.9: A, E; Fig. 8.11: A, I, J; Fig. 8.13: D, H). On some occasions this point-shaping process was performed during the initial phase, or at least before bilateral shaping, as may clearly be observed in several preforms (Fig. 8.13: H, K).

The use of large flakes to manufacture handaxes is infrequent (Table 8.2). It has been observed in six specimens; four in VR (Fig. 8.9: F, J; Fig. 8.10: C-C1; Fig. 8.12: G), and two were flakes obtained from a NQ slab (Fig. 8.8: D; Fig. 8.11: E-E1). These blanks did not require significant volumetric reduction, except in the cases of thick VR flakes. Unlike the slabs, the two main surfaces in the flakes are

Fig. 8.16 Hammerstones, and a selection of flakes close to completion and of residual bits of stone (shatter) produced when a biface is knapped on a NQ slab

secant to each other and determine a perimetral cutting edge along almost the entire contour. In these specimens, bilateral shaping had been used to balance the tool's silhouette. In general, and especially in NQ flakes, no thinning had been needed to obtain the edge, as had been the case with the slabs and cobbles.

The existence of forms of transition between knapping schemes for cleavers on flakes and handaxes is significant. One interesting transverse edge handaxe originated from a cleaver whose natural edge was transformed by bilateral knapping (Fig. 8.12: G). The silhouette of another cleaver on flake (Fig. 8.10: C-C1) was totally reshaped, so that the object was turned into a pointed handaxe. Both items confirm the versatility of the toolmakers' technological resources, which did not create separate compartments and admitted transformations from one kind of tool to another, in this case handaxes and cleavers on flake.

We recorded a total of 14 handaxes that were still in the process of being shaped (preforms), on both NQ (11 specimens) and VR cobbles (Fig. 8.13: E, F, G), which come near to the amygdaloid and ovaloid silhouettes and to the sizes of the finished specimens in the series. Their presence is a clear indicator that these tools were processed at the site itself, which proves that NQ and VR handaxes were fashioned and used right there. This is shown clearly by the use marks observed on the artifacts, especially all the rolling and wearing of the apical ends, and by the presence of fragments that broke off accidentally, particularly handaxe points.

8.3.1.1 Bifacial Shaping By-products

A significant number of the 1,421 flakes may have originated in the shaping of the handaxes and other bifaces. No characteristic flakes resulting from the knapping of handaxes have been identified. This is due to the fragility of the raw material: thin flakes fractured very easily, which makes it practically impossible to recognize them, as we have verified experimentally both for VR and NQ (Fig. 8.16). Thus most of the flakes (754) lack a butt: the area that best records the distinctive criteria of handaxe flakes (Soressi and Geneste 2011: 345, Fig. 10).

More than half of the lithic assemblage recorded in the TKLF Sector A consists of slab fragments: 3,237 specimens between 5 and 140 mm long, which were detached during the knapping process, and to a lesser degree during their use. If we consider the thickness of the handaxes, the ranges in NQ and in VR are similar, respectively 30–79 mm and 36–71 mm (Table 8.3). Whereas the thickness of the NQ artifacts could have been determined by the slab blank, this is not the case for those made on VR. It is therefore likely that the optimal thickness of the handaxes would have been between 30 and 79 mm. Therefore, at least part of the 115 NQ slab fragments with thicknesses in this range could have come from the slab blanks used to make them. The absence of features that would make identification possible is due to the fact that fractures along orthogonal planes usually occur in NQ slabs during knapping.

8.3.2 Spatial Distribution of Bifaces

M. Leakey drew attention to the fact that six of the fifteen handaxes documented in the TKLF level of Trench I (immediately south of Sector A, which we are analyzing here) lay in pairs at about 30 cm from each other (Leakey 1971: 177, Fig. 80). Sector A contains several associations of two or more handaxes, or even of a handaxe with a trihedral pick or a large scraper. However, to determine the meaning of this circumstance—for example, activities in which two or more individuals took part together—we would need an analysis of the assemblage of the archaeological aggregate, combined with traceological and taphonomic studies which would fall outside the scope of this study.

If we consider that the bifaces were abandoned in the immediate vicinity of the place where they had last been used, the main area of activity of the TK sector studied here would be in the western polygon outlined in blue (Fig. 8.15), together with the sector located just west of it. In this total area of 21 m^2, the average number of bifaces is 3.3 per m^2. It is significant that within this area there are two smaller ones (gray-line polygons), each measuring 1 m^2 and containing 10 bifaces, next to areas up to 1.6 m^2 which do not contain any of these tools at all.

The proximity of the Naibor Soit outcrop would not only have facilitated the production of NQ bifaces; it could also have made it easy enough to simply discard tools that were no longer fit for their intended purpose (e.g., blunted edges, broken points). However, in some cases the apical ends had been resharped with zenithal blows (Fig. 8.7: C-C1, D-D1, F-F1; Fig. 8.8: A-A1, E, H-H1, I, J; Fig. 8.9: J); in others, cleavers had been turned into handaxes (Fig. 8.10: C-C1; Fig. 8.12: G). All these specimens were found inside the western blue-line polygon (Fig. 8.15).

8.3.3 Functionality of the Bifaces

The presence of bifaces has been explained as related to hunting, to butchering and processing the carcasses of large herbivores (Jones 1980, 1994; Schick and Toth 1993; Bello et al. 2009; Yravedra et al. 2010), and to processing plant foods (Clark 1975; Jones 1994). Very few microwear analyses have been made of early Acheulean artifacts (Keeley 1980), perhaps because well-preserved *in situ* collections are very rare (Diez-Martín and Eren 2012). Phytolith analyses, on the other hand, are starting to yield interesting findings, and constitute a promising line of research. At the ES2-Lepolosi site, in Peninj, the presence of phytoliths on the cutting edges of three lithic artifacts suggests that these tools were used in activities that involved woodworking (Domínguez-Rodrigo et al. 2001, 2009). In any case, this evidence suggests that bifaces could have had multiple uses within a complex web of different functional contexts (Diez-Martín and Eren 2012).

Analysis of their silhouettes and sections shows that the main objective in shaping these handaxes was to produce an active point. Most of them (49 out of 54) have at least one cutting edge, but without a study of use marks we cannot rule out that the aim was not solely to achieve bilateral symmetry. It is very likely that the bifacially shaped edges (30 out of 54) had a functional purpose besides achieving bilateral symmetry. The presence of backs in 19 of the handaxes is likewise significant. The backs could be related to the tool's graspability, but because most of the backs are natural—except for six cases where they were produced by bipolar knapping (sections A1, B1 and C1)—they may also result from a selection of blanks made in order to reduce the need of modification to a minimum.

Our conclusions regarding the bifaces from the TKLF differ from those reached by other scholars on the series found by M. Leakey. De la Torre and Mora (2005) point out that, with a few exceptions, the larger tools made on flakes were not actual handaxes. These scholars reduce the shaping process to a basic scheme they call rhomboidal. We have identified this scheme in only seven handaxes, and it corresponds to the bilateral shaping phase (Inizan et al. 1992). However, they do admit that the morphology of the NQ slabs requires minimal secondary retouch, and they consider that this reduction process comprises bifacial shaping, which they recognize in a TKLF sample containing only 11 handaxes (de la Torre and Mora 2013: Fig. 8.6A).

8.4 Conclusions

Sedimentary and post-sedimentary processes preserved the original assemblages, without causing any major displacement of the centimetric items (Santonja et al. 2014). Our study enables us to conclude that here in the TKLF, façonnage and débitage were separate production processes that were also linked to differentiated exploitation of the two main raw materials used, namely NQ and VR (cf. Boëda 1991). In this techno-economic context, the bifaces, and handaxes in particular, not only present highly defined features but also constitute the most distinctive industrial component of the Lower Occupation Floor.

In the excavated area we recorded the presence of handaxe preforms and fragments that had accidentally broken off during the knapping or use of these tools. These elements make it possible to conclude that most of these tools were manufactured here, were used there, and were later abandoned. We cannot rule out the possibility that some finished bifaces could have been brought to the site from somewhere else. In the excavated sector, we found no cores for the specimens made on large NQ and VR flakes—some of them considerably larger than 10 cm (Isaac 1972). These may have been produced near the areas where the raw material was procured. They may have been selected from the items obtained through an initial volumetric reduction of the blanks, then brought to the site. This sequence, which has been considered possible in other cases as well (Goren et al. 2008), would explain the absence of large cores in the TK and at other Acheulean sites in different regions (Diez-Martín et al. 2014; Gallotti et al. 2014).

The 112 NQ items per m^2 recorded in Sector A and amounting to 494.3 kg in 51.9 m^2 (i.e., almost 10 kg per m^2) constitute a very high concentration of finds (Isaac 1981). It implies a very positive balance between the amount of raw material brought to the site and the amount of products taken away (Schick 1987: 791). This figure alone suggests that this raw material must have been available quite close to the site (cf. Hay 1976: 182–186; Stiles 1991). At the time the site was established, the closest source of tabular NQ—the Naibor Ndogo inselberg—would have been about 750 m away (Santonja et al. 2014). However, the sediments that accumulated between the site and the slopes of Naibor Ndogo are now dozens of meters deep and may conceal even closer outcrops. Easy access to the stone most employed in this lithic industry is undoubtedly an important factor in explaining the site's location.

The bifacial reduction processes we documented also suggest interesting general conclusions. The volumetric reduction phases were intense on VR cobbles and occasionally on NQ slabs as well (Fig. 8.6). In many cases, the proportions of the NQ blanks allowed the toolmaker to skip the volumetric shaping phase and start shaping the bilateral symmetry of the tools right away. This made it possible to obtain optimal symmetries, sometimes over the entire artifact, sometimes only over the apical third, especially in some of the pointed handaxes. The production of the bifaces found in the TKLF involved a real economy of raw materials. The technology was adapted to the NQ blanks and was generally based on bilateral shaping. The results are comparable to those of the technology used on VR cobbles, which was more conventional and based on bifacial reduction. As regards the handaxes' functionality, we can only say that the most evident techno-functional unit is the apical end, to which we can add the bifacially shaped cutting-edge segments of 30 specimens. The cortical bases, part of the natural backs, and above all the contour areas shaped by bipolar percussion, may have constituted comprehensive techno-functional units.

The toolmakers' effective adaptation of the processes to exploit the characteristics of the raw material with the aim of manufacturing repeated shapes attests to their command of the bifacial reduction concept in the TKLF. Though the relevant percentages are lower, their command of the process of producing cleavers on flakes has likewise been recognized at the site.

The shapes observed in the TKLF handaxes can be divided into three groups, based on their silhouettes: pointed, oval with peripheral edge, and rectangular with transverse edge (cf. Figures 8.7, 8.8, 8.9, 8.10, 8.11, and 8.12). Their sizes and weights are concentrated around clearly defined values (lengths from 177 to 278 mm, widths from 90 to 132 mm, thicknesses from 40 to 63 mm, weights from 985 to 2,251 g in 66.6% of the sample). Especially in the case of the pointed handaxes, which account for 70% of the assemblage we studied, the similarities are more evident and make it possible to believe with even greater confidence that these tools correspond to preconceived formal schemes, or "mental templates", devised according to the activities carried out at the site (Gowlett 1984: 185; Toth and Schick 2009: 272; *versus* McPherron 2000). In short, the TKLF was probably a site of specific activity, whose location not far from a lake and close to several streams would have been chosen for its easy access to food and water, and to types of stone suitable for manufacturing the tools needed. Its fluvial environment is similar to that of other areas of activity connected to the Acheulean techno-complex (Hay 1976: 181). Its toolkit is characterized by bifaces, many of them large and with very specific features that appear to be related to their functions. These tools were designed for specific tasks and would not have been as useful in a different context; thus, they were left at the site when it was abandoned. The situation observed in the TKLF argues for a close correlation between the bifaces and activities linked to this particular landscape, as proposed in the ecological hypothesis (Domínguez et al. 2005).

Our preliminary analysis of the distribution of the bifaces in the excavated sector raises an important question. These tools are not distributed evenly throughout the 51.9 m^2. They were concentrated in some areas and absent in others. This situation also occurs in the Trench I area (46.5 m^2) excavated by Leakey (1971: 177, Fig. 80b). The presence of bifaces may have been significantly different, depending on where they were found. If a smaller area had been excavated, it is possible that few or no bifaces would have been found in it. This fact would support a different interpretation of the site itself and even call its Acheulan nature into question. Only an analysis of representative areas would make it possible to formulate techno-economic interpretations of open-air Paleolithic sites. This conclusion requires one to be extremely cautious in evaluating the absence or scarce presence of bifaces in other sites of Bed II at Olduvai, in particular when the excavated areas are very small.

The dating obtained for the Tuff II^D (ca. 1.35 Ma) places the TKLF at a time that was not exactly at the beginning of the Acheulean techno-complex, a phase that today is generally limited to chronologies on the order of 1.7–1.5 Ma (c.f. Semaw et al. 2009; Diez-Martín and Eren 2012). It does, however, fall in the early Acheulean phase (Clark 1994; Schick and Toth 2001; Bar-Yosef 2006) with a temporal span that is currently considered to be between 1.5 and 1.0 Ma (Diez-Martín and Eren 2012; Sahnouni et al. 2013). In any case, the technological level of the façonnage schemes documented in the TKLF attests to a command of the bifacial shaping concept even in these early —albeit not initial—periods of the Acheulean techno-complex on the African continent. This conclusion challenges the expedience of the conventional division of the Acheulan into early, late (or middle), and terminal. Today this tripartite concept is ever more clearly perceived as being the echo of an outdated research phase (the fossil approach, Vega 2001; Monnier 2006) which takes the gradual improvement in the process of manufacturing handaxes to be Acheulean's main sequential element.

The situation we observed in the TKLF underlines the importance of understanding the sites as areas of activity. The identification of these activities and of the functionality of the site, which at present is our main objective in the different levels of the TK, could provide criteria for improving our knowledge of the evolution of behavioral patterns in the human groups that used Acheulean technology. As has been pointed out elsewhere (cf. Isaac 1984: 50; Semaw et al. 2009: 181), apart from any evaluation resulting from the apparently greater or lesser sophistication of the façonnage process, the role and usefulness of the bifaces in the whole context of activities carried out at the sites of Acheulean occupation could constitute a key element for identifying evolutionary phases or trends in the Acheulean techno-complex.

Acknowledgements We wish to thank our colleagues Rosalía Gallotti and Margherita Mussi for inviting us to take part in the conference on The Emergence of the Acheulean in East Africa, in honor of the 50th anniversary of the discovery of Melka Kunture (1963–2013), held in Rome in 2013.

Our work was carried out in the framework of The Olduvai Paleoanthropological and Paleoecological Project (TOPPP). Raquel Rojas-Mendoza made the drawings of the lithic artifacts that illustrate this paper.

We are grateful to the Spanish Ministry of Science and Technology (HAR2010-18952-C02-01 and HAR2013-45246-C3-2-P Projects) and to the Comunidad de Madrid, which, through its research and development programs (S2010/BMD-2330 R+D Project), funded this research.

References

Asfaw, B., Beyene, Y., Suwa, G., Walter, R., White, T., Wodegabriel, G., et al. (1992). The earliest Acheulean from Konso-Gardula. *Nature, 360,* 732–735.

Bar-Yosef, O. (1994). The Lower Palaeolithic of the Near East. *Journal of World Prehistory, 8,* 211–265.

Bar-Yosef, O. (2006). The known and the unknown about the Acheulean. In N. Goren-Inbar & G. Sharon (Eds.), *Axe Age. Acheulean Toolmaking from Quarry to Discard* (pp. 479–494). London: Equinox.

Bello, S. M., Parfitt, S. A., & Stringer, C. (2009). Quantitative micromorphological analyses of cut marks produced by ancient and modern handaxes. *Journal of Archaeological Science, 36,* 1869–1880.

Beyene, Y. (2003). The emergence and development of the Acheulean at Konso. *Anthropological Science, 111,* 58.

Boëda, E. (1991). Approche de la variabilité des systèmes de production lithique des industries du Paléolithique inférieur et moyen : chronique d'une variabilité attendue. *Techniques et Culture, 17–18,* 37–79.

Boëda, E. (2001). Détermination des unités techno-fonctionnelles de pièces bifaciales provenant de la couche acheuléenne C3 base du site de Barbas I. In D. Cliquet (Ed.), *Les industries à outils bifaciaux du paléolithique moyen d'Europe occidentale* (pp. 51–75). Liège: ERAUL 98.

Böeda, E., Geneste, J.-M., & Meignen, L. (1990). Identification de chaînes opératoires lithiques du Paléolithique ancien et moyen. *Paléo, 2,* 43–80.

Bower, J. R. F. (1977). Attributes of Oldowan and Lower Acheulean Tools: "tradition" and design in the Early Lower Palaeolithic. *The South African Archaeological Bulletin, 32,* 113–126.

Chavaillon, J., Chavaillon, N., Hours, F., & Piperno, M. (1979). From the Oldowan to the Middle Stone Age at Melka Kunture (Ethiopia): Understanding cultural changes. *Quaternaria, 21,* 87–114.

Chazan, M., Ron, H., Matmon, A., Porat, N., Goldberg, P., Yates, R., et al. (2008). Radiometric dating of the earlier Stone Age sequence in Excavation I at Wonderwerk Cave, South Africa: Preliminary results. *Journal of Human Evolution, 55,* 1–11.

Clark, J. D. (1970). *The Prehistory of Africa*. London: Thames and Hudson.

Clark, J. D. (1975). A comparison of the Late Acheulean industries of Africa and the Middle East. In K. Butzer & G. Isaac (Eds.), *After the Australopithecines* (pp. 605–660). The Hague: Mouton.

Clark, J. D. (1994). The Acheulean industrial complex in Africa and elsewhere. In R. Corruccini & R. Ciochon (Eds.), *Integrative Paths to the Past* (pp. 451–469). New Jersey: Prentice Hall.

Clark, J. D., & Kurashina, H. (1979). Hominid occupation of the east-central highlands of Ethiopia in the Plio-Pleistocene. *Nature, 282,* 33–39.

de la Torre, I. (2004). Estrategias tecnológicas en el Pleistoceno inferior de África oriental (Olduvai y Peninj, norte de Tanzania). Ph.D. Dissertation, Universidad Complutense.

de la Torre, I., & Mora, R. (2005). *Technological Strategies in the Lower Pleistocene at Olduvai Beds I and II.* Liège: ERAUL.

de la Torre, I., & Mora, R. (2013). The Transition to the Acheulean in East Africa: An Assessment of Paradigms and Evidence from Olduvai Gorge (Tanzania). *Journal of Archaeological Method and Theory, 21,* 781–823.

Diez Martín, F., & Eren, M. I. (2012). The Early Acheulean in Africa: Past paradigms, current ideas, and future directions. In M. Domínguez-Rodrigo (Ed.), *Stone Tools and Fossil Bones. Current Debates in the Archaeology of Human Origins* (pp. 310–357). Cambridge: Cambridge University Press.

Diez-Martín, F., Sánchez Yustos, P., Domínguez-Rodrigo, M., & Prendergast, M. E. (2011). An experimental study of bipolar and freehand knapping of Naibor Soit quartz from Olduvai Gorge (Tanzania). *American Antiquity, 76,* 690–708.

Diez-Martín, F., Sánchez-Yustos, P., Gómez de la Rúa, D., Gómez-González, J. A., de Luque, L., & Barba, R. (2014). Early Acheulean technology at Es2-Lepolosi (ancient MHS-Bayasi) in Peninj (Lake Natron, Tanzania). *Quaternary International, 322–323,* 209–236.

Domínguez-Rodrigo, M., Alcalá, L., Luque, L., & Serrallonga, J. (2005). Quelques aperçus sur les significations paléoécologique et comportementale des sites oldowayens anciens et acheuléens du Peninj (Upper Humbu Formation, Ouest du Lac Natron, Tanzanie). In M. Sahnouni (Ed.), *Le Paléolithique en Afrique. L'Histoire la plus longue* (pp. 129–156). Paris: Editions Artcom.

Domínguez-Rodrigo, M., Pickering, T. R., Baquedano, E., Mabulla, A., Mark, D. F., Musiba, C., et al. (2013). First partial skeleton of a 1.34-million-year-old *Paranthropus boisei* from Bed II, Olduvai Gorge, Tanzania. *PLoS ONE, 8,* e80347.

Domínguez-Rodrigo, M., Serrallonga, J., Juan-Treserras, J., Alcalá, L., & Luque, L. (2001). Woodworking activities by early humans: A plant residue analysis on Acheulean stone tools from Peninj, Tanzania. *Journal of Human Evolution, 39,* 421–436.

Domínguez-Rodrigo, M., Serrallonga, J., Luque, L., Diez-Martín, F., & Bushozi, P. (2009). The archaeology of the Acheulean sites from South Escarpment. In M. Domínguez-Rodrigo, L. Alcalá, & L. Luque (Eds.), *Peninj. A Research Project on the Archaeology of Human Origins (1995–2005)* (pp. 205–226). Oxford: Oxbow.

Domínguez-Rodrigo, M., Uribelarrea, D., Santonja, M., Bunn, H. T., García-Pérez, A., Pérez-González, A., et al. (2014). Autochthonous anisotropy of archaeological materials by the action of water: Experimental and archaeological reassessment of the orientation patterns at the Olduvai sites. *Journal of Archaeological Science, 41,* 44–68.

Gallotti, R., Raynal, J.-P., Geraads, D., & Mussi, M. (2014). Garba XIII (Melka Kunture, Upper Awash, Ethiopia): A new Acheulean site of the late Lower Pleistocene. *Quaternary International, 343,* 17–27.

Gibbon, R. J., Granger, D. E., Kuman, K., & Partridge, T. (2009). Early Acheulean technology in the Rieptus Formation, South Africa. *Journal of Human Evolution, 56,* 152–160.

Goren-Inbar, N., Sharon, G., Alperson-Afil, N., & Laschiver, I. (2008). The Acheulean massive scrapers of Gesher Benot Ya'aqov. A product of the biface chaîne opératoire. *Journal of Human Evolution, 55,* 702–712.

Gowlett, J. A. J. (1984). Mental abilities of early man: A look at some hard evidence. In R. Foley (Ed.), *Hominid Evolution and Community Ecology* (pp. 167–192). London: Academic Press.

Gowlett, J. A. (1986). Culture and conceptualization: The Oldowan-Acheulean gradient. In G. Bailey & P. Callow (Eds.), *Stone Age Prehistory: Studies in Memory of Charles McBurney* (pp. 243–260). Cambridge: Cambridge University Press.

Hay, R. L. (1976). *Geology of the Olduvai Gorge*. Berkeley: University of California Press.

Hay, R. L. (1990). Olduvai Gorge: A Case History in the Interpretation of Hominid Paleoenvironments in East Africa. *Geological Society of America Special Paper, 242,* 23–37.

Hodgson, D. (2009). Evolution of the visual cortex and the emergence of symmetry in the Acheulean techno-complex. *Comptes Rendus Palevol, 8,* 93–97.

Inizan, M.-L., Reduron, M., Roche, H., & Tixier, J. (1995). *Préhistoire de la pierre taillée 4.* Technologie de la pierre taillée (pp. 199). C.R.E.P.

Inizan, M.-L., Roche, H., & Tixier, J. (1992). *Technology of Knapped Stone*. Meudon: CREP, CNRS.

Isaac, G. Ll. (1972). Chronology and tempo of cultural change during the Pleistocene. In W. Bishop & J. Miller (Eds.), *Calibration of*

Human Evolution (pp. 381–430). Edinburgh: Scottish Academic Press.
Isaac, G. Ll. (1981). Stone Age Visiting Cards: Approaches to the study of early land-use patterns. In I. Hodder, G. Ll. Isaac, & N. Hammond (Eds.), *Patterns in the Past* (pp. 37–103). Cambridge: Cambridge University Press.
Isaac, G. Ll. (1984). The archaeology of human origins. Studies of the Lower Pleistocene in East Africa 1971–1981. In F. Wendorf & A. E. Close (Eds.), *Advances in World Archaeology* (pp. 1–87). Orlando (FL): Academic Press.
Jones, P. R. (1980). Experimental butchery with modern stone tools and its relevance for Palaeolithic archaeology. *World Archaeology, 12,* 153–165.
Jones, P. R. (1994). Results of experimental work in relation to the stone industries of Olduvai Gorge. In M. Leakey (Ed.), *Olduvai Gorge, 5. Excavations in Beds II, IV and the Masek Beds, 1968–1971* (pp. 254–298). Cambridge: Cambridge University Press.
Keeley, L. H. (1980). *Experimental Determination of Stone Tool Use: A Microwear Analysis*. Chicago: University of Chicago Press.
Kimura, Y. (2002). Examining time trends in the Oldowan technology at Beds I and II, Olduvai Gorge. *Journal of Human Evolution, 43,* 291–321.
Klein, R. G. (1999). *The Human Career: Human Biological and Cultural Origins* (2nd ed.). Chicago: University of Chicago Press.
Kleindienst, M. R. (1962). Components of the East African Acheulean assemblage: An analytic approach. In G. Mortelmans (Ed.), *Actes du IVème Congrès Panafricain de Préhistoire et de l'Étude du Quaternarie, 40* (pp. 81–105). Tervuren: Musée Royal de l'Afrique Centrale.
Kyara, O. A. (1999). *Lithic Raw Materials and Their Implications on Assemblage Variation and Hominid Behavior During Bed II, Olduvai Gorge, Tanzania.* New Brunswick (NJ): University of Rutgers.
Leakey, M. D. (1971). *Olduvai Gorge, vol. 3. Excavations in Beds I and II, 1960–1963.* Cambridge: Cambridge University Press.
Leakey, M. D. (1975). Cultural Patterns in the Olduvai Sequence. In K. W. Butzer & G. Ll. Isaac (Eds.), *After the Australopithecines. Stratigraphy, Ecology, and Cultural Change in the Middle Pleistocene* (pp. 477–493). Chicago: Mouton.
Leakey, M. D. (1976). The early stone industries of Olduvai Gorge, Tanzania. In D. Clark & G. Ll. Isaac (Eds.), *Les plus anciennes industries en Afrique. Union Internationales des Sciences Préhistoriques et Protohistoriques* (pp. 24–41). 9th Congrés. Nice: UISPP.
Leakey, M. D. (1978). Olduvai Gorge 1911–75: A history of the investigations. In W. W. Bishop (Ed.), *Geological Background to Fossil Man: Recent Research in the Gregory Rift Valley, East Africa* (pp. 151–155). Edinburgh: Scottish Academic Press.
Lepre, C. J., Roche, H., Kent, D. V., Harmand, S., Quinn, R. L., Brugal, J.-P., et al. (2011). An earlier origin for the Acheulean. *Nature, 477,* 82–85.
Ludwig, B. V., & Harris, J. W. K. (1998). Towards a technological reassessment of East African Plio-Pleistocene lithic assemblages. In M. D. Petraglia & R. Korisettar (Eds.), *Early Human Behaviour in Global Context. Rise and Diversity of the Lower Paleolithic Record* (pp. 84–107). London: Routledge.
McPherron, S. P. (2000). Handaxes as a measure of the mental capabilities of early hominids. *Journal of Archaeological Science, 27,* 655–663.
Manega, P. C. (1993). Geochronology, Geochemistry and Isotopic Study of the Plio-Pleistocene Hominid Sites and the Ngorongoro Volcanic Highland in Northern Tanzania. Ph.D. Dissertation, Boulder: University of Colorado.
Monnier, G. (2006). The Lower/Middle Paleolithic periodization in Western Europe. An evaluation. *Current Anthropology, 47,* 709–744.
Mourre, V. (2003). Implications culturelles de la technologie des hachereaux. Ph.D. Dissertation. University of Paris X, Nanterre.
Petraglia, M. D., & Potts, R. (1994). Water flow and the formation of early Pleistocene artifact sites in Olduvai Gorge, Tanzania. *Journal of Anthropological Archaeology, 13,* 228–254.
Quade, J., Levin, N., Semaw, S., Stout, D., Renne, P., Rogers, M., et al. (2004). Paleoenvironments of the earliest stone toolmakers, Gona, Ethiopia. *Geological Society of America Bulletin, 116,* 1529–1544.
Roche, H., & Kibunjia, M. (1994). Les sites archéologiques plio-pléistocènes de la Formation de Nachukui, West Turkana, Kenya. *Compte-reduns de l'Académie des Sciences, Paris, 318*(II), 1145–1151.
Roche, H., Brugal, J.-P., Delagnes, A., Feibel, C., Harmand, S., Kibunjia, M., et al. (2003). Les sites archéologiques plio-plésitocènes de la formation de Nachukui, Ouest-Turkana, Kenya: Bilan synthétique 1997–2001. *Comptes Rendus Palevol, 2,* 663–673.
Sahnouni, M. (1991). Étude comparative des galets taillés polyédriques, subsphériques et sphériques des gisements d'Ain Hanech (Algérie Orientale) et d'Olduvai (Tanzanie). *L'Anthropologie, 97,* 51–68.
Sahnouni, M., Semaw, S., & Rogers, M. (2013). The African Acheulean. An Archaeological Summary. In P. Mitchell & P. Lane (Eds.), *The Oxford Handbook of African Archaeology* (pp. 307–323). Oxford: Oxford University Press.
Sánchez-Yustos, P., Diez-Martín, F., Domínguez-Rodrigo, M., & Tarriño Vinagre, A. (2012). Discriminación experimental de los rasgos técnicos en la talla bipolar y a mano alzada en lascas a través de los cuarzos de Naibor Soit (Garganta de Olduvai, Tanzania). *Munibe, 63,* 5–26.
Santonja, M., Panera, J., Rubio-Jara, S., Pérez-González, A., Uribelarrea, D., Domínguez-Rodrigo, M., et al. (2014). Technological strategies and the economy of raw materials in the TK (Thiongo Korongo) lower occupation, Bed II, Olduvai Gorge, Tanzania. *Quaternary International, 322–323,* 181–208.
Schick, K. (1987). Modeling the formation of Early Stone Age artifact concentrations. *Journal of Human Evolution, 16,* 789–797.
Schick, K., & Toth, N. (1993). *Making Silent Stones Speak.* London: Phoenix.
Schick, K., & Toth, N. (2001). Paleoanthropology at the millennium. In G. Feinman & T. Price (Eds.), *Archaeology at the Millennium: A Sourcebook* (pp. 39–108). New York: Kluwer/Plenum.
Semaw, S., Rogers, M., & Stout, D. (2009). The Oldowan-Acheulean transition: Is there a "Developed Oldowan" tradition? In M. Camps & P. Chauhan (Eds.), *Sourcebook of Paleolithic Transitions. Methods, Theories and Interpretations* (pp. 173–193). New York: Springer.
Sharon, G. (2007). Acheulean Large Flake Industries: Technology, Chronology, and Significance. Oxford: Archaeopress, British Archaeological Reports International Series 1701.
Sharon, G. (2008). The impact of raw material on Acheulean large flake production. *Journal of Archaeological Science, 35,* 1329–1344.
Soressi, M., & Geneste, J.-M. (2011). The History and Efficacy of the Chaîne Opératoire Approach to Lithic Analysis: Studying Techniques to Reveal Past Societies in an Evolutionary Perspective. *PaleoAnthropology, 2011,* 334–350.
Stanistreet, I. G. (2012). Fine resolution of early hominin time, Beds I and II, Olduvai Gorge, Tanzania. *Journal of Human Evolution, 63,* 300–308.
Stiles, D. (1977). Acheulean and Developed Oldowan: The meaning of variability in the Early Stone Age. *Mila, 6,* 1–35.

Stiles, D. (1979). Early Acheulean and Developed Oldowan. *Current Anthropology, 20,* 126–129.
Stiles, D. (1991). Early hominid behaviour and culture tradition: Raw material studies in Bed II, Olduvai Gorge. *African Archaeological Review, 9,* 1–19.
Suwa, G., Asfaw, B., Haile-Selassie, Y., White, T., Katoh, S., WoldeGabriel, G., et al. (2007). Early Pleistocene *Homo erectus* fossils from Konso, Southern Ethiopía. *Anthropological Science, 11,* 133–151.
Tixier, J. (1956). Le hachereau dans l'Acheuléen Nord-Africain. Notes typologiques. *Congrès Préhistorique de France*, XVeme sessión (Poitiers), pp. 914–923.
Toth, N., & Schick, K. (2009). The Importance of Actualistic Studies in Early Stone Age Research: Some Personal Reflections. In K. Schick & N. Toth (Eds.), *The Cutting Edge: New Approaches to the Archaeology of Human Origins* (pp. 267–344). Gosport (IN): Stone Age Institute Press.
Turq, A. (2000). Paléolithique inférieur et moyen entre Dordogne et Lot. *Paléo* (Supplément 2).
Vega, L. G. (2001). Aplicación de la metodología de los programas de investigación al análisis historiográfico del Paleolítico. *Complutum, 12,* 185–215.
Wenban-Smith, F. (1998). Clactonian and Acheulean industries in Britain: Their chronology and significance reconsidered. In N. Ashton, F. Healy, & P. Pettitt (Eds.), *Stone Age Archaeology. Essays in honour of John Wymer* (pp. 90–104). Oxford: Oxbow Monograph.
Willoughby, P. R. (1987). *Spheroids and Battered Stones in the African Early Stone Age.* Oxford: Archaeopress, BAR International Series 321.
Wynn, T. (1989). *The Evolution of Spatial Competence.* Urbana: University of Illinois Press.
Wynn, T. (2002). Archaeology and cognitive evolution. *Behavioural and Brain Sciences, 25,* 389–402.
Yravedra, J., Domínguez-Rodrigo, M., Santonja, M., Panera, J., Rubio-Jara, S., Pérez-González, A., et al. (2016). The larger mammal palimpsest from TK (Thiongo Korongo), Bed II, Olduvai Gorge, Tanzania. *Quaternary International, 417,* 3–15.
Yravedra, J., Domínguez-Rodrigo, M., Santonja, M., Pérez-González, A., Panera, J., Rubio-Jara, S., et al. (2010). Cut marks on the Middle Pleistocene elephant carcass of Áridos 2 (Madrid, Spain). *Journal of Archaeological Science, 37,* 2469–2476.

Chapter 9
Faunal Change in Eastern Africa at the Oldowan – Acheulean Transition

Denis Geraads

Abstract The Early Pleistocene Transition from the Oldowan to the Acheulean in eastern Africa was roughly contemporaneous with a number of other events commonly assumed to be connected with hominin evolution. I review here the large mammal evidence, well documented in several major eastern African sites. Definite conclusions are hard to reach because of temporal gaps in the fossil record, and very patchy history of many lineages, but I conclude that, although some groups do show some turnover during this period, most of them did not change more than before or after it. We may conclude that this cultural change did not seriously impact the faunal assemblage. In addition, we may surmise that, since climate change at this period, if any, did not seriously impact the fauna, it is unlikely to have played a major role in human evolution at that time.

Keywords Mammalian assemblage • Early Pleistocene • Ethiopia • Kenya • Tanzania

9.1 Introduction

The site of Melka Kunture, about 50 km south of Addis Ababa in the Ethiopian highlands, contains a number of archaeological floors ranging from the Oldowan to the Late Stone Age, which document several major technological transitions (Gallotti 2013 and references therein; see also www.melkakunture.it/index.html). The earliest of them, the Oldowan–Acheulean transition, is illustrated by the localities of Garba IV and Gombore I (Morgan et al. 2012; Gallotti 2013, and references therein). The various localities of Melka Kunture also yield numerous faunal remains that allow reconstructing part of the faunal assemblage living nearby, even though selection by hominins and/or taphonomic biases certainly altered the representation of the biocenosis (Geraads 1979, 1985a). This co-occurrence of artifacts and faunal remains offers a valuable opportunity to document their evolutions side by side, in the frame of eastern African prehistoric and faunal evolution.

9.2 A Period of Major Changes?

During the middle part of the Early Pleistocene, from c. 1.9 to 1.5 Ma (following Gibbard et al. 2010, I regard the Pleistocene as beginning at 2.6 Ma), a number of important changes in human taxonomic composition, anatomy, and behavior are observed in eastern Africa:

(1) Succeeding ancestral *Homo habilis* (FAD 2.8 Ma, Villmoare et al. 2015), new species of the genus *Homo* appeared more or less simultaneously in the Upper Burgi Member of the Koobi Fora Formation of Kenya, at slightly more than 2 Ma (Joordens et al. 2013). These are *H. rudolfensis* and the first representatives of the *erectus* group, usually called *H. ergaster* in eastern Africa. *Homo erectus* appeared soon afterward if its earliest example is KNM-ER 2598 from just below the KBS tuff (Lepre 2014). Taxonomic assignment of these early specimens is debated, but they seem to document hominin diversification, although the possibility that they merely illustrate intraspecific variation must be considered (Lordkipanidze et al. 2013).
(2) Brain size steadily increased in hominins starting in the earliest Pliocene, with no sudden jump in hominin endocranial volume during the 1.9–1.5 Ma period (contra Maslin et al. 2014). Early *Homo* was not very different from *Australopithecus* in endocranial volume,

D. Geraads (✉)
Centre de Recherche sur la Paléobiodiversité et les Paléoenvironnements, Sorbonne Universités, MNHN, CNRS, UPMC, CP 38, 8 rue Buffon, 75231 Paris Cedex 05, France
e-mail: denis.geraads@mnhn.fr

R. Gallotti and M. Mussi (eds.), *The Emergence of the Acheulean in East Africa and Beyond: Contributions in Honor of Jean Chavaillon*, Vertebrate Paleobiology and Paleoanthropology,
https://doi.org/10.1007/978-3-319-75985-2_9

but what did happen at that time was a change in the slope of the time vs. volume curve. Appearance of some large brains was primarily linked with an increase in body mass (Ruff et al. 1997; Hublin et al. 2015).

(3) The *Australopithecus—Homo* transition was associated with a change in diet. On the whole, there was a trend toward incorporation of more and more ^{13}C-enriched food over time, but with considerable variation among regions and individual specimens (Cerling et al. 2013a; Sponheimer et al. 2013). As observed by these authors, in *Paranthropus* there is no doubt that this was linked to high reliance on C_4 plants, but in *Homo* it may be due to meat consumption of C_4-eating herbivores. The proportion of meat in the diet remains unknown, but Aiello and Wheeler (1995) and Braun et al. (2010) suggested that it was a contributing factor to increasing brain size.

(4) What we know of the cultural changes derives mostly from lithic technology, and the transition from the Oldowan to the Acheulean that took place during this interval is certainly a major one. It is first documented at Kokiselei 4 in West Turkana (Lepre et al. 2011) and Konso-Gardula in Ethiopia (Asfaw et al. 1992; Beyene et al. 2013), both at ca. 1.75 Ma; this transition is distinctly younger at Olduvai (de la Torre and Mora 2013). At Melka Kunture, it occurs at Garba IVD (Gallotti 2013).

(5) At about 1.8 Ma, hominins are first recorded in Eurasia, with Dmanisi in Georgia being the most securely dated site (Lordkipanidze et al. 2013). The beginning of the human expansion into Eurasia is often considered as a cultural event, but there is no evidence that it was, in contrast to more recent, deliberate migrations involving a human decision. The geographic ranges of all species expand and shrink for purely ecological or geographical reasons, and it is either a change of this nature or an improvement of their ecological flexibility that allowed humans to spread to Eurasia. The idea that climate change or failure to compete with Acheulean makers "forced" the makers of the Oldowan industry to leave Africa is biological nonsense: just like any other species, hominins expanded their range out of Africa as soon as they were able to do so.

(6) In addition, there were global and regional changes in the climate, vegetation, topography, and hydrological system. It is well known that global temperatures further dropped from their earliest Pleistocene values, the glaciated areas enlarged, and savannahs expanded at the expense of closed country (even though the change was neither regular, nor continuous, nor Pan-African). Maslin et al. (2014) noted that a number of eastern African lakes appeared or became deeper during this period, although Lepre (2014) questioned this dating and the correlation with other climatic events.

A recurrent issue is precise dating; there are a number of rich sites in eastern Africa, but the chronological calibration of some of the most relevant sites is still imperfect. This is true of Olduvai Bed II (in contrast to Bed I; Sanistreet 2012) or, at the time of writing, of the earliest sites of Melka Kunture. In addition, the definitions of *Homo* and its various species, or of the Acheulean, vary with authors. There is also evidence that some of the above-mentioned events are diachronic in eastern Africa (as noted, the Acheulean seems to appear later at Olduvai than at Konso and West Turkana). At least three of these events are definitely not synchronous, because the earliest *Homo* out of Africa, at Dmanisi and other sites, was still using an Oldowan technology and was still small-brained (Agustí and Lordkipanidze 2011; Lordkipanidze et al. 2013).

It is likely that these changes are interrelated to some extent and, although the hypothesis of a single triggering factor is probably an oversimplification, environmental changes are often regarded as heavily bearing on human evolution (Maslin et al. 2015, and references therein). If this is true, we may expect that the large mammal fauna of which hominins were part was similarly impacted, even if not necessarily to the same degree. We may also expect that emergence of new hominin species modified the ecosystem, including the fauna, and probably more and more so as technology and behavior allowed our ancestors to increase their impact on the environment.

Therefore, analyzing faunal changes help to shed light on various aspects on human biological and cultural evolutions. In particular, did faunas change more (or less) before or after this period than during it? If they did, were these changes synchronous with any of the other changes, and are they causally related, or at least correlated?

I shall review briefly below the main groups of large mammals, but it should be borne in mind that even though Early Pleistocene eastern African large mammal faunas are relatively well documented in sites such as Olduvai and Peninj in Tanzania, Rawi, Kanam, Koobi Fora, and West Turkana in Kenya, Omo, Konso-Gardula, and Melka Kunture in Ethiopia, a relatively complete fossil history is known for a few lineages only. Unfortunately, some rich sites (especially in the Middle Awash) are still mostly unpublished.

I have chosen not to put too much weight on frequently used observations such as first appearance datum (FAD) and last appearance datum (LAD). First, they are meaningful only for taxa that have a relatively continuous record. In addition, even for these taxa, it is usually hard to draw precise boundaries between closely related genera and species, so that in fact the only sharply defined events are FADs that correspond to immigration (e.g., *Equus*), and LADs that correspond to extinctions of a lineage. I believe that most often the use of these "data" is dictated more by the need to provide a basis to some objective treatment than by the wish to reflect biological reality.

9.3 A Review of the Evolution of the Main Groups

Carnivores are always rare, so that sampling certainly accounts for a large part of inter-site differences. If a turnover is to be recognized, it is probably c. 1.4 Ma rather than earlier, as the Okote Member of Koobi Fora is the youngest eastern African stratum with the large civet *Pseudocivetta* (also present at Garba IVE at Melka Kunture: Geraads et al. 2004a) and the otter *Torolutra* (Werdelin and Lewis 2013a). Saber-toothed felids of the genera *Megantereon* and *Homotherium*, as well as the "pseudo-machairodont" *Dinofelis* also went extinct in eastern Africa around that time (Werdelin and Peigné 2010). It is tempting to believe that the demise of these formidable predators left open a niche that *Homo* could partly fill (or, alternatively, that they failed in their competition with *Homo*), but *Homotherium* at least survived at Tighenif in Algeria until c. 1 Ma (Geraads 2016), and much later in Europe, showing that coexistence with our ancestors was possible. The only extinctions that occurred around 1.9 Ma were those of the large otters *Enhydriodon* and *Hydrictis gudho*, last recorded in the Upper Burgi Member of Koobi Fora (Werdelin and Lewis 2013a), but these taxa are quite rare. It is also around that time that the modern spotted hyena *Crocuta crocuta* replaced its smaller ancestor *C. dietrichi*, and became the most common large carnivore in Africa.

The date of appearance of the iconic African felids is not known with certainty. Cheetahs were present in the Pliocene, but it is likely that some of their highly specialized adaptations for fast running in open landscapes appeared more recently, perhaps in relation with savannah expansion (Geraads 2014). True leopards appeared in the Early Pleistocene of Olduvai Bed I, together with *Panthera leo*, although identifications at species level (Petter 1973) should perhaps be regarded with caution. They also coexisted with their saber-toothed cousins.

Werdelin and Lewis (2005, 2013b) and Lewis and Werdelin (2007) analyzed the evolution of eastern African carnivore assemblages in the Plio-Pleistocene and concluded that species richness decreased after a peak at 3 Ma, with a second, lower peak of diversity between 2 and 1.5 Ma, but acknowledged that this second peak is probably a sampling artifact due to the richly sampled Olduvai Bed I.

Ungulates make up the bulk of large mammal faunas, so that one would expect their chronological ranges to be better known, but species delineation is often imprecise. Because of their aquatic habits, hippos are among the most frequently encountered ungulates in eastern African, water-deposited open-air sites, but their species identification is difficult. This certainly explains apparent discrepancies in abundances in the Turkana basin, where they have been studied in some detail (Gèze 1985; Harris 1991a). The affinities of the most common lineage are still uncertain, so that it is better left as aff. *Hippopotamus* (Weston and Boisserie 2010). Aff. *H. protamphibius* is common throughout the Omo sequence; early representatives were hexaprotodont (with six lower incisors), but they progressively became tetraprotodont (four lower incisors) during the latest Pliocene. The canines remained relatively small. In the nearby Koobi Fora Formation, a contemporaneous tetraprotodont form has been called aff. *H. karumensis*, but we are probably dealing with the same lineage (Harris 1991b; Weston and Boisserie 2010). It evolved into a diprotodont form in the later part of the sequence, in the KBS Member of the Koobi Fora Fm, whose contemporaneous deposits are poorly fossiliferous at Omo (Shungura Fm Members H-J). Meanwhile, during the Early Pleistocene, new lineages of large hippos appeared and became dominant; they resemble the modern *Hippopotamus amphibius*, with large canines and four lower incisors. The most commonly reported Pleistocene eastern African form is *H. gorgops*; through time, its orbits and occipital became more and more elevated above the skull roof, obviously an adaptation to its aquatic life, but it is hard to understand why the closely related, contemporaneous *H. amphibius* did not develop these characteristics to the same degree. These lineages probably diverged in the Early Pleistocene, but the precise date will probably remain uncertain, as early representatives of both species cannot easily be told apart. Last, one (or more) species of pygmy hippo is known from the Turkana basin as late as the time of Shungura Member L and Okote Member, Olduvai Bed II (Harris 1991a), and Gombore I at Melka Kunture. In any case, there is no evidence of a major change in hippopotamid fauna in the 2–1.5 Ma range.

Suids have been much studied in eastern Africa (Harris and White 1978; Cooke 2007; Bishop 2010) but most lineages evolved continuously, with no major gap in morphology, and speciation events, either anagenetic or cladogenetic, are hard to date with precision, making attempts to define FADs somewhat arbitrary. In the genus *Kolpochoerus*, which some authors recognize as early as the Early Pliocene (but see Pickford 2012), the teeth remained relatively primitive, with premolars of normal size, and third molars that never reached the lengthening and degree of complexity seen in some other genera. The most common lineage is that of *K. limnetes* (or *K. heseloni* if the holotype of *K. limnetes* does not belong to this genus)—*K. olduvaiensis* (called *K. paiceae* by Souron 2012), the latter being the most derived member of the genus, with longer and more hypsodont third molars, although it was not more of a grazer (Harris and Cerling 2002) than another species, relatively common in Middle Pleistocene sites, *K. majus*, which is probably ancestral to the living giant forest hog, *Hylochoerus*. *Kolpochoerus majus* differs from the main lineage in retaining primitive features in the dentition, which looks like a mere enlarged version of Pliocene ones. This suggests an omnivorous diet, but the $\delta^{13}C$ of the only two analyzed specimens is close to 0 (Bedaso et al. 2010), as in pure grazers, showing that it probably fed mostly on grasses. The earliest record of *K. majus* is from the lower part of the Konso sequence, at c. 1.9 Ma (Suwa et al. 2003; Beyene et al. 2013), and it is perhaps also represented by fragmentary remains in the lower levels of Garba IV. Cooke (2007) and Souron (2012) place the transition from *K. limnetes* to *K. olduvaiensis* (or *K. paiceae*) at about the same period (Member J of the Shungura Fm at Omo, KBS Member of Koobi Fora). In any case, the transition between the two forms is gradual (Souron 2012, Fig. 2.18) and if *Kolpochoerus* changed its diet, it was certainly earlier than the Early Pleistocene (Souron 2012, Fig. 3.4).

The genus *Metridiochoerus* includes several species, one of which certainly gave rise to the modern warthog, *Phacochoerus*. The main species is *M. andrewsi*, known since the earliest Pleistocene, and common through most of the Early Pleistocene. Three other species, with hypsodont third molars, *M. modestus*, *M. hopwoodi*, and *M. compactus* have their FADs at about 2 Ma (Upper Burgi or KBS Members of Koobi Fora; uppermost part of Shungura Mb G). They were quite rare at that time, but may become more common later on, especially at Konso around 1.5 Ma (Suwa et al. 2014); they all survived into the Middle Pleistocene (Cooke 2007). The third molars of *M. compactus* consisted of tall simple pillars, as in the warthog, and this species was also remarkable in its huge canines. There is no doubt that these suids were, like *Kolpochoerus*, mostly grazers adapted to open environments, as also shown by their carbon isotopic values (Harris and Cerling 2002; Cerling 2015); thus, this grazing diet is at least 2 Ma old.

Ecologically, *Metridiochoerus* seems to replace *Notochoerus* that belongs to another subfamily but had a similar dental morphology, with longer and more hypsodont third molars than in *Kolpochoerus*, showing that environmental factors were perhaps less important than interspecific competition. *Notochoerus* becomes uncommon after c. 2 Ma, until its extinction c. 1.8 Ma (it is last recorded in the KBS Member of Koobi Fora and Shungura Member H; Cooke 2007). On the whole, the Suidae do show some turnover during the 2–1.5 Ma period, but none of the taxa replacements is of great magnitude.

Giraffids were not very abundant elements of the ungulate faunas in terms of number of individuals, but because of their large size they might have formed a significant part of the animal biomass. They include at least two species of *Giraffa*, of different sizes but morphologically similar to the modern giraffe, plus the genus *Sivatherium*, a very large form with buffalo-like proportions but as least twice as heavy (Harris et al. 2010a and references therein). It is known from the early Pliocene until the Middle Pleistocene at least, but it may be that it evolved toward more grazing habits (Cerling et al. 2015). Abundance data are lacking for most sites, but it was still common at Anabo Koma in Djibouti, at c. 1.5 Ma (Geraads 1985b), showing that it was not seriously affected by ecological changes in the course of the Early Pleistocene. The evolution of *Giraffa* is still imperfectly understood. *Giraffa jumae* has been reported throughout the Early Pleistocene; it may be that the transition between *G. gracilis* and the modern species *G. camelopardalis* occurred during the Early Pleistocene; it involved an increase in size and premolar molarization, but no major evolutionary change (Geraads et al. 2013).

No camel is known in eastern Africa between the earliest Pleistocene and latest Pleistocene (Harris et al. 2010b), but *Camelus* was always so rare there before historic times that this might be a sampling artifact.

The Bovidae (cattle, buffaloes, antelopes, goats, and gazelles) were by far the most common and most diversified group of large mammals in the African late Neogene, and for decades (Vrba 1995) they have repeatedly been used to examine effects of climate change. Most of them belong to tribes that mainly occur in Africa.

The Bovini are the main exception, as they are widespread in the Old World; they were chiefly represented by the genus *Syncerus* that includes the modern African buffaloes, and by its long-horned cousins of the genus *Pelorovis*. *Syncerus* was infrequent, but the history of *Pelorovis* is better known. The genus is first recorded from the top of the Hadar sequence (Geraads et al. 2012). Afterward, *P. turkanensis*, whose horns remain of moderate length, is progressively replaced by *P. oldowayensis*, whose horns may have spanned up to 3 m from tip to tip. It has been recorded from Middle Bed II of Olduvai and from the KBS Member of Koobi Fora (Gentry and Gentry 1978; Harris 1991b). At Melka Kunture, it is known from Simbiro, where it might have coexisted with *P. turkanensis*, also present in this locality although perhaps not in the same level. Besides these buffaloes, a close relative of the aurochs, of the genus *Bos*, is known from the latest Early Pleistocene of Buia in Eritrea (Martinez-Navarro et al. 2010) and the earliest Middle Pleistocene of Asbole in the Lower Awash (Geraads et al. 2004b) but this immigration of a Eurasian form remained localized.

Of the spiral-horned antelopes of the tribe Tragelaphini, large forms are better known than small ones, perhaps because their horn cores preserve better. The greater kudu, *Tragelaphus strepsiceros*, probably descended from the Hadar *T. lockwoodi*, is known throughout the Pleistocene (Gentry and Gentry 1978). Another kudu-like form, *T. gaudryi*, is known throughout the Omo sequence (Gentry 1985; Bibi and Kiessling 2015). The most significant event during the time interval considered here was that the lineage of the large *T. nakuae*, whose horns are only slightly spiraled and is abundant in the Plio-Pleistocene of Ethiopia and Northern Kenya, virtually disappeared without descent after the KBS Member of Koobi Fora (Harris 1991b); however, its extinction might be as late as Shungura Member J or K according to Bibi and Kiessling (2015). Geraads and Coppens (1995) showed that this species must have had habitat preferences similar to those of the impala, *Aepyceros*, the abundance of which seems to have been unaffected at that time. We may tentatively surmise that *T. nakuae* was replaced by the eland (*Taurotragus*), which appeared in the late Early Pleistocene (Gentry and Gentry 1978).

The smallest antelopes of the tribes Neotragini and Cephalophini (dik-diks and duikers) are so rare in most sites, certainly because of taphonomic or collecting biases, that no firm conclusion can be drawn from their presence/absence, but dik-diks may have been as common as today. This is not true of the gazelles (tribe Antilopini), a group with northern affinities, whose abundance in eastern Africa is recent; their fossil representatives in that region are not sufficiently well known to reconstruct the history of the genus *Gazella*. *Antidorcas recki* is the extinct eastern African form of the South African springbok; it is best known from Olduvai, but fluctuations of its abundance there were more related to local conditions than to chronology (Gentry and Gentry 1978). Other groups with northern affinities, the Caprini and Ovibovini, were extremely rare. One of the most commonly found antelopes is the impala, *Aepyceros*, with several species of rather uniform morphology. The modern species, *A. melampus*, prefers the woodland-savannah ecotone; it is absent from Ethiopia today, and from Melka Kunture, but there was no obvious trend in the abundance of this genus through the Pleistocene.

Reduncins include today the kobs, waterbucks, and relatives (*Kobus*) and reedbucks (*Redunca*). They are dwellers of wet grasslands, even swamps, so that their abundance is an indicator of moist, open environments, and varies therefore with the local context; no general trend can be recognized, at least until c. 1.5 Ma, after which time they were never the dominant bovid group again. No major change in taxonomic composition occurred in the first part of the Early Pleistocene, as the ancient forms *Kobus ancystrocera*, *K. sigmoidalis*, and *Menelikia* have been reported until at least the top of the Omo sequence. *Kobus sigmoidalis* has even been recorded from the Middle Pleistocene of Konso (Suwa et al. 2003) but its daughter species *K. ellipsiprymnus*, the modern waterbuck, appeared at c. 2 Ma (Gentry and Gentry 1978; Gentry 1985). The common kob *K. kob* is known throughout the period considered here, but becomes common only during its later part (see below).

Hippotragins (oryx, sable, and roan antelopes) were not very common as fossils and are not easy to identify to species. The most common one was *Hippotragus gigas*, known throughout the Olduvai sequence (Gentry and Gentry 1978) and still present at Buia at c. 1 Ma (Martinez-Navarro et al. 2004).

The alcelaphins include, among other species, the modern wildebeest, hartebeest, topi, and blesbok, living in large herds emblematic of the modern eastern African savannahs. They underwent a spectacular radiation in the Early Pleistocene, with probably at least 20 species, five or more of which being sometimes found in the same layers. However, it is only after c. 1.5 Ma that they are commonly the dominant antelopes, although relative abundances must be computed with caution because their teeth preserve better than those of other bovids (probably because they are more robust, but this important issue has never been investigated).

The wildebeest (*Connochaetes*) lineage dates from the earliest Pleistocene (Gentry 2010) and was first represented by a species with slender horn cores of more or less circular cross-section, of which a local subspecies is abundant at Garba IV at Melka Kunture, *C. gentryi leptoceras* (Geraads et al. 2004a). In the second half of the Early Pleistocene, around 1.5 Ma, it was replaced by its descendant with shorter, stouter horn cores, closer to the modern blue wildebeest, *C. taurinus* (Gentry 2010, and references therein).

Megalotragus was an extinct genus close to *Connochaetes* but of larger size and with less divergent horns, probably related to giant South African forms. With its two eastern African species, *M. kattwinkeli* and *M. isaaci*, it is known through the whole Early Pleistocene, without any obvious trend in its abundance (Harris 1991b; Gentry 2010).

Parmularius was another extinct genus, already present in the Pliocene, with a long face and rather short horn cores inserted far above the orbits, as is the case for the hartebeest. In the Pleistocene, it is best known from Olduvai, where both *P. rugosus* and *P. angusticornis* appeared in Lower Bed II, at c. 1.6 Ma, succeeding their likely ancestor *P. altidens* (Gentry and Gentry 1978; Gentry 2010).

Damaliscus remains imperfectly known, perhaps because its horn cores are less diagnostic than those of other alcelaphins. It certainly appeared in the Early Pleistocene, and two species with slightly spiraled horns were present around 1.5–1.7 Ma at Olduvai (*D. agelaius*) and Melka Kunture (*D. strepsiceras*); the "*Parmularius* cf. *pandatus*" from the lower part of the Konso sequence (Suwa et al. 2003) was perhaps related. Slightly later, after 1.5 Ma, a species with large, curved, but non-spiraled horns became relatively common until the Middle Pleistocene, *D. niro* (Suwa et al. 2003; Gentry 2010). The modern species are unknown before the Middle Pleistocene.

Beatragus, with its extremely long horns reminiscent of those of the impala, appeared at the Plio-Pleistocene boundary but is never common; it survives today as the endangered hirola, *B. hunteri* (Vrba 1997; Gentry 2010).

Numidocapra (including *Rabaticeras*) was a Pan-African but poorly known form with horn cores that were almost parallel at their bases but were slightly spiraled in a direction opposite to that of most antelopes (clockwise in the right horn). It is known at Olduvai and Anabo Koma in Djibouti (and Aïn Hanech in Algeria) at c. 1.7–1.5 Ma, but was too rare for its chronological range to be estimated (Bonis et al. 1988). It could be the ancestor of *Alcelaphus*, the hartebeests, which are now among the most abundant African antelopes but are not known before the Middle Pleistocene; the report of this genus from FwJj20 dated to c. 1.95 Ma (Braun et al. 2010) is probably erroneous.

On the whole, even though it may be that several species, especially alcelaphins, appeared at that time, the period from 2 Ma to 1.5 Ma clearly does not document a major change in the ecological preferences of bovid assemblages; changes in their taxonomic composition occur throughout this period, and if some turnover is to be recognized, it is more probably close to 1.6–1.5 Ma, i.e., after the appearance of the Acheulean.

The Perissodactyla, or odd-toed Ungulates, include the Equidae, Rhinocerotidae, and Chalicotheriidae (no tapirs are known in Africa). They passed the climax of their diversity long before the Pleistocene. Among the Equidae, the three-toed hipparions had been, together with antelopes, the dominant ungulates in the Late Miocene, but no longer played this role after the Pliocene, at which time their diversity had already decreased. Their systematics is still controversial (African forms are often called *Eurygnathohippus*, but the monophyly of this genus is disputable), and the fragmentary remains that form the bulk of the collected fossils do not help much in recognizing evolutionary events (Bernor et al. 2010, and references therein). They survived until the Early/Middle Pleistocene boundary at Gombore II at Melka Kunture (Geraads et al. 2004a) and at Konso (Suwa et al. 2003), but went extinct soon afterward. The FAD of the monodactyl equids of the genus *Equus* is the earliest Pleistocene at c. 2.2 Ma, but here again, a consensus on the species content has not been reached. After c. 2 Ma, *Equus* became more and more abundant relative to hipparions, but it would perhaps be naive to assume that the rise of *Equus* was the cause of the rarefaction and eventual extinction of the three-toed forms, which had been so successful and widespread for 10 My.

The rhinos were very similar to the modern *Diceros bicornis* ("black" rhino), a browser, and *Ceratotherium simum* ("white" rhino), a grazer, so that relative abundances of these genera provide indications on the vegetation. At the beginning of the Pleistocene, *Ceratotherium* still had relatively unspecialized teeth, prompting inclusion in a distinct species (*C. mauritanicum* in my opinion), but during the course of the Early Pleistocene hypsodonty increased and its molars became more similar to those of the modern species (Geraads 2010, and references therein). However it would be an oversimplification to link this change with an increasingly open habitat, because this evolution occurred later in North Africa although this region was probably still more open, because it was accompanied by a shortening of the metapodials unexpected in an open country form, and because isotopes show that Late Pliocene *Ceratotherium* was already a grazer (Bedaso et al. 2013; Cerling et al. 2015).

The last perissodactyl was the elusive chalicotheriid *Ancylotherium*, once called the "gorilla-horse", a survivor from the Miocene, with its long forearms and clawed paws, perhaps adapted to feeding on bark, lianas or soft trunks. Its last known occurrence is at Konso-Gardula, at c. 1.3 Ma (Suwa et al. 2003).

Of the Proboscidea, the most distinctive was *Deinotherium*, with its downturned lower incisors, no upper tusks, and lophodont molars resembling those of tapirs more than those of elephants, in accordance with a pure leaf diet. Little changed since the Miocene, it disappeared at c. 1.4 Ma in the Turkana basin (Harris et al. 1988; unpublished Omo catalogue). Elephants include the genus *Loxodonta*, to which the present-day African elephant belongs, but was a rare genus in the Pleistocene, and the much more common *Elephas* (or *Palaeoloxodon*), a relative of the present-day Asian elephant. The evolution of the Pleistocene species *Elephas recki* was chiefly marked by an increase in hypsodonty, and in the number of plates that make up the molars, allowing its use in biochronology. However, this evolution seems to have followed a gradualistic model with no recognizable bursts of evolution (Beden 1979, 1983). The species disappears from the African fossil record in the Middle Pleistocene.

Leaving aside rare groups like the hyraxes and aardvarks, the last taxon to be considered are the Primates. Apes (other than hominins) are unknown in the African Early Pleistocene but their virtual absence as fossils strongly suggests that a taphonomic factor, probably their forested environment, mitigated against fossilization, or that relevant sediments have not been found yet, but they were certainly present somewhere. The largest monkey was the terrestrial *Theropithecus oswaldi*, a relative of the modern gelada baboon of the Ethiopian highlands. Its lineage is documented since the Pliocene, with a regular evolution until its extinction toward the end of the Middle Pleistocene when it reached the size of a female gorilla (Jablonski and Leakey 2008; Jablonski and Frost 2010). Its diet incorporated more and more C_4 grasses through its history, but this change is continuous (Cerling et al. 2013b). Another species of the genus, *T. brumpti*, restricted to the Turkana basin, disappears after mid-Member G of the Shungura Fm and is unknown from the Upper Burgi Member at Koobi Fora, so its extinction is not posterior to 2 Ma (Frost 2007). Other baboons were less common and included *Parapapio*, which went extinct at c. 1.4 Ma after it gave rise to modern baboons of the genus *Papio*. It was also around that time, after the Okote Member of Koobi Fora, that the large terrestrial colobines *Cercopithecoides kimeui* and *C. williamsi* disappear, together with the arboreal *Rhinocolobus turkanensis* and the papionin *Lophocebus*, the latter now restricted to Western Africa (Jablonski and Leakey 2008; Jablonski and Frost 2010). *Paracolobus mutiwa* is a rare form known only from the Turkana basin. It is known at Omo only until Member G, but overlying sediments are poorly fossiliferous; at Koobi Fora there are only two specimens from the Upper Burgi, so that the LAD of 1.88 Ma (Frost 2007) is based upon slender evidence. *Colobus freedmani* Jablonski and Leakey 2008 is known by a few specimens from the Okote Member of Koobi Fora, c. 1.5 Ma. Thus, on the whole, although there is certainly some renewal of the cercopithecid fauna in the Early Pleistocene, it is not restricted to a particular period.

9.4 Quantitative Analysis

In addition to changes in taxonomic composition, faunal assemblages evolve through changes in abundances of the various taxa. In order to evaluate them, I have calculated the number of specimens of the main taxa of large mammals in the best-documented area, the Turkana basin, from roughly 2.2–1.3 Ma, using the available catalogues. I have divided the basin into its three main research areas and into three successive time periods (Fig. 9.1). I have excluded the carnivores that are too rare to be significant, and the hippos and proboscideans because they were not systematically collected. Still, it should be remembered that several biases may distort the results. A major one is the identification bias: some taxa have been studied in more detail in some areas than in others; some taxa were studied by different specialists who came to different conclusions; some taxa were not fully entered in the catalogue. Another one is the difference in collecting strategies, which may operate even within the same project area (Alemseged et al. 2007). Such biases certainly explain some of the most obvious gross differences, such as the seemingly high frequency of hominins in the Natoo Member, due to the virtual absence of other Primates in the database, or the high peaks of some bovid species. In addition, some of the apparent changes are restricted to some parts of the basin only and probably reflect local conditions only.

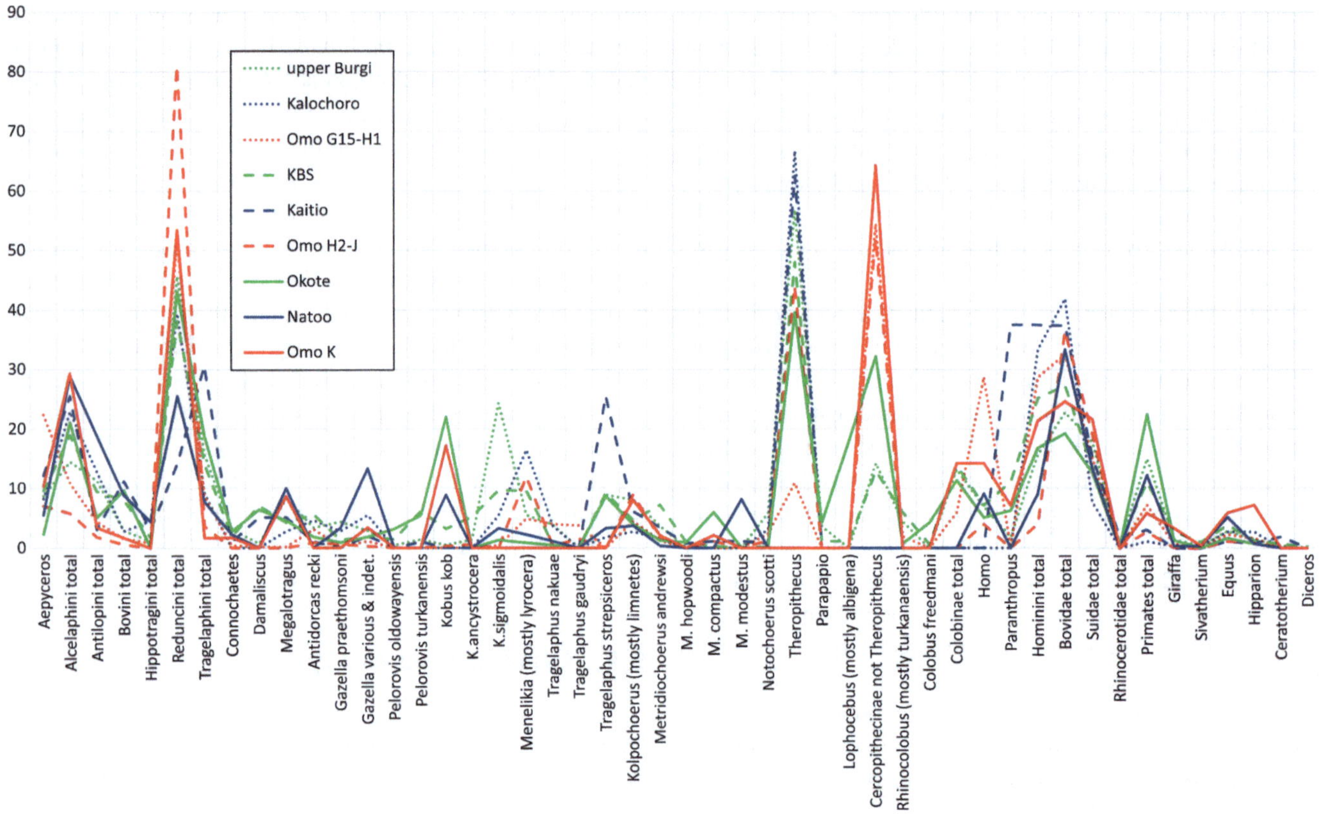

Fig. 9.1 Frequencies of the main taxa of large mammals in the Turkana basin in the Early Pleistocene. Values for individual bovid taxa are calculated as their percentage of all identified bovids; for suids, of all suids; for primates, of all primates; for other taxa (10 taxa to the right of the graph), of all large mammals excluding Proboscidea, Hippopotamidae, and Carnivora. Koobi Fora in green, West Turkana in blue, Omo Shungura in red. Each area has been divided into three successive time units that are roughly contemporaneous through the three areas; the oldest one is represented by a dotted line, the intermediate one by a dashed line, the youngest one by a continuous line, so that any trend in the whole basin should follow this order, irrespective of the color. Compiled from the Turkana database (courtesy of R. Bobe) and the Omo catalogue. The database contains no entry for non-hominin Primates in the Natoo Member, so that their number was artificially set to their average proportion in other members

Besides this, changes in abundances showing the same pattern through the whole basin are few. The sharpest one is the greater frequency of *Kobus kob* after c. 1.6–1.5 Ma, but other bovids do not display consistent patterns. Among suids, as noted above, the disappearance of *Notochoerus* in the youngest time slice is the terminal event of its earlier rarefaction; it may have made room for *Metridiochoerus compactus* that probably had similar ecological requirements. On the whole, however, Fig. 9.1 shows that even within this single basin, differences between research areas are at least as important as those that occur through time.

This lack of major change during the 2–1.5 Ma period is in agreement with some recently published results. For instance, Hakala (2012) calculated that from the Upper Burgi to the KBS and Okote Member at Koobi Fora, the proportions of the bovid tribes remained more or less the same, with only a slight increase of alcelaphins and a slight decrease of antilopins; these members are also very close in a factor analysis (Bobe et al. 2007). Similarly, Itoh et al. (2015) note that "These characteristics and size of the *K. sigmoidalis* horn cores from the Konso Formation were relatively constant during the Konso Formation range from interval 1 ($\sim$1.9 Ma) to interval 5 ($\sim$1.3 Ma)". On the whole, bovids suggest a rather steady environment over that period of time. From a detailed analysis of abundances and appearance/extinction patterns in eastern Africa, Frost (2007) concluded that the period between 3.5 and the early Middle Pleistocene, noticeable by the abundance of *Theropithecus*, is relatively stable (Frost 2007, Figs. 10, 11), except for a modest renewal at c. 2 Ma (Delson 1984; Frost 2007, Fig. 7). Cerling et al. (2015) analyzed isotopic values of a large number of mammals across the Plio-Pleistocene and placed the major change at c. 2.35 Ma. Last, Bibi and Kiessling (2015) concluded that "speciation and extinction proceeded continuously throughout the Pliocene and Pleistocene". They add that "a single origination pulse may be present at 2.0–1.75 Ma", but acknowledge that this pulse might be an artifact of the fossil record, in agreement with the above-mentioned lack of fossiliferous strata below Olduvai Bed I and the Upper Burgi Member.

9.5 Conclusion

As noted above, the quality of the fossil record is highly dependent on the existence and richness of fossil sites, so that FADs and LADS often reflect these parameters more than real evolutionary events (see, e.g., Behrensmeyer et al. 1997). In particular, appearance of several taxa at c. 2–1.8 Ma is obviously correlative of the existence of the rich deposits of the KBS Member and Olduvai Bed I, while the apparent extinctions at c. 1.5–1.4 Ma correspond with the end of the fossiliferous sequences of Omo Shungura and Koobi Fora. Attempts to estimate appearance or extinction dates and faunal turnover can note this incompleteness of the fossil record, but there is no way to reconstruct missing data.

What emerges from the above review, however, is that the transition from the Oldowan to the Acheulean is not associated with a major turnover of large mammal faunas. Kovarovic et al. (2013) have also shown that at Olduvai, faunas from the lower Bed II at c. 1.75 Ma do not much differ from those above the Lemuta tuff if the taphonomic difference in body weight profiles is taken into account; both mostly sampled woodland habitats. We may conclude, not only that this cultural transition was not triggered by a change of faunal assemblages, but also that the cultural transition did not seriously impact the large mammal assemblage as a whole. It may be, however, that the change in behavior associated with Oldowan–Acheulean transition had a significant influence on some particular taxa, and that certain ungulate species were exterminated during this period (even if this is unlikely). At a somewhat larger scale, Lewis and Werdelin (2007) link the decrease of carnivore diversity after 2 Ma to the entry of *Homo* into the carnivore guild and believe that, although our ancestors were certainly unable to compete directly with large predators, they may have interfered with them in scavenging activities (see, e.g., Ferraro et al. 2013).

Thus, it seems that there is no well-supported evidence for a link between the Oldowan–Acheulean transition and a faunal turnover. It is interesting to observe that Behrensmeyer et al. (1997) were also unable to find definite support for the widely accepted change between 2.8 and 2.5 Ma, corresponding to the emergence of *Homo*. This relative independence between the fauna and key events of human evolution is also apparent in the fact that the first known humans out of Africa were virtually unaccompanied by other large mammals, because the fauna from Dmanisi and other sites in the Caucasus, according to the few available publications (O'Regan et al. 2011; also Amirkhanov et al. 2014, and references therein), is almost totally Eurasian in character (so that it is hard to escape the conclusion that it is the ability to use stone tools that allowed *Homo* to explore this new environment).

If there is no major change in the composition of the large mammal assemblage, obviously this period is ill-suited to search for correlations between possible climate or environmental changes and biological evolution in this group, but in any case it does not provide support for an effect of external factors on its biological and cultural evolutions at that time. In this context, the challenge of finding an extrinsic cause to human diversification and biological and technological advances at that time remains open.

The impact of climate change on biological evolution is the current paradigm. A recent example is provided by Bibi and Kiessling (2015: 10626) who, without analyzing any extrinsic parameter, "conclude" that "global climate drove large mammal evolution at the million-year timescale". In fact, it is extremely hard to demonstrate. For instance, Faith and Behrensmeyer (2013) observe that "…problems of temporal and spatial scale are known to plague analyses of turnover relative to climate change, and this could explain why some studies fail to observe climatic effects on turnover". One might ask why such an elusive correlation is often taken for granted; it may be that the high concern about present-day climate change prompts researchers to overemphasize its impact on faunas, but it is also obvious that this high concern drives funding toward climate-related issues.

Acknowledgements I thank Rosalia Gallotti and Margherita Mussi for having invited me to contribute to this volume. The revised version of this chapter benefited from detailed comments by Laura Bishop and two anonymous reviewers. Thanks also to René Bobe for having allowed me to use the Turkana database.

References

Agustí, J., & Lordkipanidze, D. (2011). How "African" was the early human dispersal out of Africa? *Quaternary Science Reviews, 30*, 1338–1342.

Aiello, L. C., & Wheeler, P. (1995). The Expensive-Tissue Hypothesis the Brain and the Digestive System in Human and Primate Evolution. *Current Anthropology, 36*, 199–221.

Alemseged, Z., Bobe, R., & Geraads, D. (2007). Comparability of fossil data and its significance for the interpretation of hominin environments: a case study in the lower Omo valley, Ethiopia. In R. Bobe, Z. Alemseged, & K. Behrensmeyer (Eds.), *Hominin Environments in the East African Pliocene: An Assessment of the Faunal Evidence* (pp. 159–181). Dordrecht: Springer.

Amirkhanov, H. A., Ozherel'ev, D. V., Gribchenko, Y. N., Sablin, M. V., Semenov, V. V., & Trubikhin, V. (2014). Early Humans at the eastern gate of Europe: The discovery and investigation of Oldowan sites in northern Caucasus. *Comptes Rendus Palevol, 13*, 717–725.

Asfaw, B., Beyene, Y., Suwa, G., Walter, R. C., White, T. D., WoldeGabriel, G., et al. (1992). The earliest Acheulean from Konso-Gardula. *Nature, 360,* 732–735.

Bedaso, Z., Wynn, J. G., Alemseged, Z., & Geraads, D. (2010). Paleoenvironmental reconstruction of the Asbole fauna (Busidima Formation, Afar, Ethiopia) using stable isotopes. *Geobios, 43,* 165–177.

Bedaso, Z., Wynn, J., Alemseged, Z., & Geraads, D. (2013). Dietary and paleoenvironmental reconstruction using stable isotopes of herbivore tooth enamel from middle Pliocene Dikika, Ethiopia: Implication for *Australopithecus afarensis* habitat and food resources. *Journal of Human Evolution, 64,* 21–38.

Beden, M. (1979). Les Eléphants (*Loxodonta* et *Elephas*) d'Afrique Orientale : systématique, phylogénie, intérêt biochronologique. Ph. D. Dissertation, Université de Poitiers, France.

Beden, M. (1983). Family Elephantidae. In J. M. Harris (Ed.), *Koobi Fora Research Project, Vol. 2: The fossil Ungulates: Proboscidea, Perissodactyla, and Suidae* (pp. 40–129). Oxford: Clarendon Press.

Behrensmeyer, A. K., Todd, N. E., Potts, R., & McBrinn, G. (1997). Late Pliocene turnover in the Turkana Basin, Kenya and Ethiopia. *Science, 278,* 1589–1594.

Bernor, R. L., Armour-Chelu, M. J., Gilbert, H., Kaiser, T. M., & Schulz, E. (2010). Equidae. In L. Werdelin & W. J. Sanders (Eds.), *Cenozoic Mammals of Africa* (pp. 685–721). Berkeley: University of California Press.

Beyene, Y., Katoh, S., WoldeGabriel, G., Hart, W. K., Uto, K., Sudo, M., et al. (2013). The characteristics and chronology of the earliest Acheulean at Konso, Ethiopia. *Proceedings of the National Academy of Sciences, 110,* 1584–1591.

Bibi, F., & Kiessling, W. (2015). Continuous evolutionary change in Plio-Pleistocene mammals of Eastern Africa. *Proceedings of the National Academy of Sciences, 112,* 10623–10628.

Bishop, L. C. (2010). Suoidea. In L. Werdelin & W. J. Sanders (Eds.), *Cenozoic Mammals of Africa* (pp. 821–842). Berkeley: University of California Press.

Bobe, R., Behrensmeyer, A. K., Eck., G. G., & Harris, J. M. (2007). Patterns of abundance and diversity in late Cenozoic bovids from the Turkana and Hadar basins, Kenya and Ethiopia. In R. Bobe, Z. Alemseged, & A. K. Behrensmeyer (Eds.), *Hominin environments in the East African Pliocene: An assessment of the faunal evidence* (pp. 129–157). Dordrecht: Springer.

Braun, D. R., Harris, J. W. K., Levin, N. E., McCoy, J. T., Herries, A. I. R., Bamford, M. K., et al. (2010). Early hominin diet included diverse terrestrial and aquatic animals 1.95 Ma in East Turkana, Kenya. *Proceedings of the National Academy of Sciences, 107,* 10002–10007.

Cerling, T. E., Manthi, F. K., Mbua, E. N., Leakey, L. N., Leakey, M. G., Leakey, R. E., et al. (2013a). Stable isotope-based diet reconstructions of Turkana Basin hominins. *Proceedings of the National Academy of Sciences, 110,* 10501–10506.

Cerling, T. E., Chritz, K. L., Jablonski, N. G., Leakey, M. G., & Manthi, F. K. (2013b). Diet of *Theropithecus* from 4 to 1 Ma in Kenya. *Proceedings of the National Academy of Sciences, 110,* 10507–10512.

Cerling, T. E., Andanje, S. A., Blumenthal, S. A., Brown, F. H., Chritz, K. L., Harris, J. M., et al. (2015). Dietary changes of large herbivores in the Turkana Basin, Kenya from 4 to 1 Ma. *Proceedings of the National Academy of Sciences, 112,* 11467–11472.

Cooke, H. B. S. (2007). Stratigraphic variation in Suidae from the Shungura Formation and some coeval deposits. In R. Bobe, Z. Alemseged, & A. K. Behrensmeyer (Eds.), *Hominin Environments in the East African Pliocene: An Assessment of the Faunal Evidence* (pp. 107–127). Dordrecht: Springer.

de Bonis, L., Geraads D., Jaeger, J.-J., & Sen S. (1988). Vertébrés pléistocènes de Djibouti. *Bulletin de la Société géologique de France, sér. 8, 4,* 323–334.

de la Torre, I., & Mora, R. (2013). The Transition to the Acheulean in East Africa: An Assessment of Paradigms and Evidence from Olduvai Gorge (Tanzania). *Journal of Archaeological Method and Theory, 21,* 781–823.

Delson, E. (1984). Cercopithecid biochronology of the African Plio-Pleistocene: Correlation among eastern and southern hominid-bearing localities. *Courier Forschungsinstitut Senckenberg, 69,* 199–218.

Faith, J. T., & Behrensmeyer, A. K. (2013). Climate change and faunal turnover: Testing the mechanics of the turnover-pulse hypothesis with South African fossil data. *Paleobiology, 39,* 609–627.

Ferraro, J. V., Plummer, T. W., Pobiner, B. L., Oliver, J. S., Bishop, L. C., Braun, D. R., et al. (2013). Earliest Archaeological Evidence of Persistent Hominin Carnivory. *PLoS ONE, 8,* e62174.

Frost, S. (2007). African Pliocene and Pleistocene cercopithecid evolution and global climatic change. In R. Bobe, Z. Alemseged, & A. K. Behrensmeyer (Eds.), *Hominin Environments in the East African Pliocene: An Assessment of the Faunal Evidence* (pp. 51–76). Dordrecht: Springer.

Gallotti, R. (2013). An older origin for the Acheulean at Melka Kunture (Upper Awash, Ethiopia): Techno-economic behaviours at Garba IVD. *Journal of Human Evolution, 65,* 594–620.

Gentry, A. W. (1985). The Bovidae of the Omo group deposits, Ethiopia. *Cahiers de Paléontologie – Travaux de Paléontologie est-africaine. T.1: Périssodactyles, Artiodactyles (Bovidae)* (pp. 119–191). Paris: CNRS.

Gentry, A. W. (2010). Bovidae. In L. Werdelin & W. J. Sanders (Eds.), *Cenozoic Mammals of Africa* (pp. 741–796). Berkeley: University of California Press.

Gentry, A. W., & Gentry, A. (1978). Fossil Bovidae (Mammalia) of Olduvai Gorge, Tanzania. *Bulletin of the British Museum of Natural History (Geology), Part I, 29,* 289–446; *Part II, 30,* 1–83.

Geraads, D. (1979). La faune des gisements de Melka-Kunturé (Ethiopie): Artiodactyles, Primates. *Abbay, 10,* 21–49.

Geraads, D. (1985a). La faune des gisements de Melka Kunturé (Ethiopie). In Fondation Singer-Polignac (Ed.), *L'environnement des Hominidés au Plio-Pléistocène* (pp. 165–174). Paris: Masson.

Geraads, D. (1985b). *Sivatherium maurusium* (Pomel) (Giraffidae, Mammalia) du Pléistocène de la République de Djibouti. *Paläontologische Zeitschrift, 59,* 311–321.

Geraads, D. (2010). Rhinocerotidae. In L. Werdelin & W. J. Sanders (Eds.), *Cenozoic Mammals of Africa* (pp. 669–683). Berkeley: University of California Press.

Geraads, D. (2014). How old is the cheetah skull shape? The case of *Acinonyx pardinensis* (Mammalia, Felidae). *Geobios, 47,* 39–44.

Geraads, D. (2016). Pleistocene Carnivora (Mammalia) from Tighennif (Ternifine), Algeria. *Geobios, 49,* 445–458.

Geraads, D., & Coppens, Y. (1995). Évolution des faunes de Mammifères dans le Plio-Pléistocène de la vallée de l'Omo (Ethiopie) : apports de l'analyse factorielle. *Comptes-rendus de l'Académie des Sciences, IIa, 320,* 625–637.

Geraads, D., Eisenmann, V., & Petter, G. (2004a). The Large Mammal Fauna of the Oldowayan sites of Melka-Kunturé, Ethiopia. In J. Chavaillon & M. Piperno (Eds.), *Studies on the Early Paleolithic site of Melka Kunture, Ethiopia* (pp. 169–192). Florence: Origines.

Geraads, D., Alemseged, Z., Reed, D., Wynn, J., & Roman, D. C. (2004b). The Pleistocene fauna (other than Primates) from Asbole, lower Awash Valley, Ethiopia, and its environmental and biochronological implications. *Geobios, 37,* 697–718.

Geraads, D., Bobe, R., & Reed, K. (2012). Pliocene Bovidae (Mammalia) from the Hadar Formation of Hadar and

Ledi-Geraru, Lower Awash, Ethiopia. *Journal of Vertebrate Paleontology, 32,* 180–197.

Geraads, D., Bobe, R., & Reed, K. (2013). Pliocene Giraffidae (Mammalia) from the Hadar Formation of Hadar and Ledi-Geraru, Lower Awash, Ethiopia. *Journal of Vertebrate Paleontology, 33,* 470–481.

Gèze, R. (1985). Répartition paléoécologique et relations phylogénétiques des Hippopotamidae (Mammalia, Artiodactyla) du Néogène d'Afrique Orientale. In Fondation Singer-Polignac (Ed.), *L'environnement des Hominidés au Plio-Pléistocène* (pp. 81–100). Paris: Masson.

Gibbard, P. L., Head, M. J., Walker, M. J. C., & The Subcommission on Quaternary Stratigraphy. (2010). Formal ratification of the quaternary system/period and the Pleistocene Series/Epoch with a base at 2.58 Ma. *Journal of Quaternary Sciences, 25,* 96–102.

Hakala, S. E. (2012). Distribution of mammalian fauna during the early Pleistocene of the Koobi Fora formation, East Turkana, Kenya. Master thesis, University of Georgia.

Harris, J. M. (1991a). Hippopotamidae. In J. M. Harris (Ed.), *Koobi Fora Research Project, Vol. 3: The fossil Ungulates: Geology, fossil Artiodactyls and palaeoenvironments* (pp. 31–85). Oxford: Clarendon Press.

Harris, J. M. (1991b). Bovidae. In J. M. Harris (Ed.), *Koobi Fora Research Project, Vol. 3: The fossil Ungulates: Geology, fossil Artiodactyls and palaeoenvironments* (pp. 139–320). Oxford: Clarendon Press.

Harris, J. M., & Cerling, T. E. (2002). Dietary adaptations of extant and Neogene African suids. *Journal of Zoology, 256,* 45–54.

Harris, J. M., & White, T. D. (1978). Evolution of the Plio-Pleistocene African Suidae. *Transactions of the American Philosophical Society, 69,* 1–128.

Harris, J. M., Brown, F. H., & Leakey, M. G. (1988). Stratigraphy and palaeontology of Pliocene and Pleistocene localities west of Lake Turkana, Kenya. *Contributions in Science, Natural History Museum of Los Angeles County, 399,* 1–128.

Harris, J. M., Solounias, N., & Geraads, D. (2010a). Giraffoidea. In L. Werdelin & W. J. Sanders (Eds.), *Cenozoic Mammals of Africa* (pp. 797–811). Berkeley: University of California Press.

Harris, J. M., Geraads, D., & Solounias, N. (2010b). Camelidae. In L. Werdelin & W. J. Sanders (Eds.), *Cenozoic Mammals of Africa* (pp. 815–820). Berkeley: University of California Press.

Hublin, J.-J., Neubauer, S., & Gunz, P. (2015). Brain ontogeny and life history in Pleistocene hominins. *Philosophical Transactions of the Royal Society B, 370,* 20140062.

Itoh, M., Nakaya, H., Asfaw, B., Beyene, Y., & Suwa, G. (2015). Early Pleistocene Reduncini (Bovidae) from the Konso Formation, Southern Ethiopia. *Abstracts of the SVP 75th meeting, Dallas, 149.*

Jablonski, N. G., & Leakey, M. G. (Eds.). (2008). *Koobi Fora Research Project, Vol. 6: The fossil monkeys.* San Francisco: California Academy of Sciences.

Jablonski, N. G., & Frost, S. (2010). Cercopithecoidea. In L. Werdelin & W. J. Sanders (Eds.), *Cenozoic Mammals of Africa* (pp. 393–428). Berkeley: University of California Press.

Joordens, J. C. A., Dupont-Nivet, G., Feibel, C. S., Spoor, F., Sier, M. J., Jeroen, H. J. L., et al. (2013). Improved age control on early *Homo* fossils from the upper Burgi Member at Koobi Fora, Kenya. *Journal of Human Evolution, 65,* 731–745.

Kovarovic, K., Slepkov, R., & McNulty, K. P. (2013). Ecological continuity between Lower and Upper Bed II, Olduvai Gorge, Tanzania. *Journal of Human Evolution, 64,* 538–555.

Lepre, C. J. (2014). Early Pleistocene lake formation and hominin origins in the Turkana-Omo rift. *Quaternary Science Reviews, 102,* 181–191.

Lepre, C. J., Roche, H., Kent, D. V., Harmand, S., Quinn, R. L., Brugal, J.-P., et al. (2011). An earlier origin for the Acheulian. *Nature, 477,* 82–85.

Lewis, M. E., & Werdelin, L. (2007). Patterns of change in the Plio-Pleistocene carnivorans of Eastern Africa: Implications for hominin evolution. In R. Bobe, Z. Alemseged, & A. K. Behrensmeyer (Eds.), *Hominin Environments in the East Africa Pliocene: An Assessment of the Faunal Evidence* (pp. 77–105). Dordrecht: Springer.

Lordkipanidze, D., Ponce de León, M. S., Margvelashvili, A., Rak, Y., Rightmire, P., Vekua, A., et al. (2013). A complete skull from Dmanisi, Georgia, and the evolutionary biology of early *Homo. Science, 342,* 326–331.

Martinez-Navarro, B., Rook, L., Segid, A., Yosief, D., Ferretti, M. P., Shoshani, J., et al. (2004). The large fossil mammals from Buia (Eritrea): Systematics, biochronology and paleoenvironments. *Rivista Italiana di Paleontologia e Stratigrafia, 110 suppl.,* 61–88.

Martinez-Navarro, B., Rook, L., Papini, M., & Libsekal, Y. (2010). A new species of bull from the Early Pleistocene paleoanthropological site of Buia (Eritrea): Parallelism on the dispersal of the genus *Bos* and the Acheulian culture. *Quaternary International, 212,* 169–175.

Maslin, M. A., Brierley, C. M., Milner, A. M., Shultz, S., Trauth, M. H., & Wilson, K. E. (2014). East African climate pulses and early human evolution. *Quaternary Science Reviews, 101,* 1–17.

Maslin, M. A., Shultz, S., & Trauth, M. H. (2015). A synthesis of the theories and concepts of early human evolution. *Philosophical Transactions of the Royal Society, 370,* 20140064.

Morgan, L. E., Renne, P. R., Kieffer, G., Piperno, M., Gallotti, R., & Raynal, J.-P. (2012). A chronological framework for a long and persistent archaeological record: Melka Kunture, Ethiopia. *Journal of Human Evolution, 62,* 104–115.

O'Regan, H. J., Turner, A., Bishop, L. C., Elton, S., & Lamb, A. (2011). Hominins without fellow travellers? First appearances and inferred dispersals of Afro-Eurasian large-mammals in the Plio-Pleistocene. *Quaternary Science Reviews, 30*(11–12), 1343–1352.

Petter, G. (1973). Carnivores pléistocènes du ravin d'Olduvai. In L. S. B. Leakey, R. J. G. Savage, & S. C. Coryndon (Eds.), *Fossil Vertebrates of Africa 3* (pp. 43–100). London: Academic Press.

Pickford, M. (2012). Ancestors of Broom's Pigs. *Transactions of the Royal Society of South Africa, 67,* 17–35.

Ruff, C. B., Trinkaus, E., & Holliday, T. W. (1997). Body mass and encephalization in Pleistocene *Homo. Nature, 387,* 173–176.

Sanistreet, I. A. (2012). Fine resolution of early hominin time, Beds I and II, Olduvai Gorge, Tanzania. *Journal of Human Evolution, 63,* 300–308.

Souron, A. (2012). Histoire évolutive du genre *Kolpochoerus* (Cetartiodactyla: Suidae) au Plio-Pléistocène en Afrique Orientale. Ph.D. Dissertation, Université de Poitiers.

Sponheimer, M., Alemseged, Z., Cerling, T. E., Grine, F. E., Kimbel, W. H., Leakey, M. G., et al. (2013). Isotopic evidence of early hominin diets. *Proceedings of the National Academy of Sciences, 110,* 10513–10518.

Suwa, G., Nakaya, H., Asfaw, B., Saegusa, H., Amzaye, A., Kono, R. T., et al. (2003). Plio-Pleistocene terrestrial mammal assemblage from Konso, Southern Ethiopia. *Journal of Vertebrate Paleontology, 23,* 901–916.

Suwa, G., Souron, A., & Asfaw, B. (2014). Fossil Suidae of the Konso Formation. *The University Museum, The University of Tokyo, Bulletin, 47,* 73–88.

Villmoare, B., Kimbel, W. H., Seyoum, C., Campisano, C. J., DiMaggio, E. N., et al. (2015). Early *Homo* at 2.8 Ma from Ledi-Geraru, Afar, Ethiopia. *Science*, *347*, 1352–1355.

Vrba, E. S. (1995). The fossil record of African antelopes (Mammalia, Bovidae) in relation to human evolution and paleoclimate. In E. S. Vrba, G. H. Denton, T. C. Partridge, & L. H. Burckle (Eds.), *Paleoclimate and Evolution, with Emphasis on Human Origins* (pp. 385–424). New Haven: Yale University Press.

Vrba, E. S. (1997). New fossils of Alcelaphini and Caprinae (Bovidae: Mammalia) from Awash, Ethiopia, and phylogenetic analysis of Alcelaphini. *Palaeontologia Africana, 24,* 127–198.

Werdelin, L., & Lewis, M. E. (2005). Plio-Pleistocene Carnivora of eastern Africa: Species richness and turnover patterns. *Zoological Journal of the Linnean Society, 144,* 121–144.

Werdelin, L., & Lewis, M. E. (Eds.). (2013a). *Koobi Fora Research Project. Vol. 7: The Carnivora.* San Francisco: California Academy of Sciences.

Werdelin, L., & Lewis, M. E. (2013b). Temporal Change in Functional Richness and Evenness in the Eastern African Plio-Pleistocene Carnivoran Guild. *PLoS ONE, 8,* 1–11.

Werdelin, L., & Peigné, S. (2010). Carnivora. In L. Werdelin & W. J. Sanders (Eds.), *Cenozoic Mammals of Africa* (pp. 603–658). Berkeley: University of California Press.

Weston, E., & Boisserie, J.-R. (2010). Hippopotamidae. In L. Werdelin & W. J. Sanders (Eds.), *Cenozoic Mammals of Africa* (pp. 853–871). Berkeley: University of California Press.

Chapter 10
The Acheulean Assemblages of Asia: A Review

Robin W. Dennell

Abstract Acheulean assemblages—defined by the presence of handaxes and cleavers—are found across much of Asia. The best known are from the Levant and India and date from the Early Pleistocene. Although bifaces have been found in other parts of Asia, they are poorly dated but probably mostly Middle Pleistocene in age. In East Asia, the Movius Line as originally formulated is invalid because Acheulean, bifacial assemblages are present in China as well as the Korean Peninsula. Nevertheless, there are significant differences between some of these assemblages and those from west and south Asia. Problems of dating and differing definitions of "the Acheulean" are current impediments to establishing the spatial and temporal patterning of Acheulean assemblages in Asia. Additional major shortcomings are the lack of information on the climatic context of most Asian Acheulean assemblages, and the almost total absence of information on the identity and subsistence of their makers.

Keywords Asia • Early/Middle Pleistocene • Typology • Technology

10.1 Introduction

The discovery of a biface by the Scottish geologist Bruce Foote at Pallavaram, southern India, on May 30, 1863 (Kennedy 2003: 11), marks the beginning of our knowledge of the Asian Acheulean. Since then, there have been major discoveries of Acheulean assemblages—defined by the presence of handaxes and cleavers—in Israel, Syria, India, Saudi Arabia, and lesser ones in Turkey, Jordan, the Caucasus, and Iran. In Israel and India, the earliest Acheulean assemblages are dated in excess of 1 Ma. The discovery of bifacial assemblages in China and the Korean Peninsula that are similar to those further west has shown that the Movius Line is invalid for East Asia—at least in a strict sense of the term.

Before reviewing this evidence, three points need to be made about how assemblages are recognized as "Acheulean". The first concerns the status of surface finds. The evidence for "the Acheulean" in Asia comprises a small number of excavated Early and Middle Pleistocene localities and a large number of surface, and thus undated, finds of bifaces that are assumed from circumstantial geological evidence (e.g., being found on an eroding Middle Pleistocene surface), and/or their size and shape to be the same age as excavated specimens of Acheulean artifacts. This assumption is reasonable in areas such as the Levant and India, where several localities with Acheulean assemblages have been excavated, but is problematic (as seen below) in regions such as Central and Southeast Asia where few Paleolithic sites have been excavated. As a consequence, artifacts that are classified as Acheulean in these regions may in fact be much younger (and even Holocene), and conversely, specimens assumed to be geologically recent may be much older. The underlying problem is that, as pointed out by Brumm and Moore (2012), any technology involving bifacial flaking will result in some objects resembling Acheulean handaxes: as one example, recent rough-outs of stone axes were often mistaken for Acheulean handaxes in the early exploration of North America's prehistoric past (Meltzer 2015). Because typology is no guarantee of antiquity, statements about the geographical extent of Acheulean assemblages in Asia are at their most reliable when discussing areas where several localities have been excavated, and at their weakest when the only evidence comes from surface finds.

The second point is that researchers also disagree over the percentage or numbers of bifaces needed to classify an assemblage as Acheulean. For example, Leakey (1971) classified sites at Olduvai as Acheulean only when 40% of

R.W. Dennell (✉)
Department of Archaeology, University of Exeter, Exeter, EX4 4QE, UK
e-mail: r.w.dennell@exeter.ac.uk

R. Gallotti and M. Mussi (eds.), *The Emergence of the Acheulean in East Africa and Beyond: Contributions in Honor of Jean Chavaillon*, Vertebrate Paleobiology and Paleoanthropology,
https://doi.org/10.1007/978-3-319-75985-2_10

the artifacts in an assemblage were bifaces. If this criterion was applied to Asia, most of the evidence for the Acheulean would disappear as bifaces are usually present as a much smaller component of an assemblage. On the grounds that a biface, roughly symmetrical and with several well-directed flake removals, represents a higher level of skill than in simple, Oldowan-type core reduction, many researchers argue that only one such artifact is needed to label an assemblage as Acheulean (e.g., Kuman 1998: 175–177). She further emphasizes the necessity of large samples, as the likelihood of finding an example of the upper limits of stone flaking (in this case, an Acheulean biface) is much greater when sample sizes are large. Unfortunately, this is rarely the case in many parts of Asia, and the absence of bifaces may often simply reflect the small size of the assemblage.

The third point is that discussions of the Acheulean have been complicated by suggestions that it is primarily characterized by LCT's, or Large Cutting Tools more than 10 cm in length. As a result, some assemblages that lack handaxes and/or cleavers have been reclassified as Acheulean. Semaw et al. (2009), for example, argued that because Oldowan B assemblages in East Africa contain LCT's, they should be regarded as Acheulean. Similarly, Mishra et al. (2010) has proposed that a small assemblage from Ngebung, Sangiran, Java should be classified as a Large Flake Acheulean even though handaxes are absent. In this chapter, I concentrate on those Pleistocene assemblages defined as "Acheulean" by the presence of handaxes and/or cleavers.

Inevitably in a continent as vast and diverse as Asia, there is enormous variation in the quantity, quality, and tempo of research into the Early Paleolithic. Large areas are still largely uncharted, notably Turkey, Iran, Pakistan, Central and South East Asia, and in other areas, not a great deal of change has occurred in the last decade. The most exciting developments in recent years have occurred in Arabia, where a great deal has now been learnt; India, where the age of its early Paleolithic record has increased substantially through research at Attirampakkam; and China, where the presence of Acheulean bifacial assemblages is now clear.

This review will progress from west to east and begin by outlining the main evidence for Acheulean assemblages from West Asia. Dennell (2009) provides a fuller discussion of the Early Paleolithic of Asia, including most of the Acheulean material discussed in this paper.

10.2 West Asia

If we assume that the Acheulean originated in Africa (Lepre et al. 2011) and was adopted in Asia through immigration or cultural transmission (i.e., by copying neighboring groups), Arabia is likely the first region outside Africa to contain it.

10.2.1 *The Arabian Peninsula*

Until recently, the Arabian Peninsula—covering over a million square miles—has not played any significant role in Paleolithic discussions. This has been unfortunate because the Arabian Peninsula is the most obvious entry point to Asia from Africa, either across the Bab-el Mendab Strait at the southern end of the Red Sea, or via the Sinai Peninsula at the northern end (Beyin 2006; Derricourt 2006). Although the southern and western sides of the Arabian Peninsula are mountainous, the northern part unfolds gently toward southern Turkey, and east to the Arabian/Persian Gulf and the Zagros Mountains. It should thus be an area where populations could easily have moved over large distances.

The little that was known about the Early Paleolithic of Arabia came from surface (and therefore undatable) collections that did at least demonstrate the existence of Oldowan, Acheulean (Whalen and Peace 1990), Middle, and Upper Paleolithic assemblages, although there were no dated sequences, landscape studies, or detailed artifactual studies. Whalen and colleagues also excavated the first Acheulean sites near Saffaqah in the early 1980s (Whalen et al. 1983). Fortunately, our understanding of the Early Paleolithic of the Arabian Peninsula has been transformed in recent years thanks to international field programs, especially those led by researchers from Oxford. This research has shown that Acheulean material is widespread throughout western Arabia, but scarcer toward the Arabian/Persian Gulf. Perhaps the most remarkable discovery is that there has been a "Green Arabia" on numerous occasions throughout the Pleistocene (mainly, but not entirely during interglacials and interstadials), with the discovery of large-scale drainage systems (Fig. 10.1) and thousands of paleolakes, even in areas now as arid as the Rub al-Khali, or Empty Quarter (e.g., Shipton et al. 2014; Matter et al. 2015; Petraglia et al. 2015 and related papers in the Green Arabia volume). This means that not only was the Arabian Peninsula endowed with much greater bodies of water than now, but that dispersal into and across Arabia was considerably facilitated by the presence of large drainage systems that discharged into the Arabian/ Persian Gulf.

Acheulean assemblages were already known from previous surveys (e.g., Whalen et al. 1988, 1989; Whalen and Pease 1990). As summarized by Petraglia (2003: 171), "Acheulean sites in Arabia are numerous, widely dispersed, and occur in a variety of settings and environments including coastal zones, elevated zones in sub-mountainous regions, and in interior plains, occurring along river terraces and near lakeshores". Recent investigations have considerably enlarged this knowledge. One particularly important regional finding is that cleavers similar to those from Gesher Benot Ya'aqov, Israel, are present at several Arabian sites,

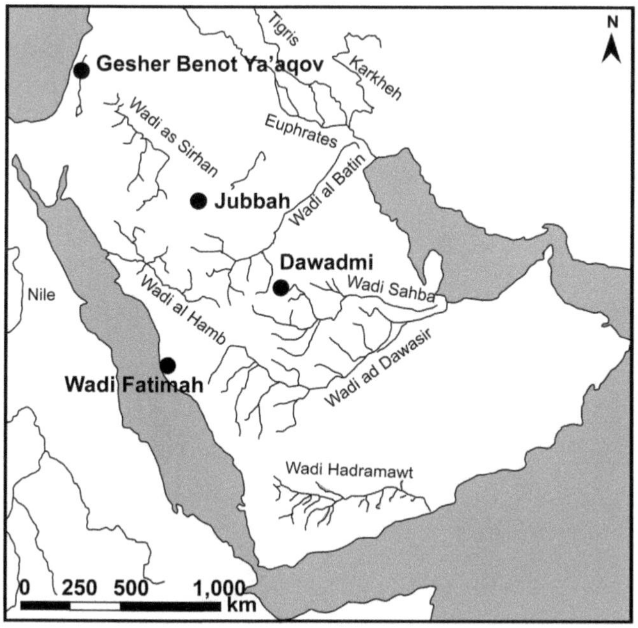

Fig. 10.1 Early–Middle Pleistocene drainage systems of Arabia (excluding palaeo-lakes), and location of Dawadmi, Wadi Fatimah, Jubbah (Saudi Arabia), and Gesher Benot Ya'aqov (GBY) (Israel). After Shipton et al. 2014, Fig. 1

implying links with both East Africa and the Levant in the early Middle Pleistocene (Petraglia et al. 2010; Shipton et al. 2014). An important breakthrough has been the realization that individual artifact sites are usually part of much larger agglomerations across large areas of the landscape (e.g., Petraglia et al. 2010). For example, there are at least 32 Acheulean sites along the Wadi Fatimah in the western part of Saudi Arabia (Whalen et al. 1988; Shipton et al. 2014). The most spectacular example is ad-Dawadmi, where there are at least 62 individual sites along a volcanic dyke covering an area ca. 100 km by 50 km, thus making it perhaps the largest Acheulean locality worldwide (Jennings et al. 2015). These sites were primarily quarries for roughing out and making bifaces from the local andesite. Dating remains a problem, but stratified sequences with Middle Paleolithic assemblages that can be linked to paleolakes have been explored in the Jubbah region of the An Nefud desert in northern Saudi Arabia (Petraglia et al. 2012; Shipton et al. 2014), and future work may be able to extend this chronological framework further back into the Pleistocene.

10.2.2 The Negev Desert (Israel)

Acheulean handaxes, picks, and other artifacts have been found that probably weathered out of the Early Pleistocene Zeheiha Formation in the Nahal Zihor. Here, a lake system developed that probably lasted 45–150 ka under a semi-arid climate, with each lake lasting between 3,000 and 10,000 years and up to 3–5 m deep and supporting viable fish populations. Cosmogenic burial data indicate the Zeheiha Formation dates from ca. 1.6 Ma (Guralnik et al. 2011). The bifaces and picks are described as similar to those from 'Ubeidiya and Latamne (see below). Younger, probably Middle Pleistocene, Acheulean assemblages are found on nearby terraces or slopes near the paleolake (Ginat et al. 2003). As in Arabia, lakes would have been a favored location for early hominins.

10.2.3 The Levant (Israel and Syria)

Several Acheulean sites are known in the Levant (Fig. 10.2) that are summarized by Bar-Yosef (1994, 1998a), Bar-Yosef and Belmaker (2011), and Sharon and Barsky (2015). Three phases are recognized and categorized variously as early, middle, and late Acheulean (e.g., Bar-Yosef and Belmaker 2011) or early, Large Flake Acheulean (known only from Gesher Benot Ya'aqov [GBY] and late Acheulean (Sharon and Barsky 2015). Those in Israel are discussed first.

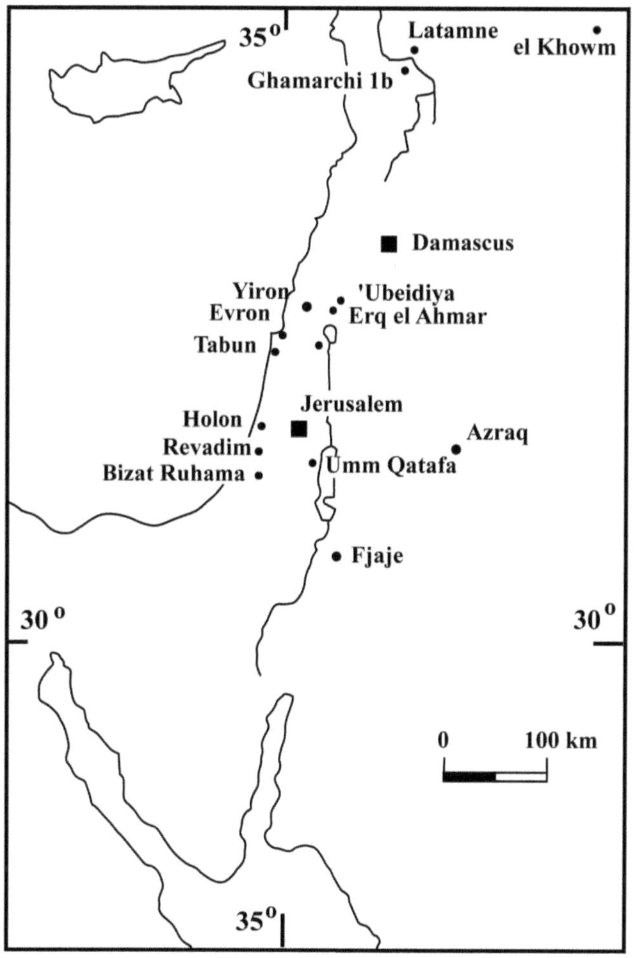

Fig. 10.2 Principal sites with Acheulean assemblages in the Levant

10.2.3.1 Israel

The two flagship Acheulean sites in Israel are ‘Ubeidiya and Gesher Benot Ya’aqov. Both lie within lacustrine formations and contain numerous assemblages with Acheulean bifaces. The only skeletal evidence for hominins from the archaeological strata from either site is a right lower incisor (UB 335) from ‘Ubeidiya site I-26a that is attributed to *Homo* sp. indet., but in terms of its probable age of 1.4 Ma may likely be derived either from an East African population of *Homo ergaster* or from a SW Asian one represented by the Dmanisi population of early *Homo erectus* (a.k.a. *H. georgicus*) (Belmaker et al. 2002).

‘Ubeidiya
The 190 m-thick ‘Ubeidiya formation comprises sediments deposited in and around a lake that underwent four major and many minor cycles of deposition (Feibel 2004). Most of the fossil and archaeological sites are found in lakeshore, marsh, and wadi deposit belonging to the lower fluvial (*fluviatile inférieur*) cycle. By comparing the faunal assemblages from ‘Ubeidiya with better-dated ones elsewhere, Tchernov (1987) proposed that the animals (and associated artifacts) at ‘Ubeidiya dated to ca. 1.0–1.4 Ma. Recent paleomagnetic work indicates that the fossil and artifact layers are somewhat earlier, and between 1.2 and 1.6 Ma (Bar-Yosef and Belmaker 2011).

Because the layers of the ‘Ubeidiya Formation have been tilted up to 60–80°, it was possible to excavate a long sequence of deposits from long trenches across the top of the deposits, but it was difficult to expose a large area of any particular surface. Artifacts were found on or in more than 80 occupation levels in three types of context: on top of a swampy layer; beach deposits that graded laterally into lake or swamp deposits; and in fluvial gravel conglomerates.

The artifact assemblages are described as similar to contemporaneous Developed Oldowan B and early Acheulean assemblages from East Africa because of similar tool-types, notably core-choppers, picks, polyhedrons, spheroids, heavy-duty scrapers and bifaces (Fig. 10.3), and attributed to hominin dispersals from Africa via the “Levantine corridor” (Bar-Yosef and Goren-Inbar 1993; Sharon and Barsky 2015). Bifaces are rare and usually <7% of the total in most assemblages, but occasionally as high as 22%; cleavers are very rare. Two aspects of these assemblages are noteworthy. The first is that there is little variation between assemblages from different sedimentary contexts and periods as the same tool-types are found in all samples. This would seem to indicate that the ‘Ubeidiya hominins did not use specific tool-kits for different types of environments or activities. The second is that different types of stone were used for making specific types of artifacts: chopping tools and flake tools were usually made from flint, bifaces from basalt, and spheroids from limestone.

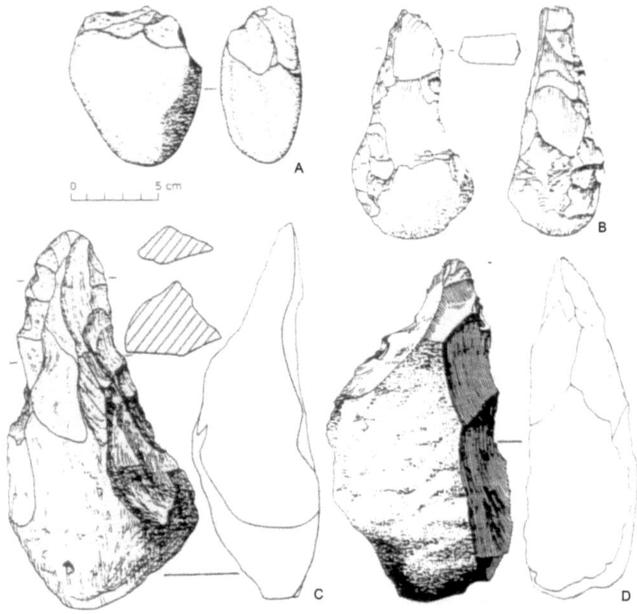

Fig. 10.3 Acheulean artifacts from ‘Ubeidiya, Israel. **A** core chopper; **B** quadrihedral (flint); **C** biface (basalt); **D** biface (flint). After Bar-Yosef 1994a, b, Fig. 6

The taphonomic history of the lithic and faunal assemblages from ‘Ubeidiya has been elucidated by Tchernov (1987) and more recently by Mallol (2006). In Tchernov’s opinion, “The co-occurrence of stone artifacts and bones appears to be a largely fortuitous association”. Mallol has largely confirmed this view. In her assessment, artifacts were deposited during dry seasons, and then reworked by low-energy fluvial action during wet seasons. Shea (1999) also pointed out that lake shores were inherently dangerous because of predators (including crocodiles), so hominins likely “raided” the foreshore for food and then discarded artifacts when carrying back food for consumption in safer locales. Over time, this type of repetitive low-level discard could be mistaken for a “living” floor where a large amount of material was discarded in a single short episode.

Gaudzinski-Windheuser (2005) analyzed the ‘Ubeidiya faunal assemblages, and summarized 17 of them (Gaudzinski 2004a, b), and found little evidence of hominin or carnivore involvement. Overall, on 6099 fossils from 17 localities, there were only 15 cutmarks and 27 tooth-puncture marks, or

0.24% and 0.43%, respectively. In some assemblages (e.g., II-23), there was evidence of carnivores hunting, scavenging, and feeding. However, two bones had cutmarks, which would seem to indicate some hominin involvement, although likely very slight. There is no evidence of marrow extraction by breaking bones, and the rarity of cutmarks could indicate a "light-touch butchery" rather than systematic defleshing.

Gesher Benot Ya'aqov (GBY)

GBY is a younger and considerably more informative site complex than 'Ubeidiya and may contain the earliest evidence for the controlled use of fire, elephant hunting, the processing of plant foods, and also the shaping of wood (Belitzky et al. 1991). At the site, the 34 m of deposits comprise calcareous and organic muds, shell layers, and conglomerates. The top and base of the local section are fluvial deposits, with intervening lake and lake-margin deposits. Because the Brunhes–Matuyama paleomagnetic boundary lies at the base of the local section, the overlying archaeological layers are safely dated to ca. 700–780 ka, and can be correlated with MIS 18–20. Because the site is water-logged, an enormous amount of plant material from 100 taxa was recovered. These included edible species such as wild grape, pistachio, water chestnut, wild olive, and plum (Goren-Inbar et al. 2000, 2002a, b). Evidence of fire comes from burnt artifacts and plant material. This appears to be clustered, and the areas where burning is localized may indicate "phantom hearths" (Goren-Inbar et al. 2004; Alperson-Afils et al. 2007; Alperson-Afils and Goren-Inbar 2010). Nut processing may also be evidenced by the presence of nut shells, and pitted stones that may have served as anvils (Goren-Inbar et al. 2002a).

All 12 of the investigated archaeological occurrences are classed as Acheulean (Goren-Inbar et al. 2000) or Large Flake Acheulean (Sharon 2011; Sharon and Barsky 2015), and most contain handaxes, cleavers, and *éclats de taille de biface* (Fig. 10.4). Limestone was used for chopping tools and percussors, flint for cores, flakes and flake tools, and basalt for handaxes and cleavers. This was locally available as large boulders and lava flows. GBY is unique in the Levant in the abundance of cleavers, and in the technique of making handaxes and cleavers from Large Flakes (i.e., those obtained from flakes >10 cm in maximum dimension), as in many African localities such as Olduvai Bed IV, Olorgesaillie, Isimila and Kalambo Falls (Sharon and Barsky 2015), and also Dawadmi and Wadi Fatimah in Arabia (Petraglia et al. 2010). Saragusti and Goren-Inbar (2001) point out that the bifacial GBY assemblage has more in common with African assemblages than with 'Ubeidiya, and this may indicate a dispersal of hominins from Africa at this time.

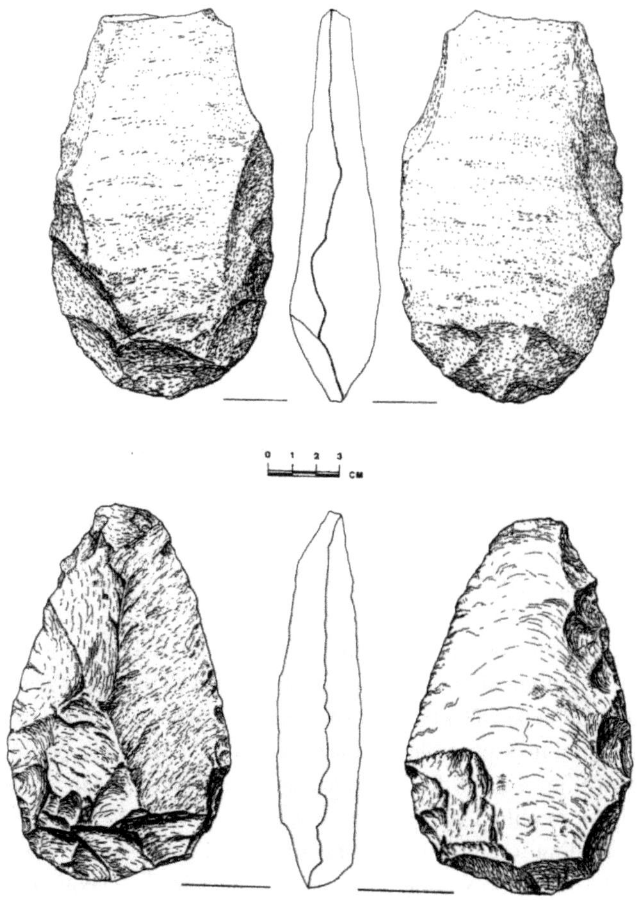

Fig. 10.4 An Acheulean cleaver and handaxe from Gesher Benot Ya'aqov (GBY), Israel, Layer II-6, level 4. After Saragusti and Goren-Inbar 2001, Fig. 5

Madsen and Goren-Inbar (2004) investigated the manufacture of bifaces through analysis of layer II-6, level 1, which contained three large basalt cores and 68 basalt bifaces. Experiments showed that a competent knapper could have made a biface in 15 min, or all 68 in 15 h. Interestingly, experiments also showed that far fewer flakes (ca. 1700) were found than the 8000 that were produced in the experimental flaking. This might indicate that hominins arrived and left with some bifaces as rough-outs that were finished elsewhere.

GBY is one of the few Asian Acheulean sites with evidence that may indicate the hunting of large animals, as opposed to butchering and scavenging them. Level II-6 contained an elephant (*Elephas antiquus*) skull, from which the palate and base had been removed and the nasal portion had been damaged. Fluvial transport was ruled out because of the presence of 37,000 artifacts <2 cm in this level that would have been winnowed out by flowing water, and because the skull was embedded in clays and silts. Although several interpretations of this find are possible, the investigators suggest that it may indicate that the elephant had been hunted (Goren-Inbar et al. 1994). This suggestion is strengthened by the analysis of the three faunal assemblages from GBY by Rabinovich et al. (2008), who concluded from the abundance and positioning of cut-, percussion- and hack-marks on the bones, and the rarity of carnivore damage that the GBY hominins were competent and systematic at defleshing fallow deer, and likely hunted them.

Late Acheulean assemblages are widespread throughout the Levant, but few sites have been adequately excavated and published. Noteworthy ones are the open-air sites of Holon (Yizraeli 1967; Porat et al. 1999), Revadim (Gvirtzman et al. 1999; Marder et al. 1999), and Berekhat Ram (Goren-Inbar 1985). Their age is unclear. ESR dates for Holon are ca. 200 ka but on typological grounds, an age of 400–500 ka is perhaps more reasonable (Bar-Yosef and Belmaker 2011). OSL dates of ca. 200–300 ka for Revadim are also probably best treated as minima. The age of Berekhat Ram is poorly constrained between 230 and 780 ka, but on typological grounds, probably 350–450 ka (Bar-Yosef and Belmaker 2011). This site is particularly noteworthy for containing a possible figurine. This was a volcanic pebble (only 35 × 25 × 21 mm) that appears to have been artificially modified to produce something similar to a figure, according to a detailed examination by d'Errico and Nowell (2000). Although Acheulean assemblages in cave sequences are rare, they are known in Israel from Tabun layer F and the equivalent of layer G from Jelinek's (1982) excavations; and also from levels D and E at Umm Qatafa. On the basis of TL dates on burnt flints, layer F at Tabun is probably >300 ka and layer G, >400 ka (Mercier et al. 1995, 2000; Mercier and Valladas 2003). An average of 213 ± 26 ka derived from three ESR dates on one tooth from level D2 at Umm Qatafa (Porat et al. 2002) is best treated as a minimum age because the Mousterian in the Levant probably began ca. 200–250 ka, and in areas where Acheulean assemblages are followed by Jabrudian ones, the latter are between 300 and 400 ka in age (Dennell 2009: 306–307).

10.2.4 Syria

The principal Acheulean sites in Syria are in the el-Kowm area and are revealed by deep sections inside wells (Le Tensorer et al. 2007; Jagher and Le Tensorer 2011). At Nadaouiyeh, there are 32 levels of a 32 m-thick section with upper (late) Acheulean material dating back to ca. 600 ka. This includes 13,000 handaxes, and a left parietal bone of *Homo erectus s.l.*, dated biostratigraphically to ca. 500 ka. Other sections are at Hummal and Umm el Tlel (Acheulean to Neolithic). There are also ca. 30 Acheulean sites in the area (Le Tensorer 2014).

One should also mention Latamne, which was excavated by Clark (1967, 1969; Bar-Yosef 1998a, pp. 249–250 for an overview) in the 1960s. Its age is uncertain but probably ca. 1.0 Ma (Bar-Yosef and Belmaker 2011). Here, over 100 m^2 of an "occupation floor" with a large early Acheulean assemblage was exposed (Fig. 10.5). Handaxes (N = 54) comprised 31.4% of all shaped tools, but cleavers (0.6%) and bifacial knives (1.6%) were very rare. In keeping with interpretations of the 1960s, Latamne was interpreted as a riverside living site that was occupied for a season or two. This interpretation probably influenced the interpretation of the nearby site of Ghamarchi 1b as a workshop or living site. Here, an upper Acheulean assemblage (N = 2129) with 140 bifaces was found on a surface of which 300 m^2 was excavated (Muhesen 1993). Both sites would repay investigation. As with 'Ubeidiya and Olduvai, it is unlikely that Latamne and Ghamarci were "living sites", and other interpretations (such as repeated low-level discard or deposition by sheet-flow) may be more reasonable.

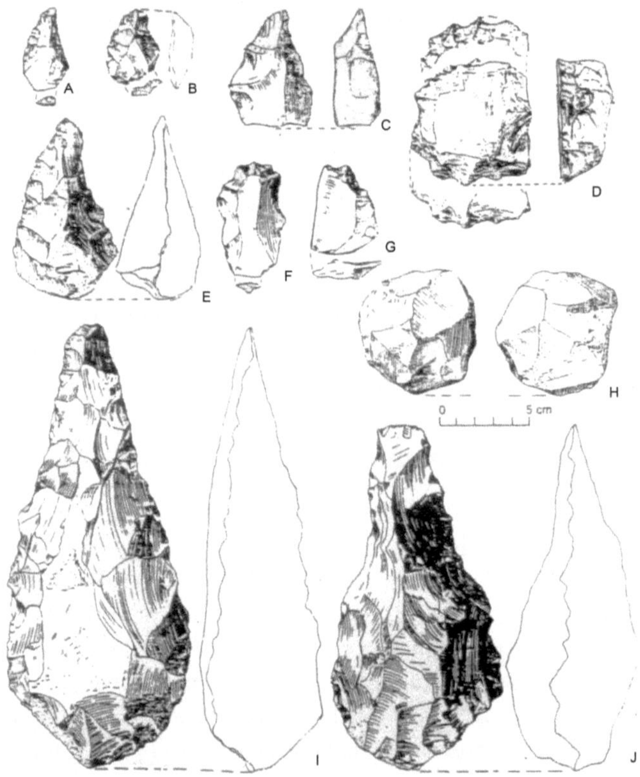

Fig. 10.5 Artifacts from Latamne, Syria: **A–D**, **F**, **G** retouched flakes; **E**, **I**, **J** bifaces; **H** spheroid. After Bar-Yosef 1994a, b, Fig. 11

10.2.5 Jordan

The Lower Paleolithic of Jordan is summarized by Bar-Yosef (1998b) and Copeland (1998). Although the Acheulean has been subdivided into Early, Middle, and Late phases on typological grounds, independent geochronological evidence is rare. There is no clear evidence of an Early Pleistocene Acheulean comparable to 'Ubeidiya. There may be some Acheulean material from Early to Middle Pleistocene contexts in the upper part of the Dauqara complex, and at Masharai'a 4 in the lower member of the Tabaqat Formation. In contrast, late Acheulean assemblages are common in lacustrine and fluvial contexts as well as from surface exposures. Good examples are an enormous spread of late Acheulean artifacts at Fjaje, where artifacts were found over a distance of ca. 20 km (Rollefson 1981); along the wadis and in the springs at Azraq, at, for example, Lion Spring and C-Spring.

10.2.6 Turkey

Our knowledge of Early Paleolithic Turkey is currently limited to undated surface material, a few open-air locations with dated artifacts, a report of some Middle Pleistocene hominin footprints (Ozansoy 1969), and the two excavated caves of Yarımburgaz (Kuhn et al. 1996) and Kara'In (Otte et al. 1998). Neither of these contained Acheulean material, although the artifact counts from Kara'In were very small for the lowest levels, so the absence of Acheulean is not necessarily proven. Kuhn (2002) provides an excellent overview of the Paleolithic of Turkey, and there is also an excellent and often updated Web site (http://tayproject.org); see TAY (Türkiye Arkeolojik Yerleşmeleri) (Archaeological Settlements of Turkey) that lists and briefly describes a large number of Paleolithic (and later) sites in Turkey.

10.2.6.1 Open-Air Locations

Acheulean bifaces and other flaked items that are probably Lower Paleolithic are known from numerous find spots and chance discoveries across Turkey. Yalçınkaya (1981) summarized what was known in the 1970s and noted that bifaces and simple flake tools were present in western Turkey, along the Black Sea littoral, and in Anatolia, particularly along the valleys of rivers such as the Euphrates. As these were all surface finds, none was dated or in context. Since then, a little more has been learnt about the early Paleolithic in western Turkey and Anatolia.

Runnels and Özdoğan (2001) discuss some Lower Paleolithic material from the Bosphorus region of Northwest Turkey. Bifaces, core-choppers, and flake tools were found at Göksu, and some Lower Paleolithic artifacts are reported from the Black Sea Coast. In Anatolia, Minzoni-Deroche and Sanlaville (1988) conducted one of the few systematic surveys in Turkey that attempted to link Acheulean artifacts to terrace sequences, in this instance, along the Euphrates and Nizip Rivers in the Gaziantep region of southern Anatolia. They identified a middle and a late Acheulean on the second and third terraces in their sequence; their middle Acheulean was compared with Latamne in Syria (see above). Albrecht and Müller-Beck (1988) discovered (typologically) late Acheulean bifaces in a gravel deposit attributed to a cool period before or (less likely) after the last interglacial in a side valley of the Euphrates. At present, the best prospects for establishing a chronological framework for the Lower Paleolithic of Turkey are in the CAVP (Central Anatolian Volcanic Plateau). At Kalatepe Deresi 3, levels IV–X, in Central Anatolia, Acheulean bifaces have been found in situ in a ravine section containing several volcanic tuff-fall deposits that it may be possible to link to dated eruption events. If so, it should be possible to narrow their present age-range of 0.16 Ma to 1.1 Ma (Tryon et al. 2009).

The lack of evidence for Acheulean (and other early Paleolithic) assemblages and sites in Turkey can be attributed to three main factors. The first is the lack of systematic fieldwork, understandable perhaps in a country so rich in cultural remains from later periods. The second is the probable extent of slope erosion and valley alluviation in western Turkey and along major river valleys. Thirdly, Anatolia has very harsh winters and would have been scarcely habitable during glacial periods, so its Paleolithic evidence is likely to be sparse (Kuhn 2002).

10.2.7 Iraq

In 1951, Bruce Howe investigated the site of Barda Barka in Iraqi Kurdistan and recovered several pebble tools, Acheulean bifaces and flake tools from an undated sand and gravel unit. This has recently been posthumously published by the Oriental Institute, Chicago (Howe 2014), and still constitutes our main evidence for the Lower Paleolithic of Iraq.

10.2.8 Iran

Apart from a few bifaces found in surface locations (Biglari and Shidrang 2006) almost nothing is known about the Early Paleolithic of Iran. Indeed, until recently, so little was known that even the discovery of a single biface merited a publication (Singer and Wymer 1978). For this reason, the discovery of Ganj Par on a terrace of the Sefirud River at the southern edge of the Caspian Sea is of great importance.

This open-air site contains an Acheulean assemblage (N = ca. 140) with handaxes, picks, and the first cleaver found in Iran. The assemblage is claimed as similar to those from 'Ubeidiya, Olduvai Upper Bed II, and Konso Gardula, but is thought to be Middle Pleistocene in age (Biglari et al. 2004; Biglari and Jahani 2011). Future surveys may hopefully illuminate the Early Paleolithic of this poorly documented part of SW Asia and help fill the enormous gap between Arabia and South Asia. At present, there is "no persuasive explanation why the Acheulian is not represented in the vast area between the Zagros and Baluchistan" (Bar-Yosef and Belmaker 2011: 1332).

10.2.9 *The Caucasus*

The evidence for Middle Pleistocene Acheulean assemblages in cave sites in the Caucasus is summarized by Ljubin and Bosinski (1995), Doronichev (2000), Lioubine (2002), and Doronichev and Golvanova (2003). Undated assemblages classified as Acheulean from open-air sites such as Eni-El and Satani Dar in Armenia are summarized by Lioubine (2002) and Fourloubey et al. (2003). Some of these open-air sites may be very ancient. According to Presnyakov et al. (2012), early Acheulean artifacts at Karakhach were in strata bracketed by volcanic ashes in which zircons were dated by SHRIMP-II U–Pb techniques to 1.942 ± 0.046 Ma, 1.804 ± 0.030 Ma, and 1.750 ± 0.020 Ma. At Kurtan, an ash layer underlying paleosols with Acheulean artifacts was dated to 1.432 ± 0.028 Ma. Further work is needed to confirm these dates.

In the cave sites, Acheulean bifaces are present but extremely rare: at Azych, only 7/289 (2.4%), and Kudaro I, only 50/5000 (1%). Tcona, lying at 2100 m a.s.l. and thus one of the highest-altitude Paleolithic sites in Asia, is an exception in that 47 artifacts out of 97 were bifaces (Doronichev and Golovanova (2003: 83), but this appears to have been a site that was briefly visited with a tool-kit already prepared. All these caves appear to have been primarily denning sites for cave bears, with only fleeting usage by hominins.

10.3 Central Asia

The five "stans" of Central Asia (Kazakhstan, Kyrgyzstan, Uzbekistan, Tajikistan, and Turkmenistan) cover an area ca. sixteen times the size of Britain, yet almost nothing is known of their Early Paleolithic. The best documented area is Tajikistan, where Ranov (1995) found a sequence of non-Acheulean assemblages in paleosols dating back to ca. 900 ka; the artifacts were made from pebbles, so the absence of bifaces may simply reflect the absence of suitably-sized stone for making them. Vishnyatsky's (1999) excellent review of the Central Asian Paleolithic mentions various sites with bifaces such as Yangadja (Turkmenistan), and Esen 2 and Karakuduk (Uzbekistan). None is dated, so it is unclear whether these finds indicate the occasional presence of Middle Pleistocene hominins with a bifacial technology, or evidence of a more recent Paleolithic record in this vast and largely desert region.

10.3.1 *Mongolia*

A chopper and a handaxe were reported from a terrace at Tsagan-dyrs that may date to the last interglacial. Handaxes, age unknown, were also found at Yarkh (Shackley 1984). Little more is known about the Early Paleolithic of this enormous country covering three million square kilometers. Given the harshness of Mongolian winters, it is unsurprising that there is hardly any indication of early Paleolithic occupation.

10.3.2 *South Asia (India and Pakistan)*

Acheulean bifaces in South Asia are found from the Soan Basin in Pakistan to NE India and from Nepal to the Kortallyar Basin in South India (Fig. 10.6). There are also some major Acheulean sites in India, several landscape surveys, and a large literature on the Early Paleolithic of South Asia. Additionally, Paddayya's (1982) monograph on the Acheulean landscape of the Hunsgi-Baichbal valleys in South India is one of the best of its kind. Nevertheless, South Asia has not played a prominent role in international discussions about the Acheulean. Reasons for this include the lack of associated faunal evidence because the soils are generally unsuitable for the preservation of faunal material; the lack of associated hominin remains, apart from a poorly dated and not particularly diagnostic cranial specimen from Hathnora in the Narmada Valley that is probably late Middle Pleistocene in age and attributed to *Homo* sp. indet. (Athreya 2007); and the problems of dating Indian (and Pakistani) early Paleolithic sites. This neglect is regrettable, because India has enormous potential for Acheulean research.

There are several excellent recent reviews of the Indian Early Paleolithic, such as Mishra (1994), Petraglia (1998, 2001), Gaillard and Mishra (2001), Pappu (2002), and Paddayya (2007). There is also a lengthy review by Dennell (2009: 336–395) of data up to 2008. Korisettar and Rajaguru (1998) provide an excellent overview of the Indian Middle Pleistocene. In the space available here, only a small selection can be made of the material.

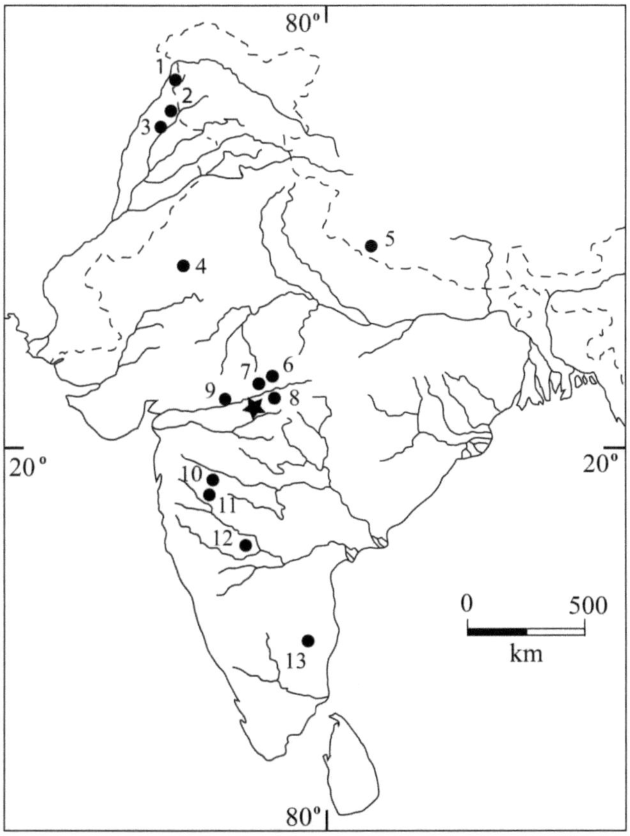

Fig. 10.6 Map of principal Acheulean sites in South Asia. Pakistan: 1, Dina; 2, Jalalpur; 3, Rohri Hills; India: 4, Singhi Talav; 5, Dang Valley (Nepal); 6, Belan Valley; 7, Paisra; 8, Bhimbetka; 9, Raisen; 10, Chirki; 11, Kukdi; 12, Hunsgi-Baichbal; 13, Attirampakkam/Kortallyar Basin. The star represents the hominin site of Hatnora

10.3.2.1 Dating

Compared with the Levant, few Indian Acheulean sites have been dated, and most of the published dates are single ^{230}Th-^{234}U determinations that were often obtained many years ago. The most important single development in recent years is the dating of the earliest Acheulean assemblages from Attirampakkam, Trench T3, layer 6, in the Kortallyar Basin, South India, to over 1 Ma (Pappu et al. 2011). Here, a 9-m section of red clays that contained two Acheulean bifaces and two cleavers (and 282 other tools) (Pappu et al. 2003: 596) were shown to lie below the Brunhes–Matuyama boundary and were thus >0.78 Ma; this was the first clear indication that the early Acheulean in India lies in the Early Pleistocene. A more precise date was obtained by applying (^{26}Al/^{10}Be) isotopic dating on six artifacts from layers 6–8; the pooled average gave a probable age of 1.51 ± 0.07 Ma. On the basis of this new evidence, it now appears that the Acheulean of South Asia has the same antiquity as that seen at 'Ubeidiya, Israel (see above; Dennell 2011).

There are several other important stratigraphic sequences with Acheulean material in India, although dating is imprecise. Relative age is usually assigned on typological grounds, with an early Acheulean defined by a low ratio of cleavers to handaxes, and an absence of the Levallois technique, and a late Acheulean defined (sometimes) by a higher cleaver to handaxe ratio and the use of Levallois and discoid core techniques (e.g., Misra 1989). Late Acheulean artifacts tend also to be thinner and found on surface exposures, whereas early Acheulean ones tend to be thicker and found in stratified sections. Three of the most carefully studied fluvial sequences are in the Son and Belan valleys (Sharma and Clark 1983; Williams and Clarke 1995), and the Narmada Valley, which has a long and complex stratigraphy from the Early to the Upper Pleistocene (Patnaik et al. 2009). The middle Son Valley is particularly important because Acheulean bifaces from Bamburi 1 and Patpara date to only 140–120 ka (i.e., the end of MIS 6) and are thus the youngest worldwide; this also implies a late survival of archaic hominins in South Asia (Haslam et al. 2011; Shipton et al. 2013).

One of the most important archaeological sequences is the site of Chirki (Corvinus 1981, 1983), which lies on a tributary of the Nevasa River in Central India. The main Acheulean evidence was found in a gravel layer overlying bedrock and dated by Th-U to >350 ka. This assemblage (N = 2524) is considered early Acheulean because of the high proportion of handaxes made from cobbles, and the absence of the Levallois technique. Another key sequence is the rock-shelter of Bhimbetka F III-23 in Central India, which was carefully excavated by Misra (1985). Layers 6–8 contained a late Acheulean assemblage (N = 18,271) (Fig. 10.7), defined by a low frequency of bifaces (2.4%), high frequencies of Levallois flakes, and a complete absence of chopper-chopping tools.

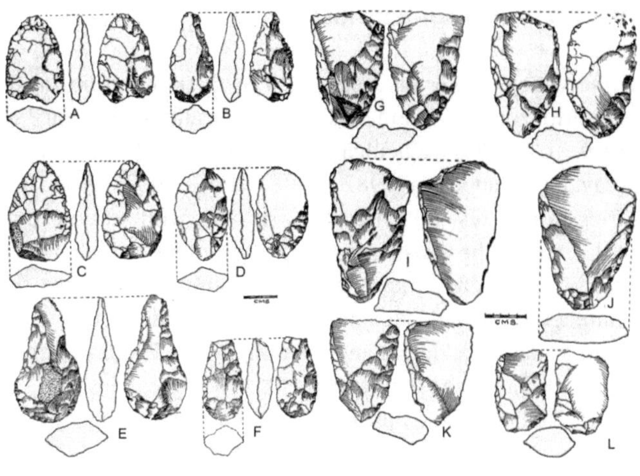

Fig. 10.7 A selection of Acheulean handaxes (**A–F**) and cleavers (**G–L**) from Bhimbetka III F-23, India. After Misra 1985, Fig. 2

10.3.2.2 Assemblage Variation

A conspicuous feature of Indian Acheulean assemblages is that both handaxes and cleavers are present, and often, cleavers are considerably commoner than handaxes. In this respect, the Indian Acheulean differs significantly from that in the Levant, where cleavers are rare apart from GBY. Because the two from Attirampakkam Trench 3 layer 6 predate those from GBY, it is possible that hominins in South Asia developed cleavers independently. On the other hand, the recent findings of cleavers in Arabian Acheulean sites (Shipton et al. 2014) leave open the possibility of diffusion along southern Asia, although there is currently no indication of any Acheulean material from southern Iran.

Ratios of handaxes to cleavers vary considerably. At Mailapur (total assemblage = 178), for example, the ratio of handaxes to cleavers is 6.6/1, but as low as 0.1/1 at Minarwala Nala, Raisen (N = 621). The percentages of bifaces in assemblages is also highly variable, from 100% in a small assemblage (N = 9) at Mudnur in the Baichbal Valley to only 1.1% in the assemblages (N = 275) from layers 3–4 at Singi Talav in the Thar Desert (Dennell 2009; Table 9.2). In the absence of faunal data, it is not possible to explain such variation.

10.3.2.3 Landscape Use

There have been several studies of Acheulean settlement patterns in India; examples are Raisen (Jacobson 1985), Paisra (Pant and Jayaswal 1991), the Kaladgi Basin (Pappu and Deo 1994; Petraglia et al. 2003); and at a sub-continental scale, a basin model developed by Korisettar (2007). The two most significant are those in the Hunsgi-Baichbal valleys and the Kortallyar Basin.

10.3.2.4 Hunsgi-Baichbal

These valleys cover an area ca. 30 by 17 km and lie in the semi-arid zone of south India. As a result of 30 years of survey, over 200 Acheulean sites were found and several excavated (Paddayya 1982, 1991, 2001, 2007). Site size varies from a few dozen artifacts to ones with several hundred. One of the most interesting is Isampur, which was a quarry site where limestone blocks were extracted for the manufacture of handaxes and may be the oldest quarry site found to date (Paddayya and Petraglia 1997; Paddayya et al. 1999, 2000). Its age is uncertain. On typological grounds, an age of 500–600 ka was expected, but an average of 10 readings from two bovid teeth gave an ESR date of 1.27 Ma (Paddayya et al. 2002). Neither age can be ruled out at present. The 15,000 artifacts from this site show all stages of lithic reduction, from initial roughing out to final flaking. As at GBY, experiments were conducted to investigate how handaxes and cleavers were made. Interestingly, different techniques would have been used: handaxes were best flaked by façonnage (i.e., by shaping a core by removing small flakes), but cleavers were made by débitage, i.e., by removing a large flake from a block and then trimming the piece by façonnage. This further implied that the Isampur hominins were able to segregate their knowledge so that two separate systems were used to make (and perhaps use) tools (Shipton et al. 2009). Data from Isampur and other sites indicate that most artifacts were discarded within 1–2 km of the source, with a few moved up to 5 km. At Isampur, it also appears that cores were transported only 250 m, hammers ca. 300 m and bifaces ca. 700 m (Petraglia et al. 2005). Paddayya (1982) also modeled seasonal land use and suggested that in the dry season, settlement was aggregated around a few springs and streams, but hominins dispersed in smaller groups in the wet season of the summer monsoon, when resources (particularly plant foods) were more plentiful.

10.3.2.5 Kortallyar Basin

This basin has been intensively researched since Bruce Foote found the first handaxe in India in 1863. Earlier research focused on establishing stratigraphic sequences, but recent work by Pappu (2001, 2007) has combined detailed geomorphological mapping with systematic surveys for material over an area of ca. 200 km^2. Although 22 mostly Middle Paleolithic sites with ca. 2000 artifacts were found, Acheulean material is scarce, largely because it is rarely found on exposed surfaces but mainly in sections (Pappu and Akhilesh 2007). For this reason, the recent dating work at Attirampakkam is exceptionally important in establishing the antiquity of the Acheulean in South Asia.

10.3.2.6 Pakistan

Almost nothing is known about the Acheulean in Pakistan, despite its size and geographical position as a potential corridor between Arabia and India. Handaxes from sections in the Jhelum Basin in north Pakistan were dated to shortly above the Brunhes–Matuyama boundary and are thus probably 600–700 ka-old (Rendell and Dennell 1985). Handaxes were also regularly noted on Middle Pleistocene exposures in the Soan Valley (personal observation). In Sindh Province in southern Pakistan, there are several large scatters that include Acheulean bifaces in the Rohri Hills; these may have been quarry sites, but most have been destroyed in recent years (Biagi and Cremaschi 1998; Biagi 2006).

10.4 East Asia (China and the Korean Peninsula)

Before discussing the evidence for Acheulean (or "Acheulean-like") assemblages from East Asia, it is first necessary to discuss the current status of the Movius Line.

10.4.1 The Movius Line

As is well known, Hallam Movius (1948) divided the Early Paleolithic Old World into two zones. Europe, Africa, Southwest and South Asia were defined by the use of bifaces, whereas East and Southeast Asia were defined by unstandardized flakes and cores. These differences were seen in cognitive terms: those hominins who used bifaces were regarded as "dynamic" and "progressive", in contrast with those "conservative" hominins that did not (see Dennell 2014a for a critique of these views). As Movius (1948: 411) stated the tools are "relatively monotonous and unimaginative assemblages of choppers, chopping tools, and hand-adzes… as early as Lower Paleolithic times, southern and eastern Asia was a region of cultural retardation …it seems very unlikely that this vast area could ever have played a vital and dynamic role in early human evolution … Very primitive forms of Early Man apparently persisted there long after types at a comparable stage of physical evolution had become extinct elsewhere". The basis of Movius's scathing dismissal of the East and Southeast Asian early Paleolithic as an area of "cultural retardation" was the fieldwork in Burma (Myanmar), in which he participated with the geologist Helmut de Terra and the Jesuit paleontologist and archaeologist Teilhard de Chardin in 1937–38. In a previous publication (Dennell 2014b), I have argued that the chronological and archaeological conclusions drawn from this fieldwork are invalid, and the Movius Line was essentially a house built on sand. In the case of China, Movius's influence was extremely negative. His sole basis for his sweeping generalization about the early Paleolithic of China was locality 1 at Chou-kou-tien (now Zhoukoudian), which was in 1948 the only Middle Pleistocene archaeological site known in China. Since then, a great deal more has been learnt about the Chinese Early Paleolithic, especially from open-air locations. Additionally, it is now apparent that the Acheulean in Africa, Europe and Asia is largely an open-air phenomenon, as bifaces are generally rare or absent in many Middle Pleistocene cave sequences. As noted already, for example, they are very rare at Kudaro I and Azych in the Caucausus, and they are absent in cave sequences where stone was locally available that could have been used for making bifaces: examples are Yarimburgaz, Turkey (N = 1675; Kuhn et al. 1996) and Gran Dolina, Atapuerca, Spain. Other examples are Truegolyna (southern Russia) and Selungur (Kyrgyzstan [N = 1417; Islamov 1990]). The absence of bifaces at Zhoukoudian does not therefore warrant the sweeping generalization that bifaces were never used in East Asia. I have therefore suggested (Dennell 2016) that we need to move beyond the Movius Line, and focus instead on the patterning of assemblages with and without bifaces across Eurasia.

10.4.2 A Movius Line sensu lato?

Although maps showing the Movius Line are still found in many current textbooks of human evolution, there is now general agreement among researchers that it is no longer valid as originally formulated. Several scholars (e.g., Petraglia and Shipton 2008; Lycett and Bae 2010; Lycett and Norton 2010) argue that a Movius Line "sensu lato" (i.e., in a loose sense) still has validity with respect to China and the Korean Peninsula. They point out, for example, that the bifaces in some East Asian assemblages are noticeably larger and thicker than those in South or West Asia; the proportions of bifaces is often lower than in some (but by no means all) West Asian assemblages; and sites with Acheulean assemblages are far rarer than in, for example, India, the Levant, Africa or Western Europe. Of these points, the most significant is the first—the bifaces in some of the East Asian Acheulean are demonstrably heavier and thicker (as in the Bose Basin, see Fig. 10.9 and below) than their western counterparts. Put another way, some Chinese bifaces would not be out of place in India or West Asia, but others would certainly stand out as anomalous. The proportion of bifaces in assemblages is not as reliable an indicator of difference. For example, the proportion of bifaces is only 1.1% at Singhi Talav layers 3a, 3b and 4 (Gaillard et al. 1983) and 1.4%, as at Attirampakkam, layer 6, T3, India (2/286; Pappu et al. 2003), or <7%, as at many locations in Ubeidiya, Israel (Bar-Yosef and Goren-Inbar 1993). The third point, that East Asian Acheulean sites are rare compared with areas further west, needs qualifying. Much of China remains unexplored regarding its Paleolithic record, and in north China, much is doubtless buried by loess on the Loess Plateau. As noted above, Acheulean material is virtually absent in Iran and is also very rare in European regions such as Greece or Portugal or in Turkey.

At the heart of these debates are the respective roles of immigration, diffusion, and independent development in the long and undoubtedly complex past of the East Asian Early Paleolithic, and also the general issue of what "the

Acheulean" is supposed to mean as a much-used but poorly-understood term. Some comments on both issues are raised at the end of this paper.

10.4.3 China

Assemblages with bifaces and LCTs are known from open-air locations in China (Fig. 10.8), as in Europe, Southwest Asia, and India. Those in North and Central China include Dingcun (Yang et al. 2014), Lantian, the Luonan Basin (Wang 2005), Yunxian level 3 (N = 9/317; de Lumley et al. 2008) and the Danjiangkou Reservoir Region (Li et al. 2014). Of these, Yunxian is late Early Pleistocene and Dingcun likely late Middle Pleistocene. At present, the best prospects for dating early Paleolithic material (including bifaces) lie in finding artifacts in the loess-paleosol sections of the Loess Plateau and adjacent regions (e.g., Li et al. 2011; Lu et al. 2011).

In some cases, reanalysis of artifact assemblages has resulted in them becoming Acheulean instead of ones composed of flakes and cores. For example, objects from Dingcun, Shanxi Province were initially classified as cores and choppers but later classed as handaxes by Yang et al. (2014). Similarly, an artifact from Pingliang, near Lantian, Shaanxi Province, was classified as a pick, or a proto-biface but is now, and with little controversy, classed as a handaxe (Wang 2005: 14).

Much attention has recently been paid to the bifacial, assemblages from the Bose Basin in South China (Hou et al. 2000; Wang et al. 2014a, b; Fig. 10.9) that are dated by tektites to ca. 803 ka, the age of the Great Australasian Tektite shower, when fragments of a meteorite or asteroid incinerated a large area of South East Asia. The reliance upon tektites as the sole means of dating this material has its detractors (e.g., Langbroek 2015) as well as its supporters (e.g., Wang and Bae 2015) and can only be resolved when other dating techniques are employed; until then, the tektite dating should be accepted. The bifaces, especially those from Fengshudao (Wang et al. 2014a, b), are larger and considerably thicker than their West and South Asian counterparts. Many pieces from the Bose Basin are unifacial, and some (but not all; see Leng 2001: 71–79) are shaped differently from the types of handaxes and cleavers seen in western Eurasia (Hou et al. 2000; Huang et al. 2001). These play a major role in debate over the validity of the Movius Line in East Asia.

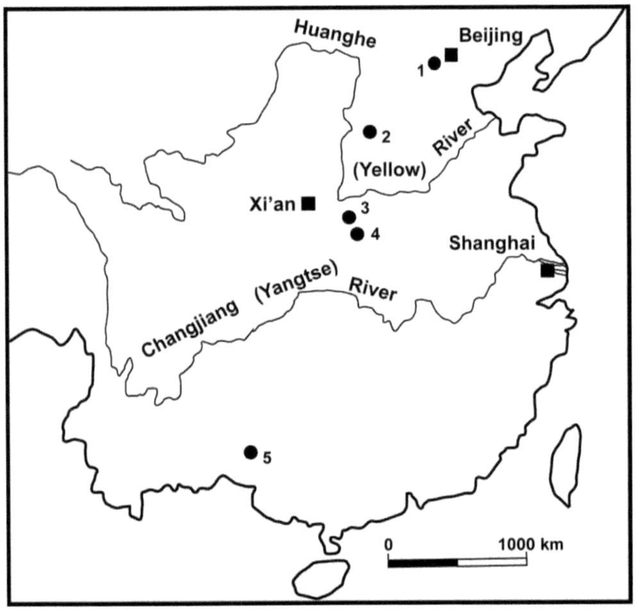

Fig. 10.8 Principal early paleolithic sites in China: (1) Zhoukoudian. Those with Acheulean bifacial assemblages, (2) Dincun; (3), Luonan Basin; (4) Yunxi; (5) Danjiangkou; (6) Bose Basin

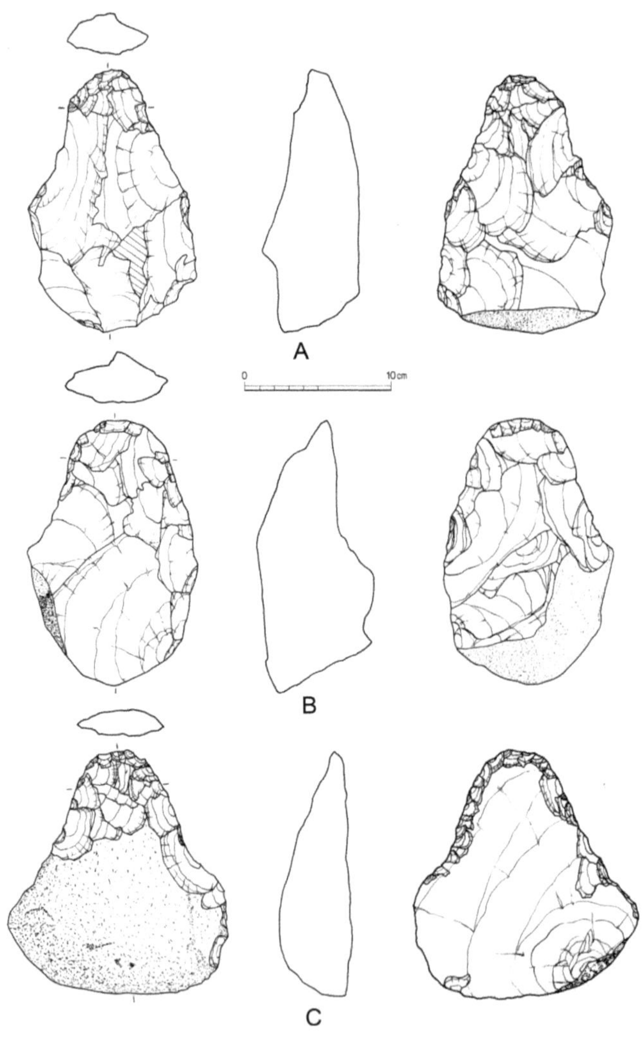

Fig. 10.9 Three bifaces from Datong, Bose Basin, China. After Huang et al. 2001, Plate 1.10

The contrast in the type of lithic materials found in caves and open-air locations in China is shown very clearly from Wang's (2005: 209) account of the material from Longyadong Cave and nearby open-air locations in the Luonan Basin. This cave sequence has recently been dated by thermally transferred OSL (TT-OSL) to 389 ± 18 and 274 ± 14 ka and can be correlated with the L4 (cold) loess and S3 (interglacial) paleosol units (Sun et al. 2013). At Longyadong, scrapers, points and burins were the only tools present, but large tools such as choppers, handaxes, cleavers (Fig. 10.10), bifacially-modified trihedrals, and spheroids were present in open-air locations. At these, the proportion of bifaces among all tools varies from 100% to 14.3%, which is a similar range to that seen at open-air sites at 'Ubeidiya, Olduvai, and other Acheulean locations. The proportion of bifaces relative to small retouched tools at Luonan likewise varies from 100% to 20%, as at 'Ubeidiya, Olduvai, and other locations. Wang (2005) explains the differences in assemblage composition between Longyadong Cave and the open-air sites in functional terms and suggests that this is representative of a larger pattern in China: "The lithic artifacts in cave sites are dominated by small flakes and small retouched tools, while the lithic assemblages in most open-air sites are characterized by large pebble tools, such as choppers, picks, spheroids, and handaxes" (Wang 2005: 13). This phenomenon, he proposes, "is taken to reflect differences in site function and hominid behaviour" (Wang 2005: 229).

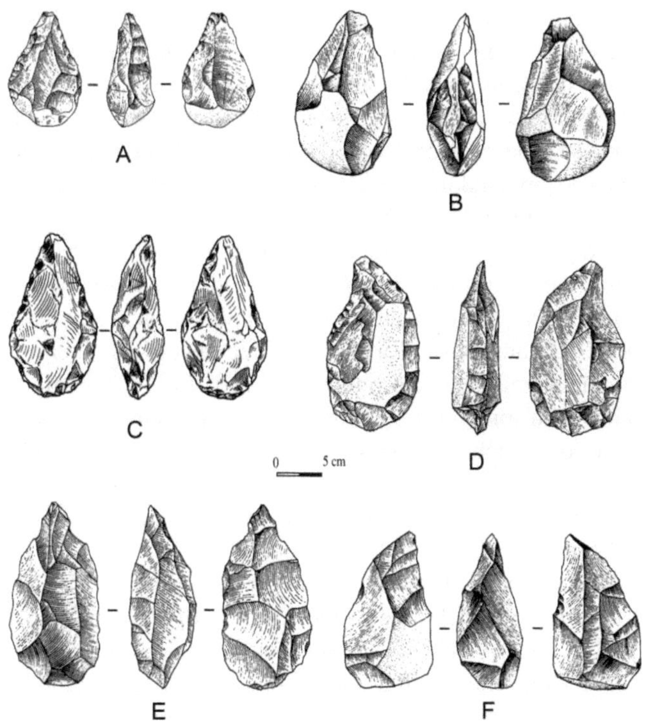

Fig. 10.10 Handaxes from open-air sites in the Luonan Basin, China. After Wang 2005, Fig. 9.2

Three further points should be kept in mind when discussing the Early Paleolithic of China. The first is that our knowledge of the Early Paleolithic of China is based largely upon a small number of areas that have been surveyed (notably Zhoukoudian, the Nihewan, Luonan, and Bose Basins), unlike Western Europe, where many areas have been explored for over 100 years. Although much has been learnt in the last decade, particularly from investigations at Yunxian (de Lumley et al. 2008) and the Danjiangkou Reservoir Region (e.g., Li et al. 2014), the Paleolithic community in China is still extremely small, especially in comparison with its size, and large areas remain unexplored in terms of Paleolithic evidence. Secondly, the presence or absence of bifaces can depend greatly upon the quality and size of flakeable nodules that are locally available. Clearly, the manufacture of a biface requires blocks, nodules, or cobbles that are larger than the finished item, and amenable to repetitive flaking. Where these conditions are not met, the absence of bifaces and large flakes may simply indicate that the local inhabitants lacked the means of making them, but not a lack of ability to do so. One example is the Nihewan Basin, China, where the rock is poor-quality crystalline that shatters easily when struck, and rarely allows the working of a large artifacts; also Tajikistan, where the raw material were pebbles ca. 3–5 cm long (Ranov 1995); Bizat Ruhama, Israel, where artifacts were typically <3 cm long; (Zaidner 2003) and Evron Quarry, Israel, where the mean core length was only 35 mm (Ronen et al. 1998). Here, it is likely the hominins were using a raw material of necessity, rather than prime choice. The Nihewan Basin is unlikely therefore to be representative of the Early Paleolithic of China in areas where high-quality stone was locally available. Thirdly, in a country as large and diverse as China, we should not be surprised to find significant differences in assemblage composition as there is ample scope for different pulses of immigration, local developments, and convergent development.

10.4.4 Korea

Norton et al. (2006) have summarized the evidence from four localities (Kumpari, Chuwoli, Kawoli, and Chongokni) in the Imjin-Hantan River Basin (IHRB) in South Korea that contain Acheulean types of bifaces. Of these, Chongokni is the best known. As a result of 11 excavations and surveys over 20 years, over 5000 heavy-duty tools, flake implements, and débitage have been recovered; however, these include only 10 handaxes (Lee 2001). Their age is uncertain but on the assumption (perhaps questionable) that constant sedimentation rates were constant, the age range of Korean bifaces appears to be from ca. 350–300 ka to the Late Pleistocene (Bae et al. 2012).

10.5 Mainland and Island SE Asia

Brumm and Moore (2012) point out that there are numerous surface finds of bifaces in both mainland and island Southeast Asia that are routinely dismissed as being "Acheulean", even though similar finds west of the Movius Line are routinely classified as Acheulean. As they state (2012: 32) "Dating African and western Eurasian surface finds as Acheulean while dismissing similar Southeast Asian and Far Eastern artifacts is a case of shifting the goalposts, one that potentially distorts Acheulean evidence in the Palaeolithic Old World". The remedy is to concentrate on finding bifaces in stratified contexts in order to determine whether or not they are Early or Middle Pleistocene in age, and not (as often alleged for SE Asia), Holocene. The best evidence of an Acheulean assemblage from SE Asia is from Ngebung 2, Sangiran, Central Java, where cleavers were found along with flakes, choppers, hammerstones, horsehooves, polyhedrons, and bolas in association with faunal remains that included *Stegodon, Axis lydekkeri,* and *Bos bubalus.* The likely age is 0.86–0.88 Ma. Handaxes are also reported from numerous localities in Sumatra, Java, Bali, Lombok, Sulawesi, and Halmahera. Their age is unknown but likely to be late Middle or early Late Pleistocene (Simanjuntak et al. 2010). It is unclear how much of this material resulted from immigration or from local convergent development. What is clear, however, is that bifaces similar to ones classified as Acheulean west of the Movius Line are far commoner in Southeast Asia than often realized.

10.6 Discussion

Having outlined the evidence for Acheulean assemblages and sites across Asia, what then can we say about "the Acheulean" in this continent? Despite the facts that we know almost nothing about who made and used these assemblages or how they lived, two major and related themes emerge.

10.6.1 *The Patchiness of the Asian Acheulean*

The first aspect of the record for Acheulean assemblages in Asia is that the Acheulean in Asia is not monolithic and continuous, but is instead thin and patchy. The only regions where a plausible case can be made for continuous occupation from the Early to late Middle Pleistocene are the Levant and probably India (although the dating needs considerable improvement in the latter case). In regions such as Turkey, Jordan, Iran, Pakistan, Central and South East Asia, the evidence for hominins in the Early and Middle Pleistocene (even when including non-bifacial assemblages) is extremely patchy, especially when considered against the time-span involved. The Arabian Peninsula may prove to be another core region of continuous settlement, but more needs to be learnt about its paleohydrological history before being sure. China south of the Qinling Mountains may prove to be another region where occupation was continuous, but too little is known at present. Even allowing for differences in the amount of fieldwork, and factors affecting the preservation, reexposure and discovery of Paleolithic material, it is unlikely that the patchiness of the Asian Acheulean will change substantially (see also Bar-Yosef and Belmaker 2011): there are a few core areas, separated by large areas where occupation records are very ephemeral. Much of "the Acheulean" world is likely to have been empty for much of the time; when we discuss "the Acheulean" in regions such as Iran or China, it is likely that it comprised brief episodes of occupation by small populations of perhaps a few hundred individuals.

Even in areas such as the Levant and India where population continuity is likely, the record is still very thin, especially when set against the time-span involved. This is particularly true of the first million years of Acheulean presence in Asia. In the Levant, for example, there is only the record from 'Ubeidiya (1.6–1.2 Ma), Latamne (ca. 1.0 Ma), Evron Quarry (ca. 1.0 Ma), GBY (0.78–0.70 Ma); in India, Attirampakkam (1.5–1.0 Ma) is the only Acheulean site so far dated to the 900 ka time-range of 1.6–0.7 Ma. Even in an area as well-studied as the Levant, the temporal gaps between dated Acheulean sites are commonly >100,000 years. Even at a local scale, therefore, it is likely that settlement records are discontinuous and marked by repeated extinction events. As I summarized elsewhere (Dennell 2009: 475) "The early hominin settlement of Asia is a repeated theme of regional expansion and contraction, colonization and abandonment, integration and isolation as rainfall increased or decreased. When viewed in closer detail much of the Asian Early Paleolithic record is likely to comprise regional discontinuities and local extinctions, rather than long-term continuity and permanent residence". As a consequence in Asia, "there is no real cultural continuity and that what we view as a constant stream of migrants, was actually interrupted many times" (Bar-Yosef and Belmaker 2011: 1332).

The primary determinants at a continental scale that determine the patchiness of the Asian Early Paleolithic (including the Acheulean) are wind and rainfall: the westerly winds from the Mediterranean that carry rain inland to western Iran and Central Asia, and the summer monsoon winds that carry rain across South and South East Asia and China. As now abundantly demonstrated (see Dennell 2009,

Chaps. 3 and 7 for a summary), these follow the same glacial–interglacial rhythms as the ice fronts of Europe: cool/cold and dry episodes alternated with warm, moist/wet periods. In semi-arid and arid areas of continental Asia, long-time continuity in occupation would always have been threatened by downturns in rainfall. Even in core areas, hominin survival would have been at risk by short-term climatic downturns operating at a timescale ranging from a few millennia to a few decades, along with random fluctuations in birth and mortality patterns and resource availability. At the level of individual groups, the vagaries of weather are of more pressing concern than the rhythms of climate change over scores of millennia. Groups using Acheulean bifaces are likely to have been small, organized in loose social networks, and highly vulnerable to vicissitudes of climate, resource availability, and random mortality. They also lacked the buffering and back-up systems of later hominins, such as storage technologies, large-scale exchange systems, and dependable social networks defined by kinship, language, or ideology. Local and sometimes regional extinction are thus likely features of Acheulean settlement histories. As Bar-Yosef and Belmaker (2011: 1332) point out "Such local and temporal extinctions are possibly one of the reasons, together with low archaeological visibility, why the number of Lower Paleolithic sites, even in well-researched areas, is still so small. Therefore the current distribution of lithic industries across Eurasia is undoubtedly incomplete … indicating that *there is no real cultural continuity and that what we view as a constant stream of migrants, was actually interrupted many times* (e.g. Dennell 2004, 2009)" (italics mine).

10.6.2 The Slow Pace of Change

The second feature of the Acheulean in Asia is the exceedingly slow rate of change: compared with later hominins of the Middle and Upper Paleolithic (or their regional equivalents), the Acheulean is unimaginably conservative. For over a million years—or 50,000 generations (assuming a generation length of 20 years)—lithic technology continued to be centered on bifaces and unstandardized flake tools. On the surface, the Acheulean appears to be a record of unbelievable monotony. Below the surface, however, some long-terms trends may be detected.

The first is the evidence of the hominin skeletal record. Brain size, as measured by cranial capacity, gradually increased from ca. 650–750 cc for early *H. erectus* (as at Dmanisi) to 1000–1200 cc (as with *H. erectus* at Zhoukoudian and *H. heidelbergensis* in Europe). Body size also increased to the modern human range. These changes were probably driven by an increased reliance on meat and carnivory as a source of high-quality protein, which in turn reflects greater skill at hunting medium- and large-sized animals on a regular basis. There was also a reduction in sexual dimorphism, which may have resulted (by analogy with other primates) in greater pair bonding and parental investment.

The second long-term trend is shown by the lithic record, which shows three indicators of long-term change. The first is a greater reliance on soft hammer techniques for flaking (e.g., the antler percussor at Boxgrove) and the gradual emergence by the late Acheulean of the Levallois technique, and a greater use of LCT's. Experimental research at Isampur indicates that hominins were able to segregate different operations for making handaxes or cleavers (see above), although current dating issues preclude understanding of when this occurred. There are also a few hints of changes in subsistence. The Levant, as example, provides three snapshots: as seen above, 'Ubeidiya at 1.6–1.2 Ma has little evidence of hunting but GBY at 780–700 ka has evidence for the hunting of fallow deer and perhaps elephant; and Qesem, at 400 ka, shows evidence of communal sharing of meat and presumably cooperative hunting. The trends are therefore of greater cooperation over longer planning periods. Fire was also another important innovation, and possibly used on a regular basis at GBY 700–780 ka and at roughly the same time, at Locality 1, Zhoukoudian (contra Weiner et al. 1998; see Dennell 2009: 411–413).

The dominant impressions from the Acheulean component of the Early Paleolithic in Asia are of group survival and risk avoidance. Given a precarious hold on survival, and with no fallback strategies for coping with short-term changes of climate, mortality, or misfortune, survival ultimately depended upon inter- and intra-group cooperation and in minimizing risk. Through natural selection, trial and error and often, good luck, those hominins using bifacial, Acheulean assemblages managed to survive their first million and a half years in Asia.

References

Albrecht, G., & Müller-Beck, H. (1988). The Paleolithic of Şehremuz near Samsat on the Euphrates River: Summary of the excavation findings and a morphology of the handaxes. *Paléorient, 14,* 76–86.

Alperson-Afil, N., Richter, D., & Goren-Inbar, N. (2007). Phantom hearths and the use of fire at Gesher Benot Ya'aqov, Israel. *PalaeoAnthropology, 2007,* 1–15.

Alperson-Afil, N., & Goren-Inbar, N. (2010). *The Acheulian Site of Gesher Benot Ya'aqov Volume II: Ancient Flames and Controlled Use of Fire*. Dordrecht: Springer.

Athreya, S. (2007). Was *Homo heidelbergensis* in South Asia? A test using the Narmada fossil from central India. In M. D. Petraglia & B. Allchin (Eds.), *The Evolution and History of Human Populations in South Asia. Inter-disciplinary Studies in Archaeology, Biological Anthropology, Linguistics and Genetics* (pp. 137–170). Dordrecht: Springer.

Bae, K., Bae, C. J., & Kim, K. (2012). The age of the Paleolithic handaxes from the Imjin-Hantan River Basins, South Korea. *Quaternary International, 281,* 14–25.

Bar-Yosef, O. (1994). The lower Palaeolithic of the Near East. *Journal of World Prehistory, 8,* 211–265.

Bar-Yosef, O. (1998a). Early colonizations and cultural continuities in the Lower Palaeolithic of western Eurasia. In M. Petraglia & R. Korisettar (Eds.), *Early human behavior in global context: The rise and diversity of the Lower Palaeolithic record* (pp. 221–279). London: Routledge.

Bar-Yosef, O. (1998b). Jordan prehistory: A view from the West. In D. O. Henry (Ed.), *The Prehistoric Archaeology of Jordan* (pp. 162–178). Oxford: British Archaeological Reports (International Series), 705.

Bar-Yosef, O., & Goren-Inbar, N. (1993). *The lithic assemblages of 'Ubeidiya: A Lower Palaeolithic site in the Jordan Valley, Jerusalem.* Jerusalem: The Hebrew University of Jerusalem.

Bar-Yosef, O., & Belmaker, M. (2011). Early and Middle Pleistocene faunal and hominins dispersals through Southwestern Asia. *Quaternary Science Reviews, 30,* 1318–1337.

Belmaker, M., Tchernov, E., Condemi, S., & Bar-Yosef, O. (2002). New evidence for hominid presence in the Lower Pleistocene of the Southern Levant. *Journal of Human Evolution, 43,* 43–56.

Belitzky, S., Goren-Inbar, N., & Werker, E. (1991). A Middle Pleistocene wooden plank with man-made polish. *Journal of Human Evolution, 20,* 349–353.

Beyin, A. (2006). The Bab al Mandab vs the Nile-Levant: An appraisal of the two dispersal routes for early modern humans out of Africa. *African Archaeological Review, 23,* 5–30.

Biagi, P. (2006). The archaeological sites of the Rohri Hills (Sindh, Pakistan): The way they are being destroyed. *Web Journal of Cultural Heritage, 1,* 77–95.

Biagi, P., & Cremaschi, M. (1998). The early Palaeolithic Sites of the Rohri Hills (Sind, Pakistan) and their environmental significance. *World Archaeology, 19,* 421–433.

Biglari, F., & Jahani, V. (2011). The Pleistocene human settlement in Gilan, South west Caspian Sea: Recent research. *Eurasian Prehistory, 8,* 3–28.

Biglari, F., & Shidrang, S. (2006). The Lower Paleolithic Occupation of Iran. *Near Eastern Archaeology, 69,* 160–168.

Biglari, F., Heydari, S., & Shidrang, S. (2004). Ganj Par: The first evidence for Lower Paleolithic occupation in the Southern Caspian Basin, Iran. *Antiquity, 78,* Project Gallery.

Brumm, A., & Moore, M. (2012). Biface distributions and the Movius Line: A Southeast Asian perspective. *Australian Archaeology, 74,* 32–46.

Clark, J. D. (1967). The Middle Acheulean occupation site at Latamne, Northern Syria (First paper). *Quaternaria, 9,* 1–68.

Clark, J. D. (1969). The Middle Acheulean occupation site at Latamne, Northern Syria (Second paper). *Quaternaria, 10,* 1–60.

Copeland, L. (1998). The Lower Palaeolithic of Jordan. In D.O. Henry (Ed.), *The Prehistoric Archaeology of Jordan* (pp. 5–21). Oxford: British Archaeological Reports (International Series), 705.

Corvinus, G. (1981). *A survey of the Pravara River system in Western Maharashtra, India. Vol. 1: The stratigraphy and geomorphology of the Pravara River system.* Tübingen: Institut für Urgeschichte, Tübinger Monographien zur Urgeschichte, 7.

Corvinus, G. (1983). *A survey of the Pravara River system in Western Maharashtra, India. Vol. 2. The excavations of the Acheulean Site of Chirki-on-Pravara, India.* Tübingen: Institut für Urgeschichte, Tübinger Monographien zur Urgeschichte, 7.

de Lumley, H., Batalla i Llasat, G., Cauche, D., Notter, O., Yanxian, L., Tianyuan, L., et al. (2008). L'industrie du Paléolithique inferieur du site de l'homme de Yunxian. In: H. de Lumley & L. Tianyuan (Eds.), *Le site de l'homme de Yunxian: Quyuanhekou, Quingpu, Yunxian, Province du Hebei* (pp. 467–583). Paris: CRNS Editions Recherche sur les Civilisations.

Dennell, R. W. (2004). Hominid dispersals and Asian biogeography during the Lower and Early Middle Pleistocene, ca. 2.0–0.5 Mya. Asian Perspectives, *43*(2), 205–226.

Dennell, R. W. (2009). *The Paleolithic settlement of Asia.* Cambridge: Cambridge University Press.

Dennell, R. W. (2011). An earlier Acheulian arrival in South Asia. *Science, 331,* 1532–1533.

Dennell, R. W. (2014a). East Asia and human evolution: From cradle of mankind to cul-de-sac. In R. W. Dennell & M. Porr (Eds.), *Southern Asia, Australia and the search for human origins* (pp. 8–20). Cambridge: Cambridge University Press.

Dennell, R. W. (2014b). Hallam Movius, Helmut de Terra, and the Line that never was: Burma, 1938. In K. Boyle, R. J. Rabett, & C. Hunt (Eds.), *Living in the landscape: Essays in honour of Graeme Barker* (pp. 11–34). Cambridge: McDonald Institute for Archaeological Research.

Dennell, R. W. (2016). Life without the Movius Line. *Quaternary International, 400,* 14–22.

d'Errico, F., & Nowell, A. (2000). A new look at the Berekhat Ram figurine: Implications for the origins of symbolism. *Cambridge Archaeological Journal, 10,* 123–167.

Derricourt, R. (2006). Getting "Out of Africa": Sea crossings, land crossings and culture in the hominin migrations. *Journal of World Prehistory, 19,* 119–132.

Doronichev, V. B. (2000). Lower Palaeolithic occupation of the northern Caucasus. In D. Lordkipanidze, O. Bar-Yosef, & M. Otte (Eds.), *Early Humans at the Gates of Europe* (pp. 67–77). Liège: ERAUL, Université of Liège, Études et Recherches Archéologiques de l'Université de Liège.

Doronichev, V., & Golovanova, L. (2003). Bifacial tools in the Lower and Middle Palaeolithic of the Caucasus and their contexts. In M. Soressi & H. Dibble (Eds.), *Multiple Approaches to the Study of Bifacial Technologies* (pp. 77–106). Philadelphia: University of Pennsylvania Museum of Archaeology and Anthropology, University Museum Monograph, 115.

Feibel, C. S. (2004). Quaternary lake margins of the Levant Rift Valley. In N. Goren-Inbar & J. D. Speth (Eds.), *Human Paleoecology in the Levantine Corridor* (pp. 21–36). Oxford: Oxbow Books.

Fourloubey, C., Beauval, C., Colonge, D., Liagre, J., Ollivier, V., & Chataigner, C. (2003). Le paléolithique en Arménie : état des connaissances acquises et données récentes. *Paléorient, 29,* 5–18.

Gaillard, C., & Mishra, S. (2001). The lower Palaeolithic in South Asia. In F. Sémah, C. Falguères, D. Grimaud-Hervé, & A.-M. Sémah (Eds.), *Origines des peuplements et chronologies des cultures paléolithiques dans le Sud-Est Asiatique* (pp. 73–91). Semananjung.

Gaillard, C., Raju, D. R., Misra, V. N., & Rajaguru, S. N. (1983). Acheulean occupation at Singi Talav in the Thar Desert: A preliminary report on 1982 excavation. *Man and Environment, 7,* 112–130.

Gaudzinski, S. (2004a). Early hominid subsistence in the Levant. Taphonomical studies at the Plio/Pleistocene 'Ubeidiya Formation (Israel): Evidence from layer II-24. In N. Goren-Inbar & J. D. Speth (Eds.), *Human Paleoecology in the Levantine Corridor* (pp. 75–88). Oxford: Oxbow Books.

Gaudzinski, S. (2004b). Subsistence patterns of Early Pleistocene hominids in the Levant—Taphonomic evidence from the 'Ubeidiya Formation (Israel). *Journal of Archaeological Science, 31,* 65–75.

Gaudzinski-Windheuser, S. (2005). Subsistenzstrategien frühpleistozäner Hominiden in Eurasien. Taphonomische Faunenbetrachtungen der Fundstellen der 'Ubeidiya Formation (Israel). *Römisch-Germanischen Zentralmuseums, 61,* 1–310.

Ginat, H., Zilberman, E., & Saragusti, I. (2003). Early Pleistocene lake deposits and Lower Palaeolithic finds in Nahal (wadi) Zihor, Southern Negev desert, Israel. *Quaternary Research, 59,* 445–458.

Goren-Inbar, N. (1985). The lithic assemblage of the Berekhat Ram Acheulean site, Golan Heights. *Paléorient, 11,* 7–28.

Goren-Inbar, N., Lister, A., Werker, E., & Chech, M. (1994). A butchered elephant skull and associated artifacts from the Acheulean site of Gesher Benot Ya'aqov, Israel. *Paléorient, 20,* 99–112.

Goren-Inbar, N., Feibel, C. S., Versoub, K. L., Melamed, Y., Kislev, M. E., Tchernov, E., et al. (2000). Pleistocene milestones on the Out-of-Africa corridor at Gesher Benot Ya'aqov, Israel. *Science, 289,* 944–947.

Goren-Inbar, N., Sharon, G., Melamed, Y., & Kislev, M. (2002a). Nuts, nut-cracking, and pitted stones at Gesher Benot Ya'aqov, Israel. *Proceedings of the National Academy of Science USA, 99,* 2455–2460.

Goren-Inbar, N., Werker, E., & Feibel, C. S. (2002b). *The Acheulean Site of Gesher Benot Ya'aqov, Israel. 1. The Wood Assemblage.* Oxford: Oxbow.

Goren-Inbar, N., Alperson, N., Kislev, M. E., Simchoni, O., Melamed, Y., Ben-Nun, A., et al. (2004). Evidence of hominin control of fire at Gesher Benot Ya'aqov, Israel. *Science, 304,* 725–727.

Guralnik, B., Matmon, A., Avni, Y., & Fink, D. (2011). ^{10}Be exposure ages of ancient desert pavements reveal Quaternary evolution of the Dead Sea drainage basin and rift margin tilting. *Earth and Planetary Science Letters, 290,* 132–141.

Gvirtzman, G., Wieder, M., Mardr, O., Khalaily, H., Rabinovich, R., & Ron, H. (1999). Geological and pedological aspects of an Early-Paleolithic site: Revadim, Central Coastal Plain, Israel. *Geoarchaeology, 14,* 101–126.

Haslam, M., Roberts, R. G., Shipton, C., Pal, J. N., Fenwick, J. L., Ditchfield, P., et al. (2011). Late Acheulean hominins at the Marine Isotope Stage 6/5e transition in north-central India. *Quaternary Research, 75,* 670–682.

Hou, Y., Potts, R., Baoyin, Y., Zhengtang, G., Deino, A., & Wei, W. (2000). Mid-Pleistocene Acheulean-like stone technology of the Bose Basin, South China. *Science, 287,* 1622–1626.

Howe, B. (2014). *Barda Balka.* Chicago: Oriental Institute Communications, 31.

Huang, W., Ne, N., & Sagawa, M. (2001). *Comparative studies on handaxes found at the Bose sites in Guangxi, China II.* Sendai: Sagawa Matatoshi, Tohoku Gakuin University.

Islamov, Y. I. (1990). Sel'oungour, un nouveau site du paléolithique inférieur en Asie centrale. *L'Anthropologie, 94,* 675–688.

Jacobson, J. (1985). Acheulean surface sites in Central India. In V. N. Misra & P. Bellwood (Eds.), *Recent advances in Indo-Pacific prehistory* (pp. 49–58). New Delhi: Oxford and IBH Publishing.

Jagher, R., & Le Tensorer, J.-M., (2011). El Khowm, a key area for the Palaeolithic of the Levant in Central Syria. In J.-M. Le Tensorer, R. Jagher, & M. Otte (Eds.), *The Lower and Middle Palaeolithic in the Middle East and neighbouring regions* (Vo. 126, pp. 197–208). Liège: ERAUL.

Jelinek, A. J. (1982). The Tabun Cave and Palaeolithic man in the Levant. *Science, 216,* 1369–1375.

Jennings, R. P., Shipton, C., Breeze, P., Cuthbertson, P., Bernal, M. A., OshanWedage, W. M. C., et al. (2015). Multi-scale Acheulean landscape survey in the Arabian Desert. *Quaternary International, 382,* 58–81.

Kennedy, K. A. R. (2003). *God-apes and fossil men: Palaeoanthropology in South Asia.* Ann Arbor: University of Michigan Press.

Korisettar, R., & Rajaguru, S. N. (1998). Quaternary stratigraphy, palaeoclimate and the Lower Palaeolithic of India. In M. Petraglia & R. Korisettar (Eds.), *Early Human Behavior in Global Context: The Rise and Diversity of the Lower Palaeolithic Record* (pp. 304–342). London: Routledge.

Korisettar, R. (2007). Toward developing a basin model for Palaeolithic settlement of the Indian subcontinent: Geodynamics, monsoon dynamics, habitat diversity and dispersal routes. In M. D. Petraglia & B. Allchin (Eds.), *The Evolution and History of Human Populations in South Asia. Inter-disciplinary Studies in Archaeology, Biological Anthropology, Linguistics and Genetics* (pp. 69–96). Dordrecht: Springer.

Kuhn, S. (2002). Palaeolithic archaeology in Turkey. *Evolutionary Anthropology, 11,* 198–210.

Kuhn, S. L., Arsebük, G., & Howell, F. C. (1996). The Middle Pleistocene lithic assemblage from Yarımburgaz Cave, Turkey. *Paléorient, 22,* 31–49.

Kuman, K. (1998). The earliest South African industries. In M. D. Petraglia & R. Korisettar (Eds.), *Early Human Behaviour in Global Context: The Rise and Diversity of the Lower Palaeolithic Record* (pp. 151–186). London: Routledge.

Langbroek, M. (2015). Do tektites really date the bifaces from the Bose (Baise) Basin, Guangxi, southern China? *Journal of Human Evolution, 80,* 175–178.

Le Tensorer, J.-M. (2014). Regional perspective of early human populations in Syria: The case of El Kowm. In N. Sanz (Ed.), *Human origin sites and the World Heritage Convention in Eurasia* (pp. 54–71). Paris: UNESCO.

Le Tensorer, J.-M., Jagher, R., Rentzel, P., Hauck, T., Ismail-Meyer, K., Pümpin, C., et al. (2007). Long-term site formation processes at natural springs Nadaouiyeh and Hummal in the El Kowm Oasis, Central Syria. *Geoarchaeology, 22,* 621–639.

Leakey, M. D. (1971). *Olduvai Gorge, Volume 3: Excavations in Beds I and II, 1960–1963.* Cambridge: Cambridge University Press.

Lee, H. W. (2001). The Paleolithic industries in Korea: Chronology and related new find spots. In S. Milliken & J. Cook (Eds.), *A Very Remote Period Indeed: Papers on the Palaeolithic presented to Derek Roe* (pp. 133–137). Oxford: Oxbow Books.

Leng, J. (2001). *Early Paleolithic Technology in Eastern and Southern Asia.* Oxford: British Archaeological Reports (International Series), 924.

Lepre, C. J., Roche, H., Kent, K. V., Harmand, S., Quinn, R. L., Brugal, J.-P., et al. (2011). An earlier origin for the Acheulian. *Nature, 477,* 82–85.

Li Hao, Li Chaorong, & Kuman, K. (2014). Rethinking the "Acheulean" in East Asia: Evidence from recent investigations in the Danjiangkou Reservoir Region, central China. *Quaternary International, 347,* 163–175.

Li Huayu, Sun, X., Wang, S., Cosgrove, R., Zhang, H., Yi, S., et al. (2011). Ages for hominin occupation in Lushi Basin, middle of South Luo River, central China. *Journal of Human Evolution, 60,* 612–617.

Lu, H., Zhang, H., Wang, S., Cosgrove, R., Sun, X., Zhao, J., et al. (2011). Multiphase timing of hominin occupations and the paleoenvironment in Luonan Basin. *Central China, Quaternary Research, 76,* 142–147.

Lioubine, V. P. (2002). *L'Acheuléen du Caucase.* Liège: ERAUL, 93.

Ljubin, V. P., & Bosinski, G. (1995). The earliest occupation of the Caucasus region. In W. Roebroeks & T. van Kolfschoten (Eds.), *The Earliest Occupation of Europe* (pp. 207–253). Leiden: University of Leiden Press.

Lycett, S. J., & Norton, C. J. (2010). A demographic model for Palaeolithic technological evolution: The case of East Asia and the Movius Line. *Quaternary International, 211,* 55–65.

Lycett, S. J., & Bae, C. J. (2010). The Movius Line controversy: The state of the debate. *World Archaeology, 42,* 521–544.

Madsen, B., & Goren-Inbar, N. (2004). Acheulean giant core technology and beyond: An archaeological and experimental case study. *Eurasian Prehistory, 2,* 3–52.

Mallol, C. (2006). What's in a beach? Soil micromorphology of sediments from the Lower Palaeolithic site of 'Ubeidiya, Israel. *Journal of Human Evolution, 51,* 185–206.

Marder, O., Gvirtzman, G., Ron, H., Khalaily, H., Wieder, M., Bankirer, R., et al. (1999). The Lower Paleolithic site of Revadim Quarry, preliminary finds. *Journal of the Israel Prehistoric Society, 28,* 21–53.

Matter, A., Neubert, E., Preusser, F., Rosenberg, T., & Al-Wagdani, K. (2015). Palaeo-environmental implications derived from lake and sabkha deposits of the southern Rub' al-Khali, Saudi Arabia and Oman. *Quaternary International, 382,* 120–131.

Meltzer, D. J. (2015). *The Great Paleolithic War: How science forged an understanding of America's ice age past.* Chicago and London: Chicago University Press.

Mercier, H., & Valladas, H. (2003). Reassessment of TL age estimates of burnt flints from the Paleolithic site of Tabun Cave, Israel. *Journal of Human Evolution, 45,* 401–409.

Mercier, N., Valladas, H., Valladas, G., Reyss, J.-L., Jelinek, A., Meignen, L., et al. (1995). TL dates of burnt flints from Jelinek's excavations at Tabun and their implications. *Journal of Archaeological Science, 22,* 495–509.

Mercier, N., Valladas, H., Froget, L., Joron, J.-L., & Ronen, A. (2000). Datation par thermoluminescence de la base du gisement paléolithique de Tabun (mont Carmel, Israel). *Comptes Rendus de l'Académie des Science, Paris (ser. 2), 330,* 731–738.

Minzoni-Deroche, A., & Sanlaville, P. (1988). Le paléolithique inferieur de la région de Gaziantep. *Paléorient, 14,* 87–98.

Mishra, S. (1994). The South Asian Lower Palaeolithic. *Man and Environment, 19,* 57–71.

Mishra, S., Gaillard, C., Hertler, C., Moigne, A.-M., & Simanjuntak, T. (2010). India and Java: Contrasting records, intimate connections. *Quaternary International, 223–224,* 265–270.

Misra, V. M. (1985). The Acheulean succession at Bhimbetka, Central India. In V. N. Misra & P. Bellwood (Eds.), *Recent advances in Indo-Pacific Prehistory* (pp. 35–47). New Delhi: Oxford and IBH Publishing.

Misra, V. N. (1989). Stone age India: An ecological perspective. *Man and Environment, 14,* 17–64.

Movius, H. L. (1948). The lower Palaeolithic cultures of southern and eastern Asia. *Transactions of the American Philosophical Society, 38,* 329–420.

Muhesen, S. (1993). L'Acheuléen récent évolué de l'Oronte. In P. Sanlaville, J. Besançon, L. Copeland, & S. Muhesen (Eds.), *Le paléolithique de la vallée moyenne de l'Oronte (Syrie)* (pp. 145–166). Oxford: British Archaeological Reports (International Series), 587.

Norton, C. J., Bae, K., Harrris, J. W. K., & Lee, H. (2006). Middle Pleistocene handaxes from the Korean Peninsula. *Journal of Human Evolution, 51,* 527–536.

Otte, M., Yalçinkaya, I., Kozlowski, J., Bar-Yosef, O., Lopez Bayón, I., & Taskiran, H. (1998). Long-term technical evolution and human remains in the Anatolian Palaeolithic. *Journal of Human Evolution, 34,* 413–431.

Ozansoy, F. (1969). Pleistocene fossil human footprints in Turkey. *Bulletin of the Mineral Research and Exploration Institute of Turkey, 72,* 146–150.

Paddayya, K. (1982). *The Acheulean culture of the Hunsgi Valley (Peninsular India): A settlement system perspective.* Poona: Deccan College Postgraduate and Research Institute.

Paddayya, K. (1991). The Acheulean culture of the Hunsgi-Baichbal Valleys, peninsular India: A processual study. *Quartär, 41–42,* 111–138.

Paddayya, K. (2001). The Acheulean culture project of the Hunsgi and Baichbal Valleys, peninsular India. In L. Barham & K. Robson-Brown (Eds.), *Human Roots: Africa and Asia in the Middle Pleistocene* (pp. 235–258). Bristol: Western Academic and Specialist Press.

Paddayya, K. (2007). The Acheulean of peninsula India with special reference to the Hunsgi and Baichbal Valleys of the lower Deccan. In M. D. Petraglia & B. Allchin (Eds.), *The Evolution and History of Human Populations in South Asia. Inter-disciplinary Studies in Archaeology, Biological Anthropology, Linguistics and Genetics* (pp. 97–119). Dordrecht: Springer.

Paddayya, K., & Petraglia, M. D. (1997). Isampur: An Acheulean workshop site in the Hunsgi Valley, Gulbarga District, Karnataka. *Man and Environment, 22,* 95–100.

Paddayya, K., Jhaldiyal, R., & Petraglia, M. D. (1999). Geoarchaeology of the Acheulean workshop at Isampur, Hunsgi Valley, Karnataka. *Man and Environment, 24,* 167–184.

Paddayya, K., Jhaldiyal, R., & Petraglia, M. D. (2000). Excavation of an Acheulean workshop at Isampur, Karnataka (India). *Antiquity, 74,* 751–752.

Paddayya, K., Blackwell, B. A. B., Jhaldiyal, R., Petraglia, M. D., Fevrier, S., Chaderton, D. A., et al. (2002). Recent findings on the Acheulean of the Hunsgi and Baichbal Valleys, Karnataka, with special reference to the Isampur excavation and its dating. *Current Science, 83,* 641–647.

Pant, P. C., & Jayaswal, V. (1991). *Paisra: The stone age settlement of Bihar.* Delhi: Agam Kala Prakashan.

Pappu, R. S. (2002). The Lower Palaeolithic culture of India. In K. Paddayya (Ed.), *Recent studies in Indian archaeology* (pp. 17–59). Delhi: Munshiram Manharlal Publishers.

Pappu, R. S., & Deo, S. G. (1994). *Man-Land relationships during Palaeolithic times in the Kaladgi Basin, Karnataka.* Pune: Deccan College Post-Graduate and Research Institute.

Pappu, S. (2001). *A re-examination of the Palaeolithic archaeological record of northern Tamil Nadu.* South India. Oxford: British Archaeological Reports (International Series), 1003.

Pappu, S. (2007). Changing trends in the study of a Palaeolithic site in India: A century of research at Attirampakkam. In M. D. Petraglia & B. Allchin (Eds.), *The Evolution and History of Human Populations in South Asia. Inter-disciplinary Studies in Archaeology, Biological Anthropology, Linguistics and Genetics* (pp. 121–135). Dordrecht: Springer.

Pappu, S., & Akhilesh, K. (2007). Preliminary observations on the Acheulean assemblages from Attirampakkam, Tamil Nadu. In N. Goren-Inbar & G. Sharon (Eds.), *Axe Age: Acheulian Tool-making from Quarry to Discard* (pp. 155–180). Oxford: Oxbow.

Pappu, S., Gunnell, Y., Taieb, M., Brugal, J.-P., & Touchard, Y. (2003). Excavations at the Palaeolithic site of Attirampakkam, South India: Preliminary findings. *Current Anthropology, 44,* 591–598.

Pappu, S., Gunnell, Y., Aklilesh, K., Braucher, R., Taieb, M., Demory, F., et al. (2011). Early Pleistocene presence of Acheulian hominins in South India. *Science, 331,* 1596–1599.

Patnaik, R., Chauhan, P. R., Rao, M. R., Blackwell, B. A. B., Skinner, A. R., Sahni, A., et al. (2009). New geochronological, palaeoclimatological and Paleolithic data from the Narmada Valley hominin locality, central India. *Journal of Human Evolution, 56,* 114–133.

Petraglia, M. D. (1998). The Lower Palaeolithic of India and its bearing on the Asian record. In M. Petraglia & R. Korisettar (Eds.), *Early Human Behavior in Global Context: The Rise and Diversity of the Lower Paleolithic Record* (pp. 343–390). London: Routledge.

Petraglia, M. D. (2001). The Lower Palaeolithic of India and its behavioural significance. In L. Barham & K. Robson-Brown (Eds.), *Human Roots: Africa and Asia in the Middle Pleistocene* (pp. 217–233). Bristol: Western Academic and Specialist Press.

Petraglia, M. D. (2003). The Lower Palaeolithic of the Arabian Peninsula: Occupations, adaptations, and dispersals. *Journal of World Prehistory, 17,* 141–179.

Petraglia, M. D., & Shipton, C. (2008). Large cutting tool variation west and east of the Movius Line. *Journal of Human Evolution, 55,* 962–966.

Petraglia, M. D., Schuldenrein, J., & Korisettar, R. (2003). Landscapes, activity, and the Acheulean to middle Palaeolithic transition in the Kaladgi Basin, India. *Journal of Eurasian Prehistory, 1,* 3–24.

Petraglia, M. D., Shipton, C., & Paddayya, K. (2005). Life and mind in the Acheulean: A case study from India. In C. Gamble & M. Porr (Eds.), *The Hominid Individual in Context: Archaeological Investigations of Lower and Middle Palaeolithic Landscapes, Locales and Artefacts* (pp. 197–219). London/New York: Routledge.

Petraglia, M. D., Drake, N., & Alsharekh, A. (2010). Acheulean landscapes and Large Cutting Tools assemblages in the Arabian peninsula. In M. D. Petraglia & J. I. Rose (Eds.), *The Evolution of Human Populations in Arabia: Paleoenvironments, Prehistory and Genetics* (pp. 103–116). Dordrecht: Springer.

Petraglia, M. D., Alsharekh, A., Breeze, P., Clarkson, C., Crassard, R., et al. (2012). Hominin dispersal into the Nefud Desert and Middle Palaeolithic Settlement along the Jubbah Palaeolake, Northern Arabia. *PLoS ONE, 7,* e49840.

Petraglia, M. D., Parton, A., Groucutt, H. S., & Alsharekh, A. (2015). Green Arabia: Human prehistory at the crossroads of continents. *Quaternary International, 382,* 1–7.

Porat, N., Zhou, L. P., Chazan, M., Noy, T., & Horwitz, L. K. (1999). Dating the lower Palaeolithic open-air site of Holon, Israel by luminescence and ESR techniques. *Quaternary Research, 51,* 328–341.

Porat, N., Chazan, M., Schwarcz, H., & Horwitz, L. K. (2002). Timing of the Lower to Middle Paleolithic boundary: New dates from the Levant. *Journal of Human Evolution, 43,* 107–122.

Presnyakov, S. L., Belyaeva, E. V., Lyubin, V. P., Rodionov, N. V., Antonov, A. V., Saltykova, A. K., et al. (2012). Age of the earliest Paleolithic sites in the northern part of the Armenian Highland by SHRIMP-II U–Pb geochronology of zircons from volcanic ashes. *Gondwana Research, 21,* 928–938.

Rabinovich, R., Gaudzinski, S., & Goren-Inbar, N. (2008). Systematic butchering of fallow deer (*Dama*) at the early Middle Pleistocene Acheulian site of Gesher Benot Ya'aqov (Israel). *Journal of Human Evolution, 54,* 134–149.

Ranov, V. A. (1995). The 'Loessic Palaeolithic' in South Tadjikistan, Central Asia: Its industries, chronology and correlation. *Quaternary Science Reviews, 14,* 731–745.

Rendell, H. M., & Dennell, R. W. (1985). Dated lower Palaeolithic artefacts from northern Pakistan. *Current Anthropology, 26,* 393.

Rollefson, G. O. (1981). The late Acheulean site at Fjaje, Wadi el-Bustan, southern Jordan. *Paléorient, 7,* 5–21.

Ronen, A., Burdukiewicz, J.-M., Laukhin, S. A., Winter, Y., Tsatkin, A., Dayan, T., et al. (1998). The lower Paleolithic site of Bizat Ruhama in the northern Negev, Israel. *Archäologisches Korrespondenzblatt, 28,* 163–173.

Runnels, C., & Özdoğan, M. (2001). The Paleolithic of the Bosphorus region, NW Turkey. *Journal of Field Archaeology, 28,* 69–111.

Saragusti, I., & Goren-Inbar, N. (2001). The biface assemblage from Gesher Benot Ya'aqov, Israel: Illuminating patterns in "Out of Africa" dispersal. *Quaternary International, 75,* 85–89.

Semaw, S., Rogers, M., & Stout, D. (2009). The Oldowan-Acheulian transition: Is there a "Developed Oldowan" artifact tradition? In M. Camps & P. Chauhan (Eds.), *Sourcebook of Paleolithic Transitions. Methods, Theories and Interpretations* (pp. 173–193). New York: Springer.

Shackley, M. (1984). Palaeolithic archaeology in the Mongolian People's Republic; A report on the state of the art. *Proceedings of the Prehistoric Society, 50,* 23–34.

Sharma, G. R., & Clark, J. D. (Eds.). (1983). *Palaeoenvironments and Prehistory in the Middle Son Valley (Madhya Pradesh, North Central India).* Allahabad: Abinash Prakashan.

Sharon, G. (2011). Large flake Acheulean. *Quaternary International, 223–224,* 226–233.

Sharon, G., & Barsky, D. (2015). The emergence of the Acheulian in Europe—A look from the east. *Quaternary International, 411,* Part B, 25–33.

Shea, J. J. (1999). Artifact abrasion, fluvial processes, and "Living Floors" from the early Paleolithic site of 'Ubeidiya (Jordan Valley, Israel). *Geoarchaeology, 14,* 191–207.

Shipton, C., Petraglia, M. D., & Paddayya, K. (2009). Stone tool experiments and reduction methods at the Acheulean site of Isampur Quarry, India. *Antiquity, 83,* 769–785.

Shipton, C., Clarkson, C., Pal, J. N., Jones, S. C., Roberts, R. G., Harris, C., et al. (2013). Generativity, hierarchical action and recursion in the technology of the Acheulean to Middle Palaeolithic transition: A perspective from Patpara, the Son Valley, India. *Journal of Human Evolution, 65,* 93–108.

Shipton, C., Parton, A., Breeze, P., Jennings, R., Groucutt, H. S., White, T. S., et al. (2014). Large Flake Acheulean in the Nefud Desert of Northern Arabia. *PaleoAnthropology, 2014,* 446–462.

Simanjuntak, T., Sémah, F., & Gaillard, C. (2010). The Palaeolithic in Indonesia: Nature and chronology. *Quaternary International, 223,* 418–421.

Singer, R., & Wymer, J. J. (1978). A hand-ax from Northwest Iran: The question of human movement between Africa and Asia in the Lower Paleolithic period. In L. G. Freeman (Ed.), *Views of the past: Essays in Old World prehistory and paleoanthropology* (pp. 13–27). The Hague: Mouton.

Sun, X., Lu, H., Wang, S., Yi, S., Chen, C., & Zhang, W. (2013). TT-OSL dating of Longyadong Middle Paleolithic site and paleoenvironmental implications for hominin occupation in Luonan Basin (central China). *Quaternary Research, 79,* 168–174.

Tchernov, E. (1987). The age of the 'Ubeidiya Formation, an Early Pleistocene hominid site in the Jordan Valley, Israel. *Israel Journal of Earth Sciences, 36,* 3–30.

Tryon, C. A., Logan, M. A. V., Mouralis, D., Kuhn, S., Slimak, L., & Balkan-Atli, N. (2009). Building a tephrostratigraphic framework for the Paleolithic of Central Anatolia, Turkey. *Journal of Archaeological Science, 36,* 637–652.

Vishnyatsky, L. B. (1999). The Paleolithic of Central Asia. *Journal of World Prehistory, 13,* 69–122.

Wang, S. (2005). *Perspectives of hominid behaviour and settlement patterns.* Oxford: British Archaeological Reports International Series, 1406.

Wang, S., Lu, H., & Xing, L. (2014a). Chronological and typo-technological perspectives on the Paleolithic archaeology in Lantian, Central China. *Quaternary International, 347,* 183–192.

Wang, W., & Bae, C. (2015). How old are the Bose (Baise) Basin (Guangxi, southern China) bifaces? The Australasian tektites question revisited. *Journal of Human Evolution, 80,* 171–174.

Wang, W., Bae, C. J., Huang, S., Huang, X., Tian, F., Mo, J., et al. (2014b). Middle Pleistocene bifaces from Fengshudao (Bose Basin, Guangxi, China). *Journal of Human Evolution, 69,* 110–122.

Weiner, S., Xu, Q., Goldberg, P., Liu, J., & Bar-Yosef, O. (1998). Evidence for the use of fire at Zhoukoudian, China. *Science, 281,* 251–253.

Whalen, N. M., & Pease, D. W. (1990). Variability in Developed Oldowan and Acheulean bifaces of Saudi Arabia. *Atlatl, 13,* 43–48.

Whalen, N. M., Sindi, H., Wahida, G., & Siraj-Ali, J. S. (1983). Excavation of Acheulean sites near Saffaqah in ad-Dawādmī 140-1982. *Atlatl, 7,* 9–21.

Whalen, N. M., Siraj-Ali, J., Sindi, H. O., Pease, D. W., & Badein, M. A. (1988). A complex of sites in the Jeddah-Wadi Fatimah area. *Atlatl, 11,* 77–85.

Whalen, N. M., Davis, W. P., & Pease, D. W. (1989). Early Pleistocene migrations into Saudi Arabia. *Atlatl, 12,* 59–75.

Williams, M. A. J., & Clarke, M. F. (1995). Quaternary geology and prehistoric environments in the Son and Belan Valleys, North Central India. In S. Wadia, R. Korisettar, & V. S. Kale (Eds.), *Quaternary environments and geoarchaeology of India* (Vol. 32, pp. 282–308). Bangalore: Geological Society of India Memoir.

Yalçınkaya, I. (1981). Le paléolithique inférieur de Turquie. In J. Cauvin & P. Sanlaville (Eds.), *Préhistoire du Levant* (pp. 207–218). Paris: Colloques Internationaux du CRNS, 598.

Yang, S. X., Huang, W. W., & Hou, Y. M. (2014). Is the Dingcun lithic assembly a "chopper-chopping tool industry", or "Late Acheulian"? *Quaternary International, 321,* 3–11.

Yizraeli, T. (1967). A lower Paleolithic site at Holon (preliminary report). *Israel Exploration Journal, 17,* 144–152.

Zaidner, Y. (2003). The use of raw material at the Lower Palaeolithic site of Bizat Ruhama, Israel. In M. Burdukiewicz & A. Ronen (Eds.), *Lower Palaeolithic small tools in Europe and the Levant* (pp. 121–132). Oxford: British Archaeological Reports (International Series), 1115.

Chapter 11
From 800 to 500 ka in Western Europe. The Oldest Evidence of Acheuleans in Their Technological, Chronological, and Geographical Framework

Marie-Hélène Moncel and Nick Ashton

Abstract This paper focuses on the early evidence of assemblages with bifacial tools, in particular their technology within the context of chronology and geography, focusing on the sites of La Noira, Arago levels P and Q and Cagny-la-Garenne I–II in France, Brandon Fields, Maidscross Hill, High Lodge and Boxgrove in the UK, and Notarchirico in Italy. Assemblages with bifacial tools, including Large Cutting Tools (LCTs), demonstrate a high diversity of technological and morphological features as early as 700 ka and are contemporary with non-handaxe assemblages. They also show specific features that contrast between northern and southern Europe, such as the use of large flakes for bifacial manufacture, or the presence of cleavers on flakes. Lack of data regarding a local origin and more elaborate bifaces in these sites indicate an early arrival of new traditions in western and southern Europe on a pre-existing hominin presence. The assemblages are compared to those without LCTs such as Happisburgh Site 3 and Pakefield in UK, Isernia La Pineta in Italy, Atapuerca Gran Dolina TD6 and Vallparadis in Spain, Pradayrol and Soleihac in France. Hypotheses on factors behind the variation, such as function, type of site, raw material constraint, and traditions of manufacture, are discussed. The period 800–500 ka is a key episode for examining behavioral changes which occurred in Europe. The discovery of hominin fossils such as the Mauer mandible in Germany led to the definition of *Homo heidelbergensis*. The emergence of new behaviors such as the ability to produce large flakes and/or large bifacial tools (handaxes, cleavers and others) leads to discussion about new skills, new social organizations, and the arrival or *in situ* evolution of hominins.

Keywords Western Europe • pre-MIS 12 • Technology

11.1 Introduction

The earliest assemblages with bifacial technology are described in East Africa from 1.76 Ma, and in the Levant and India from 1.5 Ma onward (Kleindienst 1961; Clark 1969; Bar-Yosef and Goren-Inbar 1993; Martínez-Navarro et al. 1997; Goren-Inbar et al. 2000; Carbonell et al. 2005, 2008; Lepre et al. 2011; Mgeladze et al. 2011; Pappu et al. 2011; Beyene et al. 2013; Gallotti and Mussi 2018; Semaw et al. 2018). This early Acheulean can include bifaces, cleavers, and other heavy-duty tools and sometimes Large Cutting Tools (LCTs) made on flakes over 10 cm long (see Kleindienst 1961 and Isaac 1977 for the definition). In Europe, evidence of bifacial technology is much more recent than in Africa. Recent discoveries in Spain, France, and Britain have broadened our ideas about human colonization in southern and northern parts of the continent and show the onset of this technology before 500 ka, for example, in Italy at Notarchirico at 640 ka, in France at Arago (levels P and Q) at over 550 ka and La Noira (stratum a) at 700 ka (Piperno 1999; Barsky and Lumley 2010; Lefèvre et al. 2010; Barsky 2013; Moncel et al. 2013; Falguères et al. 2015). Moreover, the recent discovery of la Boella in Spain with bifacial tools dated to 1–0.9 Ma might show the beginning of European bifacial technology, but also raises questions about its origin, whether local or introduced, and reduces the chronological gap between Asia and Europe (Vallverdú et al. 2014). Assemblages with no evidence of bifacial technology also persist after 1 Ma.

This paper mainly focuses on the period from 800–500 ka, a key phase in Europe with the co-occurrence of assemblages

M.-H. Moncel (✉)
Département de Préhistoire, UMR 7194 CNRS, Muséum national d'Histoire naturelle, 1 rue René Panhard, 75013 Paris, France
e-mail: marie-helene.moncel@mnhn.fr

N. Ashton
Department of Britain, Europe & Prehistory, British Museum, Franks House. 56 Orsman Road, London, N1 5QJ, UK
e-mail: nashton@britishmuseum.ac.uk

R. Gallotti and M. Mussi (eds.), *The Emergence of the Acheulean in East Africa and Beyond: Contributions in Honor of Jean Chavaillon*, Vertebrate Paleobiology and Paleoanthropology,
https://doi.org/10.1007/978-3-319-75985-2_11

with bifacial tools (Notarchirico, La Noira, Arago levels P-Q, Brandon Fields, Boxgrove and Cagny-la-Garenne) and those without (Pakefield, Happisburgh III, Isernia la Pineta and Atapuerca Gran Dolina TD6) (Moncel 2008, 2010). The variation in assemblages raises questions about the relationship between different hominin types, in particular whether the introduction or development of bifacial technology is linked to the dispersal of *Homo heidelbergensis* (Stringer 1996, 2012; Mounier et al. 2009; Wagner et al. 2010; Moncel et al. 2013; Ashton 2015). The few hominin fossils, dating to between 800 and 500 ka, are attributed to either *Homo antecessor* or *Homo heidelbergensis* (Gran Dolina TD6, Mauer and Boxgrove), and the diversity of anatomical features suggests possible hominin intra- or inter-diversity in Europe. Recent DNA studies and the anatomical features of the few available fossils might suggest a greater diversity of hominins and relationships between European and Asian populations. Europe is located at the end of the continent and may have recorded multiple influences or influxes of people or more local developments. Dental analysis points to longitudinal migrations of new hominin groups from Asia, and genetic data suggest speciation events in Africa or Eurasia prior to 600 ka (Martinón-Torrès et al. 2007, 2011; Stringer 2012; Bermudes de Castro and Martinón-Torrès 2013; Meyer et al. 2014).

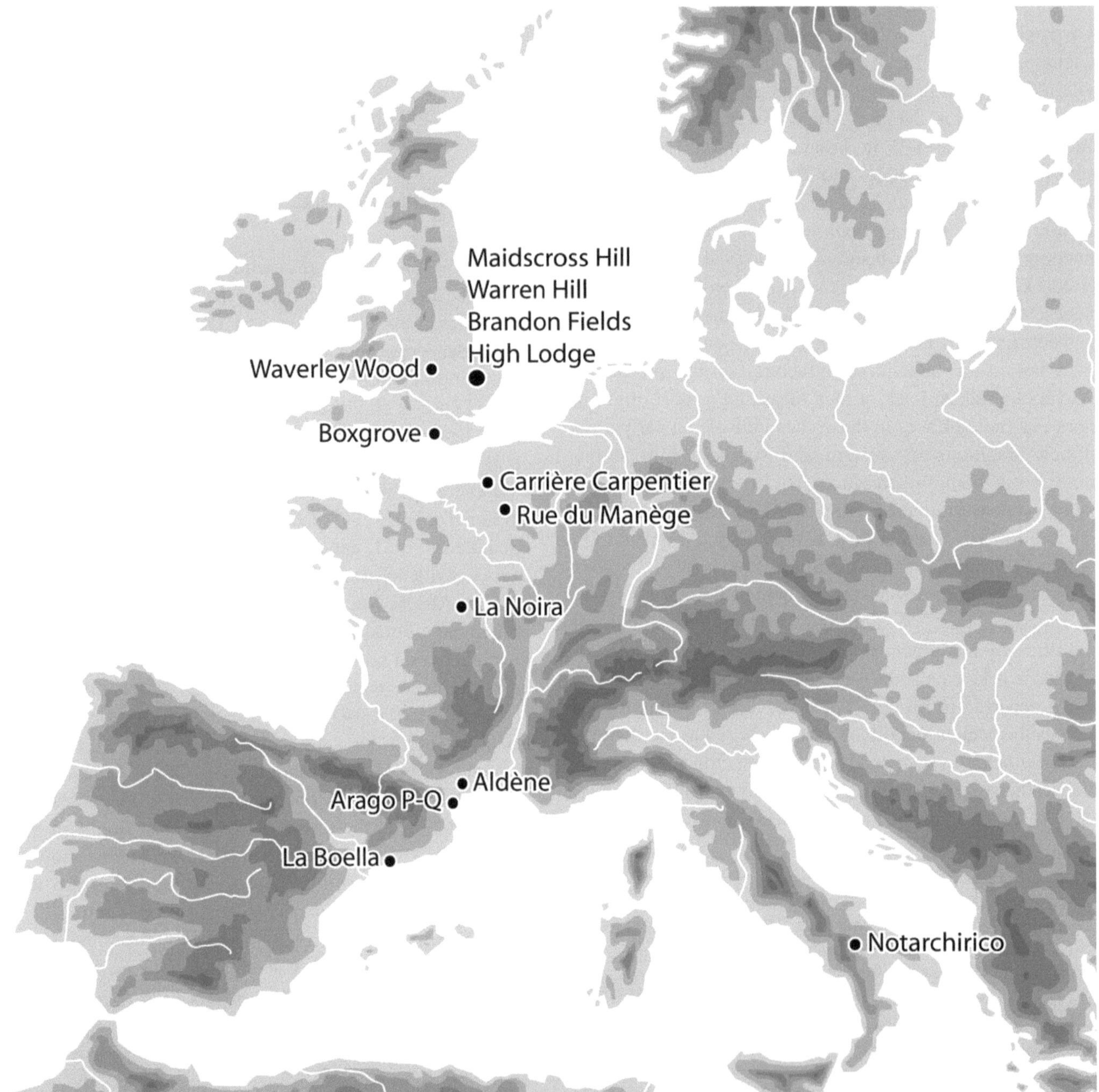

Fig. 11.1 Map of sites with bifacial tools (earlier than MIS 13 and MIS 13)

Table 11.1 Main biface and non-biface sites discussed in the text from 1 Ma

Site	Country	Date	Dating methods	Biface technology	Main reference
La Boella	Spain	1–0.9 Ma	Paleomagnetism Biostratigraphy	Yes	Vallverdú et al. (2014)
Pradayrol	France	900 ka	Biostratigraphy	No	Guadelli et al. (2012)
Atapuerca TD6	Spain	800 ka	Paleomagnetism Biostratigraphy	No	Ollé et al. (2013)
Vallparadis	Spain	800 ka	Biostratigraphy	No	Martínez et al. (2010)
Happisburgh III	UK	>800 ka	Chronostratigraphy Paleomagnetism Biostratigraphy	No	Parfitt et al. (2010)
Pakefield	UK	700 ka	Chronostratigraphy Biostratigraphy	No	Parfitt et al. (2005)
La Noira	France	700 ka	Chronostratigraphy ESR on quartz	Yes	Despriée et al. (2011)
Notarchirico	Italy	640 ka	Tephras	Yes	Piperno (1999)
Soleihac	France	700–500 ka	Tephras	No	Bracco (1991)
Isernia	Italy	600 ka	Paleomagnetism Biostratigraphy	No	Coltorti et al. (2005)
Maidscross Hill & Brandon Fields	UK	600 ka?	Chronostratigraphy ESR on quartz	Yes	Ashton and Lewis (2012)
Arago P-Q		550 ka	Biostratigraphy ESR/U-Th dating	Yes	Barsky and Lumley (2010)
Happisburgh I	UK	500 ka	Chronostratigraphy Biostratigraphy	Yes	
High Lodge	UK	500 ka	Chronostratigraphy Biostratigraphy	No	Ashton et al. (1992)
Boxgrove	UK	500 ka	Biostratigraphy	Yes	Roberts and Parfitt (1999)
Cagny La Garenne I and II	France	450 ka	Chronostratigraphy	Yes	Antoine et al. (2007)

The sporadic archaeological evidence between 800 and 500 ka prompts questions about the meaning of these assemblages, whether there were episodic arrivals of new hominin groups with biface technology, the diffusion of new ideas, or a local origin (Fig. 11.1, Table 11.1). It also calls into question the use and definition of the term Acheulean or Acheuleans, compared to East Africa where the early Acheulean has been defined. From 500 ka onward in western and southern Europe, assemblages with bifacial technology cover both southern and northern latitudes and after this time northwest Europe seems to have been occupied on a semi-continuous basis. By bifacial technology, we mean tools with bifacial shaping that managed the bifacial volume with a series of removals and peripheral shaping leading to two convergent edges and a tip.

11.2 Earliest Assemblages with Bifacial Technology

A number of sites show the use of bifacial technology before 500 ka both in northern and southern Europe. In northern Spain, La Boella (1–0.9 Ma) recently yielded some of the earliest evidence of large bifacial tools with a limited shaping with a few bifacial removals on the edges and a pointed tip (Vallverdú et al. 2014). In pit 1, level 2 (unit II), a crude schist pick and a cleaver were discovered in a butchery context with *Mammuthus meridionalis*. The débitage shows unipolar and centripetal cores. Pebble tools and some flake tools also occur. In southern Spain in the Guadix Basin, Cueva Negra del Estrecho del Quípar and Solana del Zamborino yielded lithic assemblages with bifacially shaped tools and small flakes showing centripetal exploitation (Scott and Gibert 2009). However, the suggested date of c. 0.76 Ma has been questioned (Jiménez-Arena et al. 2011) and rediscussed (Walker et al. 2013).

Notarchirico (levels B, D, F) in southern Italy documents one of the earliest occurrences of bifacial technology in southern Europe, dated to 0.64 Ma (Piperno 1999; Lefèvre et al. 2010; Santagata 2012; Nicoud 2013). The bifacial tools are poorly standardized with limited diversity and include numerous pointed chopping tools, occasional pseudo-cleavers on limestone cobbles, and some bifaces made on cobbles or flakes of quartzite, limestone and flint (Fig. 11.2). The bifaces are made with deep removals, which can be invasive or limited to the periphery, and some of them are modified by a second series of small removals on the tip. Retouch on quartz,

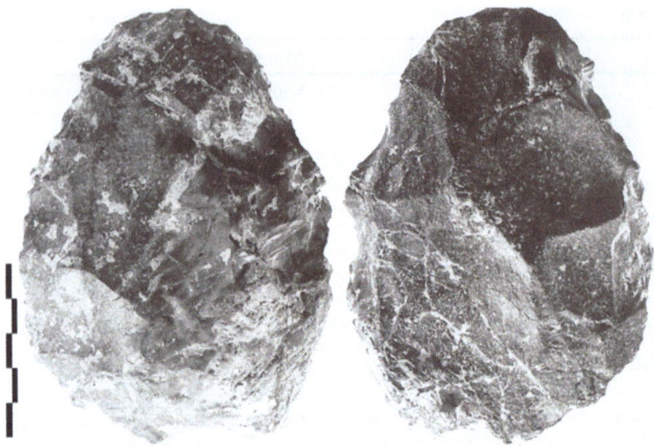

Fig. 11.2 Biface in quartzite from Notarchirico, level F (Piperno 1999)

quartzite, limestone, and flint flakes or cobbles altered the initial shape on pointed cobble tools, scrapers, notches, denticulates, and Tayac points.

In southern France at Caune de l'Arago (levels P and Q) at 0.57 Ma, several reduction processes were used to create small débitage, including controlled bipolar flaking on an anvil, discoidal working, and sometimes a hierarchy in the flaking surfaces (Barsky and Lumley 2010; Barsky 2013; Falguères et al. 2015). Discoidal working was applied to quartz or siliceous stone cobbles from 15–30 km away. Retouched flakes are abundant and mainly pointed. Bifacial tools on a variety of raw materials include well-worked bifaces of various sizes with overall management of the volume by a series of invasive bifacial removals and with final retouching (Fig. 11.3). There are also some cleavers on flakes.

In central France, La Noira dated to c. 700 ka is a site where hominins used local "millstone", a rock formed as slabs by diagenetic silicification within Oligocene lacustrine limestone (Fig. 11.4). The slabs were available in large quantities on the river's edge where they were taken for flaking or shaping (Despriée et al. 2011; Moncel et al. 2013). The site was abandoned at the next cold stage. Many slabs show a limited number of invasive removals on the slab surface or are broken by direct percussion with hard hammers, producing both small and large flakes from these probable crude cores. More structured methods of core reduction with direct hard hammer have also been identified, where some centripetal and bifacial cores show less influence from blank shape.

The tool assemblage consists of both well-worked bifaces, bifacial tools, bifacial cleavers, cleavers on flakes, crudely shaped tools, and large retouched tools. There is great variation in the shape and the invasiveness of removals, sometimes with limited or peripheral shaping and on occasion suggestions of additional functional areas. Bifaces *sensu stricto* demonstrate the pursuit of symmetrical bifacial or bilateral balance and a preconceived form. Two or three successive series of scars may be observed, made up of a first series of deep and invasive removals using face by face or alternate shaping methods, and series of shorter and thinner removals in order to finish forming the bifacial volume. Sections are plano-convex, symmetrical, or twisted.

Fig. 11.3 Biface in metamorphic schist from the Caune de l'Arago, levels P-Q (CERP Tautavel, Barsky and de Lumley 2010, on the courtesy of H. de Lumley)

Fig. 11.4 Bifaces from La Noira in millstone (stratum a) **A**, **B** pointed bifaces, **C** bifacial tool with a transversal edge (*Photograph* Marie-Hélène Moncel)

On most of the tools, unifacial retouch modified some parts of the edge and tip, but did not systematically produce regular cutting edges. These pieces have a range of diverse morphologies with cordiform, triangular, ovate, or amygdaloid shapes.

In northwest Europe several biface sites are now though to date to this period. In Britain the neighboring sites of Brandon Fields and Maidscross Hill lie on the second terrace of the Bytham River and might be as early as 600 ka (Preece 2001; Ashton and Lewis 2005; Ashton et al. 2011; Ashton and Lewis 2012). Both assemblages are made on locally available flint nodules and pebbles from fluvial gravels (Fig. 11.5). The bifaces can be split into two groups, with the first group being more rolled, crudely made with a hard hammer, sometimes on split cobbles, with few, deep bifacial removals, made face by face or alternately; while the second group is in a fresher condition and consists of ovate and cordiform bifaces made with a soft hammer with a general bifacial volume and invasive series of removals and final retouch with alternate shaping or face by face flaking leading to symmetrical or asymmetrical tools. Occasionally the latter are finished or resharpened with a tranchet removal across the tip. Both assemblages must pre-date the terrace gravels, but by how much is not known.

Happisburgh Site 1 lies on the East Anglian coast and probably dates to c. 500 ka (Ashton et al. 2008; Preece and Parfitt 2012; Fig. 11.6). It yielded a small flint assemblage of about 300 flakes, simple flake tools, cores, and one thin, ovate biface. The few cores use alternate and multiple platform techniques (Ashton et al. 2008).

Fig. 11.5 Bifaces from Brandon Fields and Maidscross Hill (*Photograph* Craig Williams)

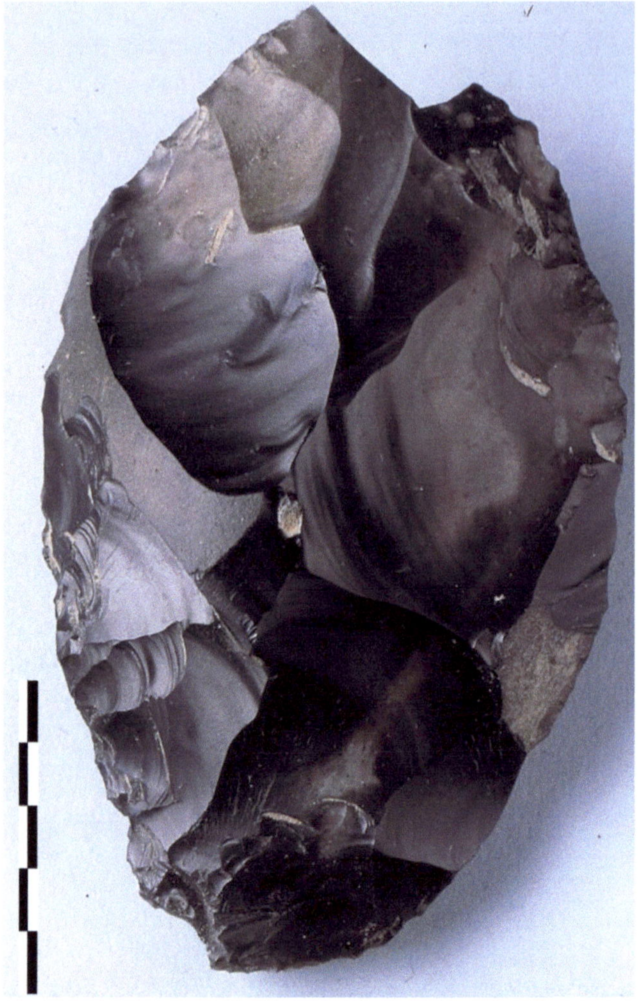

Fig. 11.6 Biface of Happisburg Site 1 (© British Museum)

On the south coast of Britain lies the site of Boxgrove, dating again to c. 500 ka based on mammalian biostratigraphy (Roberts and Parfitt 1999; Ashton 2015). Large flint scatters survive *in situ* on the former coastal mudflats with knapping mainly focused on the production of large quantities of bifaces for butchery. The bifaces are predominantly thin and ovate in shape and are made with a hard hammer, followed by shaping and final retouch with a soft hammer. They are generally symmetrical in shape and cross-section. The tips are often straight due to the frequent use of tranchet removals for probable re-sharpening. Cores are rare using single, alternate, and multiple platform techniques. The occasional flake tools are simple scrapers or minimally retouched flakes.

In northern France the Somme valley sites of Cagny-la-Garenne I and II, dating to late MIS 13 or early MIS 12, have yielded several assemblages made on locally available flint nodules (Antoine et al. 2007). The sites are interpreted as workshops with crude bifacially worked tools alongside bifaces made by series of more invasive removals (Tuffreau 1987;

Lamotte and Tuffreau 2001; Tuffreau et al. 2008; Tuffreau and Lamotte 2010). The assemblages consist of bifaces, cores, flakes, and flake tools such as notches and denticulates. The bifaces were abandoned at various shaping stages and worked both by hard and soft hammer. Generally, there is little retouch on the biface edges, apart from on the edges and tips of some elongated, lanceolate bifaces. Shaping of the tip is part of the overall shaping of the biface. There are also some bifacial cleavers. The morphologies are quite elongated (cordiform, triangular, and ovate–amygdaloid forms). Cross-sections are mainly plano-convex produced by face-on-face sequences, while some symmetrical bifaces were made by alternate flaking. The cores are mainly unifacial with some evidence of Levallois core technology at Cagny-la-Garenne I.

11.3 Assemblages Without Bifacial Technology from 800 to 500 ka

During the period of 800 to 500 ka, core and flake industries also persisted in northern and southern Europe, which poses questions about their technological links with earlier core and flake industries from perhaps 1.8 Ma. At Atapuerca in northern Spain the assemblage from Gran Dolina TD6 dates to c. 800 ka (Falguères et al. 1999; Parés and Pérez-González 1999; Parés et al. 2013). It is made from selected raw materials from surrounding areas, Neogene flint being the most abundant with large blocks flaked outside the cave (Carbonell et al. 1999). The flakes are generally small and never longer than 10 cm. Core technology is mainly multifacial and orthogonal with no preparation of the striking platform. Cobbles of various rocks with a cuboid form were also orthogonally exploited, but sometimes with unifacial and centripetal working. Bipolar exploitation on an anvil appears to have been reserved for quartz. Thick, retouched flakes, mainly on flint, amount to 6% of the artifacts (Carbonell et al. 1999; Ollé et al. 2013).

In Italy, Isernia dates to c. 600 ka, where the assemblage consists of large cobble tools on limestone and small flakes and flake tools from flint. A recent study of level 3c suggests that knapping choices and skills may not have been as opportunistic as previously described and that in some cases discoidal flint core technology was performed independently of blank shape (Gagnepain 1996; Longo et al. 1997; Peretto et al. 2004; Coltorti et al. 2005; Shao et al. 2011; Gallotti and Peretto 2015).

In Britain, Happisburgh Site 3 dates to over 800 ka. The small core and flake assemblage of c. 80 artifacts display the use of alternate and single platform techniques (Parfitt et al. 2010; Ashton et al. 2014). Several of the flakes have been modified into notches and denticulates with one example of a scraper. There is no evidence of biface manufacture, but this cannot be ruled out due to the small size of the assemblage. A similar assemblage was recovered from Pakefield dating to c. 700 ka (Parfitt et al. 2005). As the assemblage consists of only 32 artifacts, it is not known whether bifaces were also occasionally made.

A further site in Britain from this period is High Lodge, but this time probably dating to c. 500 ka based on its relationship to MIS 12 glacial sediments and by mammalian biostratigraphy (Ashton et al. 1992). The assemblages from the floodplain sediments of Beds B and C consist of cores and flakes produced by single and alternate platform techniques with refitting clearly showing the methods used (Figs. 11.7, 11.8, 11.9). Selected flakes were modified into notches and denticulates, but also unusually into very refined scrapers with invasive retouch. A predominantly ovate biface assemblage was found in the glacial outwash sand and gravel of the overlying Bed E. It is suggested that this might be derived from a different part of the floodplain, so also dating to c. 500 ka.

11.4 Assemblages With or Without Bifacial Technology

11.4.1 *A Local Origin?*

The earliest assemblages with bifacial technology appeared when there was already a foundation of core and flake working at sites such as Dmanisi, Lunery, Pirro Nord, Monte Poggiolo, Atapuerca (lower levels of Grand Dolina and Sima del Elefante), Orce, Vallonnet or Pont-de-Lavaud (Carbonell et al. 1999, 2010; Oms et al. 2000; Arzarello et al. 2006, 2015; Lumley et al. 2009; Mosquera et al. 2013; de Lombera-Hermida et al. 2015). They demonstrate that common technical rules were applied to core technology as early as 1.8 Ma at Dmanisi and 1.4 Ma in Europe, based on centripetal, orthogonal, and bipolar techniques with short, poorly structured reduction sequences and clear dependence on blank shape. Lithic technology focused on small-medium-sized flake production with a low tool ratio. The scarcity of sites over such a long period of time, also explained by geological preservation, suggests short-lived dispersal events and probably a source-sink dynamic from further south with phases of depopulation and recolonization. Northern latitudes would only have been occupied during favorable climatic periods.

The current data do not allow us to establish links between the earlier occupations and the first assemblages with bifacial technology or to assess the influence of this technology on the pre-existing methods and behaviors. For example, the production of large flakes, but less than 10 cm in length, occurred at TD6 (Gran Dolina, Atapuerca) and Orce (Oms et al. 2000; Barsky et al. 2010; Toro Moyano et al. 2011). However these flakes were never used as blanks

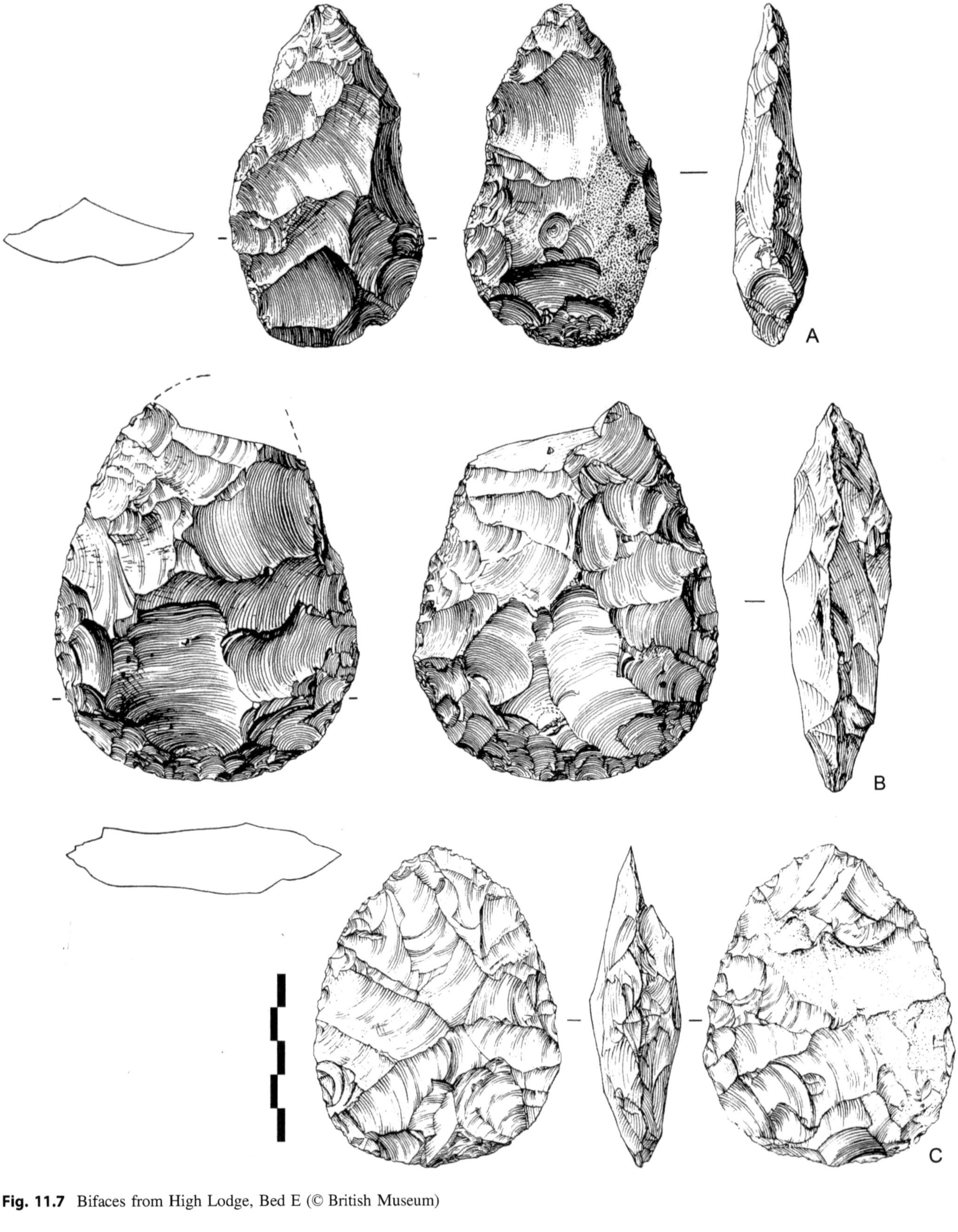

Fig. 11.7 Bifaces from High Lodge, Bed E (© British Museum)

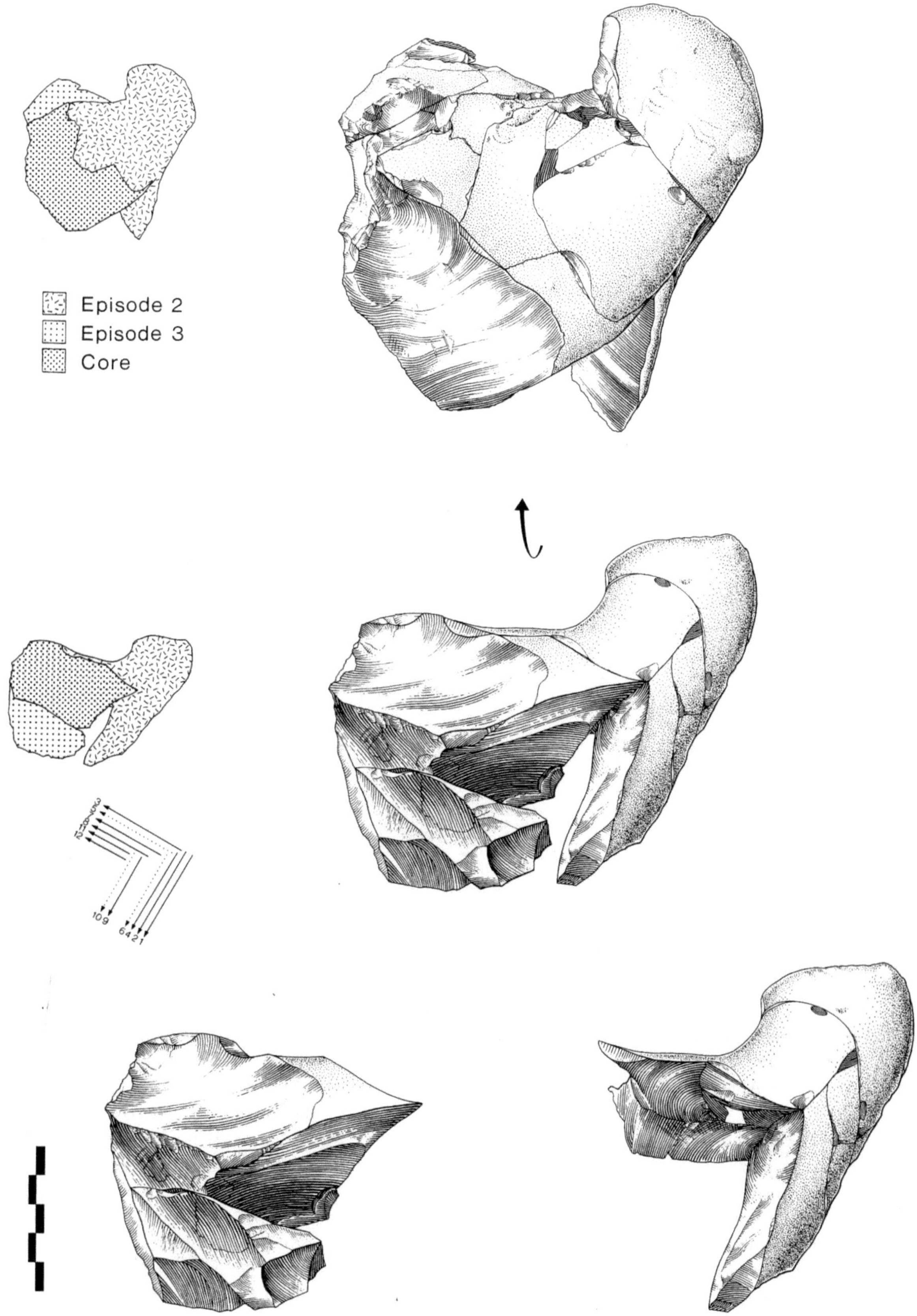

Fig. 11.8 Core refitting from High Lodge, Bed C (Illustration by Phil Dean; © British Museum)

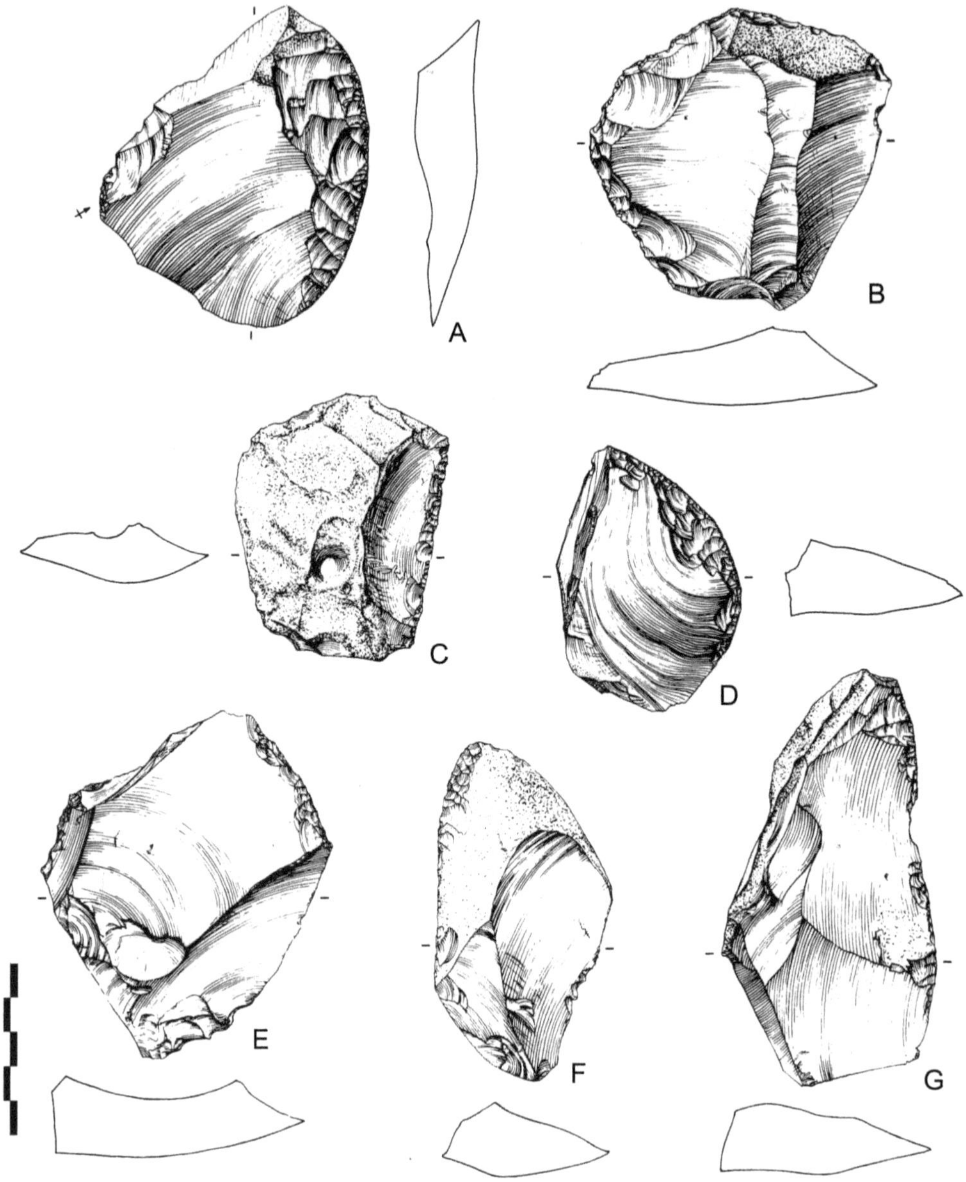

Fig. 11.9 High Lodge, Beds C, D, and E scrapers (Illustration by Phil Dean; © British Museum)

for large tool manufacture, but instead for small flake and flake tool production. Large flake production is evident from 700 ka at La Noira and Arago, and possibly at Pradayrol (Guadelli et al. 2012). It is not observed at Isernia, Notarchirico or La Boella, showing the mosaic of strategies and possibly behaviors in Europe before 500 ka.

A local European origin for bifacial working seems unlikely given that there is little evidence of *in situ* development in the form of partial bifacial working, except perhaps at La Boella. The sites of Pradayrol, Soleilhac, open-air assemblages from the Upper Roussillon and Rhône terraces in France or the site of Bogatyri (1.4–1.1 Ma) in Russia might record some attempts at bifacially worked pieces, but no real development of the technique (Bourdier 1958; Tavoso 1986; Bracco 1991; Shchelinsky et al. 2010; Guadelli et al. 2012). Some brief attempts are also noted at Sima del Elefante (1.2 Ma) in Spain (de Lombera-Hermida et al. 2015). Equally, the site of Notarchirico cannot be considered as evidence of the local origin of bifacial technology since some bifaces showing preconceived shaping of the general bifacial volume with two convergent edges and a tip occur at this site (Santagata 2012). La Boella is currently the best example of the early arrival of bifacial technology in Europe either introduced to a pre-existing foundation or as a local development (Vallverdú et al. 2014). In some ways this problem parallels the issue in East African assemblages described as Developed Oldowan or early Acheulean, where it is still unclear whether there are tangible links or a break over time (Isaac et al. 1997; de la Torre et al. 2003, 2008; Schick and Clark 2003; de la Torre and Mora, 2005; Semaw et al. 2009; de la Torre 2011).

11.4.2 Wide Diversity of Features in Lithic Assemblages from 800 to 500 ka

Lithic assemblages from 800 to 500 ka display a wide variety of features. First, as regards raw material, flint is mainly used in the north, whereas various rocks were exploited in the south. The small number of well-dated sites and their wide geographical range before 500 ka may possibly explain the diversity of strategies and assemblage composition. Each site seems to be rather different. However some comments can be made about the technological approaches despite interpretative problems relating to inter-assemblage variability.

The basic rules governing core technology are similar among the different assemblages without bifacial tools (e.g., Happisburgh 3 and Pakefield in the UK, Vallparadis and TD6 Gran Dolina in Spain) and those with bifacial tools (e.g., Notarchirico in Italy, Arago levels P-Q in France and Boxgrove in Britain; Ashton et al. 2008). Only the Arago P-Q assemblages document stone procurement from a larger area. Flaking methods are often opportunistic using unifacial, multifacial, bipolar, and discoidal flaking. However, they are not always expedient, as is the case for the earliest sites, as there is an increase in centripetal and more structured flaking which is sometimes independent of blank shape. This is more clearly observed from 700 ka onward at La Noira, with bifaces, and at Isernia, without bifaces (Longo et al. 1997; Peretto et al. 1998; Coltorti et al. 2005; Parfitt et al. 2005, 2010; Carbonell et al. 2010; Martínez et al. 2010). This could reveal a turning point in skills dated to the late Early/early Middle Pleistocene (Carbonell et al. 1999; Rodríguez 2004; Mosquera et al. 2013; Gallotti and Peretto 2015). The threshold is also connected to the ability to produce large flakes (for instance at La Noira or Arago) in association with reduction processes producing small flakes.

Aside from the ability to produce large flakes or to manage centripetal cores independently of blank shape, the mastery and management of bifacial technology is quite varied. The diversity of shaping methods and morphological results, such as sinuous or rectilinear edges, suggests varied types of use, with rectilinear edges being more efficient for cutting, while tools with irregular edges are made more ergonomic by controlling the weight and volume. The diversity of bifacial and other large tools at La Noira, some being finely made by series of removals and both hard and soft hammer, is similar to Arago levels P and Q, in terms of shape, such as ovate forms, the presence of poorly standardized cleavers on flakes, more intensively worked tools together with the methods of final shaping (Barsky and Lumley 2010; Barsky 2013). At both sites, some of the tools show management of the bifacial volume with several series of removals to obtain well-balanced surfaces. However at Arago, the large size of some tools, the range of raw material and evidence of the mobility of tools using exotic rocks from 30 km away, differentiates it from La Noira, where the full chaîne opératoire is in evidence. The distinction probably reflects site function, where La Noira is a workshop location, by contrast to the cave site of Arago with transported raw materials. This would also explain the large quantity of non-retouched flakes and crude cores or tools at La Noira, compared to the larger numbers of scrapers, notches, and points and few whole or broken cobbles at Arago.

The diversity of the bifacial tools at Notarchirico is less pronounced. Most of them are convergent chopping tools on limestone cobbles but some well-made bifaces in levels B, D, and F show the ability to master bifacial technology with series of large and short removals for managing both surfaces of the tools (Piperno 1999).

The British sites show a diversity of bifaces with poorly managed hard hammer flaking or those with well-controlled soft hammer shaping and finishing. Both are evident at the neighboring sites of Brandon Fields and Maidscross Hill, where the cruder forms are more rolled and could be older (Ashton and Lewis 2005). The more finely finished ovate and cordiform bifaces are quite similar to those at Boxgrove. At this site there is a high degree of standardization, and many of the bifaces are finished or resharpened with a tranchet blow across the tip (Roberts and Parfitt 1999).

The large number of finely made examples from Boxgrove contrasts with the contemporary or later assemblages of Arago (unit II), la Grande Vallée and Menez Dregan (level 9) in France (Monnier et al. 1994; Herisson et al. 2012; Barsky 2013) or Fontana Ranuccio in Italy with a few well-made bifaces, small pebble tools, and one tool on an elephant bone (Muttoni et al. 2009).

Contrasts can also be drawn between southern and northern Europe. In the south using flakes as blanks was common, whereas in the northwest, this technique was rare. Although large flakes could have been produced on slabs at La Noira, few were used for making bifaces or cleavers indicating perhaps that slabs were better suited to shaping. On the other hand, at Arago, reduction processes did not lead to large flake production, but flakes were still occasionally used for cleavers.

The striking feature of lithic assemblages between 800 and 500 ka is the variation in biface form, but also at times their absence. The unifying factor seems to be the increased complexity in basic knapping systems compared to those found prior to 800 ka. So are we dealing with a single phenomenon with simply a variation in biface presence and form? First we need to understand the effects of raw material and site function on assemblage variability.

11.5 Explaining Variability: Raw Materials, Site Function, and Tool-Making Traditions

Differences in raw material quality, blank form, and availability have long been recognized as a possible influence on the morphology and technology of bifaces and their prevalence within an assemblage (Villa 1981; Ashton and McNabb 1994; White 1998; Ashton and White 2003). At the macroscale, some of the variation between northwest and southern Europe can be explained by differences in the underlying geologies with Cretaceous flint widely available mainly as small- or medium-sized nodules in northern France and southern England. By contrast many of the southern sites were reliant on quartzite, igneous rocks, or even limestone slabs, for example, assemblages from the alluvial terraces of the Tarn and Garonne or at Terra Amata and Notarchirico leading to simpler forms of bifacial tools with, for some of them, a limited shaping (Piperno 1999; Lumley et al. 2009; Santagata 2012; Nicoud 2013). Moreover these rocks sometimes occurred as large blocks, making it easier to knap bifaces from large flakes. Equally, the lack of large stone nodules at Castel di Guido led to the use of bone fragments for bifacial tools composed in a large part by a bifacial tip, lateral unifacial or bifacial scraper edges, and unifacial removals on the butt (Boschian 1993). At other sites the absence of bifaces can also be explained by the lack of suitable raw material, such as at Isernia. However, at a microscale, this influence is much less obvious dependent on the site, similar modes of shaping being applied on various stones or fragments of bones on the same site or between sites in the same area.

Site function also affected biface quantity and quality. A variety of both finished and unfinished, crudely shaped forms are found at manufacturing sites such as at La Noira and Cagny la Garenne (Lamotte and Tuffreau 2001; Moncel et al. 2013, 2015, 2018b). Generally bifaces seem to be associated with butchery although other functions have also been suggested by use-wear studies (Clark and Haynes 1969; Jones 1979, 1980, 1994; Keeley 1980; Villa 1981; Schick and Toth 1993; Ashton and McNabb 1994; Mitchell 1995; Roberts and Parfitt 1999; Delagne et al. 2006; Sharon 2008; Rabinovich et al. 2008; Bello et al. 2009; de Juana et al. 2010; Garcia-Medrano et al. 2014). For example, Q1B at Boxgrove has a large number of bifaces closely associated with cut-marked bones (Roberts and Parfitt 1999). Opportunistic butchery by contrast may have led to the expedient production of crude bifaces (Piperno 1999; Lumley et al. 2004; Vallverdú et al. 2014) or the use of bone bifaces, bifacial tools, and scrapers whose function is unknown (Boschian 1993; Anzidei et al. 2012; Boschian and Sacca 2015; Ceruleo et al. 2015). Low quantities of bifaces might also reflect a wider range of site functions (Roberts and Parfitt 1999; Lhomme et al. 2003; Aureli et al. 2012, 2015).

Although much of the variability in biface form and their prevalence within an assemblage can be explained through differences in raw material and site function, there are still some sites where traditions of knapping play a strong role. For example, on many of the MIS 13 British sites, where raw material is abundant such as High Lodge (Bed E) and Boxgrove, most of the bifaces are ovate or cordiform in shape and are often finished or resharpened with a tranchet removal across the tip (giving a transverse edge like a cleaver). Where raw material is more intractable these traditions of manufacture become more difficult to discern.

11.6 Defining the Acheulean

Following decades of studies (i.e., Commont 1908; Breuil and Kelley 1954; Bordes 1961; Roe 1964, 1981; Isaac 1977), how do we define the Acheulean? Is it merely on the presence of bifaces or a wider suite of innovations? As noted above in Europe sites with or without bifaces still show an increased complexity in basic knapping systems after c. 800 ka. By comparison recent studies in Africa suggest that the Early Acheulean is not identified solely on the basis of the manufacture of LCTs, which only occurred in small quantities in assemblages till 1 Ma, or specific skills such as the ability to produce a balanced bilateral shape, the maintenance of symmetry in cross-section and plan-view and the final shaping of the tip and cutting edge. Nor is it necessarily defined simply on the basis of the systematic production of large flakes produced by various methods (e.g., Winton 2005; Sharon 2009, 2010). Other parameters are also taken into consideration. It is argued that new behaviors appear sporadically as early as 1.5 Ma with greater mobility, more selective raw material acquisition, new concepts in small and large débitage, site locations in variable ecological settings together with the curation of tools (e.g., Leakey 1951; McNabb et al. 2004; Harmand 2009; de la Torre 2011; Gallotti and Mussi 2018). Many cases are not resolved such as the Clactonian, which now is only used in Britain, but is it a non-bifacial facies of the Acheulean or a distinct cultural entity (Ashton et al. 2016)?

For Europe, La Noira and Isernia are the best examples to show that innovations also occurred in core and flake working (Gallotti and Peretto 2015). These consist of an increase in the use of different methods, in particular the centripetal method, and better overall management (volume, convexities, preparation of the striking platform). This is relevant in cultural terms, since the same technological competence was shared by hominins during specific periods of time, independent of bifacial technology. Levels P and Q at Arago also show higher mobility, more selective use of raw materials with spatial patterning to the chaîne opératoire.

The innovations in core technology, together with the problems of raw material variation and site function, suggest that the presence or absence of biface technology is not the best way to define the Acheulean, particularly given their rarity in some assemblages. It is perhaps now important to accept that there is no single definition and that there are a mosaic of material cultural expressions and behaviors that are encompassed by the term and that the biface is simply one expression of many (Barsky and Lumley 2010; Ashton et al. 2011, 2018; Moncel et al. 2013, 2015, 2018a; Ollé et al. 2013; Vallverdú et al. 2014; Ashton 2015, 2018). Within this mosaic regional signatures should be expected which would have become reinforced by the environmental backdrop in which they occurred. Links between the regions will always be difficult to demonstrate, but again should be expected, certainly within Europe, due to cyclical changes in climate and inevitable movement of populations between the south and north. Can links also be identified between Africa and Europe for the introduction of this new behavioral suite?

11.7 Episodic Arrivals Before 500 ka?

The changing climate and environment of Europe had a huge impact on the sustainability of human occupation, but also on the movement of people and transfer of new ideas. So did hominin groups have the capability to adapt to these changes? The Early Pleistocene is marked by climate cycles of 41 ka, which during the transitional periods from glacial to interglacial created temperate, more open conditions, prior to forest regeneration (Guthrie 1984; Almogi-Labin 2011; Ashton et al. 2011; Candy et al. 2011; Carrión et al. 2011; Messager et al. 2011; Rodríguez et al. 2011; Abbate and Sagri 2012; Elderfield et al. 2012; MacDonald et al. 2012; Orain et al. 2013; Garcia et al. 2014). The brief episodes of more open conditions would have been beneficial to human colonization with the availability of larger herbivore herds and easier mobility for hominins.

From the early Middle Pleistocene there is a shift to c. 100 ka climatic cycles with the Middle Pleistocene Transition (MPT) (Manzi 2004; Clark et al. 2006; Muttoni et al. 2010; Ashton and Lewis 2012). Not only were the cycles longer, but also more extreme, having a greater impact on human populations. This would have led to the successive depopulation or extinction of small groups of hominins during cold episodes in the north, and required recolonization during warmer episodes from the south (see, e.g., Dennell et al. 2011). The second climatic transition (Mid-Brunhes Event-MBE) between MIS 13 and 11 with more marked glacial–interglacial cycles might explain in part the wider diffusion of Acheulean behavior with warmer interglacials and the extension of the mammoth steppe in the north from 500 ka (Guthrie 1984; Jouzel et al. 2007; Paillard 2015).

It has also been suggested that faunal dispersals, including hominins, from Africa into southern Europe were triggered by aridification in Africa during glacial episodes, such as MIS 22 (Muttoni et al. 2010). However, the relationship between faunal turnovers across Eurasia and the introduction of bifacial technology are not easy to identify due to a lack of clear mammalian dispersals between 780 and 500 ka from Africa to Eurasia. Of significance to the success of hominin colonization of Europe were changes in the carnivore guilds around 500 ka and the demise of megahunters, such as large felids, as early as 1 Ma ago (Abbazzi et al. 2000; Stiner 2002; Belmaker 2009; Muttoni et al. 2010; Bar-Yosef and Belmaker 2011; Chapais 2011; Cuenca-Bescos et al. 2011; Madurell-Malapeira et al. 2017). These related environmental changes, in function of intensity of changes and anthropological timescale of hominins, could have promoted episodic hominin expansion into Europe, successful or otherwise, aided by new techniques, new social organizations and, as some would suggest, the occasional control of fire (McPherron 2000; Goren-Inbar and Sharon 2006; Gowlet 2006; Roebroeks and Villa 2011; Berna et al. 2012).

For northern Europe, at least, the evidence suggests occasional dispersals into the region, which also helps to explain the diversity and introduction of new behaviors. Even for southern Europe there are gaps in the record. At Atapuerca, a hiatus in occupation is apparent between 800 and 500 ka (Rodríguez et al. 2011; Mosquera et al. 2013), while in the center of France there is a lack of evidence between 1 Ma and 700 ka and from 700 to 500 ka (Despriée et al. 2011).

If the gaps in the record are real, then it seems unlikely that there is much *in situ* development of technology, ideas, and behaviors. The oldest bifacial tools are already technologically quite elaborate although rare in European assemblages in comparison to the African Acheulean dated to over 1 Ma. For the moment, there are no attributes on the tools that show a clear evolution in shaping technology or typology throughout this period; bifaces did not become increasingly more intensively shaped over time. Regardless of age, assemblages are composed of both tools made by few removals and better-shaped tools by a successive series of bifacial removals that managed the bifacial volume. Although more irregular and thick tools with large removals made by hard hammer comprise the oldest assemblages, they are associated with ovate-shaped and thinner well-worked bifaces. Moreover, some of the bifacial tools bear additional functional zones, while others are well-made bifaces with careful resharpening. Shaping is often adapted to blank shape, but there is also good management of the volume, regardless of age. Careful tip management is generally observed in the most recent assemblages, although there are exceptions such as Terra Amata dated to 450 ka (Lumley 2015). However, careful tip management is also a feature of some older assemblages. In other words there is no clear pattern showing an *in situ* development of technology. Rather it is a pattern that is more likely to reflect a much more complex system of local developments often overprinted by new ideas and transferred by people perhaps from Africa into Eurasia and almost certainly from southern into northern Europe.

Within the African record it is much easier to see a general development in the Early Acheulean. At Olduvai in Bed II (1.5 Ma), Gadeb (1.7–1.5 Ma to 0.8 Ma) or Peninj (1.6–1.5 Ma) or in South African sites there are pick-like objects, with flat or triangular cross-sections and little management of the central volume. These occur together with minimally shaped LCTs from cobbles and flakes, unifaces and an array of other heavy-duty tools with alternating or bifacial working (Leakey 1971; McPherron 2006; de la Torre et al. 2008; Endicott et al. 2010; Semaw et al. 2018). Increased refinement is evident from c. 1 Ma with the use of soft hammer working, together with a higher ratio and range of biface forms, so that by 700 ka there is much greater standardization in tool morphology (Texier and Roche 1995; Texier 2001, 2018; Roche et al. 2003; Roche 2005). Few of these developments in technology are really in evidence in the European assemblages. For example, at La Noira the developments have already occurred, such as the use of soft hammer, shaping of the volume by successive series of removals, both on the edges and the tip and secondary retouch to rectify cutting edges. Equally, blank morphology does not always influence the morphology of the bifaces, while hierarchical flaking is evident on bifacial discoidal cores.

It is difficult therefore to see the localized emergence of bifacial technology and associated behaviors at La Noira, so the inference is that they were introduced from outside Europe. However, trying to find more definite links beyond Europe is problematic. Clear signs of dispersals from Asia are difficult to demonstrate given the lack of bifaces in central Asia and their differences in form in India (Pappu et al. 2011) and China (Kuman et al. 2014), while anatomical features provide conflicting evidence (Bermudes de Castro and Martinón-Torrès 2013). Whether hominins arrived from Asia or Africa, a route around the Mediterranean seems likely (Kuhn 2002; Galanidou et al. 2013). If Levantine and African sites dated from 1–0.7 Ma are compared to European assemblages, two possible scenarios can be envisaged.

(1) Slow arrivals from the Levant of new hominins, perhaps *Homo heidelbergensis*, or new traditions which persisted for long enough in this part of the world to bring about changes (Isaac 1977; Roche et al. 1987, 1988; Texier and Roche 1995; Raynal et al. 2001, 2011; Rightmire 2001, 2009; Texier 2001; Santonja and Villa 2006; Gallotti et al. 2010; Sharon et al. 2010, 2011; Sharon 2011; Presnyakov et al. 2012; Wilkins and Chazan 2012; Gallotti 2013; Roland 2013; Garcia et al. 2014).
(2) Rapid arrivals from Africa through the Levant or North Africa (Gibraltar) by circum-Mediterranean corridors supported by early dates for the Levant and North Africa, and recent discoveries of la Boella site (Clark 1967; Bar-Yosef and Goren-Inbar 1993; Van Peer 1998; Vermeersch 2001; Gibert et al. 2003; Petraglia 2003; Boëda et al. 2004; Yasbeck 2004; Goren-Inbar and Sharon 2006; Goren-Inbar et al. 2008; Le Tensorer 2009; Raynal et al. 2010; Jagher 2011; Jagher and Le Tensorer 2011; Walker et al. 2013; Vallverdú et al. 2014; Mosquera et al. 2016).

11.8 Sustained Occupation After 500 ka

If there was only sporadic presence of humans in Europe prior to 500 ka, the record after this date seems rather different. After the MIS 12 glaciation there was probably a more sustained settlement of Europe with a stronger imprint of local development. Assemblages are often more elaborate with more distinct forms of biface, for example, the twisted ovates in Britain during MIS 11 (White 1998) or elongated cordiforms (“ficrons”) in Britain and France during MIS 9 (Wenban-Smith 2004; Lumley et al. 2015; White 2015). Even by MIS 11 some Middle Palaeolithic features emerge perhaps among Pre-Neanderthal groups (Tuffreau 1987; Villa et al. 2005; Hublin 2009; Premo and Hublin 2009; Moncel et al. 2016). The biface thickness does not decrease over time despite the more widespread use of flakes or flat blanks adapted to efficient shaping. Distinctions also seem to emerge between the volume and prehensile qualities of the biface and the functional edges; in the Lower Palaeolithic these aspects seem rather unified into a single unit with the imposition of symmetry or asymmetry, whereas in the Middle Palaeolithic the volume is merely the ergonomic instrument on which to impose a variety of functional edges through re-sharpening (Moncel 1995; Soressi and Hays 2003). This type of tool working and behavior is akin to that found on indigenous Australian stone tools (Hayden 1979).

11.9 Conclusion

Low hominin densities and distance from raw material sources would account for the European network of sites, as suggested by Lycett and Gowlett (2008) and Lycett and von Cramon-Taubadel (2008). Information is still lacking for identifying routes, and filiations cannot be established yet due to the influence of activities and environmental patterns on behavior. Despite the number of sites is low, a local origin for bifacial technology seems unlikely since clear assemblages with attempts at bifacial shaping are lacking. Sites such as La Noira attest elaborate bifaces as soon as 700 ka. There is little evidence of in situ development in the form of partial bifacial working which could have punctually happened in parallel around 1–0.9 Ma. Data seem to point more toward technological "innovations" than to "inventions" over time in Europe, as discussed in Haidle and Brauer (2011) while elsewhere both reinventions and arrivals could have happened according to local conditions. The local transformation of diverse behaviors and adaptation to local situations could be one parameter advanced to explain inter-assemblage diversity, especially if the scenario of multiple hominin arrivals (with or not potential interbreeding populations) is retained, as suggested by early and recent paleoanthropological studies (i.e., Martinón-Torrès et al. 2011; Stringer 2012; Bermúdez de Castro and Martinón-Torrès 2013; de Lumley 2015).

Acknowledgements This study is part of an ANR (Agence Nationale pour la Recherche n°2010 BLANC 2006 01) project and fieldwork and sampling at several sites in the northwest of France and England will provide new data on occupations along the paleo-Thames, Somme, and Loire Rivers before isotopic stage 12.

We would like to thank the two editors, Rosalia Gallotti and Margherita Mussi, to ask us to publish this paper and allow us developing our vision on the earliest evidence of the bifacial technology in Western Europe. We thank Craig Williams for the production of the map.

We would like to thank as well the three reviewers and again the two editors, Rosalia Gallotti and Margherita Mussi, for their relevant comments on this paper.

References

Abbate, E., & Sagri, M. (2012). Early to Middle Pleistocene *Homo* dispersals from Africa to Eurasia: Geological, climatic and environmental constraints. *Quaternary International, 267,* 3–19.

Abbazzi, L., Fanfani, F., Ferretti, M. P., Rook, L., Cattani, L., Masini, F., et al. (2000). New human remains of archaic *Homo sapiens* and Lower Palaeolithic industries from Visogliano (Duino Aurisina, Trieste, Italy). *Journal of Archaeological Science, 27,* 1173–1186.

Almogi-Labin, A. (2011). The paleoclimate of the Eastern Mediterranean during the transition from early to mid Pleistocene (900 to 700 ka) based on marine and non-marine records: An integrated overview. *Journal of Human Evolution, 60*(4), 428–436.

Antoine, P., Limondin-Lozouet, N., Chaussé, C., Lautridou, J. P., Pastre, J. F., Auguste, P., et al. (2007). Pleistocene fluvial terraces from northern France (Seine, Yonne, Somme): Synthesis and new results. *Quaternary Science Reviews, 26,* 2701–2723.

Anzidei, A. P., Bulgarelli, G. M., Catalano, P., Cerilli, E., Gallotti, R., Lemorini, C., et al. (2012). Ongoing research at the late Middle Pleistocene site of La Polledrara di Cecanibbio (central Italy), with emphasis on human-elephant relationships. *Quaternary International, 255,* 171–187.

Arzarello, M., Marcollini, F., Pavia, G., Pavia, M., Petronio, C., Petrucci, M., et al. (2006). Evidence of earliest human occurrence in Europe: The site of Pirro Nord (Southern Italy). *Naturwissenschaften, 94,* 107–112.

Arzarello, M., Peretto, C., & Moncel, M.-H. (2015). The Pirro Nord Site (Apricena, Fg, Southern Italy) in the Context of the First European Peopling: Convergences and Divergences. *Quaternary International, 389,* 255–263.

Ashton, N. M. (2015). Ecological niches, technological developments and physical adaptations of early humans in Europe: The handaxe-*heildelbergensis* hypothesis. In F. Coward, R. Hosfield, M. Pope, & F. F. Wenban-Smith (Eds.), *Settlement, Society and Cognition in Human Evolution: Landscapes in Mind* (pp. 138–153). Cambridge: Cambridge University Press.

Ashton, N. M. (2018). Landscapes of habit and persistent places during MIS 11 in Europe: A return journey from Britain. In M. Pope, J. McNabb, & C. B. Gamble (Eds.), *Crossing the Human Threshold. Dynamic Transformation and Persistent Places during the Middle Pleistocene* (pp. 142–164). London: Routledge.

Ashton, N. M., Cook, J., Lewis, S. G., & Rose, J. (1992). *High Lodge, Excavations by G. de G. Sieveking, 1962–8, and J. Cook, 1988.* British Museum Press.

Ashton, N. M., & McNabb, J. (1994). Bifaces in perspective. In N. M. Ashton & A. David (Eds.), *Stories in Stone. Occasional Paper of the Lithic Studies Society No. 4* (pp. 182–191). Oxford: Proceedings of the Anniversary Conference at St Hilda's College.

Ashton, N. M., & White, M. (2003). Bifaces and raw materials: Flexible flaking in the British Early Palaeolithic. In M. Soressi & H. Dibble (Eds.), *Multiple Approaches to the Study of Bifacial Technology* (pp. 109–124). Philadelphia: University Museum Monography, 115, University of Pennsylvania.

Ashton, N. M., & Lewis, S. G. (2005). Maidscross Hill, Lakenheath. *Proceedings of the Suffolk Institute of Archaeology and Natural History XLI* (Part 1), 122–123.

Ashton, N., & Lewis, S. G. (2012). The environmental contexts of early human occupation of northwest Europe: The British Lower Palaeolithic record. *Quaternary International, 271,* 50–64.

Ashton, N. M., Lewis, S. G., De Groote, I., Duffy, S., Bates, M., Bates, R., et al. (2014). Hominin footprints from Early Pleistocene deposits at Happisburgh, UK. *PLoS ONE, 9,* e8832.

Ashton, N. M., Parfitt, S. A., Lewis, S. G., Cooper, G. R., & Larkin, N. (2008). Happisburgh Site 1 (TG388307). In I. Candy, J. R. Lee, & A. M. Harrison (Eds.), *The Quaternary of Northern East Anglia Quaternary Research Association* (pp. 151–156). Cambridge.

Ashton, N., Lewis, J. E., & Hosfield, R. (2011). Mapping the human record: Population change in Britain during the Early Palaeolithic. In N. Ashton, J. E. Lewis, & C. Stringer (Eds.), *The Ancient Human Occupation of Britain* (pp. 39–53). Amsterdam: Elsevier.

Ashton, N. M., Lewis, S. G., Parfitt, S. A., Davis, R. J., & Stringer C. B. (2016). Handaxe and non-handaxe assemblages during Marine Isotope Stage 11 in northern Europe: Recent investigations at Barnham, Suffolk, UK. *Journal of Quaternary Science, 31,* 837–843.

Aureli, D., Contardi, A., Giaccio, B., Modesti, V., Palombo, M. R., Rozzi, R., et al. (2012). Straight-tusked elephants in the Middle Pleistocene of northern Latium: Preliminary report on the Ficoncella site (Tarquinia, central Italy). *Quaternary International, 255,* 29–35.

Aureli, D., Contardi, A., Giaccio, B., Jicha, B., Lemorini, C., Madonna, S., et al. (2015). *Palaeoloxodon* and Human Interaction: Depositional Setting, Chronology and Archaeology at the Middle Pleistocene Ficoncella Site (Tarquinia, Italy). *PLoS ONE, 10,* e0124498.
Barsky, D., & de Lumley, H. (2010). Early European Mode 2 and the stone industry from the Caune de l'Arago's archeostratigraphical levels "P". *Quaternary International, 223–224,* 71–86.
Barsky, D., Celiberti, V., Cauche, D., Grégoire, S., Lebègue, F., de Lumley, H., et al. (2010). Raw material discernment and technological aspects of the Barranco León and Fuente Nueva 3 stone assemblages (Orce southern Spain). *Quaternary International, 223–224,* 201–219.
Barsky, D. (2013). The Caune de l'Arago stone industries in their stratigraphical context. *Comptes Rendus Palevol, 1,* 305–325.
Bar-Yosef, O., & Goren-Inbar, N. (1993). *The Lithic Assemblages of Ubeidiya. A Lower Palaeolithic site in the Jordan Valley.* Jerusalem: Quedem.
Bar-Yosef, O., & Belmaker, M. (2011). Early and Middle Pleistocene Faunal and hominins dispersals through Southwestern Asia. *Quaternary Science Reviews, 30,* 1318–1337.
Bello, S. M., Parfitt, S. A., & Stringer, C. (2009). Quantitative micromorphological analyses of cut marks produced by ancient and modern handaxe. *Journal of Archaeological Science, 36,* 1869–1880.
Belmaker, M. (2009). Hominin adaptability and patterns of faunal turnover in the Lower-Middle Pleistocene transition in the Levant. In M. Camps & P. R. Chauhan (Eds.), *Sourcebook of Paleolithic Transitions: Methods, Theories, and Interpretations* (pp. 211–227). Springer.
Bermudez de Castro, J. M., & Martinón-Torrès, M. (2013). A new model for the evolution of the human Pleistocene populations of Europe. *Quaternary International, 295,* 102–112.
Berna, F., Goldberg, P., Horwitz, L. K., Brink, J., Holt, S., Bamford, M., et al. (2012). Microstratigraphic evidence of in situ fire in the Acheulean strata of Wonderwerk Cave, Northern Cape province, South Africa. *Proceedings of the National Academy of Sciences, 20,* 1215–1220.
Beyene, Y., Katoh, S., WoldeGabriel, G., Hart, W. K., Sudo, M., Kondo, M., et al. (2013). The characteristics and chronology of the earliest Acheulean at Konso, Ethiopia. *Proceedings of the National Academy of Sciences, 110,* 1584–1591.
Boëda, E., Courty, M. A., Fedoroff, N., Griggo, C., Hedley, I. G., & Muhesen, S. (2004). Le site acheuléen d'El Meirah, Syrie. In O. Auranche, M. Le Mière, & P. Sanla Ville (Eds.), *From river to the sea. The Paleololithic and the Neolithic on the Euphrates and in the Northern Levant* (pp. 165–201). Oxford: Bar International Series, 1263.
Bordes, F. (1961). *Typologie du Paléolithique ancien et moyen.* Delmas: Bordeaux.
Boschian, G. (1993). Castel di Guido - scavi 1980–1991. In *Atti della Riunione Scentifica dell'istituto Italiano di Preistoria e Protostoria, Paleosuperfici del Pleistocene e del primo Olocene in Italia. Processi di formazione e interpretazione. Venosa* (Vol. 1, pp. 167–178).
Boschian, G., & Sacca, D. (2015). In the elephant, everything is good: Carcass use and re-use at Castel di Guido (Italy). *Quaternary International, 361,* 288–296.
Bourdier, F. (1958). *Le bassin du Rhône au Quaternaire. Géologie et Préhistoire*, Université de Paris, thèse de doctorat.
Bracco, J.-P. (1991). Typologie, technologie et matières premières des industries du très ancien Paléolithique en Velay (Massif central, France) premiers résultats. In E. Bonifay & B. Vandermeersch (Eds.), *Les premiers peuplements européens* (pp. 93–100). Paris: CTHS, Congrès National des Sociétés savantes.
Breuil Abbé, H., & Kelley, H. (1954). Le Paléolithique ancien. Abbevillien. Clactonien. Acheuléen. Levalloisien. *Bulletin de la Société Préhistorique Française LI, 8,* 1–18.
Candy, I., Silva, B., & Lee, J. (2011). Climates of the Early Middle Pleistocene in Britain: Environments of the Earliest Humans. In N. Ashton, S.G. Lewis, & C. Stringer (Eds.), *The Ancient Human Occupation of Britain* (Vol. 14, pp. 11–22). Elsevier, Quaternary Science.
Carbonell, E., Garcia-Anton, M. D., Mallol, C., Mosquera, M., Ollé, A., Rodriguez, X. P., et al. (1999). The TD6 lithic industry from Grand Dolina, Atapuerca (Burgos, Spain) production and use. *Journal of Human Evolution, 37,* 653–693.
Carbonell, E., Bermùdez de Castro, J. M., Arsuaga, J. L., Allue, E., Bastir, M., Benito, A., et al. (2005). An Early Pleistocene hominin mandible from Atapuerca-TD6, Spain. *PNAS, 102,* 5674–5678.
Carbonell, E., Bermùdez de Castro, J., Parés, J., Pérez-González, A., Cuenca-Bescós, G., Olle, A., et al. (2008). The first hominin of Europe. *Nature, 452,* 456–470.
Carbonell, E., Sala Ramos, R., Rodríguez, X. P., Mosquera, M., Ollé, A., Vergès, J. M., et al. (2010). Early hominid dispersals: A technological hypothesis for "out of Africa". *Quaternary International, 223–224,* 36–44.
Carrión, J. S., Rose, J., & Stringer, C. (2011). Early Human Evolution in the Western Palaearctic: Ecological Scenarios. *Quaternary Science Reviews, 30,* 1281–1295.
Ceruleo, P., Marra, F., Pandolfi, L., Petronio, C., & Salari, L. (2015). The archaic Acheulean lithic industry of the Cretone basin (Latium, central Italy). *Journal of Archaeological Science: Reports, 3,* 480–492.
Chapais, B. (2011). The Deep Social Structure of Humankind. *Science, 331,* 1276–1277.
Clark, J. D. (1967). The Middle Acheulian occupation of Latamne, Northern Syria. *Quaternaria, 9,* 1–68.
Clark, G. (1969). *World Prehistory: A New Outline.* London: Cambridge University Press.
Clark, J. D., & Haynes, C. V. (1969). An elephant butchery site at Mwanganda's village, Karonga, Malawi, and its relevance for Palaeolithic archaeology. *World Archaeology, 1,* 390–411.
Clark, P. U., Archer, D., Pollard, D., Blum, J. D., Rial, J. A., Brovkin, V., et al. (2006). The middle Pleistocene transition: Characteristics, mechanisms, and implications for long-term changes in atmospheric pCO_2. *Quaternary Science Reviews, 25,* 3150–3184.
Coltorti, M., Feraud, G., Marzoli, A., Peretto, C., Ton-Thate, T., Voinchet, P., et al. (2005). New $^{40}Ar/^{39}Ar$, stratigraphic and paleoclimatic data on the Isernia La Pineta Lower Palaeolithic site, Molise, Italy. *Quaternary International, 131,* 11–22.
Commont, M. V. (1908). Les industries de l'ancien Saint-Acheul. *L'Anthropologie, 19,* 528–572.
Cuenca-Bescos, G., Melero-Rubio, M., Rofes, J., Martínez, I., Arsuaga, J. L., Blain, H.-A., et al. (2011). The Early-Middle Pleistocene environmental and climatic change and the human expansion in Western Europe: A case study with small vertebrates (Gran Dolina, Atapuerca, Spain). *Journal of Human Evolution, 60,* 481–491.
de Juana, S., Galan, A. B., & Domínguez-Rodrigo, M. (2010). Taphonomic identification of cutmarks made with lithic handaxes: An experimental study. *Journal of Archaeological Science, 37,* 1841–1850.
de la Torre, I., Mora, R., Domínguez-Rodrigo, M., de Luque, L., & Alcalá, L. (2003). The Oldowan industry of Peninj and its bearing on the reconstruction of the technological skills of Lower Pleistocene hominids. *Journal of Human Evolution, 44,* 203–224.
de la Torre, I., & Mora, R. (2005). *Technological Strategies in the Lower Pleistocene at Olduvai Beds I and II.* Liège: ERAUL.

de la Torre, I., Mora, R., & Martìnez-Moreno, J. (2008). The early Acheulean in Peninj (Lake Natron, Tanzania). *Journal of Anthropological Archaeology, 27*, 244–264.

de la Torre, I. (2011). The Early Stone Age lithic assemblages of Gadeb (Ethiopia) and the Developed Oldowan/early Acheulean in East Africa. *Journal of Human Evolution, 60,* 768–812.

de Lombera-Hermida, A., Bargallo, A., Terradillos-Bernal, M., Huguet, R., Vallverdú, J., García-Anton, M.-D., et al. (2015). The lithic industry of Sima del Elefante (Atapuerca, Burgos, Spain) in the context of Early and Middle Pleistocene technology in Europe. *Journal of Human Evolution, 82,* 95–106.

de Lumley, H. (2004). Habitat et mode de vie des chasseurs paléolithiques de la Caune de l'Arago (600 000–400 000 ans). *L'Anthropologie, 108,* 159–184.

de Lumley, H. (Ed.) (2009). *Terra Amata. Cadre Géographique - Historique - Contexte géologique - Stratigraphie - Sédimentologie - Datation*. Paris: CNRS Editions.

de Lumley, H., Cauche, D., El Guennouni, K., Fournier, R.-A., Grégoire, S., Lecervoisier, B., et al. (2015). *Terra Amata : Nice, Alpes-Maritimes, France. Tome IV. Fascicule 1, Les industries acheuléennes*. Paris: CNRS Editions.

de Lumley, M.-A. (2015). L'homme de Tautavel. Un *Homo erectus* européen évolué. *Homo erectus tautavelensis. L'Anthropologie, 119,* 303–348.

Delagnes, A., Lenoble, A., Harmand, S., Brugal, J.-P., Prat, S., Tiercelin, J.-J., et al. (2006). Interpreting pachyderm single carcass sites in the African Lowe rand Early Middle Pleistocene record: A multidisciplinary approach to the site of Nadung'a 4 (Kenya). *Journal of Anthropological Archaeology, 25,* 448–465.

Dennell, R. W., Martinón-Torrès, M., & Bermúdez de Castro, J. M. (2011). Hominin variability, climatic instability and population demography in Middle Pleistocene Europe. *Quaternary Science Reviews, 30,* 1511–1524.

Despriée, J., Voinchet, P., Tissoux, H., Bahain, J.-J., Falguères, C., Courcimault, G., et al. (2011). Lower and Middle Pleistocene human settlements recorded in fluvial deposits of the middle Loire River Basin, Centre Region, France. *Quaternary Science Reviews, 30,* 1474–1485.

Elderfield, H., Ferretti, P., Greaves, M., Crowhurst, S., McCave, I. N., Hodell, D., et al. (2012). Evolution of Ocean Temperature and Ice Volume Through the Mid-Pleistocene Transition. *Nature, 337,* 704–709.

Endicott, P., Ho, S. Y. W., & Stringer, C. (2010). Using genetic evidence to evaluate four palaeoanthropological hypotheses for the timing of Neanderthal and modern human origins. *Journal of Human Evolution, 59,* 87–95.

Falguères, C., Bahain, J.-J., Yokoyama, Y., Arsuaga, J. L., Bermudez de Castro, J. M., Carbonell, E., et al. (1999). Earliest humans in Europe; The age of TD6 Gran Dolina, Atapuerca, Spain. *Journal of Human Evolution, 37,* 343–353.

Falguères, C., Shao, Q., Han, F., Bahain, J. J., Richard, M., Perrenoud, C., et al. (2015). New ESR and U-series dating at Caune de l'Arago, France: A key-site for European Middle Pleistocene. *Quaternary Geochronology, 30, Part B*, 535–540.

Gagnepain, J. (1996). Etude magnétostratigraphique de sites du Plésitocène inférieur et moyen des Alpes-Maritimes et d'Italie : Grotte du Vallonnet, Marina Airport, Ca' Belvedere di Monte Poggiolo, Isernia La Pineta, Venosa Loreto. Ph.D. dissertation, Muséum national d'Histoire naturelle, Paris.

Galanidou, N., Cole, J., Iliopoulos, G., & McNabb, J. (2013). East meets West: The Middle Pleistocene site of Rodafnidia on Lesvos, Greece. *Antiquity, 336,* 87.

Gallotti, R., Collina, C., Raynal, J.-P., Kieffer, G., Geraads, D., & Piperno, M. (2010). The Early Middle Pleistocene Site of Gombore II (Melka Kunture, Upper Awash, Ethiopia) and the Issue of Acheulean Bifacial Shaping Strategies. *African Archaeological Revue, 27,* 291–322.

Gallotti, R. (2013). An older origin for the Acheulean at Melka Kunture (Upper Awash, Ethiopia): Techno-economic behaviours at Garba IVD. *Journal of Human Evolution, 65,* 594–620.

Gallotti, R., & Peretto, C. (2015). The Lower/early Middle Pleistocene small débitage productions in Western Europe: New data from Isernia La Pineta t.3c (Upper Volturno Basin, Italy). *Quaternary International, 357,* 264–281.

Gallotti, R., & Mussi, M. (2018). Before the Acheulean in East Africa: An Overview of the Oldowan Lithic Assemblages. In R. Gallotti & M. Mussi (Eds.), *The Emergence of the Acheulean in East Africa and Beyond. Contributions in Honor of Jean Chavaillon*. Cham: Springer.

García, J., Martínez, K., Cuenca-Bescós, G., & Carbonell, E. (2014). Human occupation of Iberia prior to the Jaramillo magnetochron (>1.07 Myr). *Quaternary Science Reviews, 98,* 84–99.

Garcia-Medrano, P., Ollé, A., Mosquera, M., Cáceres, I., Díez, C., & Carbonell, E. (2014). The earliest Acheulean technology at Atapuerca (Burgos, Spain): Oldest levels of the Galeria site (GII Unit). *Quaternary International, 353,* 170–194.

Gibert, J., Gibert, L., & Iglesias, A. (2003). The Gibraltar Strait: A Pleistocene door to Europe. *Journal of Human Evolution, 18,* 147–160.

Goren-Inbar, N., Feibel, C. S., Verosub, K. L., Melameb, Y., Kislev, M. E., Tchernov, E., et al. (2000). Pleistocene Milestones on the Out-of-Africa Corridor at Gesher Benot Yakov, Israel. *Science, 289,* 944–947.

Goren-Inbar, N., & Sharon, G. (2006). *Axe Age: Acheulian Tool-making from Quarry to Discard*. London: Equinox Publishing Ltd.

Goren-Inbar, N., Sharon, G., Alperson-Afil, N., & Lashiver, I. (2008). The Acheulean massive scrapers of GBY-a product of the biface chaîne opératoire. *Journal of Human Evolution, 55,* 702–712.

Gowlett, J. A. J. (2006).The early settlement of northern Europe: Fire history in the context of climate change and the social bray. Colloque: Climats-Cultures-Sociétés aux temps préhistoriques. De l'apparition des Hominidés jusqu'au Néolithique. *Palevol, 5, 1–2,* 299–311.

Guadelli, J.-L., Turq, A., & Seronie-Vivien, R. (2012). Le site de Pradayrol à Caniac-du-Causse (Lot). In A. Turq, J. Airvaux, J. Despriée, & P-J. Texier (Eds.), *La conquête de l'Ouest Il y a un million d'années en Europe. Musée des Eyzies* (pp. 136–137). Maison de l'Histoire de France.

Guthrie, R. D. (1984). Mosaics, allelochemics and nutrients: An ecological theory of Late Pleistocene megafaunal extinctions. In P. S. Martin & R. G. Klein (Eds.), *Quaternary extinctions: A prehistoric revolution* (pp. 259–298). Tucson: University of Arizona Press.

Haidle, M., & Brauer, J. (2011). From Brainwave to Tradition—How to detect Innovations in Tool Behavior, Special Issue: Innovation and the Evolution of Human Behavior. *PaleoAnthropology, 2011*, 144–153.

Harmand, S. (2009). Raw materials and techno-economic behaviors at Oldowan and Acheulean sites in the West Turkana region, Kenya. In B. Adams & B. S. Blades (Eds.), *Lithic Materials and Paleolithic Societies*. Blades: Blackwell Publishing.

Hayden, B. (1979). *Palaeolithic reflections. Lithic Technology and ethnographic excavations among Australian Aborigines*. Camberra: Australian Institute for Arboriginal Studies.

Herisson, D., Airvaux, J., Lenoble, A., Richter, D., Claud, E., & Primault, J. (2012). Le gisement acheuléen de La Grande-Vallée à Colombiers (Vienne, France) : stratigraphie, processus de formation, datations préliminaires et industries lithiques. *Paleo, 23,* 1–18.

Hublin, J.-J. (2009). The origin of Neanderthals. *PNAS, 106,* 16022–16027.

Isaac, G. (1977). *Olorgesaillie. Archaeological Studies of a Middle Pleistocene Lake Basin in Kenya*. Chicago: The University of Chicago Press.

Isaac, G., Harris, J. W. K., & Kroll, E. M. (1997). The Stone Artefact Assemblages: A Comparative Study. In G. Isaac (Ed.), *Koobi Fora Research Project* (pp. 262–362). Oxford: Oxford University Press.

Jagher, R. (2011). Nadaouiyeh Aïn Askar - Acheulian variability in the Central Syrian Desert. In J.-M. Le Tensorer, R. Jagher, & M. Otte (Eds.), *The Lower and Middle Palaeolithic in the Middle East and Neighbouring* (pp. 209–225). Liege: ERAUL.

Jagher, R., & Le Tensorer, J.-M. (2011). El Kowm, a key area for the Palaeolithic of the Levant in Central Syria. In J.-M. Le Tensorer, R. Jagher, & M. Otte (Eds.), *The Lower and Middle Palaeolithic in the Middle East and Neighbouring* (pp. 197–209). Liege: ERAUL.

Jiménez-Arenas, J. M., Santonja, M., Botella, M., & Palmqvist, P. (2011). The oldest handaxes in Europe: Fact or artefact? *Journal of Archaeological Science, 38*, 3340–3349.

Jones, P. R. (1979). Effects of raw materials on biface manufacture. *Science, 20*, 835–836.

Jones, P. R. (1980). Experimental butchery with modern stone tools and its relevance for Palaeolithic archaeology. *World Prehistory, 12*, 153–165.

Jones, P. R. (1994). Results of experimental work in relation to the stone industries of Olduvai Gorge. In M. D. Leakey & D. A. Roe (Eds.), *Olduvai Gorge. Excavations in Beds III, IV and Masek Beds, 1968–1971* (pp. 254–298). Cambridge: Cambridge University Press.

Jouzel, J., Masson-Delmotte, V., Cattani, O., Dreyfus, G., Falourd, S., Hoffmann, G., et al. (2007). Orbital and Millennial Antarctic Climate Variability over the Past 800,000 Years. *Science Reports, 317*, 793–796.

Keeley, L. H. (1980). *Experimental Determination of Stone Tool Uses*. Chicago: Chicago University Press.

Kleindienst, M. R. (1961). Variability within the late Acheulian assemblage in Eastern Africa. *South African Archaeological Bulletin, 16*, 35–52.

Kuhn, S. L. (2002). Paleolithic Archeology in Turkey. *Evolutionary Anthropology, 11*, 198–210.

Kuman, K., Li, C., & Li, H. (2014). Large cutting tools in the Danjiangkou Reservoir Region, central China. *Journal of Human Evolution, 76*, 129–153.

Lamotte, A., & Tuffreau, A. (2001). Les industries lithiques de Cagny-la-Garenne II (Somme). In A. Tuffreau (Ed.), *L'Acheuléen dans la vallée de la Somme et Paléolithique moyen dans le nord de la France : données récentes* (pp. 59–91). CERP n°6.

Leakey, L. S. B. (1951). *Olduvai Gorge. A report on the evolution of the hand-axe culture in Beds I–IV*. Cambridge: Cambridge University Press.

Leakey, M. D. (1971). *Olduvai Gorge: Excavations in Bed I and Bed II, 1960–1963*. Cambridge: Cambridge University Press.

Lefèvre, D., Raynal, J.-P., Vernet, G., Kieffer, G., & Piperno, M. (2010). Tephro-stratigraphy and the age of ancient Southern Italian Acheulean settlements: The sites of Loreto and Notarchirico (Venosa, Basilicata, Italy). *Quaternary International, 223–224*, 360–368.

Lepre, C. L., Roche, H., Kent, D. V., Harmand, S., Quinn, R. L., Brugal, J.-P., et al. (2011). An earlier origin for the Acheulian. *Nature, 47*, 82–85.

Le Tensorer, J. M. (2009). Le Paléolithique ancien de Syrie et l'importance du Golan comme voie de passage lors de l'expansion des premiers hommes hors d'Afrique. In *The International Colloquium History and Antiquities of Al-Golan 2007–2008*. Damascus: The directorate General of Antiquities and Museums.

Lhomme, V., Connet, N., & Chaussé, C. (2003). Le gisement de Soucy 6 (Yonne) et son industrie lithique dans le contexte des industries à éclats du Paléolithique inférieur en Europe du Nord-Ouest. *Bulletin de la Société préhistorique française, 100*, 241–253.

Longo, L., Peretto, C., Sozzi, M., & Vannucci, S. (1997). Artefacts, outils ou supports épuisés ? Une nouvelle approche pour l'étude des industries du Paléolithique ancien : le cas d'Isernia la Pineta (Molise, Italie Centrale). *L'Anthropologie, 101*, 579–596.

Lycett, S. J., & Gowlett, J. A. J. (2008). On questions surrounding the Acheulean "tradition". *World Archaeology, 40*, 295–315.

Lycett, S. J., & von Cramon-Taubadel, N. (2008). Acheulean variability and hominin dispersals: A model-bound approach. *Journal of Archaeological Science, 35*, 553–562.

MacDonald, K., Martinón-Torrès, M., Dennell, R. W., & Bermúdez de Castro, J. M. (2012). Discontinuity in the record for hominin occupation in south-western Europe: Implications for occupation of the middle latitudes of Europe. *Quaternary International, 271*, 84–97.

McNabb, J., Binyon, F., & Hazelwood, L. (2004). The large cutting tools from the South African Acheulean and the question of social tradition. *Current Anthropology, 45*, 653–677.

McPherron, S. P. (2000). Handaxes as a measure of the mental capabilities of early hominids. *Journal of Archaeological Science, 27*, 655–663.

McPherron, S. (2006). What typology can tell us about Acheulian handaxe production. In N. Goren Inbar & G. Sharon (Eds.), *Axe Age: Acheulian Tool-making from Quarry to Discard* (pp. 267–287). London: Equinox Publishing Ltd.

Madurell-Malapeira, J., Alba, D. M., Espigares, M.-P., Vinuesa, V., Palmqvist, P., Martínez-Navarro, B., et al. (2017). Were large carnivorans and great climatic shifts limiting factors for hominin dispersals? Evidence of the activity of *Pachycrocuta brevirostris* during the Mid-Pleistocene Revolution in the Vallparadís Section (Valles-Penedes Basin, Iberian Peninsula). *Quaternary International, 431*, 42–52.

Manzi, G. (2004). Human Evolution at the Matuyama-Brunhes boundary. *Evolutionary Anthropology, 13*, 11–24.

Martínez-Navarro, B., Turq, A., Agusti Ballester, J., & Oms, O. (1997). Fuente Nueva 3 (Orce, Granada, Spain) and the first human occupation of Europe. *Journal of Human Evolution, 33*, 611–620.

Martínez, K., Garcia, J., Carbonell, E., Agustí, J., Bahain, J.-J., Blain, H.-A., et al. (2010). A new Lower Pleistocene archaeological site in Europe (Vallparadís, Barcelona, Spain). *Proceedings of the National Academy of Sciences, 107*, 5762–5767.

Martinón-Torrès, M., Bermùdez de Castro, J. M., Gòmez-Robles, A., Arsuaga, J. L., Carbonell, E., et al. (2007). Dental evidence on the hominin dipersal during the Pleistocene. *Proceedings of the National Academy of Sciences, 104*, 13279–13282.

Martinón-Torrès, M., Dennell, R., & Bermúdez de Castro, J. M. (2011). The Denisova hominin need not be an out of Africa story. *Journal of Human Evolution, 60*, 251–255.

Messager, E., Lebreton, V., Marquer, L., Russo-Ermolli, E., Orain, R., Renault-Miskovsky, J., et al. (2011). Palaeoenvironments of early hominins in temperate and Mediterranean Eurasia: New palaeobotanical data from Palaeolithic key-sites and synchronous natural sequences. *Quaternary Science Reviews, 30*, 1439–1447.

Meyer, M., Qiaomei, Fu, Aximu-Petri, A., Glocke, I., Nickel, B., Arsuaga, J.-L., et al. (2014). A mitochondrial genome sequence of a hominin from Sima de los Huesos. *Nature, 505*, 403–406.

Mitchell, J. C. (1995). Studying biface butchery at Boxgrove: Roe deer butchery with replica handaxes. *Lithics, 16*, 64–69.

Moncel, M.-H. (1995). Biface et outil-biface du Paléolithique moyen ancien : réflexion à partir de sites d'Ardèche, Orgnac 3 et Payre. *Paleo, 7*, 157–169.

Mgeladze, A., Lordkipanidze, D., Moncel, M.-H., Despriee, J., Chagelishvili, R., Nioradze, M., et al. (2011). Hominin occupations at the Dmanisi site, Georgia, Southern Caucasus: Raw materials and technical behaviours of Europe's first hominins. *Journal of Human Evolution, 60,* 571–596.

Moncel, M-H. (2008). L'expansion humaine en Eurasie de 1,8 à 0,8 Ma : les données, les questions et les hypotheses. In J-P. Bocquet-Appel (Ed.), *La Paléodémographie* (pp. 45–59). France: Editions Errance.

Moncel, M. H. (2010). Oldest human expansions in Eurasia: Favouring and limiting factors. *Quaternary International, 223–224,* 1–9.

Moncel, M.-H., Despriée, J., Voinchet, P., Tissoux, H., Moreno, D., Bahain, J.-J., et al. (2013). Early evidence of Acheulean settlement in north-western Europe - La Noira site, a 700 000 year-old occupation in the Center of France. *PLoS ONE, 8,* e75529.

Moncel, M.-H., Ashton N., Lamotte A., Tuffreau A., Cliquet D., Despriée J. (2015). The North-west Europe early Acheulian. *Journal of Anthropological Archaeology, 40,* 302–331.

Moncel, M.-H., Arzarello, M., Peretto, C. (2016). Editorial. The Holstainian Eldorado. *Quaternary International, 409,* 1–8.

Moncel, M.-H, Arzarello, M., Boëda, E., Bonilauri, S., Chevrier, B., Gaillard, C., Forestier, H., Yinghua, Li, Sémah, F., Zeitoun, V. (2018a). Assemblages with bifacial tools in Eurasia (third part). Considerations on the bifacial phenomenon throughout Eurasia, *Comptes Rendus Palevol, 17,* 77–97.

Moncel, M.-H., Landais, A., Lebreton, V., Combourieu-Nebout, N., Nomade, S., Bazin, L. (2018b). Linking environmental changes with human occupations between 900 and 400 ka in Western Europe. *Quaternary International, 30,* 78–94.

Monnier, J. L., Hallegouët, B., Hinguant, S., Laurent, M., Auguste, P., Bahain, J.-J., et al. (1994). A new regional group of the Lower Palaeolithic in Brittany (France), recently dated by Electron Spin Resonance. *Compte Rendu de l'Académie des Sciences, 319,* 155–160.

Mosquera, M., Ollé, A., & Rodriguez, X. P. (2013). From Atapuerca to Europe: Tracing the earliest peopling of Europe. *Quaternary International, 295,* 130–137.

Mosquera, M., Ollé, A., Saladié, P., Cáceres, I., Huguet, R., Rosas, A., et al. (2016). The Early Acheulean technology of Barranc de la Boella (Catalonia, Spain). *Quaternary International, 393,* 95–111.

Mounier, A., Marchal, F., & Condemi, S. (2009). Is *Homo heidelbergensis* a distinct species? New insight on the Mauer mandibule. *Journal of Human Evolution, 56,* 219–246.

Muttoni, G., Scardia, G., Kent, D. V., Swisher, C., & Manzi, G. (2009). Pleistocene magnetochronology of early hominin sites at Ceprano and Fontana Ranuccio, Italy. *Earth and Planetary Science Letters, 286,* 255–268.

Muttoni, G., Scardia, G., & Kent, D. V. (2010). Human migration into Europe during the late Early Pleistocene climate transition. *Palaeogeography, Palaeoclimatology, Palaeoecology, 296,* 79–93.

Nicoud E. (2013). *Le paradoxe acheuléen.* Editions du CTHS.

Ollé, A., Mosquera, M., Rodríguez, X. P., de Lombera-Hermida, A., García-Antón, M. D., García-Medrano, P., et al. (2013). The Early and Middle Pleistocene technological record from Sierra de Atapuerca (Burgos, Spain). *Quaternary International, 295,* 138–167.

Oms, O., Parés, J. M., Martínez-Navarro, B., Agustí, J., Toro, I., Martínez-Fernández, G., et al. (2000). Early human occupation of Western Europe: Paleomagnetic dates for two Paleolithic sites in Spain. *Proceedings of the National Academy of Sciences, 97,* 10666–10670.

Orain, R., Lebreton, V., Russo Ermolli, E., Sémah, A.-M., Nomade, S., Shao, Q., et al. (2013). Hominin responses to environmental changes during the Middle Pleistocene in central and southern Italy. *Climate of the Past, 9,* 687–697.

Paillard, D. (2015). Quaternary glaciations: From observations to theories. *Quaternary Science Reviews, 107,* 11–24.

Pappu, S., Gunnell, Y., Akhilesh, K., Braucher, R., Taieb, M., Demory, F., et al. (2011). Early Pleistocene Presence of Acheulian Hominins in South India. *Science, 331,* 1596–1599.

Parés, J. M., & Pérez-Gonzalez, A. (1999). Magnetostratigraphy and stratigraphy at Gran Dolina section, Atapuerca (Burgos, Spain). *Journal of Human Evolution, 37,* 325–342.

Parés, J. M., Arnold, L., Duval, M., Demuro, M., Pérez-Gonzalez, A., Bermúdez de Castro, J. M., et al. (2013). Reassessing the age of Atapuerca-TD6 (Spain): New paleomagnetic results. *Journal of Archaeological Science, 40,* 4586–4595.

Parfitt, S. A., Barendregt, R. W., Breda, M., Candy, I., Collins, M. J., Coope, G. R., et al. (2005). The earliest record of human activity in northern Europe. *Nature, 438,* 1008–1012.

Parfitt, S. A., Ashton, N., Lewis, S. G., Abel, R. L., Cooper, G. R., Mike, H. F., et al. (2010). Early Pleistocene human occupation at the edge of the boreal zone in northwest Europe. *Nature, 466,* 229–233.

Petraglia, M. (2003). The lower Paleolithic of the Arabian Peninsula: Occupations, adaptations, and dispersals. *Journal of World Prehistory, 17,* 2.

Peretto, C., Amore, O., Antoniazzi, A., Bahain, J. J., Cattani, L., Cavallini, L., et al. (1998). L'industrie lithique de Cà Belvedere di Monte Poggiolo : stratigraphie, matière première, typologie, remontages et traces d'utilisation. *L'Anthropologie, 102,* 343–466.

Peretto, C., Arzarello, M., Gallotti, R., Lembo, G., Minelli, A., & Thun Hohenstein, U. (2004). Middle Pleistocene behavioural strategies: The contribution of Isernia La Pineta site. In E. Baquedano & S. Rubio Jara (Eds.), *Miscelanea en homenaje a Emiliano Aguirre,* Volumen IV, Arqueologia (pp. 369–381). Alcalá de Henares: Museo Arqueologico Regional.

Piperno, M. (Ed.). (1999). *Notarchirico. Un sito del Pleistocene medio iniziale nel bacino di Venosa.* Venosa: Edizioni Osanna.

Preece, R. C. (2001). Molluscan Evidence for differentiation of interglacials within the "Cromerian Complex". *Quaternary Science Review, 20,* 1643–1656.

Preece, R. C., & Parfitt, S. A. (2012). The Early and early Middle Pleistocene context of human occupation and lowland glaciations in Britain and northern Europe. *Quaternary International, 271,* 6–28.

Premo, L. S., & Hublin, J.-J. (2009). Culture, population structure, and low genetic diversity in Pleistocene hominins. *Proceedings of the National Academy of Sciences, 106,* 33–37.

Presnyakov, S. L., Belyaeva, E. V., Lyubin, V. P., Rodionov, N. V., Antonov, A. V., Saltykova, A. K., et al. (2012). Age of the earliest Paleolithic sites in the northern part of the Armenian Highland by SHRIMP-II U-Pb geochronology of zircons from volcanic ashes. *Gondwana Research, 21,* 928–938.

Raynal, J-P., Sbihi-Alaoui, F. Z., Geraads, D., Magoga, L., & Mohib, A. (2001). The earliest occupation of North Africa: The Maroccan perspective. *Quaternary International, 75,* 65–77.

Rabinovich, R, Gaudzinski-Windheuser, S., Goren-Inbar, N. (2008). Systematic butchering of fallow deer (*Dama*) at the early middle Pleistocene Acheulian site of Gesher Benot Ya'aqov (Israel). *Journal of Human Evolution, 54,* 134–149.

Raynal, J.-P., Sbihi-Alaoui, F.-Z., Mohib, A., El Graoui, M., Lefèvre, D., Texier, J.-P., et al. (2010). Hominid Cave at Thomas Quarry I (Casablanca, Morocco): Recent findings and their context. *Quaternary International, 223–224,* 369–382.

Raynal, J.-P., Sbihi-Alaoui, F. Z., Michib, A., El Graoui, M., Lefèvre, D., et al. (2011). Contextes et âge des nouveaux restes dentaires humains du Pleistocène moyen de la carrière Thomas I à Casablanca (Maroc). *Bulletin de la Société Préhistorique Française, 108,* 645–671.

Rightmire, G. P. (2001). Patterns of hominid evolution and dispersal in the Middle Pleistocene. *Quaternary International, 75,* 77–84.

Rightmire, G. P. (2009). Middle and later Pleistocene hominins in Africa and Southwest Asia. *Proceedings of the National Academy of Sciences, 106,* 16046–16050.

Roberts, M., & Parfitt, S. (1999). *Boxgrove: A Middle Pleistocene Hominid site at Eartham Quarry.* West Sussex, England: English Heritage.

Roche, H. (2005). From Simple Flaking to Shaping: Stone-knapping Evolution among Early Hominids. In V. Roux & B. Brill (Eds.), *Stone knapping: the necessary conditions for a uniquely hominin behavior* (pp. 35–53). MacDonald Institute Monograph.

Roche, H., Brugal, J. P., Lefèvre, D., & Texier, P.-J. (1987). Premières données sur l'Acheuléen des hauts plateaux kényans : le site d'Isenya (district de Kajiado). *Comptes Rendus de l'Académie des Sciences, 305,* 529–532.

Roche, H., Brugal, J. P., Lefèvre, D., Ploux, S., & Texier, P.-J. (1988). Isenya : état des recherches sur un nouveau site acheuléen d'Afrique orientale. *The African Archaeological Review, 6,* 27–55.

Roche, H., Brugal, J.-Ph., Delagnes, A., Feibel, C., Harmand, S., Kibunjia, M., et al. (2003). Les sites archéologiques plio-pléistocènes de la formation de Nachukui, Ouest-Turkana, Kenya : bilan synthétique 1997–2001. *Comptes Rendus Palevol, 2,* 663–673.

Rodríguez, X. P. (2004). *Technical Systems of Lithic Production in the Lower and Middle Pleistocene of the Iberian Peninsula.* Oxford: British Archaeological Research International Series, 1323.

Rodríguez, J., Burjachs, F., Cuenca-Bescós, G., García, N., van der Made, J., Pérez-González, A., et al. (2011). One million years of cultural evolution in a stable environment at Atapuerca (Burgos, Spain). *Quaternary Science Reviews, 30,* 1396–1412.

Roe, D. A. (1964). The British Lower and Middle Palaeolithic: Some problems, methods of study and preliminary results. *Proceeding Prehistoric Society, 30,* 245–267.

Roe, D. A. (1981). *The lower and middle Palaeolithic periods in Britain.* London: Routledge and Kegan Paul.

Roebroeks, W., & Villa, P. (2011). On the earliest evidence for habitual use of fire in Europe. *Proceedings of the National Academy of Sciences, 108,* 5209–5214.

Rolland, N. (2013). The Early Pleistocene human dispersals in the Circum-Mediterranean Basin and initial peopling of Europe: Single or multiple pathways? *Quaternary International, 316,* 59–72.

Santagata, C. (2012). L'utilisation de roches autres que le silex au Paléolithique ancien et moyen. Choix économiques, techniques et fonctionnels sur la base de l'étude des gisements de Saint-Anne I (Haute-Loire, France) (MIS 5 et 6) et de Notarchirico (Basilicata, Italia) (MIS 14 à 17). PhD Dissertation, Université de Bordeaux 1.

Santonja, M., & Villa, P. (2006). The Acheulian of western Europe. In N. Goren-Inbar & G. Sharon (Eds.), *Axe Age. Acheulian Tool-making from Quarry to Discard* (pp. 429–478). London: Equinox.

Shchelinsky, V. E., Dodonov, A. E., Baigusheva, V. S., Kulakov, S. A., Simakova, A. N., Tesakov, A. S., et al. (2010). Early Palaeolithic sites on the Taman Peninsula (Southern Azov Sea region, Russia): Bogatyri/Sinyaya Balka and Rodniki. *Quaternary International, 223–224,* 28–35.

Schick, K. D., & Toth, N. (1993). *Making Silent Stones Speak: Human Evolution and the Dawn of Technology.* New York: Simon and Schuster.

Schick, K., & Clark, J. D. (2003). Biface technological development and variability in the Acheulean industrial complex in the Middle Awash Region of the Afar Rift, Ethiopia. In M. Soressi & H. L. Dibble (Eds.), *Multiples Approaches to the Study of Bifacial Technologies* (pp. 1–30). Philadelphia: University of Pennsylvania, Museum of Archaeology and Anthropology.

Scott, G. R., & Gibert, L. (2009). The oldest hand-axes in Europe. *Nature, 461,* 82–85.

Semaw, S., Rogers, M., & Stout, D. (2009). The Oldowan-Acheulian Transition: Is there a "Developed Oldowan" artifact tradition? In M. Camps & P. Chauhan (Eds.), *Sourcebook of Paleolithic Transitions* (pp. 173–194). New York: Springer.

Semaw, S., Rogers, M. J., Cáceres, I., Stout, D., & Leiss, A. (2018). The Early Acheulean 1.6–1.2 Ma from Gona, Ethiopia: Issues on the Emergence of the Acheulean in Africa. In R. Gallotti & M. Mussi (Eds.), *The Emergence of the Acheulean in East Africa and Beyond. Contributions in Honor of Jean Chavaillon.* Cham: Springer.

Shao, Q., Bahain, J.-J., Falguères, C., Peretto, C., Arzarello, M., Minelli, A., et al. (2011). New ESR/U-series data for the early Middle Pleistocene site of Isernia la Pineta, Italy. *Radiation Measurements, 46,* 847–852.

Sharon, G. (2008). The impact of raw material on Acheulian large flake production. *Journal of Archaeological Science, 35,* 1329–1344.

Sharon, G. (2009). Acheulian Giant-Core Technology. *Current Anthropology, 50,* 335–367.

Sharon, G. (2010). Large flake Acheulian. *Quaternary International, 223–224,* 226–233.

Sharon, G. (2011). Flakes Crossing the Straits? Entame Flakes and Northern Africa-Iberia Contact During the Acheulean. *African Archaeological Review, 28,* 125–140.

Sharon, G., Feibel, C., Alperson, N., Harlavan, Y., Feraud, G., Ashkenazi, S., et al. (2010). New evidence for the Northern Dead Sea Rift Acheulian. *PaleoAnthropology, 2010,* 79–99.

Sharon, G., Alperson-Afil, N., & Goren-Inbar, N. (2011). Cultural conservatism and variability in the Acheulian sequence of Gesher Benot Ya'aqov. *Journal of Human Evolution, 60,* 387–397.

Soressi, M., & Hays, M. A. (2003). Manufacture, transport and use of Mousterian bifaces: A case study from the Périgord (France). In M. Soressi & D. L. Dibble (Eds.), *Multiple Approaches to the Study of Bifacial Technologies* (pp. 125–147). Philadelphia: University of Pennsylvania Museum of Archaeology and Anthropology.

Stiner, M. C. (2002). Carnivory, Coevolution, and the Geographic Spread of the Genus *Homo. Journal of Archaeological Research, 10,* 1–63.

Stringer, C. (1996). The Boxgrove Tibia: Britain's Oldest Hominid and its place in the Middle Pleistocene Record. In C. Gamble & A. J. Lawson (Eds.), *The English Heritage Reviewed* (pp. 53–56). Wesses Archaeology.

Stringer, C. (2012). The Status of *Homo heidelbergensis* (Schoetensack 1908). *Evolutionary Anthropology, 21,* 101–107.

Tavoso, A. (1986). Le Paléolithique inférieur et moyen du Haut-Languedoc. Gisements des terrasses alluviales du Tarn, du Dadou, de l'Agout, du Sor et du Fresquel. Marseille: Université de Provence, Laboratoire de Paléontologie et de Préhistoire.

Texier, P.-J. (2001). Du seuil de l'hominisation aux premières cultures : potentialité et apports de la taille expérimentale. *Praehistoria, 2,* 51–61.

Texier, P-J., & Roche, H. (1995). The impact of predetermination on the development of some Acheulean chaînes opératoires. In J. Bermùdez de Castro, J. L. Arsuaga, & E. Carbonell (Eds.), *Evolucion humana en Europa y los yacimientos de la Sierra de Atapuerca* (pp. 403–420). Vallaloid: Junta de Castilla y Leon.

Texier, P.-J. (2018). Technological Assets for the Emergence of the Acheulean? Reflexions on the Kokiselei 4 Lithic Assemblage and its place in the Archaeological Context of West Turkana, Kenya. In R. Gallotti & M. Mussi (Eds.), *The Emergence of the Acheulean in East Africa and Beyond. Contributions in Honor of Jean Chavaillon.* Cham: Springer.

Toro Moyano, I. B. D., Cauche, D., Celiberti V., Gregoire, S., Lebegue, F., Moncel, M. H., et al. (2011). The archaic stone tool industry from Barranco León and Fuente Nueva 3, (Orce, Spain): Evidence of the earliest hominin presence in southern Europe. *Quaternary International*, 1–12.

Tuffreau, A. (1987). *Le Paléolithique inférieur et moyen du nord de la France (Nord, Pas-de-Calais, Picardie) dans son cadre stratigraphique*. Ph.D. doctorat d'état, l'Université de Lille.

Tuffreau, A., Lamotte, A., & Goval, E. (2008). Les industries acheuléennes de la France septentrionale. *L'Anthropologie, 112,* 104–139.

Tuffreau, A., & Lamotte, A. (2010). Oldest Acheulean settlements in Northern France. *Quaternary International, 223–224,* 455.

Vallverdú, J., Saladié, P. S., Rosas, A., Huguet, R., Cáceres, I., Mosquera, M., et al. (2014). Age and Date for Early Arrival of the Acheulian in Europe (Barranc de la Boella, la Canonja, Spain). *PLoS ONE, 9,* e103634.

Van Peer, P. (1998). The Nile Corridor and the Out-of-Africa Model. An examination of the Archaeological Record. *Current Anthropology, 39,* 115–139.

Vermeersch, P. M. (2001). "Out of Africa" from an Egyptian point of view. *Quaternary International, 75,* 103–112.

Villa, P. (1981). Matières premières et provinces culturelles dans l'Acheuléen français. *Quaternaria*, 19–35.

Villa, P., Soto, E., Santonja, M., Pérez-González, A., Mora, R., Parcerisas, J., et al. (2005). New data from Ambrona: Closing the hunting versus scavenging debate. *Quaternary International, 126–128,* 223–250.

Wagner, G. A., Krbetschek, M., Degering, D., Bahain, J.-J., Shao, Q., Falguères, C., et al. (2010). Radiometric dating of the type-site for *Homo heidelbergensis* at Mauer, Germany. *Proceedings National Academy of Sciences, 107,* 19726–19730.

Walker, M. J., López-Martínez, M., Carrión-García, J. S., et al. (2013). Cueva Negra del Estrecho del Río Quípar (Murcia, Spain): A late Early Pleistocene hominid site with an "Acheuleo-Levalloiso-Mousteroid" assemblage. *Quaternary International, 294,* 135–159.

Wenban-Smith, F. F. (2004). Handaxe typology and Lower Palaeolithic cultural development: Ficrons, cleavers and two giant handaxes from Cuxton. *Lithics, 25,* 11–21.

White, M. J. (1998). Twisted ovate bifaces in the British Lower Palaeolithic: Some observations and implications. In N. M. Ashton, F. Healy, & P. B. Pettitt (Eds.), *Stone Age Archaeology: Essays in Honour of John Wymer* (pp. 98–104). Oxford: Oxbow Books, Lithic Studies Occasional Paper 6.

White, M. J. (2015). 'Dancing to the ryththms of the biotidal zone': Settlement history and culture history in Middle Pleistocene Britain. In F. Coward, R. Hosfield, M. Pope, & F. F. Wenban-Smith (Eds.), *Settlement, Society and Cognition in Human Evolution: Landscapes in Mind* (pp. 154–173). Cambridge: Cambridge University Press.

Wilkins, J., & Chazan, M. (2012). Blade production at 500 thousand years ago at Kathu Pan 1, South Africa: Support for a multiple origins hypothesis for early Middle Pleistocene blade technologies. *Journal of Archaeological Science, 39,* 1883–1900.

Winton, V. (2005). An investigation of knapping-skill development in the manufacture of Palaeolithic handaxes. In V. Roux & B. Bril (Eds.), *Stone knapping: the necessary conditions for a uniquely hominin behavior* (pp. 109–119). MacDonald Institute Monograph.

Yazbeck, C. (2004). Le Paléolithique du Liban : bilan critique. *Paléorient, 30,* 111–126.

Index

R. Gallotti and M. Mussi (eds.), *The Emergence of the Acheulean in East Africa and Beyond: Contributions in Honor of Jean Chavaillon*, Vertebrate Paleobiology and Paleoanthropology,
https://doi.org/10.1007/978-3-319-75985-2

MIX
Papier aus verantwortungsvollen Quellen
Paper from responsible sources
FSC® C105338

If you have any concerns about our products,
you can contact us on
ProductSafety@springernature.com

In case Publisher is established outside the EU,
the EU authorized representative is:
Springer Nature Customer Service Center GmbH
Europaplatz 3, 69115 Heidelberg, Germany

Printed by Libri Plureos GmbH
in Hamburg, Germany